Empirical Research in Accounting

This textbook provides the foundation for a course that takes PhD students in empirical accounting research from the very basics of statistics, data analysis, and causal inference up to the point at which they conduct their own research. Starting with foundations in statistics, econometrics, causal inference, and institutional knowledge of accounting and finance, the book moves on to an in-depth coverage of the core papers in capital market research. The latter half of the book examines contemporary approaches to research design and empirical analysis, including natural experiments, instrumental variables, fixed effects, difference-in-differences, regression discontinuity design, propensity-score matching, and machine learning. Readers of the book will develop deep data analysis skills using modern tools. Extensive replication and simulation analyses are included throughout.

Key Features:

- Extensive coverage of empirical accounting research over more than 50 years.
- Integrated coverage of statistics and econometrics, institutional knowledge, and research design.
- Numerous replications and a dozen simulation analyses to immerse readers in papers and empirical analysis.
- All tables and figures in the book can be reproduced by readers using the included code.
- Easy-to-use templates facilitate hands-on exercises and introduce reproducible research concepts. (Solutions available to instructors.)

Ian D. Gow is a professor at the University of Melbourne, where he teaches several courses, including courses based on this book. Ian previously served on the faculties of Harvard Business School, Northwestern University, and Yale. Ian's recent research focuses on causal inference and empirical methods. Ian has a PhD from Stanford, an MBA from Harvard and BCom and LLB degrees from the University of New South Wales.

Tongqing (Tony) Ding is a senior lecturer at the University of Melbourne, where he teaches courses on data analytics, financial statement analysis, and corporate reporting. Tony's research focuses on corporate governance, financial reporting and disclosure, ESG, and data analytics. Tony has PhD and MS degrees from the University of Colorado and degrees from Shanghai Jiao Tong University.

CHAPMAN & HALL/CRC Series on Statistics in Business and Economics

Recently Published Titles

Empirical Research in Accounting
Tools and Methods
Ian D. Gow and Tongqing Ding

For more information about this series, please visit: https://www.routledge.com/Chapman-and-HallCRC-Series-on-Statistics-in-Business-and-Economics/book-series/CHSBE?srsltid=AfmB OoprnfkrPO9MQw1mM8n4HF9PT87ONW4uUIQMUvc1r9QS1ufKCH2G

Empirical Research in Accounting
Tools and Methods

Ian D. Gow and Tongqing Ding

CRC Press
Taylor & Francis Group
Boca Raton London New York

CRC Press is an imprint of the
Taylor & Francis Group, an **informa** business

A CHAPMAN & HALL BOOK

Designed cover image: © Ian D. Gow and Tongqing Ding

First edition published 2025
by CRC Press
2385 NW Executive Center Drive, Suite 320, Boca Raton FL 33431

and by CRC Press
4 Park Square, Milton Park, Abingdon, Oxon, OX14 4RN

CRC Press is an imprint of Taylor & Francis Group, LLC

© 2025 Ian D. Gow and Tongqing Ding

ISBN: 978-1-032-58650-2 (hbk)
ISBN: 978-1-032-59783-6 (pbk)
ISBN: 978-1-003-45623-0 (ebk)

DOI: 10.1201/9781003456230

Typeset in Latin Modern font
by KnowledgeWorks Global Ltd.

Publisher's note: This book has been prepared from camera-ready copy provided by the authors.

Access the Support Material: https://www.routledge.com/Empirical-Research-in-Accounting-Tools-and-Methods/Gow-Ding/p/book/9781032586502?srsltid=AfmBOoq8itQrR-_lBxJ4_mT4H64wVhkh5V-bha1Tr4LeZK7Z-GD6jEdr

Contents

Preface

This book provides a course on financial accounting research that begins at an upper-undergraduate ("honours") or introductory PhD level. One goal of the course, like most PhD courses, is to prepare PhD students to take further research courses and to go on to do their own research. Another goal of the course is to provide students with a set of skills that is useful in other domains, such as consulting or finance. This second goal stems from the origins of parts of this course in a joint honours-PhD course at the University of Melbourne, where the honours students are undergraduates completing an additional year of study with a focus on research. While some honours students progress to PhD studies, most elect to take jobs in industry, such as consulting, auditing, or public service.

Features of this book

Some features of this book mean that a course based on it will be distinguished from a more traditional PhD-level course in a number of respects that we discuss here.

Pedagogically driven selection of papers

Many syllabuses for PhD courses in accounting focus on recent papers with a view to giving students a sense of the current themes and trends in research to help students spot gaps in the literature that they can fill with their own research. We view such courses as complementary to this course, but take a different approach.

Aiming to provide a more fundamental understanding of accounting research, the selection of papers in this course is driven more by pedagogical goals than an attempt to represent the current state of play in accounting research. In some cases, this means covering older papers (e.g., Ball and Brown, 1968), but in other cases we use a recent paper that features a core idea or approach.

Incorporation of data analysis skills

A second feature that distinguishes this course from most PhD courses in accounting is an emphasis on data analysis skills, which are woven into the course throughout. While it might have been possible to make this a course focused exclusively on such skills, it is our view that these skills are best learned through applying them to real research questions. Conversely, we believe that being able to pull data, run simulations, and get more involved with critical elements of the research process engenders a better understanding of research.

We build data analysis and computing skills into the course at each step in a systematic way. In practice, research computing skills are the bread and butter of a researcher's toolkit but are generally neglected in PhD programs' formal curricula. The prevailing ethos seems to be that research computing skills are acquired informally from other students, through research assistantships and collaboration with faculty members, and so on.

In some doctoral programs, this approach may work to a degree. But in many doctoral programs, such informal learning fails to prepare students adequately. For example, if students' collaboration with faculty members is informal and related to collaboration on papers where the students do the data analysis with limited or no hands-on guidance from the faculty member, then the opportunity for clear and comprehensive guidance is limited.

Greater emphasis on research design and methods

Accounting research is overwhelmingly an empirical discipline seeking to draw causal inferences and, as such, significant research training should be focused on research design issues. Part III of the book examines **causal inference** in depth, including natural experiments, regression discontinuity designs, instrumental variables, and fixed effects. However, we will see that there are flaws in the common belief that it is these *techniques* that offer hope for warranted causal inference, rather than the *settings* in which they can be deployed. Throughout the course, we will offer a broader set of tools for making inferences about real-world phenomena.

In this course, we strive to give students the skills that are needed to conduct the analyses that we cover. We believe that understanding of statistical and econometric techniques by accounting researchers is more likely to be enhanced by hands-on simulation analysis than by analysis of consistency and asymptotic variance of estimators. By building in the data analysis skills needed to perform such simulations, we hope that this course provides a platform for accounting researchers to think more carefully about the properties of their estimators. A core commitment we make in this book is that *every* analysis we present can be conducted by the reader using the code found herein.

Prerequisites

We presume prior knowledge of some topics and access to certain computing resources. We have endeavoured to keep these requirements to a minimum.

1. **Knowledge of accounting and business.** In terms of accounting, we assume a solid understanding of the content of an introductory financial accounting course and enough understanding of business to make sense of accounting.

2. **Prior exposure to statistics and econometrics.** Some familiarity with the elements of statistical inference and ordinary least-squares (OLS) regression will be helpful. While we do provide some introductory material on elements of these in Chapters 3–5, this is selective and it may help to use a textbook to brush up on these topics as you work through the book.

3. **Access to academic journals.** The course makes extensive use of papers in academic journals. If you are a faculty member, researcher, or student at an academic institution, then you should be able to access the papers we use through your library. Some universities provide a service (perhaps for a fee) for alumni access to academic journals. Unfortunately, if you cannot access the papers, it will be difficult to make full use of this course book beyond Part I.

4. **Access to a computer and the internet.** To accommodate a broader audience and minimize set-up costs, we assume nothing of the reader other than access to a computer and the internet and basic proficiency in using these.

5. **Ability to install R and RStudio.** In this course, while we focus on R, a popular open-source programming language for statistics and data science, we do not assume any knowledge of R in this course. Chapter 2 provides an introductory tutorial on R and RStudio. We hope that this provides the required platform for later chapters whether you are a complete novice or coming to R from another statistical software system. Throughout the book, we direct the reader to additional resources for learning more detail than we can fit in this book. With the occasional detour to resources like *R for Data Science*, working through this book should provide the reader with a strong set of data science skills.[1] Our hope is that these skills will prove useful whether you continue down the path of academic research or pursue a position in practice.

6. **Access to WRDS data.** Because it is difficult to go very far in accounting research without WRDS data, this book is targeted at the reader who has a WRDS account. If you do not have a WRDS account but are eligible for one (e.g., you are a graduate student, researcher, or faculty member at a WRDS-subscribing institution), then you should apply for such an account.[2]

A guide for readers

The book is written so as to be fairly accessible to a novice reading independently (subject to the prerequisites outlined above). We recommend that such readers work through the first few chapters in order, including running the code, completing the exercises, and thinking about the discussion questions. That said, some elements of the exercises and discussion questions are subtle and having an instructor or someone to discuss these with will help you to get the full value from this material.

We hope that this book will be useful to a variety of readers, learners, and instructors beyond the novices. Below we discuss possible approaches for some hypothetical readers.

- **I am interested in learning more about issues related to research design and causal inference.** You might find you can dive into Chapters 2 and 4, then move to Chapter 17 and subsequent chapters.

- **I am interested in learning more about issues related to research design and causal inference, but I don't really want to learn R.** The plan for the hypothetical reader in the previous bullet point likely works. Even if you aren't interested in learning R, we think that running the code helps solidify understanding and that what the code is doing is sufficiently clear that copy-pasting the code into your own computer should be enough to get the gist of what is going on.

- **I have heard about R and would like to learn more about it.** Chapters 2 and 3 cover some of the basics. But if you're already proficient in something like SAS or Stata, you may find it pretty easy to skip those chapters (after meeting the prerequisites above)

---

[1] *R for Data Science* is available at https://r4ds.hadley.nz/
[2] Go to https://wrds-www.wharton.upenn.edu/register to do so.

and go to a chapter that aligns with your research interests, and see if you can figure out what the code is doing as you work through it. We have deliberately written the book so that, apart from initial set-up in Section 1.2, code in each chapter is independent of that in the others.

Acknowledgements

While this book draws on materials we have been using for many years, writing this book began in earnest in early 2021. Since then we have received help from many others, ranging from supplying code and data, suggestions on content, feedback on drafts, and simply encouragement to persist with the project. We would like to recognize the help of Ulrich Atz, Andrew Baker, Ray Ball, Jeremy Bertomeu, Stu Black, Mark Bradshaw, Philip Brown, Patty Dechow, Jenny Zha Giedt, Lucy Gow, Amy Hutton, Rohit Kattamuri, James Kavourakis, David Larcker, Changju Lee, Andy Leone, Ying Liang, Christian Leuz, Miguel Minutti-Meza, Casey Mulligan, Matt Pinnuck, Steve O'Byrne, Shiva Rajgopal, Mario Schabus, Stefan Schantl, Richard Sloan, Dan Taylor, Jake Thomas, Jake Thornock, Stephen Walker, Charlie Wang, Yihong Wang, Eddie Watts, and Anastasia Zakolyukina.

We also thank the many students who suffered through earlier versions of the materials here, including students at Deakin, Harvard, Melbourne, Michigan, and Wharton. Quinn Swanquist taught a PhD class at the University of Alabama using an earlier version of the book. We thank Quinn and his students—Susan Rykowski, Amy Mathews, and Jack Archer—for providing detailed feedback.

Some notes on style

We follow British (hence Australian) conventions for the most part. Reflecting the enduring influence of a *Pocket Oxford Dictionary* one of us received at age seven, we tend to use "-ize" spellings instead of "-ise" spellings (in any case, these are more familiar to American readers). Also we likely use the Oxford comma more often than not. One benefit of our choice is that we do not have to follow the prescription of American English that commas and full stops (periods) *always* go inside quotes and can instead put them where they naturally belong (i.e., where speakers of languages other than American English put them) even if this produces sentences that may look odd to some American readers. (It's hard to disagree with Hadley Wickham—the lead author of the Tidyverse[3]—on this point: "That is literally the stupidest rule in American English and I refuse to follow it."[4])

For code, we largely follow the Tidyverse style guide for R code,[5] except that we often put the first item after the assignment operator (`<-`) on a new line.

---

[3] https://www.tidyverse.org
[4] https://twitter.com/hadleywickham/status/1456350268677935108
[5] https://style.tidyverse.org

Part I

Foundations

1

Introduction

1.1 Structure of the book

The book is organized into four parts.

Part I: Foundations covers a variety of topics, including research computing, statistics, causal inference, and some details of data sets commonly used in accounting research. This part of the book covers material often not included in the formal coursework of a PhD in accounting. For example, material related to statistics and causal inference is often assumed to be covered in coursework in statistics and econometrics rather than in the accounting-specific courses. Material on research computing and detailed investigation of data sets is generally not covered in PhD coursework at all, with the typical approach being for these skills and knowledge to be picked up informally.

Assuming very little in terms of prior knowledge, *Part I: Foundations* covers core concepts and skills in data analysis, statistics, and causal inference.

- Chapter 1 provides an introduction to the book, including a reading guide and instructions for setting up your computer.

- Given the centrality of data skills to getting the full value out of this book, we provide a fast-paced tutorial-style introduction to R in Chapter 2.

- As we assume very little knowledge of statistics and regression analysis, we provide an introduction to the basics of regression analysis in Chapter 3.

- Chapter 4 builds on Chapter 3 to provide an introduction to elements of causal inference.

- Chapter 5 provides an introduction to statistical inference, which is a core part of empirical accounting research.

Part I: Foundations introduces key data sets frequently used in empirical accounting research.

- Chapters 6 and 8 provide an introduction to Compustat and accessing data through WRDS.

- Chapter 7 discusses the linking of data sets from different providers with a focus on linking financial statement data from Compustat with stock return data from CRSP.

- We wrap up Part I with Chapter 9, which provides additional data skills useful for both later chapters and (we hope) readers' own research efforts.

Part I provides the foundations for the remaining parts of the book. Depending on the preferences of readers and instructors, one could either continue with *Part II: Capital Markets*

DOI: 10.1201/9781003456230-1

Research or skip ahead to *Part III: Causal Inference.* While some parts of Part III draw on skills and concepts covered in Part II, we flag such instances in each case.

The material of Part I could be covered in a number of ways. One approach would be to cover this material in a standalone introductory course or "boot camp". A reader will notice that Chapter 2 actually incorporates by reference significant portions of *R for Data Science*,[1] which could easily be a course in its own right (and one highly complementary to the material covered here), so there is plenty of material here for a full-fledged course for a program willing to devote class time to these skills.

Another approach might be to assign *Part I: Foundations* to students on a self-study basis, perhaps with select portions being covered when they are most relevant for later portions of the book. For example, for a course based on *Part II: Capital Markets Research* of the book, Chapter 7 covers the important topic of correctly linking databases—not often encountered in PhD courses—and could be assigned as background work as and when relevant to material from Part II.

Part II: Capital Markets Research provides the basis for a PhD-level course focused on capital markets research. This part alone easily provides materials for about eight weeks of coursework. For a ten- or twelve-week course, an instructor could draw on materials from other parts of the book, or could easily supplement using other materials. Part II is deliberately focused on more "classical" material and thus could easily complement related material that focuses on more contemporary work in financial accounting research. Part II starts with research from the 1960s—such as Fama et al. (1969), Ball and Brown (1968), and Beaver (1968)—and covers some of the most important studies of subsequent decades, including Bernard and Thomas (1989), Sloan (1996), and key earnings management papers of the 1980s and 1990s.

Part III: Causal Inference provides the basis for a PhD-level course focused on causal inference in empirical accounting research. Part III has a more contemporary orientation and is not focused on capital markets research.

Depending on the needs of students in a given program, Part III could be taught as a standalone course with elements of Part I being drawn upon as needed. Topics in Chapter 19 draw on materials in Part I, with extensive discussion of causal diagrams (Chapter 4), standard errors (Chapter 5), linking databases (Chapter 7), using regular expressions (Chapter 9), and two-step regressions (drawing on materials covered in Chapter 3).

While there are connections between Part II and Part III (e.g., Chapter 19 covers measures of accruals and earnings management that are covered in Chapters 15 and 16), these do not seem to rise to the level of considering Part II a prerequisite for Part III. Chapter 19 focuses on earnings management, which is the topic of an entire chapter in Part II (Chapter 16). While the material of Part III might typically be covered later in the coursework of an accounting PhD program, we have endeavoured to present this material in a way that is fairly self-contained and therefore accessible to students earlier in their PhD studies (perhaps using materials from Part I to fill in gaps). There may even be merit in covering most of Part III before Part II, as it will allow students to read Part II materials (mostly older papers) through a more contemporary lens.

Part IV: Additional Topics provides chapters on topics such as matching, handling extreme values, selection models, and statistical (machine) learning. While these are important

---

[1] https://r4ds.hadley.nz

topics, we believe they are less closely related than the materials of Parts II and III. Instructors could easily incorporate chapters from Part IV in courses based on Part II or Part III of this book, or as standalone material for courses not based on this book.

1.2 Setting up your computer

Assuming that you have the ability to install software and a WRDS account, setting up your computer so that you can run the code in this book is straightforward and takes just a few minutes. We list the required steps below and also provide a video demonstrating these steps online.[2]

1. Download and install R. R is available for all major platforms (Windows, Linux, and MacOS) at https://cloud.r-project.org.

2. Download and install RStudio. An open-source version of RStudio is available on the Posit website.[3]

3. Install the required packages from CRAN.[4] CRAN stands for "Comprehensive R Archive Network" and is the official repository for **packages** (also known as **libraries**) made available for R. In this course, we will make use of a number of R packages. These can be installed easily by running the following code in RStudio.[5]

```
install.packages(c("DBI", "MASS", "MatchIt", "RPostgres",
    "arrow", "car", "duckdb", "farr", "fixest", "furrr",
    "glmnet", "httr2", "kableExtra", "lmtest", "modelsummary",
    "optmatch", "pdftools", "plm", "rdrobust", "robustbase",
    "rpart", "rpart.plot", "sandwich", "tidyverse"))
```

Note that `farr` is an R package one of us created just for this course (Gow, 2022). (As the package is related to the course *Financial Accounting Research* at the University of Melbourne, `farr` stands for "Financial Accounting Research with R".)

4. Set up R to connect to the WRDS PostgreSQL database. To actually use much of the code from Chapter 6 on, you will need to tell R how to access WRDS data stored in its PostgreSQL database by running the following line within RStudio.

```
Sys.setenv(PGHOST = "wrds-pgdata.wharton.upenn.edu",
        PGPORT = 9737L,
        PGDATABASE = "wrds",
        PGUSER = "your_WRDS_ID",
        PGPASSWORD = "your_WRDS_password")
```

Obviously, you should replace `your_WRDS_ID` and `your_WRDS_password` with your actual WRDS ID and WRDS password, respectively. This code will need to be run each time you

---

[2]https://www.youtube.com/watch?v=xRY6Y8qXUJ8
[3]https://www.rstudio.com/products/rstudio/download/#download
[4]https://cran.r-project.org
[5]You can copy and paste the code into the "Console" in RStudio.

open RStudio to access WRDS data in the code examples below. But once you have run this code, you do not need to run it again during the same session (i.e., until you close and reopen RStudio).

If the only PostgreSQL database you access is the WRDS database, you could put the values above in .`Renviron`, a special file that is opened every time you open R (see here[6] for more information on this file).[7] The contents of this file would look something like this:

```
PGHOST = "wrds-pgdata.wharton.upenn.edu"
PGPORT = 9737L
PGDATABASE = "wrds"
PGUSER = "your_WRDS_ID"
PGPASSWORD = "your_WRDS_password"
```

We discuss alternative approaches to setting up the WRDS database connection in Section 6.1, but we recommend this approach as it keeps the user-specific aspects of the code separate from the parts of the code that should work for everyone. By using environment variables, we ensure that the code in the book works for you if you copy it and paste it in your R console.

Note that we have striven to make the code in each chapter independent of the code in other chapters. So, if you feel comfortable with using R and have fulfilled the requirements listed above, you could easily jump ahead to a chapter of interest and start running code.

---

[6]https://go.unimelb.edu.au/ucd8

[7]We put our passwords in a special password file, as described in the PostgreSQL documentation at https://www.postgresql.org/docs/current/libpq-pgpass.html, so we don't need to set `PGPASSWORD`. It's obviously not a good idea to put your password in code.

2

Describing data

This chapter aims to immerse the active reader in applied data analysis by providing a brief, fast-paced introduction to both R and basic statistics. As discussed in Appendix C, R is both a programming language and a software environment focused on statistical computing and graphics. Like a natural language,[1] the best way to learn a computer language is to dive in and use it. Unlike a natural language, one can achieve a degree of fluency in a matter of hours. Also as a learner, you do not need to find a native speaker of R willing to tolerate your bad grammar; you can practise R without leaving your computer and get immediate feedback on your attempts.[2]

As we assume very little here in terms of prior exposure to R or statistics, some points in this chapter may seem a little tedious. But we expect these parts can be covered quickly. We highly recommend that you treat this chapter as a tutorial and work through the code that follows, either by typing or by copying and pasting into the R console (this is what we mean by an "active reader" above). Once you have completed the steps detailed in Section 1.2, the code below should allow you to produce the output below on your own computer in this fashion.[3]

Some parts of the code may seem a little mysterious or unclear. But just as one does not need to know all the details of the conditional tense in French to use *Je voudrais un café, s'il vous plaît*, you don't need to follow every detail of the commands below to get a sense of what they mean and the results they produce (you receive a coffee). For one, most concepts covered below will be examined in more detail in later chapters.

In addition, you can easily find help for almost any task in R's documentation, other online sources, or in the many good books available on statistical computing (and we provide a guide to further reading at the end of this chapter). Many of the R commands in the online version of this book[4] include links to online documentation.

We believe it is also helpful to note that the way even fluent speakers of computer languages work involves internet searches for things such as "how to reorder column charts using ggplot2" or asking questions on StackOverflow.[5] To be honest, some parts of this chapter were written with such an approach. The path to data science fluency for many surely involves years of adapting code found on the internet to a problem at hand.

As such, our recommendation is to go with the flow, perhaps playing with the code to see the effect of changes or trying to figure out how to create a variant that seems interesting to you. To gain the most from this chapter, you should try your hand at the exercises

---

[1] https://en.wikipedia.org/wiki/Natural_language

[2] Unfortunately, as with any computer language, R can be a stickler for grammar.

[3] With a decent internet connection, the steps outlined in Section 1.2 should take less than ten minutes. The idea of allowing the reader to produce *all* the output seen in this book by copying and pasting the code provided is a core commitment to the reader that we make.

[4] https://iangow.github.io/far_book/

[5] https://stackoverflow.com

DOI: 10.1201/9781003456230-2

included in the chapter. As with the remainder of the book, these exercises often fill in gaps
deliberately left in the text itself.

2.1   Introduction to R

2.1.1   Functions

R contains a multitude of **functions**. For example, the `exp()` function "computes the
exponential function":

```
exp(1)
```

```
[1] 2.718282
```

We can access the documentation for `exp()` by typing `help(exp)` or `? exp` at the console.
Looking at that documentation, we see that it has a single argument: x. So we could instead
type:

```
exp(x = 1)
```

```
[1] 2.718282
```

But with a single argument, there is no reason to do so. In contrast, the documentation for
`log()` shows the form `log(x, base = exp(1))`. This tells us that `log()` takes up to two
arguments. There is no default for the first argument x, but the second argument (`base`)
has a default of `exp(1)`. So, by default, `log()` will return the natural logarithm[6] of the
value supplied as a first argument.

Function arguments can be matched either by position or by name. In the first and third
examples below, we use positional matching; in the second and fourth, we name the arguments. In the first and second examples, by not supplying a second argument, R uses the
default of `exp(1)`.

```
log(8)
```

```
[1] 2.079442
```
```
log(x = 8)
```

```
[1] 2.079442
```
```
log(100, 10)
```

```
[1] 2
```
```
log(x = 100, base = 10)
```

```
[1] 2
```

To draw random numbers from the normal distribution, we would use the `rnorm()` function,
which has the form `rnorm(n, mean = 0, sd = 1)`, where n is the size of the sample and
`mean` and `sd` are the mean and standard deviation of the distribution we are sampling from.

---

[6]https://en.wikipedia.org/wiki/Natural_logarithm

If we want to draw a single random number, use the default value for `mean`, but set `sd` to 10, we would need to name the second argument.

```
rnorm(1, sd = 10)
```

```
[1] -1.2246
```

(Note that you may get a different value here because of the "randomness" inherent in the `rnorm()` function.)

The base R functions can be supplemented by creating new functions or by loading **packages** (we will discuss packages shortly). Creating functions is quite straightforward, as can be seen in the following example.[7]

```
square <- function(x) {
  x * x
}
```

We can now call our function like any other function:

```
square(2)
```

```
[1] 4
```

```
square(64)
```

```
[1] 4096
```

Each function provided by R or a package has documentation that can be accessed using the `help` function (or its shorthand equivalent ?). Here are a couple of examples that you should try:

```
help(exp)
? sin
```

2.1.2 Operators

We have already introduced one kind of operator, the **assignment operator** `<-`. In other languages (e.g., Python), this operator is written as =. In many contexts, one can use = in place of `<-`. While you may see R code in the wild using = for assignment, we always use `<-` in this book. In RStudio, one can use `Alt` + - (or `Cmd` + - on MacOS) to get `<-`. For help with assignment operators, run the following code in your console.

```
? `<-`
```

As would be expected of any programming language, especially one focused on statistics, R has a full set of **arithmetic operators**, including +, *, and /.

```
2 + 2
```

```
[1] 4
```

```
4 * 3
```

---

[7]We are glossing over a lot of details here, such as the fact that we used the *assignment operator* to store our function in a *variable* `square` and that the return value for the function is the last calculation performed. Keep reading to learn about these. We could be explicit about the returned value using `return(x * x)`, but this is not necessary.

```
[1] 12
5 / 2
```

```
[1] 2.5
```

As an aside, these operators are actually functions behind the scenes and one can access the function-style version by enclosing the operator in backticks:

```
`*` (4, 3)
```

```
[1] 12
```

Knowing this can be helpful for getting help on operators (the help for * provides information on other arithmetic operators including ^, %%, and %/%, so we recommend you look at the results of the following command).

```
? `*`
```

Another class of operators comprises the **relational operators**, such as >, >=, ==, and !=. As can be seen below, these operators yield either TRUE or FALSE.[8]

```
6 > 4
```

```
[1] TRUE
6 < 4
```

```
[1] FALSE
6 == 4
```

```
[1] FALSE
6 != 4
```

```
[1] TRUE
6 == 3 + 3
```

```
[1] TRUE
FALSE == FALSE
```

```
[1] TRUE
```

For help with relational operators, run the following code in your console.

```
? `>`
```

Another class of operators in R are **logical operators**, including ! (not), & (and), and | (or).

```
!(6 == 4)
```

```
[1] TRUE
!TRUE
```

```
[1] FALSE
```

---

[8]When either side of the comparison is missing (NA), the result is also NA. We discuss missing values in Section 2.3.2.

```
TRUE & FALSE
```

```
[1] FALSE
```
```
TRUE | FALSE
```

```
[1] TRUE
```

For help on the logical operators, run the following code in your console.

```
? `!`
```

2.1.3 Variables

A **variable** can be used to store an **object**, such as numerical values. We can use the assignment operator `<-` to store an object in a variable. For example, we can store 3 in the variable v as follows.

```
v <- 3
```

We can then recover and manipulate this value later.

```
v
```

```
[1] 3
```
```
v * 2.5
```

```
[1] 7.5
```

Not all objects are simple values like this. We discuss some of the more complex objects that can be created in R below.

2.1.4 Data types

Values in R draw on a number of different data types, including the following:

- Logical: `TRUE` or `FALSE`
- Integer: for example, `1` or `2`
- Numeric: for example, `1.0` or `256.3`
- Character: for example, `"ABC"` or `"hello"` or `"123"`

Numeric and integer values may appear alike, but we can specify that we mean an integer by appending `L` to the literal value we type.

```
1L
```

```
[1] 1
```

We can use the `class()` function to determine the data type of a value.

```
class(1L)
```

```
[1] "integer"
```
```
class(3.1415)
```

```
[1] "numeric"
```

In addition to these basic data types, there are more elaborate types such as dates, timestamps, and so on.

2.1.5   Data structures

2.1.5.1   Vectors

The most basic data structure in R is the **vector**, which we can think of as an ordered collection of values of a certain data type. Note the value stored in v above and the results of the various calculations we have performed so far are actually vectors. One way to construct multiple-element vectors is using the c() function (here c stands for "combine" or "concatenate").

```
c(0, 1, 4, 9, 16, 25)
```

```
[1]  0  1  4  9 16 25
```

Scalar addition and multiplication work as they do with vectors in linear algebra.[9] That is adding a **scalar** (i.e., a real number) to a vector adds that scalar to each element of the vector; multiplying a vector by a scalar multiplies each element of the vector by the scalar.

Multiplying or adding two vectors means multiplying or adding each element.

```
v <- c(1, 2, 4, 6)
v
```

```
[1] 1 2 4 6
```

```
v * 2.5
```

```
[1]  2.5  5.0 10.0 15.0
```

```
v + 3
```

```
[1] 4 5 7 9
```

```
w <- c(1, 2, 3, 4)
v + w
```

```
[1]  2  4  7 10
```

```
v * w
```

```
[1]  1  4 12 24
```

When we mix values of different data types in a single vector, R converts them to the type that can accommodate all values automatically. So, a logical value can be interpreted as an integer (1L for TRUE and 0L for FALSE), an integer can be represented as a numeric value, and a number (e.g., 3.1) can be thought of as a character ("3.1"). The following code demonstrates this: going from one line to the next, we add an element of a different type and the automatic conversion of the previous elements is evident from the output.

```
c(TRUE)
```

```
[1] TRUE
```

```
c(TRUE, 2L)
```

```
[1] 1 2
```

```
c(TRUE, 2L, 3.1)
```

```
[1] 1.0 2.0 3.1
```

---

[9]See Appendix A for more on linear algebra.

```
c(TRUE, 2L, 3.1, "frog")
```

```
[1] "TRUE" "2"    "3.1"  "frog"
```

We can repeat the same idea but using the class() function to identify the data type of a vector explicitly:

```
class(c(TRUE))
```

```
[1] "logical"
```

```
class(c(TRUE, 2L))
```

```
[1] "integer"
```

```
class(c(TRUE, 2L, 3.1))
```

```
[1] "numeric"
```

```
class(c(TRUE, 2L, 3.1, "frog"))
```

```
[1] "character"
```

A similar conversion is often performed when we apply functions to vectors. For example, the application of the sum() function to a numeric vector behaves as we would expect.

```
sum(1:10)
```

```
[1] 55
```

When the sum() function is applied to a logical vector, the vector is converted to an integer type.

```
v <- c(TRUE, TRUE, FALSE, FALSE, FALSE)
sum(v)
```

```
[1] 2
```

```
class(v)
```

```
[1] "logical"
```

```
class(sum(v))
```

```
[1] "integer"
```

```
!v
```

```
[1] FALSE FALSE  TRUE  TRUE  TRUE
```

```
sum(!v)
```

```
[1] 3
```

We can access a specific element of a vector using []. For example,

```
v[1]
```

```
[1] TRUE
```

```
v[4]
```

```
[1] FALSE
```

2.1.5.2 Lists

Another key data structure in R is the **list**, which is a general form of vector in which the various elements need not be of the same type and are often themselves vectors or lists. Lists provide a convenient way to, say, return the results of statistical computations.

Rather than using [] as we did above, we use the [[]] operator to access elements of a list.[10] The following code confirms that the first element is still of `logical` data type, and the fourth element is of type `character`.

```
l <- list(TRUE, 1L, 2.1, "frog")
l[[1]]
```

```
[1] TRUE
```
```
l[[4]]
```

```
[1] "frog"
```
```
class(l[[1]])
```

```
[1] "logical"
```
```
class(l[[4]])
```

```
[1] "character"
```

Elements of a list can be named, which is very useful in many contexts:

```
players <- list(name = c("M3GAN", "Magnus", "Hikaru"),
                yob = c(2022, 1990, 1987),
                rating = c(3000, 2859, 2678))

players[["name"]]
```

```
[1] "M3GAN"  "Magnus" "Hikaru"
```

Alternatively, we can access elements of a list using the $ operator:

```
players$yob
```

```
[1] 2022 1990 1987
```

To access the help for [], [[]], and $, use the following command:

```
? `[`
```

2.1.5.3 Other data structures

While vectors and lists are the most basic data structures in R, there are others including:

- **matrices** (more generally **arrays**) are multi-dimensional generalizations of vectors.
- **data frames** are matrix-like structures in which the columns can be of different types. A data frame can also be thought of as a list of vectors.

Data structures are a subset of the more general class of **objects** in R. Another kind of object in R is the function, which we saw earlier.

---

[10]Note that l[1] and l[4] work here, but each returns a single-element list, rather than the underlying single-element vector stored in the respective position.

2.1.6 Packages

When you open R, a number of functions are automatically made available, including `exp()`, `log()`, and the operators discussed above. But in practice much of the functionality of R is provided by **packages**.

Some packages are automatically installed with R; others require separate installation. For example, the following code installs the `dplyr` package (and you would have run a command like this when installing packages to satisfy the requirements listed in Section 1.2).

```
install.packages("dplyr")
```

Once a package has been installed, it can be loaded into R using the `library()` function.

```
library(dplyr)
```

Here you might see a mysterious message something like this:

```
## The following objects are masked from 'package:stats':
##
##      filter, lag
```

What does this mean? Basically, both the `stats` package (loaded by default when you start R) and the newly loaded `dplyr` package contain functions named `filter()` and `lag()`. The message tells that the newly loaded functions "mask" the ones from the `stats` package, so that when you call `filter()` now, you get the function supplied by the `dplyr` package.

This illustrates one reason for putting functions in packages: avoiding name conflicts between functions by only loading the packages you need. You can specify `warn.conflicts = FALSE` in calling `library` to suppress this message.

```
library(dplyr, warn.conflicts = FALSE)
```

While we generally suppress these messages in this book, you should pay attention to them in practice to avoid the possibility of calling a function different from that you intended to call. Of course, R allows you to specify the package from which you want to get a function by appending the package name and `::` before the function name (e.g., `stats::lag()` to get the `lag()` function from the `stats` package). In certain contexts (e.g., writing your own R packages), this approach will be highly advisable. For example, see the code[11] for the `get_size_rets_monthly()` function from the `farr` package, which we use in Chapter 15.

2.1.7 Data frames

Because the **data frame** is such an important data structure in this book, we devote a little more time to it here. Some kind of data frame is a basic feature of any statistical software system. A data frame is a data structure that organizes data into spreadsheet-like tables in which each column represents a variable and each row represents an observation.

Some data frames are made available automatically in R. The data frame `iris` represents a famous data set used in early research by Ronald Fisher (use `? iris` to learn more about this data set).

---

[11] https://github.com/iangow/farr/blob/main/R/get_size_rets_monthly.R

Using the `dim()` function, we see that the `iris` data set has 150 rows and 5 columns.

```
dim(iris)
```

```
[] 150   5
```

Because simply typing `iris` at the command line would show all 150 rows, we use the `head()` function to display the first six rows of the data.

```
head(iris)
```

```
  Sepal.Length Sepal.Width Petal.Length Petal.Width Species
1          5.1         3.5          1.4         0.2  setosa
2          4.9         3.0          1.4         0.2  setosa
3          4.7         3.2          1.3         0.2  setosa
4          4.6         3.1          1.5         0.2  setosa
5          5.0         3.6          1.4         0.2  setosa
6          5.4         3.9          1.7         0.4  setosa
```

We mentioned above that a data frame can be considered as a list of vectors. We can use this to access the first 6 rows of the second column as follows:

```
iris[[2]][1:6]
```

```
[1]  3.5 3.0 3.2 3.1 3.6 3.9
```

We can make this more readable by referring to the second column by its name:

```
iris[["Sepal.Width"]][1:6]
```

```
[1]  3.5 3.0 3.2 3.1 3.6 3.9
```

Additionally, just as with a list, we can use the `$` operator to access the second column:

```
iris$Sepal.Width[1:6]
```

```
[1]  3.5 3.0 3.2 3.1 3.6 3.9
```

Each column of a data frame is a vector, and each vector can be of a different type from the others. The second column is a numeric vector.

```
class(iris$Sepal.Width)
```

```
[1] "numeric"
```

From visual inspection, we might expect `iris$Species` to be a character vector, but it is actually of type `factor`.

```
class(iris$Species)
```

```
[1] "factor"
```

We can use the `table()` function to get the frequency of each unique value in the `Species` variable of `iris`.

```
table(iris$Species)
```

```
    setosa versicolor  virginica
        50         50         50
```

One innovation made available through the `dplyr` package is the **tibble**, which is best viewed as a more modern version of the base R data frame. (`data.frame`). The Tidyverse website[12] provides more information about tibbles. Most of the data frames we will use in this book are actually tibble objects. We can use the `as_tibble()` function to create a tibble version of `iris`:

```
iris_tbl <- as_tibble(iris)
iris_tbl
```

```
# A tibble: 150 x 5
   Sepal.Length Sepal.Width Petal.Length Petal.Width Species
          <dbl>       <dbl>        <dbl>       <dbl> <fct>
 1          5.1         3.5          1.4         0.2 setosa
 2          4.9         3            1.4         0.2 setosa
 3          4.7         3.2          1.3         0.2 setosa
 4          4.6         3.1          1.5         0.2 setosa
 5          5           3.6          1.4         0.2 setosa
 6          5.4         3.9          1.7         0.4 setosa
 7          4.6         3.4          1.4         0.3 setosa
 8          5           3.4          1.5         0.2 setosa
 9          4.4         2.9          1.4         0.2 setosa
10          4.9         3.1          1.5         0.1 setosa
# i 140 more rows
```

A tibble does pretty much all the things that a `data.frame` does, but has certain benefits, such as nicer default printing (no need for `head()` and the type of each column is indicated at the top).

2.2 Exploring data

In this section, we use data on the financial performance and stock price behaviour of Australian banks to introduce some additional basic concepts of R, provide a brief overview of some basic statistics, and showcase some of the data visualization functionality of R.

In addition to `dplyr` (already loaded above), we will use the `farr` package, which provides the data we will work with; the `ggplot2` package, which provides the plotting functions we use below; the `forcats` package, which allows us to work with factors (see discussion below); and `tidyr`, which we use to reshape data frames.

```
library(dplyr)
library(ggplot2)
library(farr)
library(forcats)   # For fct_inorder()
library(tidyr)     # For pivot_wider()
```

The `farr` package includes three data frames related to Australian banks. The first is `aus_banks`, which contains the identifiers (`gvkey`), tickers, and names of 10 banks.

---

[12]https://tibble.tidyverse.org

```
aus_banks
```

```
# A tibble: 10 x 3
   gvkey  ticker co_name
   <chr>  <chr>  <chr>
 1 014802 NAB    National Australia Bank Ltd
 2 015362 WBC    Westpac Banking Corp
 3 015889 ANZ    ANZ Bank New Zealand Limited
 4 024512 CBA    Commonwealth Bank of Australia
 5 200051 ADB    Adelaide Bank Ltd
 6 200575 BOQ    Bank of Queensland Ltd
 7 200695 BEN    Bendigo and Adelaide Bank Limited
 8 203646 SGB    St. George Bank Ltd
 9 212631 BWA    Bank Of Western Australia
10 312769 BBC    BNK Banking Corp Ltd
```

The second data frame is `aus_bank_funds`, which contains selected annual financial statement data for Australian banks. Investors often refer to such information as **fundamental** data, as it relates to the operating performance of the business that is used in "fundamental" valuation analysis.[13]

```
aus_bank_funds
```

```
# A tibble: 283 x 7
   gvkey  datadate        at    ceq    ib    xi    do
   <chr>  <date>       <dbl>  <dbl> <dbl> <dbl> <dbl>
 1 024512 1987-06-30  43887.  2469.  198.     0     0
 2 024512 1988-06-30  50445.  2693.  359.     0     0
 3 024512 1989-06-30  60650.  3055   476.     0     0
 4 024512 1990-06-30  67030.  3888.  524.     0     0
 5 024512 1991-06-30  89292.  4353.  883.     0     0
 6 024512 2008-06-30 487572  25619  4791    NA    NA
 7 024512 1992-06-30  88340.  5471.  409.     0     0
 8 024512 1993-06-30  90995.  5569.  443.     0     0
 9 024512 1994-06-30  91321.  5965.  682.     0     0
10 024512 1995-06-30  99595   6568.  983.     0     0
# i 273 more rows
```

The third data frame is `aus_bank_rets`, which contains monthly data on stock returns and market capitalization for Australian banks. Note that in `aus_bank_rets`, as in many financial data sets, a value of 0.148 in `ret` means that the stock had a 14.8% return for that month.

```
aus_bank_rets
```

```
# A tibble: 3,047 x 4
   gvkey  datadate       ret mkt_cap
   <chr>  <date>       <dbl>   <dbl>
 1 014802 1985-12-31 NA       1527.
 2 014802 1986-01-31  0.0833  1655.
 3 014802 1986-02-28  0.0977  1828.
 4 014802 1986-03-27  0.148   2098.
```

---

[13]The `NA` values are missing values, which we discuss in Section 2.3.2.

```
 5 014802 1986-04-30  0.0660   2237.
 6 014802 1986-05-30 -0.0464   2133.
 7 014802 1986-06-30 -0.0913   1891.
 8 014802 1986-07-31 -0.0696   1771.
 9 014802 1986-08-29  0.0551   1868.
10 014802 1986-09-30 -0.0299   1813.
# i 3,037 more rows
```

Descriptions of the variables found in the three data frames are provided in Table 2.1.

TABLE 2.1
Variables in the Australian bank data

Item	Description
gvkey	Firm identifier
ticker	Stock exchange ticker
co_name	Bank's name
ret	Stock return (monthly)
mkt_cap	Market capitalization
ni	Net income
ib	Income before extraordinary items
xi	Extraordinary items
do	Discontinued items

One thing to note about the three data frames is that, while each covers the same Australian banks, they differ significantly in their number of rows, which is a function of how many times each bank is represented in each data frame.

In `aus_banks`, a bank is shown only once, so there are only 10 rows.[14] In contrast, `aus_bank_funds` has data for each bank for a number of financial years, so it has 283 rows. And `aus_bank_rets` has data for each bank for a number of months giving it 3047 rows.

2.2.1 Data visualization

Suppose we are interested in making a plot of the market capitalization of Australian banks using the latest available data in `aus_bank_rets`. We can apply the `max()` function to the `datadate` column of aus_bank_rets to determine the meaning of "the latest available data":

```
max(aus_bank_rets$datadate)
```

```
[1] "2022-10-31"
```

This suggests that we want to take a subset of the `aus_bank_rets` data, having `datadate` equal to that value (assuming that the latest available date is the same for all the banks in the table). There are a number of ways to do this, but in this book we will focus on approaches based on `dplyr`.

According to its home page:[15] "`dplyr` is a grammar of data manipulation, providing a consistent set of verbs that help you solve the most common data manipulation challenges:

---

[14]Note that a bank will not always have exactly the same name or ticker over time, but this is an issue that we defer to Chapter 7.

[15]https://dplyr.tidyverse.org

- `mutate()` adds new variables that are functions of existing variables
- `select()` picks variables based on their names
- `filter()` picks cases based on their values
- `summarise()` reduces multiple values down to a single summary
- `arrange()` changes the ordering of the rows"

These `dplyr` verbs are functions that take a data frame (whether `data.frame` or `tibble`) as an input and return a data frame as output. As such, the `dplyr` verbs are often used in conjunction with the **pipe** (`|>` below).[16] The following code takes `aus_bank_rets` and uses the pipe (`|>`) to pass this to the `filter()` function, which returns only those rows satisfying the condition supplied (`datadate == max(datadate)`).

```
aus_bank_rets |>
  filter(datadate == max(datadate))
```

```
# A tibble: 7 x 4
  gvkey   datadate      ret   mkt_cap
  <chr>   <date>      <dbl>    <dbl>
1 014802  2022-10-31  0.125  102247.
2 015362  2022-10-31  0.168   84412.
3 015889  2022-10-31  0.121   76422.
4 024512  2022-10-31  0.154  177567.
5 200575  2022-10-31  0.168    4765.
6 200695  2022-10-31  0.157    5118.
7 312769  2022-10-31  0.116      74.2
```

Having identified the data we want for our plot, we next need to pass this to our plotting function. In this book, we will focus on the `ggplot2` package for plots. According to its webpage,[17] "`ggplot2` is a system for declaratively creating graphics, based on 'The Grammar of Graphics' (Wilkinson, 2005). You provide the data, tell `ggplot2` how to map variables to aesthetics, what graphical primitives to use, and it takes care of the details."

In this case, we create a filtered data frame (`latest_mkt_cap`)

```
latest_mkt_cap <-
  aus_bank_rets |>
  filter(datadate == max(datadate))
```

We then pipe this data frame to the `ggplot()` function. As discussed above, we need to tell `ggplot()` "how to map variables to aesthetics" using the `aes()` function inside the `ggplot()` call. In this case, we put `gvkey` on the x-axis, `mkt_cap` on the y-axis, and also map `gvkey` to `fill` (the colour of the elements) so that each firm has a different colour. We next add (using +) a "layer" using a **geom** function, in this case `geom_col()`, which gives us a column graph.[18]

```
latest_mkt_cap |>
  ggplot(aes(x = gvkey, y = mkt_cap, fill = gvkey)) +
  geom_col()
```

---

[16]This "native pipe" (`|>`) was added to R relatively recently and you will often see code using `%>%`, the pipe from the `magrittr` package that is made available via `dplyr`. Both pipes behave similarly in most respects, but there are minor differences.

[17]https://ggplot2.tidyverse.org

[18]The + behaves in many respects like the pipe operator. For discussion of why `ggplot2` uses + instead of `|>` or `%>%`, see https://community.rstudio.com/t/why-cant-ggplot2-use/4372/7.

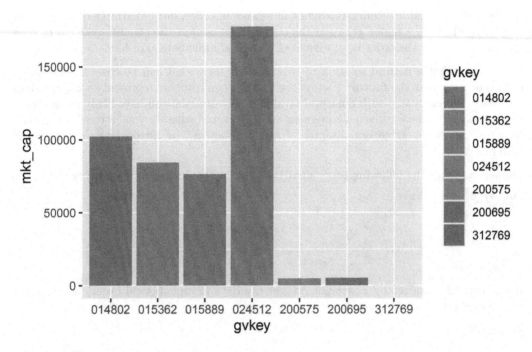

FIGURE 2.1
Market capitalization of Australian banks (initial version)

So we have a plot—Figure 2.1—but a few problems are evident:

- Why are the columns shown in a seemingly random order? Plots are much easier to interpret when information is presented in a sensible manner (e.g., bars increasing or decreasing in size).
- Which banks are represented here? Few people have any idea which bank is represented by the GVKEY 014802.
- Why are the GVKEYs shown in two places? Surely, we only need either the x-axis labels or the legend, not both.

Fortunately, all of these problems can be solved.

We start with the first problem. We can use `arrange()` to see how R thinks about the natural order of our data. The `arrange()` verb orders the rows of a data frame by the values of selected columns (in this case `gvkey`).

```
latest_mkt_cap |>
  arrange(gvkey)
```

```
# A tibble: 7 x 4
  gvkey  datadate      ret  mkt_cap
  <chr>  <date>      <dbl>    <dbl>
1 014802 2022-10-31  0.125  102247.
2 015362 2022-10-31  0.168   84412.
3 015889 2022-10-31  0.121   76422.
4 024512 2022-10-31  0.154  177567.
5 200575 2022-10-31  0.168    4765.
6 200695 2022-10-31  0.157    5118.
7 312769 2022-10-31  0.116     74.2
```

We can now see that the columns are not in a random order, but are arranged in the order of gvkey. The `<chr>` under gvkey in the output tells us that the column gvkey is of type "character" and so the order is an extended version of alphabetical order.

It turns out that a natural solution to this problem uses what can be considered another data type known as the **factor**. Factors are useful when a vector represents a **categorical variable**, especially one with a natural ordering. For example, you have likely encountered a survey where you were given a statement and then asked whether you "strongly disagree", "disagree", "neither agree nor disagree" (neutral), "agree", or "strongly agree" with the statement.

Suppose we have the following (very small) data set with the results of a survey:

```
survey_df <- tibble(
  participant = 1:10,
  response = c("disagree", "strongly agree", "strongly agree",
               "strongly disagree", "neutral",  "disagree",
               "neutral", "agree", "agree", "strongly disagree"))
```

If we sort this data set by `response`, we get the data in alphabetical order, which is not very meaningful.

```
survey_df |>
  arrange(response)
```

```
# A tibble: 10 x 2
   participant response
         <int> <chr>
1            8 agree
2            9 agree
3            1 disagree
4            6 disagree
5            5 neutral
6            7 neutral
7            2 strongly agree
8            3 strongly agree
9            4 strongly disagree
10          10 strongly disagree
```

One option might be to convert the responses to a numerical scale (e.g., "strongly disagree" equals 1 and "strongly agree" equals 5). But this would mean the loss of the underlying descriptive values that are more meaningful for humans. R offers a better approach, which is to encode `response` as a **factor** using the `factor()` function and to indicate the order using the `levels` argument.

We can use the `mutate()` verb to replace the existing version of `response` with a new factor version. The `mutate()` function can be used to add new variables to a data frame, but new variables will replace existing variables that have the same name. Applying this approach to our `survey_df` data frame produces data organized in a more meaningful way:

```
survey_df |>
  mutate(response = factor(response,
                           levels = c("strongly disagree", "disagree",
```

```
                                         "neutral", "agree",
                                         "strongly agree"))) |>
  arrange(response)
```

```
# A tibble: 10 x 2
   participant response
         <int> <fct>
 1           4 strongly disagree
 2          10 strongly disagree
 3           1 disagree
 4           6 disagree
 5           5 neutral
 6           7 neutral
 7           8 agree
 8           9 agree
 9           2 strongly agree
10           3 strongly agree
```

A logical way to organize the columns of our plot is by market capitalization. We can use `arrange(desc(mkt_cap))` to order the data in descending (`desc`) order of `mkt_cap`, and then use the function `fct_inorder()` to create a factor version of `gvkey` in the same order.

Moving to the next problem, the immediate issue is that we are getting our data from `aus_bank_rets`, which has no information about the names (or tickers) of the banks in our data. While `aus_banks` has names and tickers, it lacks data on market capitalization.

In this case, we need to use one of the **two-table verbs** offered by `dplyr`. These verbs take two tables as inputs and return as output a single table based on matching values in selected variables across the two tables. We briefly describe four of these verbs here:

- `inner_join(x, y)` only includes observations that match in both `x` and `y`.
- `left_join(x, y)` includes all observations in `x`, regardless of whether they have a match in `y` or not.
- `right_join(x, y)` includes all observations in `y`. It is roughly equivalent to `left_join(y, x)`.
- `full_join(x, y)` includes all observations from `x` and `y`.

Our immediate needs are met with `inner_join()`, but later in the chapter we will have a use for `left_join()`. Note that we will generally use the pipe to rewrite (for example) `inner_join(x, y)` as `x |> inner_join(y)`.

The variables to be used in the match are specified using the `by` argument. As our matching variable across tables is the firm identifier `gvkey`, we specify `by = "gvkey"`.

Putting all these pieces together we create a new version of `latest_mkt_cap` below. Because we don't need all the variables in our plot, we can use the `select()` function to keep just the variables we will use. We use the `fct_inorder()` function to create a factor version of `ticker` with the levels ordered based on their order in the data, which is in descending order of `mkt_cap` because of the preceding `arrange(desc(mkt_cap))` line.

```
latest_mkt_cap <-
  aus_banks |>
  inner_join(aus_bank_rets, by = "gvkey") |>
```

```
    filter(datadate == max(datadate)) |>
    select(ticker, co_name, mkt_cap) |>
    arrange(desc(mkt_cap)) |>
    mutate(ticker = fct_inorder(ticker))
```

Having created our data frame, we can pass this to `ggplot()` again to create Figure 2.2. Note that we specify `theme(legend.position = "none")` to delete the redundant legend.

```
latest_mkt_cap |>
    ggplot(aes(x = ticker, y = mkt_cap, fill = ticker)) +
    geom_col() +
    theme(legend.position = "none")
```

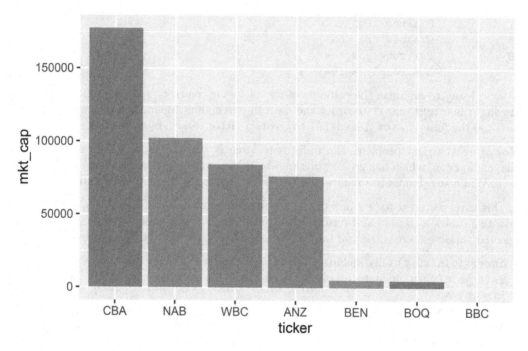

FIGURE 2.2
Market capitalization of Australian banks

2.3 Basic data analysis and statistics

Having covered (quickly!) some ground in terms of manipulating data frames and making plots in R, we now move on to some basic statistics. Although we assume little prior knowledge of statistics, our focus in this section is on conducting statistical analysis using R.

2.3.1 Mean

The mean of a variable x in a sample of n observations is $\bar{x} = \frac{\sum_{i=1}^{n} x_i}{n}$, where x_i is the value for observation i. We can calculate the sum of a vector using the sum() function and length() gives the number of elements in a vector.[19]

```
sum(aus_bank_funds$ib)
```

```
[1] 520938.7
```

So we can calculate the mean of ib "by hand" as follows:

```
sum(aus_bank_funds$ib) / length(aus_bank_funds$ib)
```

```
[1] 1840.773
```

Of course, R has a built-in mean() function, which we can see produces the same answer.

```
mean(aus_bank_funds$ib)
```

```
[1] 1840.773
```

2.3.2 Missing values

At this point we take a moment for a brief digression about **missing values**. One thing that you may notice regarding aus_bank_rets is that the first value of ret is NA. The value NA is a special value that indicates a missing value. Values can be missing for many different reasons. In aus_bank_rets, we calculated returns by comparing end-of-month stock prices with beginning-of-month stock prices and adding the value of distributions (e.g., dividends). But there will always be a first month for any given stock, which means we will have missing values for ret in that first month.

Missing values might also arise in survey responses if participants do not complete all questions. Later in this chapter, we see that missing values can also arise when merging different data sets.

The correct handling of missing values is important in many statistical contexts. For example, we might ask participants to indicate their weight (in kilograms). Anne says 51, Bertha says 55, but Charlie does not complete that question. In estimating the average weight of participants, we might ignore the missing case and simply take the average of 51 and 55 (i.e., 53).

```
survey_data <- list(name = c("Anne", "Bertha", "Charlie"),
                    weight = c(51, 55, NA))
```

But the reality is that we "don't know" what the average weight of our participants is and in some sense NA is an alternative way of saying "don't know". With this interpretation, a single "don't know" means that aggregated statistics such as sums and means are also "don't know". Reflecting its origins in statistics, R is careful about these issues, which we see in its handling of NA values. When asked the mean of weight, R's response is (in effect) "don't know".

```
mean(survey_data$weight)
```

```
[1] NA
```

---

[19]Recall that variables in data frames are actually vectors.

In the setting above, we might be concerned that Charlie chose not to answer the question about weight because he really let himself go during Covid lockdowns (and thus is above average weight) and feels too embarrassed to answer. In such a case, we would expect a biased estimate of the average weight of participants if we just ignored Charlie's non-response. But, sometimes we might be willing to make assumptions that make it appropriate to ignore missing values and we can do so by specifying `na.rm = TRUE`. We will also see a paper studying missing values in Section 8.5.

```
mean(survey_data$weight, na.rm = TRUE)
```

```
[1] 53
```

In other contexts, we may use the `is.na()` function to check whether a value is `NA`. For example, to create a data frame based on `aus_bank_rets`, but excluding cases with missing `ret` values, we could use the following code (note the use of ! to mean "not").

```
aus_bank_rets |>
  filter(!is.na(ret))
```

```
# A tibble: 3,037 x 4
   gvkey  datadate        ret mkt_cap
   <chr>  <date>        <dbl>   <dbl>
 1 014802 1986-01-31  0.0833    1655.
 2 014802 1986-02-28  0.0977    1828.
 3 014802 1986-03-27  0.148     2098.
 4 014802 1986-04-30  0.0660    2237.
 5 014802 1986-05-30 -0.0464    2133.
 6 014802 1986-06-30 -0.0913    1891.
 7 014802 1986-07-31 -0.0696    1771.
 8 014802 1986-08-29  0.0551    1868.
 9 014802 1986-09-30 -0.0299    1813.
10 014802 1986-10-31  0.0692    1938.
# i 3,027 more rows
```

2.3.3 Median (and other quantiles)

The median of a variable can be described as the middle value of that variable. When we have an odd number of observations, the median is clear:

```
median(1:7)
```

```
[1] 4
```

When there is an even number of observations, some kind of averaging is needed:

```
median(1:8)
```

```
[1] 4.5
```

```
median(aus_bank_funds$ib)
```

```
[1] 427
```

The median is actually a special case of the more general idea of **quantiles**:

```
quantile(aus_bank_funds$ib, probs = 0.5)
```

```
50%
427
```

As can be seen in the output above, quantiles are expressed as **percentiles** (i.e., 0.5 is represented as 50%). In describing a variable, the most commonly cited percentiles (apart from the median) are the 25th and the 75th, which are also known as the first and third quartiles, respectively.

```
quantile(aus_bank_funds$ib, probs = c(0.25, 0.50, 0.75))
```

```
   25%       50%        75%
52.558   427.000   2882.000
```

By default, the summary() function produces these three quantiles, the mean, minimum (min()), and maximum (max()) values.

```
summary(aus_bank_funds$ib)
```

```
   Min.   1st Qu.   Median     Mean   3rd Qu.      Max.
-1562.40    52.56   427.00  1840.77   2882.00   9928.00
```

The summary() function can also be applied to a data frame.

```
summary(aus_bank_funds)
```

```
   gvkey              datadate              at                  ceq
Length:283        Min.   :1987-06-30   Min.   :       0.4   Min.   :    0.33
Class :character  1st Qu.:1998-04-30   1st Qu.:    9227.0   1st Qu.:  538.76
Mode  :character  Median :2005-09-30   Median :   76143.1   Median : 4980.00
                  Mean   :2006-02-12   Mean   :  230875.2   Mean   :14293.54
                  3rd Qu.:2014-06-30   3rd Qu.:  332403.0   3rd Qu.:18843.50
                  Max.   :2022-09-30   Max.   : 1215260.0   Max.   :78713.00


      ib                  xi                  do
Min.   :-1562.40   Min.   :-6068.00   Min.   :-6068.00
1st Qu.:   52.56   1st Qu.:    0.00   1st Qu.:    0.00
Median :  427.00   Median :    0.00   Median :    0.00
Mean   : 1840.77   Mean   :  -31.71   Mean   :  -29.41
3rd Qu.: 2882.00   3rd Qu.:    0.00   3rd Qu.:    0.00
Max.   : 9928.00   Max.   : 2175.00   Max.   : 2175.00
                   NA's   :117        NA's   :117
```

A bank's income is likely to be an increasing function of shareholders' equity. Banks' return on assets generally come in the form of interest on loans to businesses and consumers, so more assets generally means more income. Assets equal liabilities plus shareholders' equity and bank regulators normally insist on limits on the amount of assets based on the amount of shareholders' equity. For these reasons, a common measure of bank performance is return on equity, which we can calculate as roe = ib / ceq.[20]

```
aus_bank_roes <-
  aus_bank_funds |>
  mutate(roe = ib/ceq) |>
  filter(!is.na(roe))
```

[20]In practice, it is more common to use either average or beginning shareholders' equity in the denominator, but this would add complexity that is unhelpful for our current purposes.

```
summary(aus_bank_roes$roe)
```

```
   Min.  1st Qu.   Median     Mean  3rd Qu.     Max.
-0.23966  0.08563  0.12934  0.11884  0.15560  0.23052
```

The output from `summary()` suggests a wide variation in ROE for our banks.

2.3.4 Histograms

While the output from `summary()` above is helpful, a more comprehensive representation of the data is provided by the **histogram**. To produce a histogram for a variable x, we divide the range of values of x into bins, count the number of observations falling into each bin, and then plot these counts. The histogram is made available via the `geom_histogram()` function from `ggplot2`.

Below we specify the width of the bins (`binwidth = 0.01`) and `ggplot2` does the rest. Note that we only specify the `x` aesthetic because the values on both axes are created from that one data series.

```
aus_bank_roes |>
  ggplot(aes(x = roe)) +
  geom_histogram(binwidth = 0.01)
```

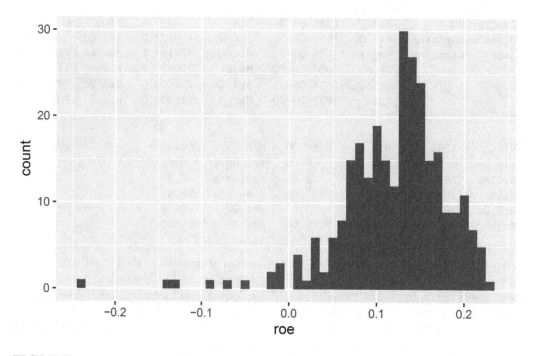

FIGURE 2.3
Distribution of ROEs for Australian banks

From Figure 2.3, we see that most observations have positive ROE, but there is a long tail of negative ROE firm-years.

2.3.5 Variance

The **variance** of a variable in a sample of n observations is $\sigma^2 = \frac{\sum_{i=1}^{n}(x_i - \bar{x})^2}{n-1}$, where $\bar{x}$ is the mean of x. Note that the denominator is $n-1$ rather than n for statistical reasons that are not important for now.[21]

We can use this formula to calculate the variance of `ib` "by hand":

```
sum((aus_bank_funds$ib - mean(aus_bank_funds$ib))^2) /
  (length(aus_bank_funds$ib) - 1)
```

```
[1] 6584270
```

However, it would be easier to put this formula into a function and use that function.

```
var_alt <- function(x) {
  sum((x - mean(x))^2) / (length(x) - 1)
}

var_alt(aus_bank_funds$ib)
```

```
[1] 6584270
```

We stored our function as `var_alt()` so that our function doesn't mask the built-in function `var()`:

```
var(aus_bank_funds$ib)
```

```
[1] 6584270
```

The **standard deviation** is the square root of the variance.

```
sqrt(var(aus_bank_funds$ib))
```

```
[1] 2565.983
```

Naturally, R has a built-in function for the standard deviation (`sd()`).

```
sd(aus_bank_funds$ib)
```

```
[1] 2565.983
```

Note that one benefit of the standard deviation is that it is expressed in the same units as the underlying variable (e.g., dollars rather than the dollars-squared in which the variance of `ib` is expressed).

2.3.6 Covariance and correlation

We will soon see that a lot of research focuses not on how variables vary alone, but on how they vary along with other variables.

For example, we might be interested in how the stock returns for our two largest banks (`CBA` and `NAB`) vary together. Here we create a focused data set `rets_nab_cba` for this purpose.

---

[21] In some texts, you will see n as the numerator. We use $n-1$, as that is what is used by R's `var()` function. An intuitive explanation for using $n-1$ is that one **degree of freedom** was used in calculating the mean, so we should subtract one from the denominator.

```
rets_nab_cba <-
  aus_bank_rets |>
  inner_join(aus_banks, by = "gvkey") |>
  filter(ticker %in% c("CBA", "NAB")) |>
  select(ticker, datadate, ret)

rets_nab_cba
```

```
# A tibble: 817 x 3
   ticker datadate          ret
   <chr>  <date>          <dbl>
 1 NAB    1985-12-31 NA
 2 NAB    1986-01-31   0.0833
 3 NAB    1986-02-28   0.0977
 4 NAB    1986-03-27   0.148
 5 NAB    1986-04-30   0.0660
 6 NAB    1986-05-30  -0.0464
 7 NAB    1986-06-30  -0.0913
 8 NAB    1986-07-31  -0.0696
 9 NAB    1986-08-29   0.0551
10 NAB    1986-09-30  -0.0299
# i 807 more rows
```

Note that we see a new operator (`%in%`) that relates to the `match()` function (type `? match` in R for documentation) and whose meaning should be clear from the context.

First, we look at the variance of `ret`. Running the line below, we should not be surprised that we get the result `NA` (see discussion of missing values above).

```
var(rets_nab_cba$ret)
```

```
[1] NA
```

We can `na.rm = TRUE` to calculate a variance omitting missing values.

```
var(rets_nab_cba$ret, na.rm = TRUE)
```

```
[1] 0.00357181
```

We can also use the `summarize()` function, which takes a data frame as input and creates a new data frame with summarized values.

```
rets_nab_cba |>
  summarize(var_ret = var(ret, na.rm = TRUE))
```

```
# A tibble: 1 x 1
  var_ret
    <dbl>
1 0.00357
```

In practice, we are probably less interested in the variance of the returns of all banks considered as a single series than in the variance of returns on a per-stock basis. Fortunately, the `group_by()` verb from the `dplyr` package makes this easy. The `group_by()` function takes an existing data frame and converts it into a **grouped data frame** so that subsequent operations using `summarize()` are performed by group. The groups created by `group_by()`

can be removed by applying `ungroup()` (or by including `.groups = "drop"` in the call to `summarize()`).

```
rets_nab_cba |>
  group_by(ticker) |>
  summarize(var_ret = var(ret, na.rm = TRUE)) |>
  ungroup()
```

```
# A tibble: 2 x 2
  ticker var_ret
  <chr>    <dbl>
1 CBA    0.00326
2 NAB    0.00384
```

At this stage, readers familiar with SQL will recognize parallels between the verbs from **dplyr** and elements of **SQL**, the most common language for interacting with relational databases. We explore these parallels in more depth in Appendix B.

A core concept in measuring how two variables vary together is **covariance**:

$$\text{cov}(x,y) = \frac{\sum_{i=1}^{n}(x_i - \overline{x})(y_i - \overline{y})}{n-1}$$

An issue with the data frame above is that the returns for NAB and CBA for (say) 2022-07-29 are in completely different rows, making it difficult to calculate the covariance between these two sets of returns. Fortunately, we can use the `pivot_wider()` function from the `tidyr` package to reorganize the data.

The following code creates two new variables `NAB` and `CBA`, representing the returns for NAB and CBA, respectively. The column names come from `ticker`, and the values from `ret`. The column `datadate` serves as the identifier for each row (hence, `id_cols = "datadate"`).

```
rets_nab_cba_wide <-
  rets_nab_cba |>
  pivot_wider(id_cols = datadate,
              names_from = ticker,
              values_from = ret) |>
  drop_na()

rets_nab_cba_wide
```

```
# A tibble: 368 x 3
   datadate        NAB      CBA
   <date>        <dbl>    <dbl>
 1 1991-10-31  0.105    0.106
 2 1991-12-31  0.0544   0.0703
 3 1992-01-31 -0.0476  -0.0758
 4 1992-02-28 -0.0171   0.0164
 5 1992-03-31 -0.0147  -0.0121
 6 1992-04-30  0.0435   0.0576
 7 1992-05-29  0.0417   0.0159
 8 1992-06-30 -0.00861 -0.0677
 9 1992-07-30  0.0376   0.00559
10 1992-08-31 -0.0663  -0.0472
# i 358 more rows
```

We will see that calculating the covariance of returns using data in this form will be straight-forward. We can easily make a function for calculating the covariance of two variables by translating the formula above into R code (we call it `cov_alt()` to avoid masking the built-in `cov()` function).

```
cov_alt <- function(x, y) {
  sum((x - mean(x)) * (y - mean(y))) / (length(x) - 1)
}
```

We can now calculate the covariance of the returns of NAB and CBA:

```
cov_alt(rets_nab_cba_wide$NAB, rets_nab_cba_wide$CBA)
```

```
[1] 0.00249763
```

And we can see that this yields the same result as we get from the built-in `cov()` function:

```
cov(rets_nab_cba_wide$NAB, rets_nab_cba_wide$CBA)
```

```
[1] 0.00249763
```

We can even apply `cov()` to a data frame (note that we drop `datadate` here using `select(-datadate)`):

```
rets_nab_cba_wide |>
  select(-datadate) |>
  cov()
```

```
          NAB         CBA
NAB 0.003641643 0.002497630
CBA 0.002497630 0.003279127
```

A concept that is closely related to covariance is **correlation**. The correlation of two variables x and y is given by the formula

$$\text{cor}(x, y) = \frac{\text{cov}(x, y)}{\sigma_x \sigma_y}$$

where σ_x and σ_y are the standard deviations of x and y, respectively. The correlation between any two variables will range between -1 and 1.

We can calculate the correlation between the returns of CBA and NAB just as we did for the covariance.

```
cor(rets_nab_cba_wide$NAB, rets_nab_cba_wide$CBA)
```

```
[1] 0.7227704
```

Like `cov()`, `cor()` can be applied to a data frame. Here we calculate the correlation between the returns for all stocks found on `latest_mkt_cap`:

```
aus_banks |>
  filter(ticker %in% latest_mkt_cap$ticker) |>
  inner_join(aus_bank_rets, by = "gvkey") |>
  pivot_wider(id_cols = datadate,
              names_from = ticker,
              values_from = ret) |>
  select(-datadate) |>
  cor(use = "pairwise.complete.obs")
```

```
            NAB         WBC        ANZ        CBA        BOQ         BEN         BBC
NAB  1.0000000  0.74844288  0.76604166  0.72277038  0.5728578  0.51585079  0.16398454
WBC  0.7484429  1.00000000  0.76320108  0.73093523  0.5578418  0.44736669  0.04194391
ANZ  0.7660417  0.76320108  1.00000000  0.70294937  0.4779466  0.48603484  0.08941787
CBA  0.7227704  0.73093523  0.70294937  1.00000000  0.5288250  0.47925147  0.04485547
BOQ  0.5728578  0.55784177  0.47794661  0.52882504  1.0000000  0.63651079  0.11605958
BEN  0.5158508  0.44736669  0.48603484  0.47925147  0.6365108  1.00000000  0.04766109
BBC  0.1639845  0.04194391  0.08941787  0.04485547  0.1160596  0.04766109  1.00000000
```

From the above, you can see that the correlation of any variable with itself is always 1.

2.3.7 Linking stock prices to fundamentals

Suppose that we want to create a plot of the market-to-book ratio of each bank, where we define this ratio as `mb = mkt_cap / ceq`. Clearly, we need to use one of the two-table verbs because `mkt_cap` is on `aus_bank_rets` and `ceq` is on `aus_bank_funds`.

Here is one approach:

```
aus_bank_rets |>
  inner_join(aus_bank_funds, by = c("gvkey", "datadate")) |>
  select(gvkey, datadate, mkt_cap, ceq) |>
  mutate(mb = mkt_cap / ceq)
```

```
# A tibble: 169 x 5
   gvkey  datadate   mkt_cap    ceq     mb
   <chr>  <date>       <dbl>  <dbl>  <dbl>
 1 014802 1987-09-30   3382.  2848.   1.19
 2 014802 1988-09-30   5076.  4097.   1.24
 3 014802 1991-09-30   8806.  7700.   1.14
 4 014802 1992-09-30   9145.  7995.   1.14
 5 014802 1993-09-30  16519.  8816.   1.87
 6 014802 1994-09-30  14125.  9852    1.43
 7 014802 1996-09-30  19671. 12519    1.57
 8 014802 1997-09-30  30063. 12579    2.39
 9 014802 1998-09-30  29516. 15028    1.96
10 014802 1999-09-30  33289. 15845    2.10
# i 159 more rows
```

The problem with this approach is that we have just one ratio per year. If you look at the financial pages of a newspaper or financial website, you will see that these provide values of ratios like this as frequently as they provide stock prices. They do this by comparing the latest available data for both fundamentals (here `ceq`) and stock prices (here `mkt_cap`). For example, on 28 January 2023, the website[22] of the Australian Stock Exchange (ASX) listed a stock price for CBA of $109.85, and earnings per share of $5.415 for a price-to-earnings ratio of 20.28. The earnings per share number is the diluted earnings per share from continuing operations reported on the face of CBA's income statement[23] for the year ended 30 June 2022.

We can start the process of constructing something similar by using `left_join()` in place of the `inner_join()` we used above.

---

[22]https://www2.asx.com.au/markets/company/cba
[23]https://go.unimelb.edu.au/hzw8

```
aus_bank_rets |>
  left_join(aus_bank_funds, by = c("gvkey", "datadate")) |>
  filter(datadate >= "1987-09-30") |>
  select(gvkey, datadate, mkt_cap, ceq) |>
  mutate(mb = mkt_cap / ceq) |>
  print(n = 15)
```

```
# A tibble: 2,984 x 5
   gvkey  datadate   mkt_cap   ceq    mb
   <chr>  <date>       <dbl> <dbl> <dbl>
 1 014802 1987-09-30   3382. 2848.  1.19
 2 014802 1987-10-30   2507.    NA  NA
 3 014802 1987-11-27   2712.    NA  NA
 4 014802 1987-12-31   2558.    NA  NA
 5 014802 1988-01-29   2617.    NA  NA
 6 014802 1988-02-29   2677.    NA  NA
 7 014802 1988-03-31   3034.    NA  NA
 8 014802 1988-04-29   3165.    NA  NA
 9 014802 1988-05-31   3716.    NA  NA
10 014802 1988-06-30   4100.    NA  NA
11 014802 1988-07-29   4445.    NA  NA
12 014802 1988-08-31   4689.    NA  NA
13 014802 1988-09-30   5076. 4097.  1.24
14 014802 1988-10-31   4900.    NA  NA
15 014802 1988-11-30   4873.    NA  NA
# i 2,969 more rows
```

The issue we see now is that most rows have no value for `ceq`, hence no value for `mb`. Following the logic applied on the ASX's website, we want to carry forward the value of `ceq` until it is updated with new financial statements. To do this, we will use the `fill()` function from the `tidyr` package. Once we `group_by(gvkey)` (to indicate that we only want to "fill" data within a company) and `arrange(datadate)` (so that R knows the applicable order for the data), we can call `fill(ceq, direction = "down")` to produce data like the following, which appears to be what we want (compare the output below with that above).

```
aus_bank_rets |>
  inner_join(aus_banks, by = "gvkey") |>
  filter(ticker %in% latest_mkt_cap$ticker) |>
  left_join(aus_bank_funds, by = c("gvkey", "datadate")) |>
  select(ticker, datadate, mkt_cap, ceq) |>
  group_by(ticker) |>
  arrange(datadate) |>
  fill(ceq, .direction = "down") |>
  ungroup() |>
  mutate(mb = mkt_cap / ceq) |>
  filter(!is.na(mb)) |>
  arrange(ticker, datadate) |>
  print(n = 15)
```

```
# A tibble: 2,364 x 5
   ticker datadate   mkt_cap   ceq    mb
```

```
    <chr>   <date>        <dbl> <dbl> <dbl>
 1  ANZ     1987-09-30    3708. 3139. 1.18
 2  ANZ     1987-10-30    2708. 3139. 0.863
 3  ANZ     1987-11-30    2555. 3139. 0.814
 4  ANZ     1987-12-31    2569. 3139. 0.819
 5  ANZ     1988-01-29    2495. 3139. 0.795
 6  ANZ     1988-02-29    2650. 3139. 0.844
 7  ANZ     1988-03-31    2933. 3139. 0.934
 8  ANZ     1988-04-29    3016. 3139. 0.961
 9  ANZ     1988-05-31    3478. 3139. 1.11
10  ANZ     1988-06-30    3180. 3139. 1.01
11  ANZ     1988-07-29    3429. 3139. 1.09
12  ANZ     1988-08-31    4100. 3139. 1.31
13  ANZ     1988-09-30    4406. 3903. 1.13
14  ANZ     1988-10-31    4572. 3903. 1.17
15  ANZ     1988-11-30    4505. 3903. 1.15
# i 2,349 more rows
```

Now that the code appears to be working, we can store the result in `aus_bank_mb`.

```
aus_bank_mb <-
  aus_bank_rets |>
  inner_join(aus_banks, by = "gvkey") |>
  filter(ticker %in% latest_mkt_cap$ticker) |>
  left_join(aus_bank_funds, by = c("gvkey", "datadate")) |>
  select(ticker, datadate, mkt_cap, ceq) |>
  group_by(ticker) |>
  arrange(datadate) |>
  fill(ceq, .direction = "down") |>
  ungroup() |>
  mutate(mb = mkt_cap / ceq) |>
  filter(!is.na(mb))
```

Then we can plot the evolution of market-to-book ratios for our banks using `geom_line()`, as shown in Figure 2.4.

```
aus_bank_mb |>
  ggplot(aes(x = datadate, y = mb,
             linetype = ticker, color = ticker)) +
  geom_line()
```

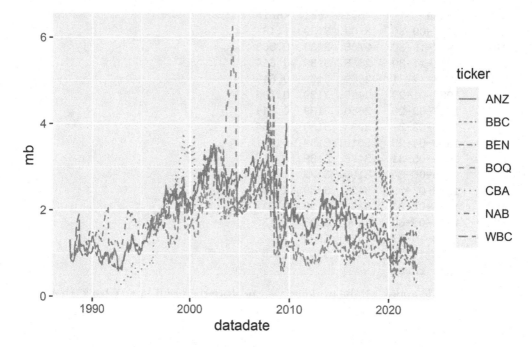

FIGURE 2.4
Time-series of market-to-book for Australian banks

2.4 Reproducible research

A topic that has received increased attention in the research community in recent years is **reproducible research**. This is a complex topic and there are various ideas of what it means for research to be reproducible.

One notion is that the results disseminated by a research team should be reproducible by other research teams running similar experiments. If results do not reproduce in this way, their **generalizability** may be questioned.

Another notion of reproducibility is more narrow. Typically results disseminated by a research team carry the implicit claim that the team took data from certain sources (sometimes these are experiments, but in accounting research, these are often from databases accessible to other researchers) and conducted certain data manipulations and statistical analyses to produce the disseminated results. The applicable notion here is that other researchers should be able to verify the data and code used to produce results, ensuring that the code functions as claimed and produces the same results. This notion of reproducibility underlies policies for sharing code and data, which will be discussed further in Chapter 19.

We bring up the issue of reproducibility so early in the course because we believe an orientation to reproducibility is a habit best formed early in one's career. Additionally, the tools provided with modern software such as R make the production of reproducible analyses easier than ever.

We also believe that reproducibility is important not only for other researchers, but also for individual researchers over time. Six months after running some numbers in Excel, you may need to explain how you got those numbers, but important steps in the analysis may have been hard-coded or undocumented.

Alternatively, you might run an analysis today and then want to update it in a year's time as more data become available (e.g., stock prices and fundamentals for 2023). If you have structured your analysis in a reproducible way, this may be as simple as running your code again. But if you copy-pasted data into Excel and used point-and-click commands to make plots, those steps would have to be repeated manually again.

Reproducibility is arguably also important in practice. Checking data and analysis is often a key task in auditing. Reproducible research leaves an audit trail that manual analyses do not.

Reproducibility also fosters learning and collaboration. It is much easier to get feedback from a more experienced researcher or practitioner if the steps are embedded in code. And it is much easier for collaborators to understand what you are doing and suggest ideas if they can see what you have done.

To make it easier for you to get into reproducible research, we have created a Quarto template—available on GitHub[24]—that you can use to prepare your solutions to the exercises below. For reasons of space, we do not provide details on using Quarto, but the Quarto website[25] provides a good starting point and we direct you to other resources in the next section.

2.5 Further reading

Hopefully, the material above provides the fast-paced introduction to data analysis and visualization promised at the opening of the chapter.

Note that our approach here emphasizes approaches using the `tidyverse`[26] packages, such as `dplyr`, `ggplot`, `tidyr`, and `forcats`. We emphasize these packages in this course because we believe they provide a cleaner, more consistent, and more modern interface for accomplishing many data tasks than so-called "Base R" approaches do.

Given this emphasis, a natural place to look for further material is the second edition of *R for Data Science*[27] ("R4DS") which also emphasizes `tidyverse` approaches. The content above overlaps with the material from a number of parts of R4DS. Chapters 3[28] and 5[29] of R4DS provide more details on the primary `dplyr` verbs and reshaping data. We briefly introduced functions, which get a fuller treatment in Chapter 27[30] of R4DS. We also discussed factors (Chapter 16[31] of R4DS), missing values (Chapter 18[32] of R4DS), and joins (Chapter 19[33] of R4DS).

---

[24] https://raw.githubusercontent.com/iangow/far_templates/main/r-intro.qmd

[25] https://quarto.org

[26] https://www.tidyverse.org

[27] https://r4ds.hadley.nz

[28] https://r4ds.hadley.nz/data-transform.html

[29] https://r4ds.hadley.nz/data-tidy.html

[30] https://r4ds.hadley.nz/functions.html

[31] https://r4ds.hadley.nz/factors.html

[32] https://r4ds.hadley.nz/missing-values.html

[33] https://r4ds.hadley.nz/joins.html

We introduced elements from Base R not covered until much later in *R for Data Science*. For example, the use of [[and $ to select elements of a list or a data frame are first seen in Chapter 14[34] of R4DS. Chapter 27[35] of R4DS provides an introduction to important elements of Base R.

In accounting and finance research, data visualization seems relatively underutilized. We speculate that there are two reasons for this. First, making plots is relatively difficult with the tools available to most researchers. For example, many researchers appear to use Excel to make plots, but Excel's plotting functionality is relatively limited and requires manual steps. Second, data visualization is most useful when the goal is to gain genuine insight into real-world phenomena and less useful for finding the statistically significant "results" that are so often sought (we cover this issue in more depth in Chapter 19).

We agree with the authors of *R for Data Science* that thinking about data visualization forces the analyst to think about the structure of the data in a careful way. Chapters 1,[36] 9,[37] 10,[38] and 11[39] of R4DS provide more depth on data visualization. We also try to use data visualization throughout this book. An excellent book-length treatment of the topic is provided by Healy (2018).

We discussed reproducible research briefly above and provided a link to a Quarto template for readers to use in completing the exercises below. For reasons of space, we do not discuss Quarto in detail here. Fortunately, Chapter 28[40] of R4DS provides an introduction to Quarto.

Our expectation and hope is that, by the end of this book, you will have covered almost all the topics addressed by R4DS. But given that R4DS often provides more in-depth coverage than we can here, we recommend that you read that book carefully as you work through this book and we provide guidance to assist the reader in doing this.

Moving beyond the Tidyverse, Adler (2012) continues to provide a solid introduction to R oriented to Base R. A more programming-oriented pre-Tidyverse introduction to R is provided by Matloff (2011). An introduction to programming using R is provided by Grolemund (2014).

2.6 Exercises

We have created Quarto[41] templates that we recommend that you use to prepare your solutions to the exercises in the book. These templates are available at https://github.com/iangow/far_templates/blob/main/README.md. Download the file r-intro.qmd and make sure to save it with a .qmd extension so that when you open it in RStudio, it is recognized as a Quarto document.

1. Create a function cor_alt(x, y) that uses cov_alt() and var_alt() to calculate the correlation between x and y. Check that it gives the same value as the

---

[34] https://r4ds.hadley.nz/strings.html#sec-other-languages
[35] https://r4ds.hadley.nz/base-r
[36] https://r4ds.hadley.nz/data-visualize.html
[37] https://r4ds.hadley.nz/layers.html
[38] https://r4ds.hadley.nz/eda
[39] https://r4ds.hadley.nz/communication.html
[40] https://r4ds.hadley.nz/quarto.html
[41] https://quarto.org

built-in function `cor()` for the correlation between `ret_nab` and `ret_cba` from `rets_nab_cba_wide`.

2. If we remove the `drop_na()` line used in creating `rets_nab_cba_wide`, we see missing values for CBA. There are two reasons for these missing values. One reason is explained here,[42] but the other reason is more subtle and relates to how values are presented in `datadate`. What is the first reason? (*Hint:* What happened to CBA in 1991?) What is the second reason? How might we use `lubridate::ceiling_date(x, unit = "month")` to address the second reason? Does this second issue have implications for other plots?

3. Adapt the code used above to calculate the *correlation* matrix for the returns of Australian banks to instead calculate the *covariance* matrix. What is the calculated value of the variance of the returns for NAB?

4. From the output above, what is the value for the variance of NAB's returns given by the `cov()` function applied to `rets_nab_cba_wide`? Why does this value differ from that you calculated in the previous question?

5. What do the two-table verbs `semi_join()` and `anti_join()` do? In what way do they differ from the two-table verbs listed above? How could we replace `filter(ticker %in% latest_mkt_cap$ticker)` (see above) with one of these two verbs?

6. In calculating ROE above, we used `ib` rather than a measure of "net income". According to WRDS,[43] "ni [net income] only applies to Compustat North America. Instead use: `ni = ib + xi + do`." Looking at the data in `aus_bank_funds`, does this advice seem correct? How would you check this claim? (*Hint:* You should probably focus on cases where both `xi` and `do` are non-missing and checking more recent years may be easier if you need to look at banks' financial statements.)

7. Figure 2.4 is a plot of market-to-book ratios. Another measure linking stock prices to fundamentals is the **price-to-earnings ratio** (also known as the **PE ratio**). Typically, PE ratios are calculated as

$$PE = \frac{\text{Stock price}}{\text{Earnings per share}}$$

where

$$\text{Earnings per share} = \frac{\text{Net income}}{\text{Shares outstanding}}$$

So we might write

$$PE = \frac{\text{Stock price} \times \text{Shares outstanding}}{\text{Net income}}$$

What critical assumption have we made in deriving the last equation? Is this likely to hold in practice?

8. Calculating the PE ratio using `pe = mkt_cap / ib`, create a plot similar to Figure 2.4, but for the PE ratios of Australian banks over time.

---

[42] https://en.wikipedia.org/wiki/Commonwealth_Bank
[43] https://go.unimelb.edu.au/ocd8

9. Suppose you wanted to produce the plots in the test (market capitalization; market-to-book ratios; histogram of ROE) using Excel starting from spreadsheet versions of the three data sets provided above? Which aspects of the task would be easier? Which would be more difficult? What benefits do you see in using R code as we did above?

10. Using the documentation from the `farr` package, describe the contents of the `by_tag_year` data frame (type `help(by_tag_year)` or `? by_tag_year` after loading the `farr` package).

11. Using `by_tag_year`, create a plot that displays the total number of questions asked across all languages over time.

12. Produce a plot like the one above, but focused on questions related to R.

13. If we want to know the popularity of R relative to other languages, we're probably more interested in a percentage, instead of just the counts. Add a new variable that is the fraction of all questions asked in each year with a specific tag to the data set and plot this variable focused on questions related to R.

14. Two popular R packages we have used in this chapter—`dplyr` and `ggplot2`—also have Stack Overflow tags. Perform the same steps that you did for R above for these two tags to see whether they are growing as well.

15. Produce a plot that depicts the relative popularity of R, Python, SAS, and Stata over time.

16. Which language among R, SAS, and Stata has triggered the most questions in the history of Stack Overflow? (*Hint*: Use the `dplyr` verbs `summarize()` and `group_by()`.)

3

Regression fundamentals

In this chapter, we provide a short introduction to the fundamentals of regression analysis with a focus on ordinary least-squares (OLS) regression. Our emphasis here is on helping the reader to build intuition for the mechanics of regression. We demonstrate different ways of achieving various regression results to strengthen the reader's intuition for what regression is doing. In this spirit, we close the chapter with a brief discussion of the Frisch-Waugh-Lovell theorem, which provides a way of representing multivariate regression coefficients as the result of a single-variable regression.

While we motivate some of our regressions with a data set that prompts a number of causal questions, we largely sidestep the issue of when OLS regression does or does not produce valid estimates of causal effects. Thus, while we hint at possible causal interpretations of results in this chapter, the reader should be cautious about these interpretations. We begin our formal analysis of causal inference with Chapter 4.

Additionally, while we note that OLS will provide noisy estimates of **estimands**, we do not address issues regarding how precise these estimates are or how to assess the statistical significance of results.[1] A few p-values and t-statistics will crop up in regression analysis shown in this chapter, but we ignore those details for now. We begin our study of statistical inference in Chapter 5.

> **Tip**
>
> The code in this chapter uses the packages listed below. For instructions on how to set up your computer to use the code found in this book, see Section 1.2 (note that Step 4 is not required as we do not use WRDS data in this chapter). Quarto templates for the exercises below are available on GitHub.[a]
>
> ---
> [a]https://github.com/iangow/far_templates/blob/main/README.md

```
library(dplyr)
library(ggplot2)
library(farr)
library(modelsummary)
library(tidyr)              # pivot_wider()
library(fixest)             # feols()
```

---

[1]An estimand is simply a fancy way of saying the thing we're trying to estimate.

DOI: 10.1201/9781003456230-3

3.1   Introduction

Suppose we have data on variables y, x_1 and x_2 for n units and we conjecture that there is a linear relationship between these variables of the following form:

$$y_i = \beta_0 + \beta_1 \times x_{i1} + \beta_2 \times x_{i2} + \epsilon_i$$

where $i \in 1, \ldots, n$ denotes the data for a particular unit. We can write that in matrix form as follows:

$$\begin{bmatrix} y_1 \\ y_2 \\ \ldots \\ y_{n-1} \\ y_n \end{bmatrix} = \begin{bmatrix} 1 & x_{11} & x_{12} \\ 1 & x_{21} & x_{22} \\ \ldots & \ldots & \ldots \\ 1 & x_{n-1,1} & x_{n-1,2} \\ 1 & x_{n,1} & x_{n,2} \end{bmatrix} \times \begin{bmatrix} \beta_0 \\ \beta_1 \\ \beta_2 \end{bmatrix} + \begin{bmatrix} \epsilon_1 \\ \epsilon_2 \\ \ldots \\ \epsilon_{n-1} \\ \epsilon_n \end{bmatrix}$$

And this can be written even more compactly as:

$$y = X\beta + \epsilon$$

where X is an $n \times 3$ matrix and y is an n-element vector. It is conventional to denote the number of columns in the X matrix using k, where $k = 3$ in this case.[2]

In a regression context, we call X the **regressors**, y the **regressand**, and ϵ the **error term**. We assume that we observe X and y, but not ϵ. If we did observe ϵ, then we could probably solve for the exact value of the coefficients β with just a few observations. Lacking such information, we can produce an **estimate** of our **estimand**, β. Our estimate ($\hat{\beta}$) is likely to differ from β due to noise arising from the randomness of the unobserved ϵ and also possibly bias. There will usually be a number of **estimators** that we might consider for a particular problem. We will focus on the **ordinary least-squares regression** (or **OLS**) estimator as the source for our estimates in this chapter.

OLS is a mainstay of empirical research in the social sciences in general and in financial accounting in particular. In matrix notation, the **OLS estimator** is given by

$$\hat{\beta} = (X^{\mathsf{T}}X)^{-1}X^{\mathsf{T}}y$$

Let's break this down. First, X^{T} is the **transpose** of X, meaning the $k \times n$ matrix formed by making the rows of X into columns. Second, $X^{\mathsf{T}}X$ is the product of the $k \times n$ matrix X^{T} and the $n \times k$ matrix X, which results in a $k \times k$ matrix. Third, the -1 exponent indicates the **inverse** matrix. For a real number x, x^{-1} denotes the number that when multiplied by x gives 1 (i.e., $x \times x^{-1} = 1$). For a **square matrix** Z (here "square" means the number of rows equals the number of columns), Z^{-1} denotes the square matrix that when multiplied by Z gives the **identity matrix**, $\mathbf{I}$ (i.e., $Z \times Z^{-1} = \mathbf{I}$).

---

[2]For a quick overview of some details of matrix algebra, see Appendix A.

The 3×3 identity matrix looks like this

$$\begin{bmatrix} 1 & 0 & 0 \\ 0 & 1 & 0 \\ 0 & 0 & 1 \end{bmatrix}$$

Note that, just as there is no meaningful way to calculate the inverse of 0, it's not always possible to take the inverse of a matrix. But, so long as no column of X is a linear combination of other columns and $n > k$, then we can calculate $(X^{\mathsf{T}}X)^{-1}$ (and there are standard algorithms for doing so). Now, $X^{\mathsf{T}}y$ is the product of a $k \times n$ matrix (X^{T}) and a vector with n elements (this can be thought of as an $n \times 1$ matrix), so the result will be a $k \times 1$ matrix. Thus, the product of $(X^{\mathsf{T}}X)^{-1}$ and $X^{\mathsf{T}}y$ will be a vector with k elements, which we can denote as follows:

$$\begin{bmatrix} \hat{\beta}_1 \\ \hat{\beta}_2 \\ \hat{\beta}_3 \end{bmatrix}$$

So $X^{\mathsf{T}}X$ is a $k \times k$ matrix (as is its inverse), and $X^{\mathsf{T}}y$ is the product of a $k \times N$ matrix times an $N \times 1$ matrix (i.e., a vector). So $\hat{\beta}$ is a k-element vector. If we have a single regressor x, then X will typically include the constant term, so $k = 2$.

For this chapter, we will assume that the model $y = X\beta + \epsilon$ is a **structural (causal) model**. What this means is that if we could somehow increase the value of x_{i1} by 1 unit without changing any other part of the system, we would see an increase in the value of y_i equal to β_1. This model is causal in the sense that a unit change in x_{i1} can be said to *cause* a β_1 change in y_i.

To make this more concrete, let's consider some actual (though not "real") data. The following code uses R functions to generate random data. Specifically, we generate 1000 observations with $\beta = 1$ and $\sigma = 0.2$. This is the first of at least a dozen **simulation analyses** that we will consider in the book.

To make our analysis easier to replicate, we include **set.seed(2021)** to set the **random-number generator** used by **rnorm()** to the same point, so that we can reproduce the analysis ourselves later and so that others can reproduce it too. For more on random-number generation, type **? set.seed** in the R console. Note that the value 2021 is arbitrary and represents nothing more than the year in which this material was first written. Any value would work here, e.g., 2024, 42, or 20240215 could be used.

```
set.seed(2021)
N <- 1000
x <- rnorm(N)
e <- rnorm(N, sd = 0.2)
y <- x * 1 + e
```

We next construct X as a matrix comprising a column of ones (to estimate the constant term) and a column containing x.

```
X <- matrix(c(rep(1, N), x), ncol = 2)
head(X)
```

```
     [,1]       [,2]
[1,]    1 -0.1224600
[2,]    1  0.5524566
```

```
[3,]     1  0.3486495
[4,]     1  0.3596322
[5,]     1  0.8980537
[6,]     1 -1.9225695
```

Naturally, R has built-in matrix operations. To get X^T, the transpose of the matrix X, we use t(X). To multiply two matrices, we use the matrix multiplication operator %*%. And to invert $(X^\mathsf{T}X)$ to get $(X^\mathsf{T}X)^{-1}$, we use the solve() function. Thus, the following calculates the OLS estimator $\hat{\beta} = (X^\mathsf{T}X)^{-1}X^\mathsf{T}y$.

```
b <- solve(t(X) %*% X) %*% t(X) %*% y
b
```

```
             [,1]
[1,] 0.007539896
[2,] 0.997646586
```

3.2 Running regressions in R

According to the documentation[3] for lm():

> The basic function for fitting ordinary multiple models is lm(), and a streamlined version of the call is as follows:

```
fitted.model <- lm(formula, data = data.frame)
```

For example,

```
fm1 <- lm(y ~ x1 + x2, data = production)
```

would fit a multiple regression model of y on x1 and x2 using data from the data frame production. We use the lm() function—part of base R—to estimate regressions in this chapter.

Note that R was developed by statisticians and thus works in a way consistent with that history. For example, if we run the following code, we actually estimate coefficients on the constant term (the intercept), x1, x2, and the product of x1 and x2.

```
fm2 <- lm(y ~ x1 * x2, data = production)
```

In contrast to other statistical packages, with lm() there's no need to calculate the product of x1 and x2 and store it as a separate variable and there's no need to explicitly specify the "main effect" terms (i.e., x1 and x2) in the regression equation. R (and the lm() function) takes care of these details for us. The third argument to lm() is subset, which allows us to specify as condition that each observation needs to satisfy to be included in the regression.

Here the first argument to lm() (formula) is "an object of class *formula* ... a symbolic description of the model to be fitted." As we are regressing y on x, we use the formula y ~ x here; we will soon see more complicated formula expressions. The value to the data argument is normally a data frame, so below we put x and y into a tibble. We then call the

---

[3]https://cran.r-project.org/doc/manuals/r-release/R-intro.html#Linear-models

`lm()` function, store the returned value in the variable `fm`, then show some of the contents of `fm`:

```
df <- tibble(y, x)
fm <- lm(y ~ x, data = df)
fm
```

```
Call:
lm(formula = y ~ x, data = df)

Coefficients:
(Intercept)              x
    0.00754        0.99765
```

From this output, we see that we get the same results using `lm()` as we do using matrix algebra.

The data sets we will focus on next are `test_scores` and `camp_attendance`, both of which are part of the `farr` package. The `test_scores` data frame contains data on test scores for 1000 students over four years (grades 5 through 8).

```
test_scores
```

```
# A tibble: 4,000 x 3
      id grade score
   <int> <int> <dbl>
1      1     5  498.
2      1     6  513.
3      1     7  521.
4      1     8  552.
5      2     5  480.
6      2     6  515.
# i 3,994 more rows
```

The `camp_attendance` data set contains data on whether a student attended a science camp during the summer after sixth grade.

```
camp_attendance
```

```
# A tibble: 1,000 x 2
      id camp
   <int> <lgl>
1      1 TRUE
2      2 TRUE
3      3 FALSE
4      4 FALSE
5      5 FALSE
6      6 TRUE
# i 994 more rows
```

We can also see that exactly half the students in the sample attended the science camp.

```
camp_attendance |>
  summarize(mean(camp)) |>
  pull()
```

```
[1] 0.5
```

```
camp_scores <-
  test_scores |>
  inner_join(camp_attendance, by = "id") |>
  rename(treat = camp) |>
  mutate(post = grade >= 7)
```

The question we might be interested in is whether attending the science camp improves test performance. The natural first thing to do would be plot the data, which we do in Figure 3.1.

```
camp_scores |>
  group_by(grade, treat) |>
  summarize(score = mean(score),
            .groups = "drop") |>
  ggplot(aes(x = grade, y = score,
             linetype = treat, colour = treat)) +
  geom_line()
```

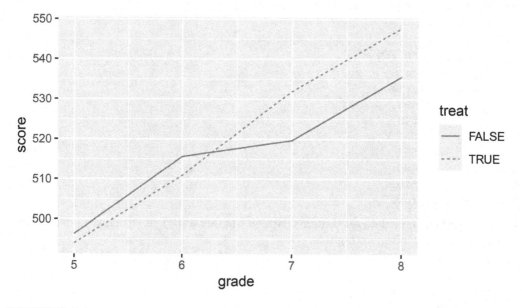

FIGURE 3.1
Student test scores by grade and camp participation

Figure 3.1 shows that the students who went to the camp had lower scores in grades 5 and 6 (i.e., before the camp), but stronger performance in grades 7 and 8. This provides *prima facie* evidence of a positive effect of the science camp on scores.

While the ideal approach might be to *randomly* assign students to the camp and then compare scores after the camp, in this case, it appears that students with lower scores went to camp. Given our lack of contextual information, there is no *obvious* story for what we see in Figure 3.1 that is more plausible than (or at least as simple as) one that attributes a positive effect on test scores of going to the science camp.

A question that we might now have is: What is the best way to test for an effect of the science camp on test scores? One reasonable hypothesis is that the summer camp has a biggest effect on seventh-grade scores and that we might compare seventh-grade scores with sixth-grade scores to get the best estimate of the effect of camp on scores.

In the code below, we use `filter()` to limit our analysis to data related to sixth and seventh grades. We then calculate mean scores for each value of (`post`, `treat`). We then use `pivot_wider()` to put `score` values for different levels of `post` in the same row so that we can calculate `change`.

```
test_summ <-
  camp_scores |>
  filter(grade %in% 6:7L) |>
  group_by(post, treat) |>
  summarize(score = mean(score), .groups = "drop") |>
  pivot_wider(names_from = post, values_from = score) |>
  rename(post = `TRUE`, pre = `FALSE`) |>
  mutate(change = post - pre)
```

You may find it helpful in understanding code examples such as this one to look at the intermediate output along the line of pipes. For example, if you highlight the text from `test_scores |>` through to the end of the `summarize()` line just *before* the pipe at the end of the last line, and click "Run Selected Lines" in the "Code" menu of RStudio (or hit CTRL + Enter on Windows or Linux or ⌘ + Enter on MacOS) you will see what is coming out at that point of the pipe (it should look like the following):

```
camp_scores |>
  filter(grade %in% 6:7L) |>
  mutate(post = grade >= 7) |>
  group_by(post, treat) |>
  summarize(score = mean(score), .groups = "drop")
```

```
# A tibble: 4 x 3
  post  treat score
  <lgl> <lgl> <dbl>
1 FALSE FALSE  516.
2 FALSE TRUE   511.
3 TRUE  FALSE  520.
4 TRUE  TRUE   532.
```

We strongly recommend that you use this approach liberally when working through this book (and when debugging your own pipelines).

We stored the results of our analysis in `test_summ`, which we show in Table 3.1.

```
test_summ
```

TABLE 3.1
Initial analysis of student scores

treat	pre	post	change
FALSE	515.539	519.657	4.118
TRUE	510.810	531.938	21.128

We see in Table 3.1 that the scores of the treated students (i.e., those who went to the summer camp) increased by 21.128, while the scores of the control students increased by 4.118.

One approach is to view the outcome of the control students as the "but-for-treatment" outcome that the treated students would have seen had they not gone to summer camp. With this view, the effect of going to camp is the difference in the difference in scores, or 17.010. This is the **difference-in-differences** estimator of the causal effect.

We can also recover this estimator using the `lm()` function:

```
fm_dd <- lm(score ~ treat * post, data = camp_scores,
        subset = grade %in% 6:7L)
coefficients(fm_dd)
```

```
    (Intercept)         treatTRUE              postTRUE treatTRUE:postTRUE
     515.539023         -4.728960              4.118259          17.009708
```

Note that we did not need to specify the inclusion of the **main effects** of `treat` and `post`; R automatically added those when we requested their interaction (`treat * post`). Also note that we did not need to convert the logical variables `treat` and `post` so that `TRUE` is 1 and `FALSE` is 0; in effect, R also did this for us.

A natural question might be whether we could do better using all four years of data. There is no simple answer to this question unless we have a stronger view of the underlying causal model. One causal model might have it that students have variation in underlying talent, but that there is also variation in industriousness that affects how students improve over time. From the perspective of evaluating the effect of the camp, variation in industriousness is going to add noise to estimation that increases if we are comparing performance in fifth grade with that in eighth grade.

Another issue is that the effects of the summer camp might fade over time. As such, we might get a larger estimated effect if we focus on seventh-grade scores than if we focus on (or also include) eighth-grade scores. But from a policy perspective, we might care more about sustained performance improvement and actually prefer eighth-grade scores.

However, if we were willing to assume that, in fact, scores are a combination of underlying, **time-invariant** individual talent, the *persistent* effects (if any) of summer camp, and random noise, then we'd actually do better to include all observations.[4]

```
fm_dd_all <- lm(score ~ treat * post, data = camp_scores)
coefficients(fm_dd_all)
```

```
    (Intercept)         treatTRUE              postTRUE treatTRUE:postTRUE
     505.887759         -3.536433             21.764629          15.735916
```

Another possibility is that scores are a combination of underlying, time-invariant individual talent, the persistent effects (if any) of summer camp, and random noise *and* the grade in which the test is taken. For example, perhaps the test taken in seventh grade is similar to that taken in sixth grade but, with an extra year of schooling, students might be expected to do better in the higher grade assuming the scores are not scaled in any way within grades. (Recall that we just have this data set without details that would allow us to rule out such ideas, so the safest thing to do is to examine the data.) The easiest way to include grade is as a **linear trend**, which means that grade is viewed as a number (e.g., 5 or 8):

```
fm_dd_trend <- lm(score ~ treat * post + grade, data = camp_scores)
coefficients(fm_dd_trend)
```

```
    (Intercept)         treatTRUE              postTRUE              grade
```

---

[4]The idea here being that more data is better than less. This idea will show up again in later chapters.

```
        412.335651              -3.536433            -12.254320             17.009474
treatTRUE:postTRUE
        15.735916
```

Note that we have exactly the same coefficients on `treat` and `treat * post` as we had before (−3.5364 and 15.7359, respectively). The easiest way to understand the estimated coefficients is to plug in some candidate X values to get the fitted values.

Suppose we have a student who went to summer camp. In grade 6, this student's predicted score would be

$$412.3357 + -3.5364 + 6 \times 17.0095 = 510.8561$$

In grade 7, this student's predicted score would be

$$412.3357 + -3.5364 + 7 \times 17.0095 + 15.7359 = 543.6015$$

An alternative approach that allows for `grade` to affect scores would be to estimate a separate intercept for each level that `grade` takes on. That is, we'd have a different intercept for `grade==6`, a different intercept for `grade==7`, and so on. While we could achieve this outcome by creating variables using an approach such as `mutate(grade7 = grade == 7)`, it is easier to use the R's support for factors.

As discussed in Chapter 2, factors are a type that is useful for representing **categorical variables**, which often have no meaningful numerical representation (e.g., "red" or "blue", or "Australia" or "New Zealand") or where we want to move away from a simple numerical representation (e.g., grade 7 may not be simply 7/6 times grade 6).[5] Rather than simply adding a factor version of `grade` to the model above, let's run a simpler regression.

```
fm_grade <- lm(score ~ factor(grade), data = test_scores)
coefficients(fm_grade)
```

```
  (Intercept) factor(grade)6 factor(grade)7 factor(grade)8
    495.06454       18.11000       30.73311       46.64206
```

This model estimates **fixed effects** for each `grade` without other **covariates**. Table 3.2 provides the mean scores by grade for comparison.

```
test_scores |>
  group_by(grade) |>
  summarize(score = mean(score),
            .groups = "drop")
```

TABLE 3.2

Mean scores by grade

grade	score
5	495.065
6	513.175
7	525.798
8	541.707

---

[5] *R for Data Science* has a whole chapter devoted to factors: https://r4ds.hadley.nz/factors.html.

The idea of fixed effects is that there are time-invariant factors that have a constant effect on the outcome (hence *fixed* effects). In some settings, we would posit fixed effects at the level of the individual. Here we are positing fixed effects at the grade level. Working through the exercises should provide additional insights into what we are doing here. For more on fixed effects, see Cunningham (2021, pp. 391–392) and also Chapter 21.

Now, let's estimate **fixed effects** for both `grade` *and* student (`id`). This will yield more fixed effects than we have students (we have 1000 students), so we suppress the coefficients for the fixed effect in the regression output, which is shown in Table 3.3.

```
fm_id <- lm(score ~ treat * post + factor(grade) + factor(id),
            data = camp_scores, x = TRUE)


modelsummary(fm_id,
             estimate = "{estimate}{stars}",
             coef_omit = "^factor",
             gof_map = "nobs",
             stars = c('*' = .1, '**' = 0.05, '***' = .01))
```

TABLE 3.3
Test-score regressions with `grade` and `id` fixed effects

	(1)
(Intercept)	495.370***
	(2.527)
treatTRUE	−2.235
	(3.570)
postTRUE	38.774***
	(0.276)
treatTRUE × postTRUE	15.736***
	(0.319)
Num.Obs.	4,000

Note that we specified `x = TRUE` so that the X matrix used in estimation was returned by `lm()`. The size of this matrix is given by the `dim()` function:

```
dim(fm_id$x)
```

```
[1] 4000 1006
```

We have 1006 columns because we have added so many fixed effects. This means that $(X^\mathsf{T} X)^{-1}$ is a 1006×1006 matrix. As we add more years and students, this matrix could quickly become quite large and inverting it would be computationally expensive (even more so for some other operations that would need even larger matrices). To get a hint as to a less computationally taxing approach, let's see what happens when we "demean" the variables in a particular way.

```
demean <- function(x) x - mean(x)


camp_scores_demean <-
  camp_scores |>
  group_by(id) |>
  mutate(score = demean(score)) |>
```

```
  group_by(grade) |>
  mutate(score = demean(score))

fm_demean <- lm(score ~ treat * post,
                data = camp_scores_demean, x = TRUE)
```

Results from this analysis are shown in Table 3.4.

```
modelsummary(fm_demean,
             estimate = "{estimate}{stars}",
             coef_omit = "^factor",
             gof_map = "nobs",
             stars = c('*' = .1, '**' = 0.05, '***' = .01))
```

TABLE 3.4
Regression with demeaned test scores

	(1)
(Intercept)	3.934***
	(0.138)
treatTRUE	−7.868***
	(0.195)
postTRUE	−7.868***
	(0.195)
treatTRUE × postTRUE	15.736***
	(0.276)
Num.Obs.	4,000

The size of the X matrix is now 4000×4. This means that $(X^T X)^{-1}$ is now a much more manageable 4×4 matrix. While we had to demean the data, this is a relatively fast operation.

3.2.1 Exercises

1. In using `pivot_wider()` in Chapter 2, we supplied a value to the `id_cols` argument, but we omitted that in creating `test_summ`. If we wanted to be explicit, what value would we need to provide for that argument in the code creating `test_summ`?

2. What is the relation between the means in Table 3.2 and the regression coefficients in `fm_grade`?

3. Why is there no estimated coefficient for `factor(grade)5` in `fm_grade`?

4. Now let's return to our earlier regression specification, except this time we include fixed effects for `grade` (see code below and output in Table 3.5). We now have *two* fixed effects omitted: `factor(grade)5` and `factor(grade)8`. Why are we now losing two fixed effects, while above we lost just one? (*Hint*: Which variables can be expressed as linear combinations of the `grade` indicators?)

```
fm_dd_fe <- lm(score ~ treat * post + factor(grade),
               data = camp_scores)
```

```
modelsummary(fm_dd_fe,
             estimate = "{estimate}{stars}",
             gof_map = "nobs",
             stars = c('*' = .1, '**' = 0.05, '***' = .01))
```

TABLE 3.5
Regression results with **grade** fixed effects

	(1)
(Intercept)	496.833***
	(0.274)
treatTRUE	−3.536***
	(0.316)
postTRUE	38.774***
	(0.388)
factor(grade)6	18.110***
	(0.316)
factor(grade)7	−15.909***
	(0.316)
treatTRUE × postTRUE	15.736***
	(0.448)
Num.Obs.	4,000

5. In words, what are we doing to create `camp_scores_demean`? Intuitively, why might this affect the need to use fixed effects?

6. Can you relate the coefficients from the regression stored in `fm_demean` to the numbers in Table 3.6? Which of these estimated coefficients is meaningful? All of them? Some of them? None of them?

```
camp_scores_demean |>
  group_by(grade, treat) |>
  summarize(score = mean(score), .groups = "drop")
```

TABLE 3.6
Demeaned test scores by **grade** and **treat**

grade	treat	score
5.0000	False	3.3377
5.0000	True	−3.3377
6.0000	False	4.5302
6.0000	True	−4.5302
7.0000	False	−3.9746
7.0000	True	3.9746
8.0000	False	−3.8933
8.0000	True	3.8933

7. The `feols()` function from the `fixest` package offers a succinct syntax for adding fixed effects and uses computationally efficient algorithms (much like our

demeaning approach above) in estimating these. What is the same in the results below and the two specifications we estimated above? What is different? Why might these differences exist? What is the `I()` function doing here? What happens if we omit it (i.e., just include `post * treat`)?

```
fefm <- feols(score ~ I(post * treat) | grade + id, data = camp_scores)
coefficients(fefm)
```

```
I(post * treat)
       15.73592
```

3.3   Frisch-Waugh-Lovell theorem

The Frisch-Waugh-Lovell theorem states that the following two regressions yield identical regression results in terms of both the estimate $\hat{\beta}_2$ and residuals.

$$y = X_1\beta_1 + X_2\beta_2 + \epsilon$$

and

$$M_{X_1}y = M_{X_1}X_2\beta_2 + \eta$$

where M_{X_1} is the "residual maker" for X_1 or $I - P_{X_1} = I - X_1(X_1^\mathsf{T}X_1)^{-1}X_1^\mathsf{T}$, y is a $n \times 1$ vector, and X_1 and X_2 are $(n \times k_1)$ and $(n \times k_2)$ matrices.

In other words, we have two procedures that we can use to estimate $\hat{\beta}_2$ in the regression equation above. First, we could simply regress y on X_1 and X_2 to obtain estimate $\hat{\beta}_2$. Second, we could take the following more elaborate approach:

1. Regress X_2 on X_1 (and a constant term) and store the residuals (ϵ_{X_1}).
2. Regress y on X_1 (and a constant term) and store the residuals (ϵ_y).
3. Regress ϵ_y on ϵ_{X_1} (and a constant term) to obtain estimate $\hat{\beta}_2$.

The Frisch-Waugh-Lovell theorem tells us that only the portion of X_2 that is **orthogonal** to X_1 affects the estimate $\hat{\beta}_2$. Note that the partition of X into $[X_1 X_2]$ is quite arbitrary, which means that we also get the same estimate $\hat{\beta}_1$ from the first regression equation above and from estimating

$$M_{X_2}y = M_{X_2}X_1\beta_1 + \upsilon$$

To verify the Frisch-Waugh-Lovell theorem using some actual data, we draw on the data set `comp` from the `farr` package and a regression specification we will see in Chapter 24.

As our baseline, we run the following linear regression and store it in `fm`.

```
fm <- lm(ta ~ big_n + cfo + size + lev + mtb +
            factor(fyear) * (inv_at + I(d_sale - d_ar) + ppe),
         data = comp, na.action = na.exclude)
```

Here the dependent variable is `ta` (total accruals), `big_n` is an indicator variable for having a Big N auditor (see Chapter 25.1) and the other variables are various controls (use help(comp) or ? comp for descriptions of these variables). We again use the I() function we saw above and interact factor(fyear) with three different variables.

We then run two **auxiliary regressions**: one of ta on all regressors *except* cfo (we store this in fm_aux_ta) and one of cfo on all regressors *except* cfo (we store this in fm_aux_cfo). We then take the residuals from each of these regressions and put them in a data frame under the names of the original variables (ta and size respectively). Finally, using the data in aux_data, we regress ta on size.

```
fm_aux_ta <- lm(ta ~ big_n + size + lev + mtb +
                factor(fyear) * (inv_at + I(d_sale - d_ar) + ppe),
            data = comp, na.action = na.exclude)

fm_aux_cfo <- lm(cfo ~ big_n + size + lev + mtb +
                 factor(fyear) * (inv_at + I(d_sale - d_ar) + ppe),
             data = comp, na.action = na.exclude)

aux_data <- tibble(ta = resid(fm_aux_ta),
                   cfo = resid(fm_aux_cfo))
fm_aux <- lm(ta ~ cfo, data = aux_data)
```

The Frisch-Waugh-Lovell theorem tells us that the regression in fm_aux will produce exactly the same coefficient on cfo and the same residuals (and very similar standard errors) as the regression in fm, as can be seen in Table 3.7. Here we use modelsummary() from the modelsummary package to produce attractive regression output. We use coef_omit = "(fyear|ppe|inv_at|d_sale)" to focus on coefficients of greater interest.

```
modelsummary(list(fm, fm_aux),
             estimate = "{estimate}{stars}",
             coef_omit = "(fyear|ppe|inv_at|d_sale)",
             gof_map = "nobs",
             stars = c('*' = .1, '**' = 0.05, '***' = .01))
```

TABLE 3.7

Demonstration of FWL theorem

	(1)	(2)
(Intercept)	−0.017	0.000
	(0.028)	(0.004)
big_nTRUE	0.022*	
	(0.011)	
cfo	0.141***	0.141***
	(0.008)	(0.008)
size	0.000	
	(0.002)	
lev	−0.066***	
	(0.013)	
mtb	0.000***	
	(0.000)	
Num.Obs.	8,850	8,850

The Frisch-Waugh-Lovell theorem is an important result for applied researchers to understand, as it provides insights into how multivariate regression works. A side-benefit of the result is that it allows us to reduce the relation between two variables in a multivariate regression to a bivariate regression without altering that relation. For example, to understand the relation between `cfo` and `ta` embedded in the estimated model in `fm`, we can plot the data.

We produce two plots. The first—Figure 3.2—includes all data, along with a line of best fit and a smoothed curve of best fit. However, Figure 3.2 reveals extreme observations of the kind that we will study more closely in Chapter 24 (abnormal accruals more than 5 or less than 5 times lagged total assets!).

So we trim the values of `ta` at −1 and +1 and produce a second plot.[6] In Figure 3.3, there is no visually discernible relation between `size` and `ta` and the line of best fit is radically different from the curve. If nothing else, hopefully these plots raise questions about the merits of blindly accepting regression results with the messy data that we often encounter in practice.

```
aux_data |>
  filter(!is.na(cfo), !is.na(ta)) |>
  ggplot(aes(x = cfo, y = ta)) +
  geom_point() +
  geom_smooth(method = "lm", formula = y ~ x, se = FALSE, linetype = 2,
              colour = "red") +
  geom_smooth(method = "gam", formula = y ~ s(x, bs = "cs"), se = FALSE)
```

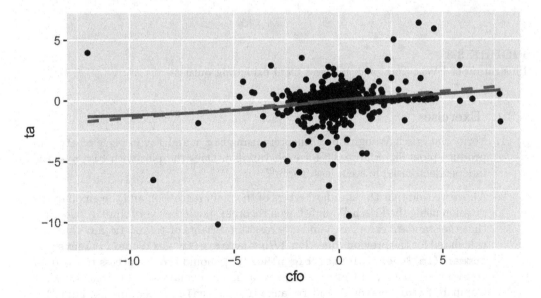

FIGURE 3.2
Illustration of FWL: Total accruals and CFO

---

[6]That is, values less than −1 are set to −1 and values greater than +1 are set to +1. This is similar to winsorization, which we discuss in Chapter 24.

```
aux_data |>
  filter(!is.na(cfo), !is.na(ta), abs(ta) < 1) |>
  ggplot(aes(x = cfo, y = ta)) +
  geom_point() +
  geom_smooth(method = "lm", formula = y ~ x, se = FALSE, linetype = 2,
              colour = "red") +
  geom_smooth(method = "gam", formula = y ~ s(x, bs = "cs"), se = FALSE)
```

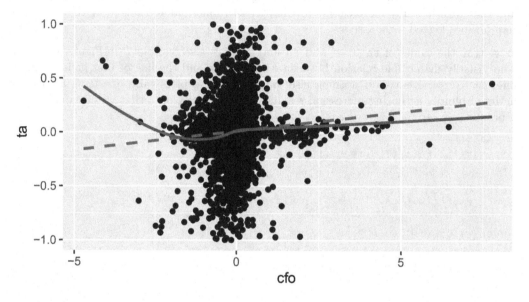

FIGURE 3.3
Illustration of FWL: Total accruals and CFO excluding outliers

3.3.1 Exercises

1. Verify the Frisch-Waugh-Lovell theorem using `big_n` and `lev` in place of `cfo` and produce plots like Figure 3.2 for each variable. Does the plot with `big_n` as the independent variable seem less helpful?

2. Above we said that the standard errors of the main regression and the auxiliary regression using the Frisch-Waugh-Lovell theorem should be "very similar". Confirm that the standard errors are similar across the variants of `fm` and `fm_aux` that you calculated for the previous question. (*Hint*: `summary(fm_aux)$coefficients` and `summary(fm)$coefficients["big_nTRUE", ]` should provide access to the data you want to compare.) Can you guess what might explain any differences? (*Hint*: Compare `fm$df.residual` and `fm_aux$df.residual` and perhaps use `sqrt()`.)

3. In words, what effect does converting `fyear` into a factor and interacting it with `inv_at`, `I(d_sale - d_ar)` and `ppe` have? (*Hint:* It may be helpful to visually inspect the more complete regression output produced without `coef_omit = "(fyear|ppe|inv_at|d_sale)"`.)

3.4 Further reading

This chapter provides a bare minimum introduction to running regressions in R plus some concepts that help develop intuition about what's going on in OLS regression. Treatments that provide a similar emphasis on intuition, but go deeper into the details include Angrist and Pischke (2008) and Cunningham (2021). Any econometrics textbook will offer a more rigorous treatment of OLS and its properties.

4

Causal inference

The importance of causal inference in accounting research is clear from the research questions that accounting researchers seek to answer. Many questions in accounting research are causal, including examples from this book:

- Does short-selling affect firms' level of earnings management?
- Does lowering the cost of disclosure increase the supply of disclosure?
- Do Big Four auditors increase the quality of financial statements?
- Do managerial incentives lead to managerial misstatements in financial reports?

The taxonomy for empirical research papers provided by Gow et al. (2016) comprises four categories:

- **descriptive** research papers
- papers focused on **prediction**
- papers that focus on **measurement** of some construct
- papers that seek—whether explicitly or not—to draw **causal inferences**

The survey conducted by Gow et al. (2016) suggests that most original research papers in the top three accounting research journals use observational data and that about 90% of these papers seek to draw causal inferences.

Accounting researchers' focus on causal inference is consistent with the view that "the most interesting research in social science is about questions of cause and effect" (Angrist and Pischke, 2008, p. 3). While you may hear people talk about "interesting associations" at times, the reality is that associations (or correlations) are only interesting if there's an interesting possible causal explanation for the associations. Associations that are mere coincidence, such as those found on the Spurious Correlations website[1] are amusing, but unlikely to be of genuine interest to researchers or policy-makers.

At times, authors appear to disclaim any intention to draw causal inferences. Bertrand and Schoar (2003) is fairly typical: "There is no such thing as a random allocation of top executives to firms. Therefore, we are not hoping in this section to estimate the causal effect of managers on firm practices. Instead, our objective is more modest. We want to assess whether there is any evidence that firm policies systematically change with the identity of the top managers in these firms."

There are at least two issues with this claim. First, why would anyone be interested in "evidence that firm policies systematically change with the identity of the top managers" if such changes come with no understanding as to *why* they change? Second, this claim is a bit of a pretence. The title of the paper is, after all, "Managing with style: The *effect* of managers on firm policies" and the first sentence of the abstract says "this paper investigates whether and how individual managers *affect* corporate behavior and performance" (emphasis added).

---

[1]http://tylervigen.com/spurious-correlations

DOI: 10.1201/9781003456230-4

As another example, suppose a researcher argues that a paper that claims that "theory predicts X is associated Y and, consistent with that theory, we show X is associated with Y" is merely a descriptive paper that does not make causal inferences. However, theories are invariably causal in that they posit how exogenous variation in certain variables leads to changes in other variables. Further, by stating that "consistent with ... theory, X is associated with Y", the clear purpose is to argue that the evidence tilts the scale, however slightly, in the direction of believing the theory is a valid description of the real world: in other words, a causal inference is drawn. A paper that argues that Z is a common cause of X and Y and claims to find evidence of this is still making causal inferences (i.e., that Z causes X and Z causes Y).

Making causal inferences requires strong assumptions about the causal relations among variables. For example, as discussed below, estimating the causal effect of X on Y requires that the researcher has controlled for variables that could confound estimates of such effects.

Recently, some social scientists have argued that better research designs and statistical methods can increase the credibility of causal inferences. For example, Angrist and Pischke (2010) suggest that "empirical microeconomics has experienced a credibility revolution, with a consequent increase in policy relevance and scientific impact." Angrist and Pischke (2010, p. 26) argue that such "improvement has come mostly from better research designs, either by virtue of outright experimentation or through the well-founded and careful implementation of quasi-experimental methods."

> **Tip**
>
> The code in this chapter uses the packages listed below. For instructions on how to set up your computer to use the code found in this book, see Section 1.2 (note that Step 4 is not required for this chapter as we do not use WRDS data). Quarto templates for the exercises below are available on GitHub.[a]
>
> ---
> [a]https://github.com/iangow/far_templates/blob/main/README.md

```
library(dplyr, warn.conflicts = FALSE)
library(ggplot2)
library(modelsummary)
```

4.1 Econometrics

Empirical financial accounting research can be viewed as fundamentally a highly specialized area of applied microeconomics. Financial accounting researchers typically take classes (at some level) in microeconomics, statistics, and econometrics, either before or in parallel with more specialized classes in accounting research. This model has two significant gaps that budding researchers need to address.

The first gap is the translation of ideas from econometrics and microeconomics into what researchers do from day to day. As we will see, sometimes things get lost in this translation. One goal of this book is to reduce the gap in the domain of empirical methods.

But it's the second gap that we want to address in this chapter, and this gap affects not only financial accounting researchers but applied economic researchers more generally. This gap is between what accounting researchers want to do, which is causal inference, and what econometrics textbooks talk about, such as *consistency*, *unbiasedness*, and *asymptotic variance*.[2]

4.1.1 What is econometrics?

Angrist and Pischke (2014, p. xi) suggest that "economists' use of data to answer cause-and-effect questions constitutes the field of applied econometrics." Given that "applied" here is to in contradistinction to "theoretical" econometrics, this is perhaps one working definition of econometrics for the purposes of accounting research, a field with little direct connection to the more theoretical end of econometrics. Given accounting researchers focus on "cause-and-effect questions" (discussed above), this definition seems consistent with what empirical accounting researchers do.

For example, we might posit that an economic phenomenon y is best modelled as follows:

$$y = X\beta + \epsilon \qquad (4.1)$$

where y and ϵ are N-element vectors, X is an $N \times K$ matrix, and β is a K-element vector of coefficients. This single-equation model "is still the workhorse in empirical economics" (Wooldridge, 2010, p. 49) and this is no doubt true in empirical accounting research.

A very important point about Equation 4.1 is that it should be viewed as a **structural (causal) model**. Wooldridge (2010, p. 49) writes "Goldberger (1972) defines a structural model as one representing a causal relationship, as opposed to a relationship that simply captures statistical associations. A structural equation can be obtained from an economic model, or it can be obtained through informal reasoning."

If we were only interested in estimating this equation to generate conditional expectation function linear in the parameters, then we would not be concerned with whether $\mathbb{E}[X, \epsilon] \neq 0$, whether there were "bias" in the estimates $\hat{\beta}$, or whether we had sufficiently "controlled for" variables in the right-hand side of the equation. Such notions only apply if we are viewing this equation as a structural model.

Viewing the equation above as a structural model, if we knew β, we could understand how changes in X cause changes in y. For example, if x_{ik} goes from 0 to 1, then we expect y_i to increase by β_k. In other words, β_k can be viewed as the (causal) effect of x_k on y. Of course, in reality, we don't *know* β; we need to estimate it using data. Much of econometrics is devoted to explaining how (and when) we can estimate β accurately and efficiently.

An important point to make is that, in viewing the equation above as a structural model, we depart slightly from more recent usage, which has tended to view a structural model as something deriving from an economic model, typically one starting with preferences and technologies for production of economic goods and information and deriving predictions about economic phenomena from that model. This narrower view of structural models does not seem to fit the reality that almost all empirical research proceeds without access to such models, notwithstanding a focus on causal inference.

Sometimes the reasoning behind the use of particular models is poor, and models implied by what researchers do can be poorer than they need to be. But models are always

[2]We provide some intuitive discussion of these ideas in Chapter 5.

simplifications of reality and thus always in some sense "wrong". And, even if we accept a model in a given setting, it is often the case that we "know" that our estimators are unlikely to provide unbiased estimates of the model's true parameter values. Yet that does not change what researchers are trying to do when they conduct empirical analysis using econometric techniques.

If you look at the indexes of some standard textbooks from just a few years ago, you will see no entries for *causal, causation,* or similar terms.[3] But we would argue that econometrics properly conceived is the social science concerned with *estimation of parameters of (structural) economic models.* We would further argue that any time we believe we can read of an estimate of a causal effect from an econometric analysis, we are estimating a structural model, even if the model does not involve advanced economic analysis. For example, if we randomly assign observations to treatment and control, then most would agree that we can (under certain assumptions) read causal effects off a regression of the outcome of interest on the treatment indicator. So this is a "structural" model, albeit a very simple one.

Recently, some econometric textbooks have been more explicit about causal inference as the goal of almost all empirical research in economics. One example is *Mostly Harmless Econometrics* (Angrist and Pischke, 2008) and another is *Causal Inference: The Mixtape* (Cunningham, 2021).

The definition of econometrics that we provide here is not vacuous; not everything that accounting researchers do with econometric techniques could be described as econometric analysis. At times, researchers run regressions without giving any thought as to the model they are trying to estimate. As we will see later in this book, there are settings where accounting researchers have drawn (causal) inferences from estimated coefficients about economic phenomena whose connection to the empirical models used is very far from clear. In other settings, researchers will implicitly assume that the null hypothesis implies a zero coefficient on some variable without doing any modelling of this. One of the goals of this book is to encourage researchers to keep in mind what we are trying to do when we conduct econometric analysis.

4.1.2 Econometrics: The case of conditional conservatism

Conservatism has long been regarded as a hallmark of financial reporting. Many accounting standards treat losses and gains differently, with a greater willingness to recognize losses than gains. For example, if the present value of expected future cash flows from an asset declines below the asset's carrying value, several accounting standards (e.g., IAS 36) require the recognition of a commensurate loss. However, if the present value of expected future cash flows from an asset increases above the asset's carrying value, those standards may defer any recognition of a gain until those future cash flows are realized. In this way, conservatism manifests as **asymmetric timeliness** in the recognition of losses and gains. Because this form of conservatism is conditional on news about the value of assets, it is also known as **conditional conservatism**.[4]

Basu (1997) developed a measure of asymmetric timeliness that is derived from estimation of the following regression:

$$X_{it}/P_{it} = \alpha_0 + \alpha_1 D_{it} + \beta_0 R_{it} + \beta_1(R_{it} \times D_{it}) + \epsilon_{it},$$

---

[3]You might find an entry on *Granger causality*, but with some caveat that this is a misnomer.

[4]In contrast, the non-recognition of most internally generated intangible assets is not conditional on news, and it therefore considered **unconditional conservatism**.

where X_{it} is the earnings per share for firm i in fiscal year t, P_{it} is the price per share for firm i at the beginning of fiscal year t, R_{it} is the return on firm i from 9 months before fiscal year-end t to three months after fiscal year-end t, and D_{it} is an indicator for $R_{it} < 0$.

Using examples and intuition, Basu (1997) argues that $\hat{\beta}_1$ is a measure of asymmetric timeliness. An extensive literature has exploited the Basu (1997) measure to understand the role of conservatism in accounting. For example, Ball et al. (2000, p. 22) "propose that common-law accounting income is more asymmetrically conservative than code-law, due to greater demand for timely disclosure of economic losses" and provide evidence based on differences in the Basu (1997) measure. Jayaraman and Shivakumar (2012, p. 95) "find an increase in [the Basu (1997) measure] after the passage of antitakeover laws for firms with high contracting pressures."

This literature proceeded notwithstanding the absence of any demonstration that the Basu (1997) measure provides an unbiased and efficient estimator of an underlying conservatism parameter. While Ball et al. (2013) suggests the promise of covering the "econometrics of the Basu asymmetric timeliness coefficient" in its title, it contains little analysis of the kind seen in an econometrics text.

A more conventional econometric analysis might posit conditional conservatism as something that can be parameterized in some way. For example, specifying θ as the key parameter—with higher values of θ representing higher levels of conservatism—it might then be demonstrated how estimates of this underlying parameter θ can be made—or at least how measures that are increasing in θ can be constructed.

In an effort to establish some econometric foundations for the Basu (1997) measure, Ball et al. (2013) "model conditional conservatism as recognition of y in the current period only when it is sufficiently bad news ... [and assume current period recognition] if $y_t < c$ and zero otherwise, where c is a threshold below which current-period recognition occurs." Focusing on the case where $c = 0$, Ball et al. (2013) show that the expected Basu (1997) coefficient will be positive, but when there is no conditional conservatism (e.g., $c = -\infty$ or $c = \infty$), the expected Basu (1997) coefficient will be zero. But Ball et al. (2013) do not even propose an underlying parameterization of conservatism on which conventional econometrics could be brought to bear. The only parameter here is c, which does not have a monotonic relationship with conservatism (Ball et al., 2013 point out that both $c = -\infty$ and $c = \infty$ are consistent with the absence of conservatism).

In effect, Ball et al. (2013) show that if conservatism is defined as an asymmetric relationship between returns and income around zero returns, then an OLS estimator that captures an asymmetric relationship between returns and income around zero returns (such as the Basu (1997) measure) will *detect* conservatism.[5] Ball et al. (2013, p. 1073) state that "we show that, holding other things constant, the Basu regression identifies conditional conservatism only when it exists." Apart from the heavy lifting being done by the words "holding other things constant", this claim belies how the Basu (1997) measure has been used in research. The two papers cited above—Ball et al. (2000) and Jayaraman and Shivakumar (2012)—interpret higher levels of the Basu (1997) measure as capturing greater conservatism. Dietrich et al. (2022, p. 2150) point out that "differences in conservatism" across firms and time "is the explicit focus of nearly all conservatism research." But Ball et al. (2013) provide no

---

[5] Ball et al. (2013) was in large part a response to results in Dietrich et al. (2007) that showed "bias" in the Basu (1997) measure. The careful reader might wonder how bias can arise if it is not clear what underlying parameter is being estimated. But Dietrich et al. (2007) choose their words carefully and merely seek to demonstrate bias in the test statistics in favour of a hypothesis of conservatism when none exists rather than bias in an estimate coefficient relative to its true value.

elaboration of the concept of "greater conservatism" let alone how the Basu (1997) measure captures it.[6]

4.1.3 Econometrics: A brief illustration

While this is a text book on accounting research, let's consider a stylized example from labour economics. Suppose that we posit the following *structural* model:

$$y_i = \beta_0 + \beta_1 x_{i1} + \beta_2 x_{i2} + \beta_3 x_{i3} + \epsilon_i$$
$$x_{i1} = \alpha_0 + \alpha_2 x_{i2} + \alpha_3 x_{i3} + \eta_i$$

where i subscripts denote individual i. In the first equation, y_i is income at age 30, x_{i1} is years of education, x_{i2} is a measure of industriousness, x_{i3} is a measure of intelligence, and ϵ_i can be interpreted as random factors that affect y_i independent of X_i. In the second equation, we add coefficients, $\alpha := (\alpha_0, \alpha_1, \alpha_2)$, and η_i, which can be interpreted as random factors that affect x_{i1} independent of x_{i2} and x_{i3}.[7]

As researchers, we can postulate a model of the form above and we might obtain data on (y, X) that we could use to estimate $\beta := (\beta_0, \beta_1, \beta_2, \beta_3)$. But we don't know that the model is correct and even if we did, we don't know the values in β.

As we saw in Chapter 3, the OLS regression estimator can be written

$$\hat{\beta} = \left(X^\mathsf{T} X\right)^{-1} \left(X^\mathsf{T} y\right)$$

and it can be shown mathematically that OLS has good properties—such as unbiasedness and efficiency—under certain conditions. But in a world of cheap computing, we don't need to break out our pencils to do mathematics. Instead, we can "play God" in some sense and fix parameter values, simulate the data, then examine how well a researcher would do in estimating the parameter values that we set.

```
set.seed(2021)

n <- 100000

df <- tibble(
  industry = rnorm(n),
  intelligence = rnorm(n),
  education = 3 * intelligence + 4 * industry + rnorm(n),
  income = 10 + 5 * education + 6 * intelligence + 7 * industry + rnorm(n))
```

Now we can estimate three different models and store them in a list named `fms`.

```
fms <- list(lm(income ~ education, data = df),
            lm(income ~ education + intelligence + industry, data = df),
            lm(income ~ intelligence + industry, data = df))
```

The results from these models are presented in Table 4.1.

---

[6]Dietrich et al. (2022) make some tentative steps in the direction of a model in which conservatism is parameterized by considering two different levels of conservatism in addition to the no-conservatism regime.
[7]Here := means "defined as ...".

```
modelsummary(fms,
             estimate = "{estimate}",
             statistic = NULL,
             gof_map = c("nobs", "r.squared"))
```

TABLE 4.1
Income regressed on education and other variables

	(1)	(2)	(3)
(Intercept)	9.997	10.003	10.019
education	6.767	4.999	
intelligence		6.003	21.008
industry		7.000	27.007
Num.Obs.	100,000	100,000	100,000
R2	0.996	0.999	0.978

4.1.4 Exercises

1. Looking at the simulation code, what are the true values of $\beta := (\beta_0, \beta_1, \beta_2, \beta_3)$ and $\alpha := (\alpha_0, \alpha_1, \alpha_2)$?
2. Do any of the three equations reported in Table 4.1 provide good estimates of β?
3. Consider regression (3) in Table 4.1. With regard to the first of the two equations, are there any issues with regard to estimating β? What (if any) OLS assumption is violated?
4. What happens if you substitute the second equation (for x_{i1}) into the first equation (for y_i)? Does this equation satisfy OLS assumptions in some way?
5. Using the structural equations, what happens if we arbitrarily increase the value of industry (x_{i3}) by one unit? What happens to education (x_{i1})? What happens to income (y_i)?
6. Can you read the effect sizes from the previous question off any of the regression results in Table 4.1? If so, which one(s)?

4.2 Basic causal relations

The section provides a brief introduction to **causal diagrams**.

Figures 4.1–4.3 illustrate the basic ideas of causal diagrams and how they can be used to facilitate thinking about causal inference. Each figure depicts potential relationships among three *observable* variables. In each case, we are interested in understanding how the presence of a variable Z impacts the estimation of the causal effect of X on Y. The only difference between the three graphs is the direction of the arrows linking either X and Z, or Y and Z. The boxes (or **nodes**) represent random variables and the arrows (or **edges**) connecting boxes represent hypothesized causal relations, with each arrow pointing from a cause to a variable assumed to be affected by it.

Pearl (2009a) shows that we can estimate the causal effect of X on Y by conditioning on a set of variables, Z, that satisfies certain criteria. These criteria imply that very different conditioning strategies are needed for each of the causal diagrams (see Gow et al. (2016) for a more formal discussion).

While conditioning on variables is much like the standard notion of "controlling for" such variables in a regression, there are critical differences. First, conditioning means estimating effects for each distinct level of the set of variables in Z. This concept of **nonparametric conditioning** on Z is more demanding than simply including Z as another regressor in a linear regression model.[8] Second, the inclusion of a variable in Z may not be an appropriate conditioning strategy. Indeed, it can be that the inclusion of Z results in biased estimates of causal effects.

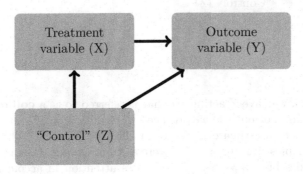

FIGURE 4.1
Z is a confounder

Figure 4.1 is straightforward. In this case, we need to condition on Z in order to estimate the causal effect of X on Y. Note the notion of "condition on" again is more general than just including Z in a parametric (linear) model.[9] The need to condition on Z arises because Z is what is known as a **confounder**.

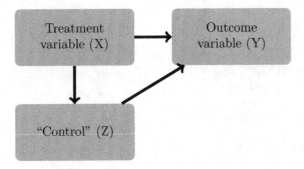

FIGURE 4.2
Z is a mediator

Figure 4.2 is a bit different. Here Z is a **mediator** of the effect of X on Y. No conditioning is required in this setting to estimate the total effect of X on Y. If we condition on X and Z, then we obtain a different estimate, one where the "indirect effect" of X on Y via Z

---

[8]Including variables in a linear regression framework is a form of *parametric* conditioning and "controls for" these variables only under strict assumptions, such as linearity in the relations between X, Y, and Z.
[9]Inclusion of Z blocks the "back-door" path from Y to X via Z.

is captured in the coefficient on Z, leaving only the "direct effect" to be reflected in the coefficient of X.[10]

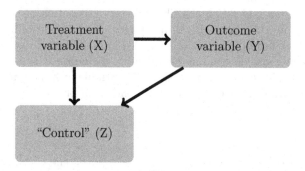

FIGURE 4.3
Z is a collider

Finally in Figure 4.3, we have Z acting as what is referred to as a **collider** variable (Pearl, 2009b, p. 17).[11] Again, not only do we not need to condition on Z, but that we *should* not condition on Z to get an estimate of the total effect of X on Y. While in epidemiology, the issue of "collider bias ...can be just as severe as confounding" (Glymour and Greenland, 2008, p. 186), collider bias appears to receive less attention in accounting research than confounding. Many intuitive examples of collider bias involve selection or stratification. Admission to university could be a function of combined test scores (T) and interview performance (I) exceeding a threshold, i.e., $T + I \geq C$. Even if T and I are unrelated unconditionally, a regression of T on I conditioned on admission to university is likely to show a negative relation between these two variables. To see this, we can generate some data following this simple structure.

```
set.seed(2021)

n <- 100000

admissions <- tibble(
  test = rnorm(n),
  interview = rnorm(n),
  score = test + interview,
  cutoff = quantile(score, .90),
  admitted = score >= cutoff)
```

We plot these data in Figure 4.4.

```
admissions |>
  ggplot(aes(x = test, y = interview, color = admitted)) +
  geom_point() +
  theme_bw()
```

---

[10]The quotes around the terms "direct effect" and "indirect effect" reflect that the meaning of these terms is very much contingent on what variables are included in the model.

[11]The two arrows from X and Y "collide" in Z.

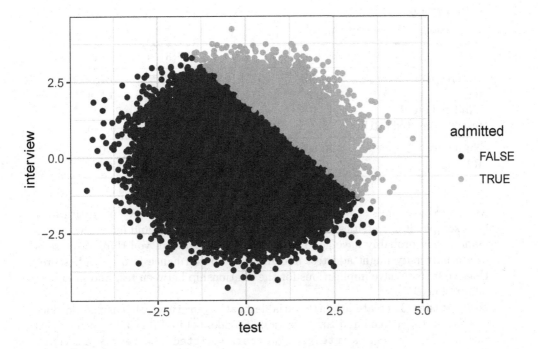

FIGURE 4.4
Admission status, interview scores, and test scores

We can also fit some models to the data we have generated. In the following code, we estimate four models and store the estimated models in a list `fms` (for "fitted models"). Models (3) and (2) capture subsets of data where applicant are or are not admitted, respectively. Models (1) and (4) use all observations, but differ in terms of the regression terms included.

```
fms <- list(lm(interview ~ test, data = admissions),
            lm(interview ~ test, data = admissions, subset = !admitted),
            lm(interview ~ test, data = admissions, subset = admitted),
            lm(interview ~ test * admitted, data = admissions))
```

The results from these models are presented in Table 4.2.

```
modelsummary(fms,
             estimate = "{estimate}{stars}",
             statistic = NULL,
             gof_map = c("nobs", "r.squared"),
             stars = c('*' = .1, '**' = 0.05, '***' = .01))
```

4.2.1 Exercises

1. Imagine that, while you understand the basic idea that both tests and interviews affect admissions, you only have access to the regression results reported in Table 4.2. Which of the four models seems to best "describe the data"? Which of the four models seems to do worst?

2. How are the coefficients reported in Table 4.2 for model (4) related to those for models (2) and (3)? Is this coincidence? Or would we always expect these relations to hold?

TABLE 4.2
Regression with collider variables

	(1)	(2)	(3)	(4)
(Intercept)	0.001	−0.161***	2.137***	−0.161***
test	−0.008***	−0.174***	−0.719***	−0.174***
admittedTRUE				2.298***
test × admittedTRUE				−0.545***
Num.Obs.	100,000	90,000	10,000	100,000
R2	0.000	0.031	0.515	0.227

3. As applied economists, we should be alert to the "endogeneity" of institutions. For example, if universities admitted students based on test and interview performance, they probably have good reasons for doing so. Can you think of reasons why a university might add test and interview performance into a single score? Does your story have implications for the relationship between test and interview performance?

4. Using `mutate()`, create a fourth variable `test_x_admitted` that is the product of `test` and `admitted` and run a version of model (4) in Table 4.2 that uses this variable. That is regress `interview` on `test`, `admitted`, and `test_x_admitted`. Do you get the same results as are shown above for model (4) in Table 4.2? Why or why not?

4.3 Causal diagrams: Formalities

Here we provide a brief formal treatment of some of the ideas on causal diagrams discussed above. See Pearl (2009a) for more detailed coverage.

4.3.1 Definitions and a result

We first introduce some basic definitions and a key result from Pearl (2009a). While this material may seem forbiddingly formal, the ideas are actually fairly straightforward and we will cover a few examples to demonstrate this. We recommend that in your first pass through the material, you spend just a little time on the two definitions and the theorem and then go back over them after reading the guide we provide just after the theorem.

Definition 4.1. A path p is said to be *d*-**separated** (or **blocked**) by a set of nodes Z if and only if

    1. p contains a chain $i \rightarrow m \rightarrow j$ or a fork $i \leftarrow m \rightarrow j$ such that the middle node m is in Z, or

    2. p contains an inverted fork (or *collider*) $i \rightarrow m \leftarrow j$ such that the middle node m is not in Z and such that no descendant of m is in Z.

Definition 4.2. A set of variables Z satisfies the **back-door criterion** relative to an ordered pair of variables (X, Y) in a **directed acyclic graph (DAG)** G if:

1. no node in Z is a descendant of X; and
2. Z blocks every **back-door path** between X and Y, i.e., every such path that contains an arrow into X.[12]

Given this criterion, Pearl (2009a, p. 79) proves the following result.

Theorem 4.1. *If a set of variables Z satisfies the back-door criterion relative to (X, Y), then the causal effect of X on Y is identifiable and is given by the formula*

$$P(y|x) = \sum_z P(y|x, z)P(z),$$

where $P(y|x)$ stands for the probability that $Y = y$, given that X is set to level $X = x$ by external intervention.[13]

In plainer language, Theorem 4.1 tells us that we can estimate the causal effect of X on Y if we have a set of variables Z that satisfies the back-door criterion. Then Definition 4.2 tells us that Z needs to block all back-door paths while not containing any descendant of X. Finally, Definition 4.1 tells what it means to block a path.

4.3.2 Application of back-door criterion to basic diagrams

Applying the back-door criterion to Figure 4.1 is straightforward and intuitive. The set of variables $\{Z\}$ or simply Z satisfies the criterion, as Z is not a descendant of X and Z blocks the back-door path $X \leftarrow Z \rightarrow Y$. So by conditioning on Z, we can estimate the causal effect of X on Y. This situation is a generalization of linear model in which $Y = X\beta + Z\gamma + \epsilon_Y$ and ϵ_Y is independent of X and Z, but X and Z are correlated. In this case, it is well known that omission of Z would result in a biased estimate of β, the causal effect of X on Y, but by including Z in the regression, we get an unbiased estimate of β. In this situation, Z is a **confounder**.

Turning to Figure 4.2, Z does not satisfy the back-door criterion, because Z (a mediator) is a descendant of X. However, $\emptyset$ (i.e., the empty set) *does* satisfy the back-door criterion. Clearly, $\emptyset$ contains no descendant of X. Furthermore, the only path other than $X \rightarrow Y$ that exists is $X \rightarrow Z \rightarrow Y$, which does not have a back-door into X. Note that the back-door criterion implies not only that we *need* not condition on Z to obtain an unbiased estimate of the causal effect of X on Y, but that we *should* not condition on Z to get such an estimate.

Finally in Figure 4.3, we have Z acting as a collider. Again, we see that Z does not satisfy the back-door criterion because Z is a descendant of X. However, $\emptyset$ again satisfies the back-door criterion. First, it contains no descendant of X. Second, the only path other than $X \rightarrow Y$ that exists is $X \rightarrow Z \leftarrow Y$, which does not have a back-door into X. Again, the back-door criterion implies not only that we need not condition on Z, but that we should not condition on Z to get an unbiased estimate of the causal effect of X on Y.

4.3.3 Exercises

1. Draw the DAG for the structural model above relating intelligence, industriousness, education, and income. For each x variable, identify the sets of conditioning variables that satisfy the back-door criterion with respect to estimating a causal effect of the variable on y.

---

[12]The "arrow into X" is the portion of the definition that explains the "back-door" terminology.

[13]How the quantities $P(y|x)$ map into estimates of causal effects is not critical to the current discussion, it suffices to note that in a given setting, it can be calculated if the needed variables are observable.

2. For any valid set of conditioning variables not considered in Table 4.2, run regressions to confirm that these indeed deliver good estimates of causal effects.

4.4 Discrimination and bias

Let's examine a real-world example related to possible gender discrimination in labour markets.[14] When critics claimed that Google systematically underpaid its female employees, Google responded[15] that when "location, tenure, job role, level and performance" are taken into consideration, women's pay was basically identical to men's. In other words, controlling for characteristics of the job, women received the same pay.

But what if stereotyping means men are given roles that are paid better? In this case, naive comparisons of wages by gender "controlling for" occupation would understate the presence of discrimination. Let's illustrate this with a DAG based on a simple occupational sorting model with unobserved heterogeneity.

Note that there is in fact no effect of being female (F) on earnings (Y) except through discrimination (D). Thus, if we could control for discrimination, we'd get a coefficient of zero on F. In this example, we aren't interested in estimating the effect of being female on earnings *per se*; we are interested in estimating the effect of discrimination. Note also that discrimination is not directly observed, but given our DAG, we can use F as a proxy for D, as we have assumed that there is no relation between F and Y (or between F and O, which denotes occupational assignment) except through D.

In this DAG, there are two paths between D and Y:

$$D \to Y$$
$$D \to O \to Y$$

Neither path is a back-door path between D and Y because neither has an arrow pointing *into* D. Conditioning on O causes the estimated effect of discrimination on income to be biased because O is a descendant of D, and a set of conditioning variables Z satisfies the back-door criterion only if no node in Z is a descendant of D.

To check our understanding of the application of the back-door criterion to this setting, we can generate some data based on a simulation that matches the causal diagram in Figure 4.5.

```
n <- 100000

set.seed(2023)
df <- tibble(
  female = runif(n) >= 0.5,
  discrimination = female,
  occupation = 1 + 0 * female - 2 * discrimination + rnorm(n),
  salary = 1 - 1 * discrimination + 2 * occupation + rnorm(n)
)
```

We run two regressions on these data and present results in Table 4.3.

---

[14]This subsection is adapted from Cunningham (2021), available at https://mixtape.scunning.com.
[15]https://gizmodo.com/google-employees-organize-their-own-study-of-gender-pay-1802767010

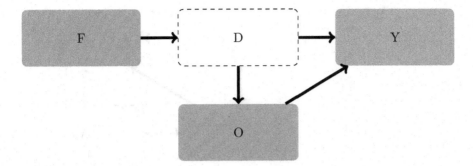

FIGURE 4.5
Discrimination

```
fms <- list(lm(salary ~ female, data = df),
            lm(salary ~ female + occupation, data = df))

modelsummary(fms,
             estimate = "{estimate}{stars}",
             statistic = NULL,
             gof_map = c("nobs", "r.squared"),
             stars = c('*' = .1, '**' = 0.05, '***' = .01))
```

TABLE 4.3
Discrimination regressions

	(1)	(2)
(Intercept)	3.013***	1.003***
femaleTRUE	−5.005***	−0.996***
occupation		1.999***
Num.Obs.	100,000	100,000
R2	0.556	0.911

4.4.1 Exercises

1. Given the equations for `occupation` and `salary` in the simulation above, what is the direct effect of discrimination on salary? What is the indirect effect (i.e., the effect via `occupation`) of discrimination on salary? What is the total effect of discrimination on salary? How do each of these effects show up in the regression results reported in Table 4.3? (*Note*: Because of sampling variation, the relationships will not be exact.)

2. Consider the possibility of an additional unobserved variable, ability (A), that affects role assignment (O) and also affects income (Y) directly. A DAG that captures this is provided in Figure 4.6.

What would be the correct conditioning strategy in this case? Would it now make sense to condition on O if the goal is to estimate the total effect of discrimination on Y? (*Hint*: In answering this question, it may help to adapt the simulation above to generate an `ability` variable and to incorporate that in the model using code like the following.)

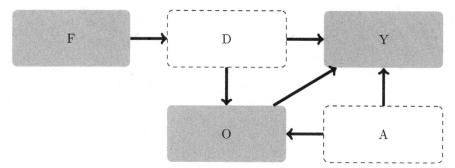

FIGURE 4.6
Discrimination with ability

```
set.seed(2023)
ability = rnorm(n)
occupation = 1 + 0 * female - 2 * discrimination + 2 * ability + rnorm(n)
salary = 1 - 1 * discrimination + 2 * occupation + 0.3 * ability + rnorm(n)
```

3. Consider the additional possibility of different occupational preferences between males and females, as depicted in Figure 4.7.

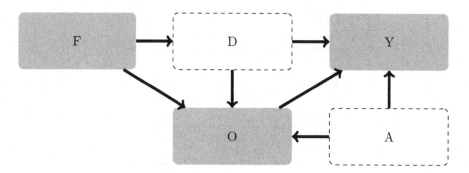

FIGURE 4.7
Discrimination with ability and job preferences

Given the DAG in Figure 4.7, is it possible to identify a set of conditioning variables that allow you to estimate the total effect of discrimination on salary? What roles does O have in this DAG? (*Hint*: Replace the 0 coefficient on `female` in the `occupation` equation used in the last question with a value of either -1 or +1. Does the sign of this coefficient affect the sign of bias, if any?) Is it possible to estimate the *direct* effect of discrimination on salary? If so, how? If not, why?

4.5 Causal diagrams: Application in accounting

Many papers in accounting research include many variables to "control for" potential confounding effects. If these variables are truly confounders, then this is appropriate. However,

Gow et al. (2016) suggest that researchers do not give consideration to the possibility that these variables are mediators or colliders. Inclusion of "controls" that are mediators or colliders will generally lead to bias, making it important to rule these possibilities out.

Larcker et al. (2007) discuss these distinctions, albeit in different language, in the context of corporate governance research and a regression model of the form:[16]

$$Y = \alpha + \sum_{r=1}^{R} \gamma_r Z_r + \sum_{s=1}^{S} \beta_s X_s + \epsilon \tag{4.2}$$

Larcker et al. (2007) suggest that

"One important feature in the structure of Equation 4.2 is that the governance factors [X] are assumed to have no impact on the controls (and thus no indirect impact on the dependent variable). As a result, this structure may result in conservative estimates for the impact of governance on the dependent variable. Another approach is to only include governance factors as independent variables, or:

$$Y = \alpha + \sum_{s=1}^{S} \beta_s X_s + \epsilon \tag{4.3}$$

The structure in Equation 4.3 would be appropriate if governance impacts the control variables and both the governance and control variables impact the dependent variable (i.e., the estimated regression coefficients for the governance variables will capture the total effect or the sum of the direct effect and the indirect effect through the controls)."

But Larcker et al. (2007) gloss over some subtle issues. If some elements of Z_r are mediators and other elements are confounders, then biased estimates will arise from both equations. The estimator based on Equation 4.3 will be biased because confounders have been omitted, while that based on Equation 4.2 will be biased because mediators have been included. Additionally, the claim that the estimates are "conservative" assumes that the indirect effect via mediators is of the same sign as the direct effect. Without this assumption, the relation between the direct effect and the total effect is unclear in both magnitude and even sign.

Additionally, Larcker et al. (2007) do not address possible colliders, which plausibly exist in their setting. For example, corporate governance plausibly affects firms' choice of leverage, while performance also plausibly affects leverage. If so, "controlling for" leverage might induce associations between governance and performance even when there is no causal relation between these variables.[17] While there is intuitive appeal to with-and-without-controls approach suggested by Larcker et al. (2007), selecting controls "requires careful thinking about the plausible causal relations between the treatment variables, the outcomes of interest, and the candidate control variables" (Gow et al., 2016, p. 485).

---

[16]We alter the mathematical notation and equation numbers of Larcker et al. (2007) to conform with notation and numbering we use here.

[17]Note that Larcker et al. (2007) do not "control for" leverage in regressions where performance is a dependent variable.

4.6   Further reading

Gow et al. (2016) covers several topics not covered in this chapter. But parts of this book (especially Part III) go into those topics more deeply than Gow et al. (2016) do.

More recent econometrics textbooks, such as Wooldridge (2010), more explicitly address causal inference than do older texts. In addition, a number of more recent textbooks—such as Angrist and Pischke (2008) and Cunningham (2021)—place causal inference front and centre.

Fuller coverage of casual diagrams can be found in Morgan and Winship (2014); Pearl (2009a) offers a more advanced treatment. Huntington-Klein (2021) uses causal diagrams as an organizing framework and spends much of the book examining causal inference before introducing the econometric tools that are the primary focus of most textbooks. A more accessible introduction to ideas in Pearl (2009a), including more recent developments, is provided by Pearl and Mackenzie (2019).

Our discussion of structural models glossed over some subtle, but important, issues that get some coverage in Chapter 5 of Pearl (2009a).

5

Statistical inference

In this chapter, we study statistical inference in the context of **null hypothesis significance testing** (NHST), which is the predominant paradigm in empirical accounting research. We examine NHST in fairly procedural terms without spending too much time justifying it. One reason for doing so is that, even if we simply accept NHST as a sensible approach to science, the way NHST is used in practice is profoundly different from how it was expected to be used when its epistemological foundations were set.[1] So real problems with NHST exist long before reaching those foundations.

Another issue with NHST is that statistical inference in most settings is based on **limit theorems**—results from mathematical statistics that can require fairly advanced mathematics to prove—such as **laws of large numbers** and **central limit theorems**.[2] To minimize the mathematical requirements here, we use simulations to illustrate ideas rather than trying to provide mathematically precise results. For readers looking to go deeper on either dimension, we provide a guide to further reading at the end of the chapter.

> **Tip**
>
> The code in this chapter uses the packages listed below. For instructions on how to set up your computer to use the code found in this book, see Section 1.2 (note that Step 4 is not required as we do not use WRDS data in this chapter). Quarto templates for the exercises below are available on GitHub.[a]
>
> ---
>
> [a]https://github.com/iangow/far_templates/blob/main/README.md

```
library(dplyr)
library(ggplot2)
library(farr)
library(modelsummary)
library(kableExtra)    # add_header_above()
library(purrr)         # map(), map_vec(), map_df()
library(tidyr)         # nest()
library(sandwich)      # NeweyWest(), vcovHC()
library(plm)           # pmg(), vcovNW()
library(fixest)        # se(), pvalue()
```

---

[1]We examine some of these issues in Chapter 19.

[2]Referring to these results in the plural ("laws" and "theorems") without the definite article ("the") and in lower case is intended to indicate that these results are actually sets of closely related results exhibiting the general pattern that stronger assumptions (imposed on the random variables in question) lead to "stronger" results.

DOI: 10.1201/9781003456230-5

5.1   Some observations

Students of econometrics in PhD programs often spend much of their time learning about the **asymptotic properties** of estimators, which might be understood as describing how estimators behave as sample sizes tend towards infinity. However, there are two gaps in this pedagogical approach. First, it is often not made clear *why* we care about asymptotic properties of estimators. A practical student might question the relevance of asymptotic properties when sample sizes in practice can at times be quite large, but are rarely infinite.

Second, in practice, researchers soon forget that the estimators they use are justified by their asymptotic properties and rarely exploit what they learned in econometrics classes to study new estimators. For example, in this chapter we will devote some time to the study of **standard errors**. Almost all estimators of standard errors are motivated by asymptotic properties and thus can perform poorly in actual research settings. Yet Gow et al. (2010) show that estimators can gain currency among researchers without any formal analysis of their properties, asymptotic or otherwise.

One reason for studying the asymptotic properties of estimators is that they are mathematically tractable for a well-trained econometrician. For a practical researcher, knowing that an estimator has good asymptotic properties might be considered more or less *necessary*. An estimator that is not **consistent** is apt to produce significantly biased estimates in finite samples (one can think of *consistency* as the asymptotic equivalent of unbiasedness).[3] And good asymptotic properties might be *practically sufficient* for the estimator to behave well in finite samples. An estimator that is *asymptotically efficient* and *asymptotically normal* might have a good chance of being an efficient estimator in finite samples and having a distribution that is sufficiently close to the normal distribution for practical statistical inference.

In this chapter, we will not derive asymptotic properties of estimators. Instead, we direct readers to econometrics texts for these details. Additionally, while we provide an intuitive introduction to laws of large numbers and central limit theorems, we do not even state these formally, let alone derive them. Instead, in keeping with a theme of this book, we will use simulations to give the reader an intuitive feel for where these results from mathematical statistics come from and what they mean.

5.2   Data-generating processes

Like causal inference, thinking about statistical inference generally involves quite abstract reasoning. Before getting into the statistics, it is helpful to imagine that the data we observe are generated by an underlying process (hence a **data-generating process** or **DGP**). A DGP will often look like a **model**, but in principle these are distinct ideas. The DGP is a feature of reality that we generally do not observe directly, while a model is a guess or useful simplification of a conjectured DGP that humans use to understand the DGP.

---

[3] If speaking to a careful econometrician, one should be careful about conflating consistency and unbiasedness, as there exist estimators that are always biased in finite samples, but are consistent and estimators that are unbiased, but not consistent.

In some settings, we may believe that our model captures all the essential details of the DGP. While the results of a coin toss will depend on the rotational momentum imparted to the coin, the vertical velocity of the coin, imperfections on the coin's surface, prevailing humidity, temperature, wind velocity, and so on, we may be happy in simply using a model in which coin tosses are independent, and equally likely to be one outcome as the other.

Part of the potential for confusion between these two ideas is that when discussing a DGP, we inevitably use a model, as we do here. Suppose that the height in centimetres of adult individuals is drawn from a normal distribution with mean of $\mu = 165$ and standard deviation of $\sigma = 8$.

```
hgt_mean <- 165
hgt_sd <- 8
```

This statement could be interpreted as the process for generating the heights of the billions of adult people who happen to inhabit Earth at this time. In a sense, the people who actually exist are a non-deterministic sample of the potential people who could exist.[4]

Alternatively, we might take the current inhabitants of the world as given and posit that the empirical distribution of heights is actually a normal distribution with parameters μ and σ, and focus instead on taking random sample from that population. Looking ahead, we might use a random sample to make inferences about the distribution of heights of adult humans who currently inhabit the world or about the data-generating process for adult heights. In the context of causal inference, it is often the latter kind of inference that we want to make, but for present purposes, we focus on the former kind.

There is no doubt that a two-parameter normal distribution is an oversimplification of the actual DGP, hence more of a model than an actual DGP. For example, we make no allowance for heights varying around the world, for the tendency of heights to increase over time (hence the reference to "current" above), or the tendency for men to be taller than women, each of which may undermine the applicability of a simple assumed DGP. But for current purposes, we ignore such quibbles and simply assume that heights come from a normal distribution (μ, σ).

We can simulate the taking of a random sample of size $n = 100$ by using the `rnorm()` function we saw in Chapter 2.

```
hgt_sample <- rnorm(100, mean = hgt_mean, sd = hgt_sd)
```

Given that we *know* that $\mu = 165$ and that $\sigma = 8$, we can calculate expected values for the mean and standard deviation of the heights in our sample. Here the sample mean is calculated as the sum of the observed heights $(\{x_i : i \in (1, \ldots, n)\})$ divided by n, where $n = 100$.

$$\overline{X}_n = \frac{\sum_{i=1}^{n} x_i}{n}$$

---

[4]We eschew the word "random" here lest it be misinterpreted in an implausible "equal probability of existing" sense.

$$\mathbb{E}[\overline{X}_n] = \mathbb{E}\left[\frac{\sum_{i=1}^{n} x_i}{n}\right]$$

$$= \frac{\sum_{i=1}^{n} \mathbb{E}[x_i]}{n}$$

$$= \frac{n\mu}{n}$$

$$= \mu$$

To calculate the variance of $\overline{X}_n$, we note that the variance of ax, where a is a constant and x is a random variable, is $\mathrm{Var}[ax] = a^2\mathrm{Var}[x]$ and the variance of the sum of independent and identically distributed variables x_i, $i \in (1,\dots,n)$, each with variance σ^2, is simply the sum of their variances:

$$\mathrm{Var}\left[\sum_{i=1}^{n} x_i\right] = \sum_{i=1}^{n} \mathrm{Var}[x_i]$$

$$= n\sigma^2$$

The variance of x/k for some constant $k \neq 0$ is

$$\mathrm{Var}\left(\frac{x}{k}\right) = \sum_{i=1}^{n} \frac{1}{n}\left(\frac{x_i - \overline{x}}{k}\right)^2$$

$$= \frac{1}{k^2}\sum_{i=1}^{n}\frac{1}{n}(x_i - \overline{x})^2$$

$$= \frac{1}{k^2}\mathrm{Var}(x)$$

From the above, we can calculate the following:

$$\mathrm{Var}[\overline{X}_n] = \mathrm{Var}\left[\frac{\sum_{i=1}^{n} x_i}{n}\right]$$

$$= \frac{1}{n^2}\mathrm{Var}\left[\sum_{i=1}^{n} x_i\right]$$

$$= \frac{\sigma^2}{n}$$

The standard deviation of the sample mean is the square root of its variance: $\frac{\sigma}{\sqrt{n}}$.

> ⚠ **Warning**
>
> This section may seem a little bewildering on a first reading because we cover a lot of concepts in R quickly. Our recommendation is to play around with the code yourself as you work through this section and focus more on the inputs and outputs than the finer details of the code at this point.

Because in this section, we are in an idealized world in which we know the underlying DGP, we can test our calculations by generating data. We can incorporate the code we used to make a sample above in a function `get_hgt_sample()` that takes a single argument (`n`) for the sample size.

```
get_hgt_sample <- function(n) {
  rnorm(n, mean = hgt_mean, sd = hgt_sd)
}
```

Calling `get_hgt_sample` with `n = 10`, we get a sample of 10 random numbers. We use `set.seed(2023)` so that we get the same random numbers each time. The value 2023 is completely arbitrary (it was the current year when we first wrote this chapter).[5]

```
set.seed(2023)
```

```
get_hgt_sample(10)
```

```
 [1] 164.3297 157.1365 149.9995 163.5108 159.9321 173.7264 157.6902 173.0131
 [9] 161.8059 161.2550
```

But we can easily generate *ten-thousand* samples. In many programming languages, we would use a `for` loop to do this and R offers such functionality, as seen in the following code (Chapter 27[6] of *R for Data Science* discusses `for` loops). In R, it is natural to use a list to store results from multiple runs of a function, so we initialize `hgt_samples` as an empty list and then use the `for` command to repeat the `get_hgt_sample(100)` call 10,000 times, storing each sample in a slot in `hgt_samples` (we discussed lists in Chapter 2).

```
n_samples <- 10000
```

```
set.seed(2023)

hgt_samples <- list()
for (i in 1:n_samples) {
  hgt_samples[[i]] <- get_hgt_sample(100)
}
```

We can take a peek at the first ten observations in the 463rd sample like this:

```
hgt_samples[[463]][1:10]
```

```
 [1] 168.4070 164.0360 163.2143 155.2887 163.0420 170.6284 146.7339 168.8120
 [9] 172.7608 156.1855
```

We can calculate the mean of the 463rd sample like this:

```
mean(hgt_samples[[463]])
```

```
[1] 164.4264
```

We could use a `for` loop to calculate the mean for each of the samples in one go, storing the results in a new list, `hgt_sample_means`.

```
hgt_sample_means <- list()
for (i in 1:length(hgt_samples)) {
  hgt_sample_means[[i]] <- mean(hgt_samples[[i]])
}
```

---

[5]Type `? set.seed` in R to learn more.
[6]https://r4ds.hadley.nz/base-r

We can peek at the first 5 values of `hgt_sample_means` like this:

```
hgt_sample_means[1:5]
```

```
[[1]]
[1] 165.5401

[[2]]
[1] 166.6662

[[3]]
[1] 163.7082

[[4]]
[1] 165.0866

[[5]]
[1] 164.1906
```

For many purposes, it will be easier to work with `hgt_sample_means` if it is a vector rather than a list and we can use the `unlist()` function to convert it to a vector:

```
unlist(hgt_sample_means[1:5])
```

```
[1] 165.5401 166.6662 163.7082 165.0866 164.1906
```

Additionally, we could use one of the `map()` functions, such as `map_vec()`, from the `purrr` library rather than a `for` loop. Chapter 26[7] of *R for Data Science* covers the `purrr` library. The `map(x, f)` function maps the function `f()` to each element of the vector `x` and returns a list containing the results. In our case, `x` will be a column of a data frame and, because we want the result of each calculation here to be a vector, we use `map_vec()`, which is like `map()` but with the output simplified to a vector.[8]

```
hgt_sample_means <- map_vec(hgt_samples, mean)
```

We can see we have a nice vector of means from a single line of code.

```
hgt_sample_means[1:10]
```

```
[1] 165.5401 166.6662 163.7082 165.0866 164.1906 164.9117 164.3328 164.5542
[9] 164.1957 164.8957
```

In fact, we could have used `map()` rather than a `for`-loop to create `hgt_samples` in the first place.

Note that, to get `n_samples` samples, we want to `map()` the `get_hgt_sample()` function to the vector `1:n_samples`. But we have a small problem: `map()` will provide each value of that vector to `get_hgt_sample()`, and `get_hgt_sample()` simply does not expect that value. One approach to addressing this is to use an **anonymous function** as we do below.

```
set.seed(2023)
hgt_samples <- map(1:n_samples, function(x) get_hgt_sample(100))
```

---

[7]https://r4ds.hadley.nz/iteration

[8]There are base R equivalents to the functions provided by `purrr`, such as `Map()` and `lapply()`. Perhaps the main benefit of using the `purrr` functions is that they are easier to learn because there is greater consistency across functions.

We call `function(x) get_hgt_sample(100)` an anonymous function because we do not give it a name. Note that we could also use a shortcut form that allows us to write an anonymous `function(x) get_hgt_sample(100)` as simply `\(x) get_hgt_sample(100)`. While we could give the function a name (by storing it in a variable), this means creating a new variable (`get_hgt_sample_alt`) without producing code that is easier to read.

```
set.seed(2023)
get_hgt_sample_alt <- function(x) get_hgt_sample(100)
hgt_samples <- map(1:n_samples, get_hgt_sample_alt)
```

Another approach would be to make a version of `get_hgt_sample()` that does expect an argument related to the index of the sample being generated. In fact, in some settings, it might be useful to retain that information, so we have `get_hgt_sample_df()` return a data frame (tibble) containing the sample index in the first column (`i`) and the actual sample in the second column (`data`).

```
get_hgt_sample_df <- function(i, n) {
  tibble(i, data = list(rnorm(n, mean = hgt_mean, sd = hgt_sd)))
}
```

To see what `get_hgt_sample_df()` is doing, we call it once to get a single (small) sample.

```
one_sample <- get_hgt_sample_df(i = 1, n = 10)
```

Looking at the data frame `one_sample`, we can see that it has two columns and only one of these—**data**—is a list column.

```
one_sample
```

```
# A tibble: 1 x 2
      i data
  <dbl> <list>
1     1 <dbl [10]>
```

Looking at the first value in the **data** column, we see that it's a vector of ten values.

```
one_sample$data[[1]]
```

```
[1] 159.2986 172.3488 170.3377 172.4086 161.2063 171.0710 164.7035 183.3683
[9] 152.9049 171.6578
```

So let's get **n_samples** using this approach.

```
set.seed(2023)
hgt_samples <- map(1:n_samples, get_hgt_sample_df, n = 100)
```

While we can access the underlying data in `hgt_samples`, it might seem that we are adding complexity for no particular reason, especially if we compare the code for looking at the 463rd sample below to the equivalent code in our first attempt above.

```
unlist(hgt_samples[[463]]$data)[1:10]
```

```
[1] 168.4070 164.0360 163.2143 155.2887 163.0420 170.6284 146.7339 168.8120
[9] 172.7608 156.1855
```

One reason for this complexity is that currently `hgt_samples` is a list of tibbles. We can use the function `list_rbind()` to bind them together into a single data frame.

```
hgt_samples_df <- list_rbind(hgt_samples)
```

Note that `hgt_samples_df` is starting to get fairly large:

```
print(object.size(hgt_samples_df), units = "auto")
```

```
8.2 Mb
```

While the size of `hgt_samples_df` does not present a problem for a modern computer, more complicated analyses might produce objects sufficiently large that we would want to pay attention to their sizes. We can now use `map_vec()` to extract statistics from `hgt_samples_df`. To keep our results small, we then drop the `data` column from our table of statistics (`hgt_samples_stats`). (We could always use the column i to join `hgt_samples_stats` with `hgt_samples_df` if we wanted to recover these data.)

```
hgt_samples_stats <-
  hgt_samples_df |>
  mutate(n = map_vec(data, length),
         mean = map_vec(data, mean),
         se = map_vec(data, sd) / sqrt(n)) |>
  select(-data)
```

We can now assess how well our formulas worked out in our simulations. We calculate the mean and standard deviation of the means, as well as the mean of the standard errors.

```
hgt_stats_summ <-
  hgt_samples_stats |>
  summarize(n = mean(n),
            mean_mean = mean(mean),
            sd_mean = sd(mean),
            mean_se = mean(se),
            sd_se = sd(se))
```

```
hgt_stats_summ
```

```
# A tibble: 1 x 5
      n mean_mean sd_mean mean_se   sd_se
  <dbl>     <dbl>   <dbl>   <dbl>   <dbl>
1   100      165.   0.810   0.799  0.0569
```

The mean of the estimated means is indeed pretty close to the μ parameter of 165 and the standard deviation of those estimated means is also very close to the value $\frac{\sigma}{\sqrt{n}}$ (0.8).

We can examine the errors in our estimates using `transmute()`, which acts like `mutate()` except that `transmute()` does not retain columns not calculated in the function call.

```
hgt_stats_summ |>
  transmute(mean_error = mean_mean - hgt_mean,
            se_error = sd_mean - hgt_sd / sqrt(n),
            se_est_error = mean_se - sd_mean)
```

```
# A tibble: 1 x 3
  mean_error se_error se_est_error
       <dbl>    <dbl>        <dbl>
1     0.0110   0.0100      -0.0112
```

5.2.1 Discussion questions

1. Explain what "error" each of `mean_error`, `se_error`, and `se_est_error` is capturing.

2. What effects do you expect changes in the values of `n_samples` or `n` to have on `mean_error`? Do you expect changing `n_samples` or changing `n` to have a greater effect? Verify your conjectures using the simulation code. (*Hint:* Consider 100,000 draws of samples of 100 and 10,000 draws of samples of 1,000. In each case, you should `set.seed(2023)` afresh and effectively replace data in `hgt_samples`.)

5.3 Hypothesis testing

Of course, in reality we generally do not observe parameters like `hgt_mean` and `hgt_sd` and our task as statisticians is to form and test hypotheses about those unobserved parameters. With **null hypothesis significance testing**, the researcher starts with a **null hypothesis** and asks whether the data are "sufficiently unusual" to justify rejection of that hypothesis.

As alluded to above, there are many deep issues lurking in this approach that we largely ignore. One such issue that we ignore is the source for null hypotheses. It is often implied that null hypotheses somehow represent the prior or default beliefs of "science". For example, in the context of pharmaceuticals, the null hypothesis is usually that a potential drug has zero (or negative) effect on the target outcome (say, curing cancer) relative to a placebo and this null hypothesis has to be rejected for a drug to receive approval from regulators.[9] In other contexts, the source for—and meaning of—the null hypothesis is less clear.

5.3.1 Normal distribution, known standard deviation

Suppose we have a null hypothesis that the mean height of adult individuals is $\mu_0 = 170$ and that we *know* that heights are drawn from a normal distribution with standard deviation of $\sigma = 8$ centimetres.

```
hgt_mean_null <- 170
```

Now we can get a sample of data to test our hypothesis:

```
set.seed(2023)
sample_size <- 100
test_sample <- get_hgt_sample(sample_size)
mean_est <- mean(test_sample)
mean_est
```

```
[1] 165.5401
```

Here we have a sample mean of 165.54. How unusual is 165.54 under the null hypothesis?

Given we know that the sample mean will be normally distributed with standard deviation equal to $\sigma/\sqrt{n}$, we can calculate the probability of observing a mean this far from the null hypothesis *assuming the null hypothesis is true* using `pnorm()`:

---

[9]We also ignore the various semantic debates, such as whether one "accepts" alternative hypotheses and "rejects" null hypotheses or whether other words better describe what is going on.

```
pnorm(mean_est, mean = hgt_mean_null, sd = hgt_sd / sqrt(sample_size))
```

```
[1] 1.238951e-08
```

In practice, it is more common to transform the mean estimate into a **z-statistic** that has a **standard normal distribution** (denoted by $N(0,1)$), with mean zero and standard deviation of one.

$$z = \frac{\bar{x} - \mu_0}{\sigma/\sqrt{n}}$$

```
z_stat <- (mean_est - hgt_mean_null) / (hgt_sd / sqrt(sample_size))
```

We can see that using the z-statistic (here -5.57) yields exactly the same p-value.

```
pnorm(z_stat)
```

```
[1] 1.238951e-08
```

It is generally posited that the criterion for rejection of the null hypothesis is established prior to analysing the data. The set of possible outcomes is partitioned into two disjoint sets that together comprise the event space. One of these sets is called the **rejection region**. The **critical value** represents the threshold between these two sets expressed in terms of the **test statistic**.

The critical value is generally calculated based on a probability threshold (also known as the **significance level** or **size of the test**) below which results are deemed **statistically significant**. A widely used significance level is 5%. Given that our null hypothesis is H_0 : $\mu = 170$, we could reject this null hypothesis because the test statistic is either too low or too high, so we need to account for both possibilities when determining the critical value.

As such we set the critical value for "surprisingly low" test statistics at the point on the distribution where 2.5% (5% ÷ 2) of values are more extremely negative given the null hypothesis. And we set the critical value for "surprisingly high" test statistics at the point where 2.5% (5% ÷ 2) of values are more extremely positive given the null hypothesis.

```
qnorm(0.025)
```

```
[1] -1.959964
```

```
qnorm(1 - 0.025)
```

```
[1] 1.959964
```

We see that these critical values are identical in magnitude and only differ in sign.[10] We can therefore set the critical value as $c = 1.96$ and identify the **rejection region** as $\{z : |z| > c\}$. As the absolute value of our observed z-statistic (5.57) clearly exceeds the critical value, we can reject the null hypothesis.

We can also calculate the p-value of our observed result. Again we need to recognize that values as extreme as those observed could be positive or negative, so the probability of finding data at least as surprising as that we have is actually twice this probability calculated above $p = 2.4779025 \times 10^{-8}$. (That the observed p-value is below 5% is equivalent to $|z| > c$.)

---

[10]This is because of the symmetry of the standard normal distribution around zero.

5.3.2 Normal distribution, unknown standard deviation

In reality, a researcher does not simply *know* the standard deviation. Fortunately, we can use our data to estimate it.

$$\hat{\sigma}^2 = \frac{1}{n-1} \sum_{i=1}^{n} (x_i - \overline{x})^2$$

We can use our estimate $\hat{\sigma}$ to calculate a **t-statistic**, which is a close analogue of the z-statistic calculated above.

$$t = \frac{\overline{x} - \mu_0}{\hat{\sigma}/\sqrt{n}}$$

It turns out that with normally distributed and independent variables, this t-statistic has a known distribution that is close to the normal distribution and that gets closer as n increase.

```
sd_est <- sd(test_sample)
t_stat <- (mean_est - hgt_mean_null)/(sd_est / sqrt(sample_size))
t_stat
```

```
[1] -5.614045
```

We can calculate the critical values for a test based on the t-statistics as follows:

```
qt(0.025, df = sample_size - 1)
```

```
[1] -1.984217
```

```
qt(1 - 0.025, df = sample_size - 1)
```

```
[1] 1.984217
```

The t-distribution is also symmetric, so we can express c as $c = 1.98$ and identify the **rejection region** as $\{t : |t| > c\}$. We can also calculate a p-value for our t-statistic.

```
pt(t_stat, df = sample_size - 1) * 2
```

```
[1] 1.808489e-07
```

5.3.3 Unknown distribution: Asymptotic methods

But what if we don't know that the underlying distribution for heights is normal? Perhaps we even have good reason to believe that it isn't normal. In this kind of situation, which is actually quite common, there are two basic approaches.

The first approach is to invoke a **central limit theorem** to claim that the distribution of means is *asymptotically* normal and thus approximately normal in finite samples.

To make things interesting, we replace our underlying normal distribution of heights with a new bimodal distribution created by drawing heights from one of two normal distributions chosen at random. We embed this distribution in the function `bimodal_dgf()`.

```
bimodal_dgf <- function(n) {
  short <- sample(c(TRUE, FALSE),
                  size = n,
                  prob = c(0.3, 0.7),
                  replace = TRUE)

  short_sample <- rnorm(n, mean = 155, sd = 3)
```

```
  tall_sample <- rnorm(n, mean = 171, sd = 3)
  if_else(short, short_sample, tall_sample)
}
```

We can invoke `bimodal_dgf()` to generate a sample of the desired size from our distribution.

```
set.seed(2023)
bimodal_sample <- bimodal_dgf(1000)
```

In Figure 5.1, we can see the sample we drew from this (unrealistic) distribution of heights has two *modes* and is clearly not normal.

```
tibble(height = bimodal_sample) |>
  ggplot(aes(x = height)) +
  geom_histogram(binwidth = 1)
```

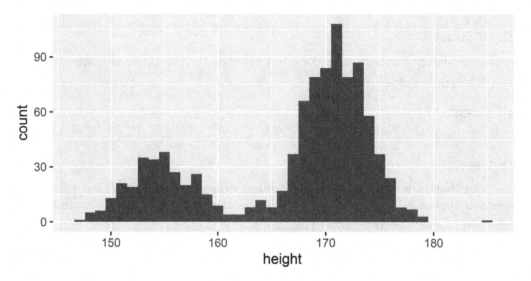

FIGURE 5.1
Histogram for the bimodal distribution

Now we make a function to draw multiple (`n_samples`) samples (each of `sample_size` size) from the distribution supplied as the `dgf` argument and calculate the mean for each sample.

```
get_means <- function(sample_size, n_samples, dgf) {
  tibble(i = 1:n_samples,
         sample = map(i, function(x) dgf(sample_size)),
         mean = map_vec(sample, mean),
         se = map_vec(sample, function(x) sd(x) / sqrt(length(x) - 1))) |>
    select(-sample) |>
    mutate(z = mean / se)
}
```

Here we draw 10,000 samples of 1000 from `bimodal_dgf()`. (You may be pleasantly surprised to learn that we can just pass one R function—`bimodal_dgf()`—into another!)

```
set.seed(2023)
bimodal_sample_stats <- get_means(sample_size = 1000,
                                  n_samples = 10000,
                                  dgf = bimodal_dgf)
```

Now we make a small function `make_clt_plot()` to plot the histogram for the variable `mean` in the supplied data frame `df`, along with the normal distribution matching the mean and standard deviation of `mean`.

```
make_clt_plot <- function(df) {
  df |>
    ggplot(aes(x = mean)) +
    geom_histogram(aes(y = after_stat(density)), fill = "green",
                   binwidth = sd(df$mean) / 10) +
    geom_vline(aes(xintercept = mean(mean), color = "red")) +
    stat_function(fun = dnorm,
                  args = list(mean = mean(df$mean),
                              sd = sd(df$mean)),
                  color = "blue") +
    theme(legend.position = "none")
}
```

Using this function with `bimodal_sample_stats`, we see in Figure 5.2 that, while the underlying distribution is bimodal, the distribution of means is very close to normal.

```
make_clt_plot(bimodal_sample_stats)
```

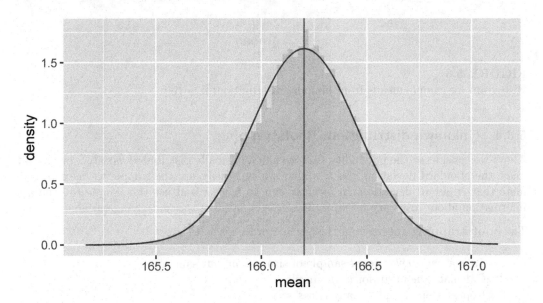

FIGURE 5.2
Histogram for sample means from bimodal distribution

We can combine the `get_means()` and `make_clt_plot()` functions into a single function `clt_demo()` that allows us to make a plot of the distribution of means with an overlaid

normal distribution so that we can see how well the latter approximates the former for a
given sample size (`sample_size`) and DGF (`dgf`).

```
clt_demo <- function(sample_size, n_samples, dgf) {
  get_means(sample_size, n_samples, dgf) |>
    make_clt_plot()
}
```

We use `clt_demo()` for Figure 5.3, which is like Figure 5.2, but for `sample_size` = 100.

```
clt_demo(sample_size = 100, n_samples = 10000, bimodal_dgf)
```

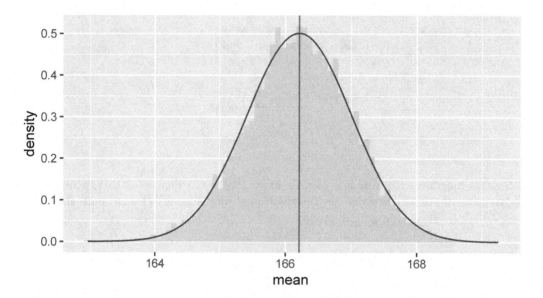

FIGURE 5.3
Histogram for sample means from bimodal distribution: $n = 100$

5.3.4 Unknown distribution: Bootstrapping

Above we considered the possibility that we had a normally distributed variable, but didn't
know the standard deviation. We saw that we can easily use the sample we have to estimate that standard deviation. In fact, we can go further and use the sample to generate
information about arbitrary features of the distribution of the underlying variable.

The **bootstrapping** approach in this setting proceeds as follows:

1. Take a *with-replacement* sample of size n from our sample.
2. Calculate the mean for that sample.
3. Repeat steps 1 and 2 many times.
4. Use the distribution from step 3 to calculate estimates of relevant parameters
   (e.g., the standard deviation).

We can use the sample we created above (`bimodal_sample`) for this purpose. Here we
calculate 10,000 samples. We first construct the `get_bs_mean()` function to perform steps
1 and 2.

```
get_bs_mean <- function(data) {
  bs_sample <- sample(data, size = length(data), replace = TRUE)
  mean(bs_sample)
}
```

We use then `map_vec()` to perform step 3.

```
set.seed(2021)
bs_stats <-
  tibble(i = 1:n_samples) |>
  mutate(mean = map_vec(i, function(x) get_bs_mean(data = bimodal_sample)))
```

We might simply use the results of our bootstrapping procedure to estimate the standard error of our mean estimates.

```
se_est <- sd(bs_stats$mean)
se_est
```

```
[1] 0.2489343
```

But we could also use these results to evaluate the appropriateness of the normal approximation. Figure 5.4 suggests that the approximation is quite good in this case.

```
make_clt_plot(bs_stats)
```

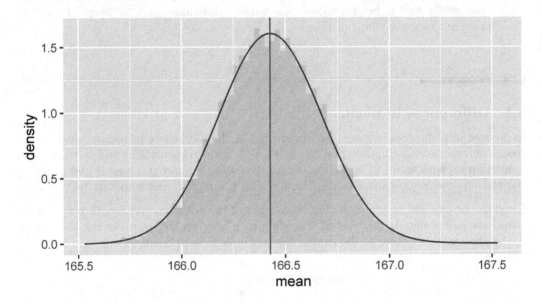

FIGURE 5.4
Histogram for sample means from bimodal distribution

Here we have barely scratched the surface of what can be done with bootstrapping analysis. In some settings, we can use bootstrapping to read off critical values without needing to rely on the distribution approximating a normal distribution. We will see bootstrapping approaches again below and we discuss them in more detail in subsequent chapters.

5.3.5 Exercises

1. What is the relation between critical values and p-values using z-statistics and t-statistics? (*Hint:* Use R functions such as `qt(p/2, df = sample_size - 1)` and `qnorm(0.025)` for various significant levels and sample sizes.)

2. Explain how the output of the following code relates to the statistical tests above. Why does this relationship exist?

```
df <- tibble(x = test_sample)
summary(lm((x - hgt_mean_null) ~ 1, data = df))
```

3. Examine `clt_demo(sample_size, n_samples = 10000, bimodal_dgf)` for different values of `sample_size`. At what value of `sample_size` does the underlying non-normality of `bimodal_dgf()` become apparent?

4. Using the value for `sample_size` you calculated in the previous question, what effect does varying `n_samples` have on the distribution? How do you interpret the pattern here?

5. Create your own data-generating function (say, `my_dgf()`) like `bimodal_dgf()`. This function should embody the distribution of a random variable and it should have a single required argument `n` for the size of the sample that it returns. Examine the output of `clt_demo()` for your function. (*Hint*: Like `bimodal_dgf()`, you might find it helpful to use R's built-in functions, such as `sample()` and `rnorm()` to make your function.)

5.4 Differences in means

As we saw in Chapter 3, we are often interested in the differences in means between two groups of observations. Suppose that the first group has n_0 observations and is drawn from a distribution with mean of $\overline{x}_0$ and standard deviation σ_0, and the second group has n_1 observations and is drawn from a distribution with mean of $\overline{x}_1$ and standard deviation σ_1.

Given the maintained assumption of independence of the observations here, the two sample means are independently distributed. The difference between the means of the two groups $\overline{x}_1 - \overline{x}_0$ has variance equal to the sum of the variances of each mean.

$$\text{var}(\overline{x}_1 - \overline{x}_0) = \frac{\sigma_1^2}{n_1} + \frac{\sigma_0^2}{n_0}$$

Taking the square root of this gives the standard error of the difference in means.

$$\text{se}(\overline{x}_1 - \overline{x}_0) = \sqrt{\frac{\sigma_1^2}{n_1} + \frac{\sigma_0^2}{n_0}}$$

Again, with large-enough samples, we can often assume that the difference in means will be approximately normally distributed.

One important point to note is that if the null hypothesis is $\mu_0 = \mu_1 = 0$, then it may be tempting to think that if the null hypothesis $\mu_1 \leq 0$ is rejected (in favour of $H_1 : \mu_1 > 0$)

while the null hypothesis $\mu_0 \leq 0$ is not, then we can conclude that the null hypothesis $\mu_1 \leq \mu_0$ can thereby be rejected (in favour of $H_1 : \mu_1 > \mu_0$). This reasoning is erroneous, as we will demonstrate.

Imagine that our sampling procedure will draw 1,000 observations from each of two groups and that we will compare the means of these two sub-samples. We will construct 10,000 such samples; we do this by constructing them in turn using the `get_means()` function from above with `dgf = rnorm`.

```
set.seed(2023)
sample_0_stats <- get_means(sample_size = 1000,
                            n_samples = 10000,
                            dgf = rnorm)

sample_1_stats <- get_means(sample_size = 1000,
                            n_samples = 10000,
                            dgf = rnorm)
```

We then merge these two samples by the sample identifier (`i`) and for each sample, we calculate the difference in means and the standard error of the difference using the formula above. From these values, we can calculate the z-statistic for each difference.

```
merged_stats <-
  sample_0_stats |>
  inner_join(sample_1_stats, by = "i", suffix = c("_0", "_1")) |>
  mutate(mean_diff = mean_1 - mean_0,
         se_diff = sqrt(se_0^2 + se_1^2),
         z_diff = mean_diff / se_diff)
```

Let's assume that the size of our test is $\alpha = 0.05$. Because we are using two-tailed tests, we set the critical value to the absolute value of $\Phi^{-1}(\alpha/2)$.

```
crit_value <- abs(qnorm(0.025))
```

The correct way to evaluate difference in means is to consider the difference significant if (1) $|z_{\text{diff}}| > c$. However, some researchers will consider the difference significant if either (2) $|z_1| > c$ while $|z_0| \leq c$ or (3) $|z_0| > c$ while $|z_1| \leq c$. And there is a risk that researchers will take the opportunity to declare the difference "statistically significant" if any of the conditions (1), (2), or (3) is met.

Below we calculate the proportion of samples for which the null hypothesis is rejected based on condition (1) (`prop_sig_diff`), based on conditions (2) or (3) (`prop_sig_diff_alt`), or based on conditions (1), (2), or (3) (`prop_sig_diff_choice`).

```
merged_stats |>
  mutate(sig_diff = abs(z_diff) > crit_value,
         sig_1 = abs(z_1) > crit_value,
         sig_0 = abs(z_0) > crit_value) |>
  summarize(prop_sig_diff = mean(sig_diff),
            prop_sig_diff_alt = mean((sig_1 & !sig_0) | (sig_0 & !sig_1)),
            prop_sig_diff_choice = mean(sig_diff | (sig_1 & !sig_0) |
                                        (sig_0 & !sig_1)))
```

TABLE 5.1

Null hypothesis rejection rates for different approaches

prop_sig_diff	prop_sig_diff_alt	prop_sig_diff_choice
0.0508	0.0967	0.1161

5.4.1 Discussion question

What is the issue implied by the statistics reported in Table 5.1? What is the correct approach implied by these statistics? Why might researchers prefer to have a choice regarding the test to be used?

5.5 Inference with regression

In practice, most statistical inference in accounting research relates to the hypotheses about regression coefficients. Fortunately, most of the concepts we have seen above with inference about means carry over to inference in regression settings.

When the OLS estimator is unbiased, the expected value of $\hat{\beta}$ will be β, but $\hat{\beta}$ is just an estimate and will typically differ from the true value, as we can see by substituting the value of y into the equation for $\hat{\beta}$ and doing some algebra. Note that below we use the facts that $\mathbb{E}\left[\hat{\beta}-\beta\right]$ is zero (implied by the expression below) and that the variance of a zero-mean vector Z equals $\mathbb{E}\left[ZZ'\right]$.

$$\hat{\beta} = (X^\mathsf{T}X)^{-1}X^\mathsf{T}y$$
$$= (X^\mathsf{T}X)^{-1}X^\mathsf{T}(X\beta+\epsilon)$$
$$= (X^\mathsf{T}X)^{-1}X^\mathsf{T}X\beta + (X^\mathsf{T}X)^{-1}\epsilon$$
$$= \beta + (X^\mathsf{T}X)^{-1}X^\mathsf{T}\epsilon$$
$$\hat{\beta}-\beta = (X^\mathsf{T}X)^{-1}X^\mathsf{T}\epsilon$$
$$\mathrm{var}(\hat{\beta}|X) = \mathbb{E}\left[(X^\mathsf{T}X)^{-1}X^\mathsf{T}\epsilon\epsilon'X(X^\mathsf{T}X)^{-1}|X\right]$$
$$= (X^\mathsf{T}X)^{-1}X^\mathsf{T}\mathbb{E}\left[\epsilon\epsilon'|X\right]X(X^\mathsf{T}X)^{-1}$$

Now if $\mathbb{E}\left[\epsilon_i\epsilon_j\right]=0, \forall i\neq j$, then $\mathbb{E}\left[\epsilon\epsilon'|X\right]$ will be a diagonal matrix with each element on the main diagonal equal to σ_i^2, the variance of ϵ for observation i (i.e., $\mathrm{var}(\epsilon_i)$). And if $\mathbb{E}[\sigma_i^2|X]=\sigma^2$ for all i, then we have:

$$\Sigma := \mathrm{var}(\hat{\beta}|X)$$
$$= \sigma^2(X^\mathsf{T}X)^{-1}X^\mathsf{T}X(X^\mathsf{T}X)^{-1}$$
$$= \sigma^2(X^\mathsf{T}X)^{-1}$$

This is the familiar result under the "standard" OLS assumptions, including the assumption that $\mathbb{E}[\sigma_i^2|X]=\sigma^2$ for all i, which is known as **homoskedasticity**.

One issue is that we do not observe σ^2, but we can estimate it using the standard formula

$$\hat{\sigma}^2 = \frac{1}{N-k}\sum_{i=1}^{N}\hat{\epsilon}_i^2$$

where $\hat{\epsilon}_i$ is the regression residual for observation i and $N - k$ incorporates a **degrees-of-freedom** correction. (Note that $k = 2$ when we have one regressor in addition to a constant term.) Putting the above together, we have

$$\hat{\Sigma} = \hat{\sigma}^2 (X^{\mathsf{T}} X)^{-1}$$

This is also the standard approach to estimating the variance of coefficients in statistical software, including the lm() function in R. To confirm this, let's create a test data set, calculate regression statistics "by hand" and then compare results with those provided by the lm() function.

To make this more concrete, let's consider some simulated data. Specifically, we generate 1000 observations with $\beta = 1$ and $\sigma = 0.2$.

```
set.seed(2021)
N <- 1000
x <- rnorm(N)
e <- rnorm(N, sd = 0.2)
y <- x * 1 + e
```

We next construct X as a matrix comprising a column of ones (to estimate the constant term) and a column containing x.

```
X <- matrix(c(rep(1, N), x), ncol = 2)
```

We then calculate β and residuals by hand.

```
b <- solve(t(X) %*% X) %*% t(X) %*% y
resid <- y - X %*% b
```

We can now calculate $\hat{\sigma}^2$ and then $\hat{\Sigma}$. This allows us to calculate the estimated standard errors of the estimated coefficients as the square root of the diagonal elements of $\hat{\Sigma}$ and then the t-statistics for each coefficient by dividing each element of $\hat{\beta}$ by its estimated standard error. Results from the following code are shown in Table 5.2.

```
resvar <- sum(resid^2) / (N - 2)
Sigma <- resvar * solve(t(X) %*% X)
se <- sqrt(diag(Sigma))
tibble(b = b[, 1], se,
       t = b / se,
       p = 2 * (1 - pt(t, df = N - 2)))
```

TABLE 5.2
Regression output calculated "by hand"

b	se	t	p
0.007540	0.006440	1.171	0.242
0.997647	0.006321	157.829	0.000

Now, let's compare the results seen in Table 5.2 with those in Table 5.3, which are obtained from the following code, which applies summary() to a model estimated using lm().

```
fm <- lm(y ~ x)
coefficients(summary(fm))
```

TABLE 5.3

Regression output calculated using `lm()`

	Estimate	Std. Error	t value	Pr(>|t|)
(Intercept)	0.007540	0.006440	1.171	0.242
x	0.997647	0.006321	157.829	0.000

You may have noticed that these approaches to testing hypotheses about regression coefficients are analogous to those used to test hypotheses about means in Sections 5.3.1 and 5.3.2. Similarly, we might use asymptotic methods or bootstrapping to test hypotheses about regression coefficients. In fact, the assumptions required to yield regression coefficients that follow the t distribution—such as normally distributed error terms—are generally implausible in most research settings and in practice most inferences are justified based on asymptotic results and assumptions of approximately normal distributions in the finite samples of actual research. In fact, all the approaches discussed in Section 5.6 are justified asymptotically.

In the remainder of this section, we discuss how bootstrapping can be applied to regression analysis. The bootstrapping approach in this setting proceeds as follows:

1. Take a *with-replacement* sample of size n from our sample of data.
2. Calculate the regression coefficients for that sample.
3. Repeat steps 1 and 2 many times.
4. Use the distribution above to calculate estimates of relevant parameters (e.g., the standard error of regression coefficients).

To demonstrate this, we put our y and x variables above into a data frame.

```
df <- tibble(y, x)
```

We create a function (`get_bs_coefs()`) that performs steps 1 and 2.

```
get_bs_coefs <- function(data) {
  rows <- 1:nrow(data)
  bs_rows <- sample(rows, size = length(rows), replace = TRUE)
  bs_sample <- data[bs_rows, ]
  fm <- lm(y ~ x, data = bs_sample)
  fm$coefficients
}
```

We then use `map_df()` to repeat these steps 10,000 times.

```
set.seed(2021)
bs_coefs <-
  tibble(i = 1:n_samples) |>
  mutate(map_df(i, function(x) get_bs_coefs(data = df)))
```

We can then calculate the standard deviation of the estimated regression coefficients as an estimate of their standard errors. Here we use `across()` to summarize multiple columns in one step and display results in Table 5.4.[11]

---

[11]The `across()` function is covered in some detail in Chapter 26 of *R for Data Science*.

```
bs_coefs |>
  select(-i) |>
  summarize(across(everything(), sd))
```

TABLE 5.4
Bootstrapped standard errors

(Intercept)	x
0.0064542	0.0064964

5.6 Dependence

In generating data in Chapter 3, we assumed that the errors and regressors were independent. However, most of the empirical accounting research uses panel data sets, typically repeated observations on a set of firms over time. In these settings, it is reasonable to assume that regressors (X) and errors (ϵ) are correlated both across firms (i.e., cross-sectional dependence) and over time (i.e., time-series dependence).

While White (1980) standard errors are consistent in the presence of heteroskedasticity, it is well known that both OLS and White produce misspecified test statistics when either form of dependence is present.

Gow et al. (2010) demonstrate that accounting researchers use a variety of approaches to address the issues of cross-sectional and time-series dependence widely understood to be present in common research settings. The literature survey in Gow et al. (2010) identified a number of common approaches: Fama-MacBeth, Newey-West, the "Z2" statistic, and standard errors clustered by firm, industry, or time.

5.6.1 Newey-West

The White (1980) approach was generalized by Newey and West (1987) to give a covariance matrix estimator robust to both heteroskedasticity and serial correlation. While the Newey-West procedure was developed in the context of a single time series, Gow et al. (2010) show it is frequently applied in panel data settings, where its use is predicated on cross-sectional independence. Assessing the performance of Newey-West in the presence of both cross-sectional and time-series dependence, Gow et al. (2010) find that it produces misspecified test statistics with even modest levels of cross-sectional dependence.

5.6.2 Fama-MacBeth

The Fama-MacBeth approach (Fama and MacBeth, 1973) is addresses concerns about cross-sectional correlation. The original Fama-MacBeth approach—which Gow et al. (2010) label "FM-t"—involves estimating T cross-sectional regressions (one for each period) and basing inferences on a t-statistic calculated as

$$t = \frac{\overline{\beta}}{se(\overline{\beta})}, \text{ where } \overline{\beta} = \frac{1}{T}\sum_{t=1}^{T}\hat{\beta}_t$$

and $se(\hat{\beta})$ is the standard error of the coefficients based on their empirical distribution. When there is no cross-regression (i.e., time-series) dependence, this approach yields consistent estimates of the standard error of the coefficients as T goes to infinity. However, cross-regression dependence in errors and regressors causes Fama-MacBeth standard errors to be understated (Cochrane, 2009; Schipper and Thompson, 1983).

To illustrate the Fama-MacBeth approach, we will generate a data set using the simulation approach of Gow et al. (2010), which is available using the `get_got_data()` function found in the `farr` package. The parameters Xvol and Evol relate to the cross-sectional dependence of X and ϵ, respectively. The parameters rho_X and rho_E relate to the time-series dependence of X and ϵ, respectively. We get 10 years of simulated data on 500 firms and store the data in the data frame named `test`.

```
set.seed(2021)
test <- get_got_data(N = 500, T = 10,
                     Xvol = 0.75, Evol = 0.75,
                     rho_X = 0.5, rho_E = 0.5)
```

We first estimate use `nest()` from `tidyr` to organize the relevant data by `year`. We then use the `map()` function from the `purrr` package twice. We first use `map()` to estimate the regression by year and put the fitted models in a column named `fm`. We then use `map()` again to extract the coefficients from the `fm` column and put them in the `coefs` column.[12]

```
results <-
  test |>
  group_by(year) |>
  nest() |>
  mutate(fm = map(data, \(df) lm(y ~ x, data = df)),
         coefs = map(fm, coefficients)) |>
  ungroup()
```

We then use the `unnest_wider()` function from the `tidyr` package to extract the data in `coefs` into separate columns in the data frame `coefs_df`.

```
coefs_df <-
  results |>
  unnest_wider(coefs) |>
  select(year, `(Intercept)`, x)
```

The data frame `coefs_df` contains the results of the first step of the Fama-MacBeth approach, which are shown in Table 5.5.

```
coefs_df
```

The second step of the Fama-MacBeth approach involves calculating the mean and standard error of the estimated coefficients. As we see in the exercises, regression of the time-series of coefficients on a constant produces the required mean and standard error.

```
fms <- list(lm(`(Intercept)` ~ 1, data = coefs_df),
            lm(x ~ 1, data = coefs_df))
```

Two common variants of the Fama-MacBeth approach appear in the accounting literature. The first variant, which Gow et al. (2010) label "FM-i", involves estimating firm- or

---

[12]Note that we use the shortcut form that allows us to write `function(df) lm(y ~ x, data = df)` as simply `\(df) ~ lm(y ~ x, data = df)`.

TABLE 5.5

Step 1 of Fama-MacBeth: Coefficients by year

year	(Intercept)	x
1	1.6356	0.9514
2	2.7017	0.9899
3	−1.1093	0.9078
4	−0.3491	1.0952
5	−0.8629	1.0035
6	−0.4451	0.9104
7	−0.2011	1.0507
8	1.5738	0.8011
9	−2.7727	0.9517
10	0.8072	0.9843

portfolio-specific time-series regressions with inferences based on the cross-sectional distribution of coefficients. FM-i had been used extensively in the accounting literature prior to Gow et al. (2010).

FM-i is appropriate if there is time-series dependence but not cross-sectional dependence. However, it is difficult to identify such circumstances in real research settings and we do not discuss FM-i further.[13]

The second common variant of the FM-t approach, which Gow et al. (2010) label FM-NW, is intended to correct for serial correlation in addition to cross-sectional correlation. FM-NW modifies FM-t by applying a Newey-West adjustment in an attempt to correct for serial correlation. While a number of studies claimed that FM-NW produces a conservative estimate of statistical significance, they did so without any formal evaluation of FM-NW.

Implementing FM-NW is straightforward using the `sandwich` package. In `get_nw_vcov()`, we simply apply `NeweyWest()` from the `sandwich` package to the fitted FM-t model to recover the FM-NW covariance matrix.[14]

```
get_nw_vcov <- function(fm, lag = 1) {
  NeweyWest(fm, lag = lag, prewhite = FALSE, adjust = TRUE)
}
```

In Table 5.6, we tabulate the Fama-MacBeth coefficients twice: once with the default standard errors and once with Newey-West standard errors.

```
modelsummary(dvnames(c(fms, fms)),
             coef_rename = c('(Intercept)' = 'Estimate'),
             vcov = c(map(fms, vcov),
                      map(fms, get_nw_vcov)),
             estimate = "{estimate}{stars}",
             output = "kableExtra",
             coef_omit = "^factor",
             gof_map = c("nobs", "r.squared"),
```

[13]In fact, Gow et al. (2010) found that FM-i was most frequently used when cross-sectional dependence was the primary issue, such as when returns are the dependent variable.

[14]The options here are chosen to line up with the `NeweyWestPanelStata()` function provided by Gow et al. (2010) at https://go.unimelb.edu.au/dzw8. This function in turn was verified to line up with the `newey.ado` function from Stata.

```
        stars = c('*' = .1, '**' = 0.05, '***' = .01)) |>
  add_header_above(c(" " = 1, "FM-t" = 2, "FM-NW" = 2))
```

TABLE 5.6
Standard errors using FM-t and FM-NW

	FM-t		FM-NW	
	(Intercept)	x	(Intercept)	x
Estimate	0.098	0.965***	0.098	0.965***
	(0.506)	(0.026)	(0.457)	(0.020)
Num.Obs.	10	10	10	10
R2	0.000	0.000	0.000	0.000

Gow et al. (2010) show that FM-NW fails to address serial correlation and offer two explanations for this failure. "First, FM-NW generally involves applying Newey-West to a limited number of observations, a setting in which Newey-West is known to perform poorly. Second, FM-NW applies Newey-West to a time-series of *coefficients*, whereas the dependence is in the underlying *data*" (Gow et al., 2010, p. 488). For example, without the underlying data, it is not possible to discern whether the sequence {2.1, 1.8, 1.9, 2.2, 2.3} represents highly auto-correlated draws from a mean-zero distribution or independent draws from a distribution with mean 2. As such, we do not examine FM-NW any further.

To estimate FM-t, we can use the `pmg()` function from the `plm` package. We simply specify the variable containing the relevant index (i.e., the t in FM-t).

```
fm_fmt <- pmg(y ~ x, data = test, index = "year")
```

Because `pmg` objects created using the `plm` package do not support the `modelsummary()` function out of the box, we implement `tidy()` and `glance()` functions so that `modelsummary()` knows how to handle these.[15]

```
tidy.pmg <- function(x, ...) {
   res <- summary(x)
   tibble(
     term      = names(res$coefficients),
     estimate  = res$coefficients,
     std.error = se(res),
     statistic = res$coefficients / se(res),
     p.value   = pvalue(res))
}


glance.pmg  <- function(x, ...) {
   res <- summary(x)
   tibble(
     r.squared       = res$rsqr,
     adj.r.squared   = res$r.squared,
     nobs            = length(res[[2]]))
}
```

---

[15]See the "extension and customization" page of the `modelsummary` documentation for details: https://go.unimelb.edu.au/vzw8.

In Table 5.7, we can see that the results using `pmg()` line up with those we calculated "by hand" above.

```
modelsummary(fm_fmt,
             estimate = "{estimate}{stars}",
             coef_omit = "^factor",
             gof_map = c("nobs", "r.squared"),
             stars = c('*' = .1, '**' = 0.05, '***' = .01))
```

TABLE 5.7
FM-t using `pmg()`

	(1)
(Intercept)	0.098
	(0.506)
x	0.965***
	(0.026)
Num.Obs.	5,000
R2	0.786

5.6.3 Z2 statistic

According to Gow et al. (2010, p. 488), the Z2 statistic first appeared in 1994 and was used in a number of subsequent studies in accounting research. The Z2-t statistic is calculated from the t-statistics from separate period-specific cross-sectional regressions as:

$$Z2 = \frac{\bar{t}}{se(\hat{t})}, \text{ where } \bar{t} = \frac{1}{T}\sum_{t=1}^{T}\hat{t}_t$$

$se(\hat{t})$ is the standard error of the t-statistics based on their empirical distribution, and T is the number of time periods in the sample.[16] Several studies claimed that Z2 adjusts for cross-sectional *and* serial correlation, but offered no basis for this claim. Gow et al. (2010) found no prior formal analysis of the properties of the Z2 statistic and found that Z2 suffers from cross-regression dependence in the same way as the Fama-MacBeth approach does.

5.6.4 One-way cluster-robust standard errors

A number of studies in the survey by Gow et al. (2010) use cluster-robust standard errors, with clustering either along a cross-sectional dimension (e.g., analyst, firm, industry, or country) or along a time-series dimension (e.g., year). Gow et al. (2010) refer to the former as CL-i and the latter as CL-t. Cluster-robust standard errors (also referred to as Huber-White or Rogers standard errors) were proposed by White (1984) as a generalization of the heteroskedasticity-robust standard errors of White (1980). With observations grouped into G clusters of N_g observations, for g in $1, \dots, G$, the covariance matrix is estimated using the following expression:

$$\hat{V}(\hat{\beta}) = (X'X)^{-1}\hat{B}(X'X)^{-1}, \text{ where } \hat{B} = \sum_{g=1}^{G} X'_g u_g u'_g X_g \tag{5.1}$$

---

[16]As discussed in Gow et al. (2010), just as there are FM-i and FM-t statistics, there were Z2-i and Z2-t statistics. We only discuss the Z2-t variant here.

where X_g is the $N_g \times K$ matrix of regressors, and u_g is the N_g-vector of residuals for cluster g. If each cluster contains a single observation, Equation 5.1 yields the White (1980) heteroskedasticity-consistent estimator.

While one-way cluster-robust standard errors permit correlation of unknown form within each cluster, they assume a lack of dependence across clusters. Gow et al. (2010) demonstrate that t-statistics for CL-t are overstated in the presence of time-series dependence and t-statistics for CL-i are overstated in the presence of cross-sectional dependence. So when both forms of dependence are present, CL-t and CL-i will both produce inflated t-statistics.

Because cluster-robust methods assume independence across clusters, clustering by, say, industry-year does *not* produce standard errors robust to either within-industry or within-year dependence. In fact, this requires that there be neither time-series nor cross-industry dependence (i.e., no dependence between observations for industry j in year t and those in industry j in year $t + 1$ *or* those in industry k in year t). More concretely, clustering by firm-year when the observations are firm-years simply produces White standard errors.

5.6.5   Two-way cluster-robust standard errors

Thompson (2011) and Cameron et al. (2011) developed an extension of cluster-robust standard errors that allows for clustering along more than one dimension.

In contrast to one-way clustering, two-way clustering (CL-2) allows for *both* time-series *and* cross-sectional dependence. For example, two-way clustering by firm and year allows for within-firm (time-series) dependence *and* within-year (cross-sectional) dependence (e.g., the observation for firm j in year t can be correlated with that for firm j in year $t + 1$ and that for firm k in year t). To estimate two-way cluster-robust standard errors, the expression in Equation 5.1 is evaluated using clusters along each dimension (for example, clustered by industry and clustered by year) to produce estimated covariance matrices V_1 and V_2. The same expression is calculated for the "intersection" clusters (in our example, observations within an industry-year) to yield the estimated covariance matrix V_I. The estimated two-way cluster-robust covariance matrix V is then calculated as $V = V_1 + V_2 - V_I$. Packages available for R make it straightforward to implement two-way cluster-robust standard errors and we demonstrate their use below.

5.6.6   Calculating standard errors

In the following code, we use functions from three libraries (`sandwich`, `plm`, and `fixest`) to estimate seven models. The `fixest` library provides the `feols()` function as an extension to `lm()` that allows for fixed effects, instrumental variables, and clustered standard errors. We demonstrated basic usage of `feols()` in Chapter 3. We will discuss the use of `feols()` with fixed effects and instrumental variables in later chapters. Here we just use `feols()` to get clustered standard errors.

```
fms <- list(OLS = lm(y ~ x, data = test),
            White = lm(y ~ x, data = test),
            NW = plm(y ~ x, test, index = c("firm", "year"),
                     model = "pooling"),
            `FM-t` = pmg(y ~ x, test, index = "year"),
            `CL-i` = feols(y ~ x, vcov = ~ firm, data = test),
            `CL-t` = feols(y ~ x, vcov = ~ year, data = test),
            `CL-2` = feols(y ~ x, vcov = ~ year + firm, data = test))
```

Note that the coefficient estimates will be equal for every model except the `FM-t` model.[17] The only difference between the six models other than `FM-t` will be the estimated standard errors.

For most of the models, we can produce the covariance matrix using `vcov()`. However, for `White`, we need to replace the default covariance matrix (so far there is no difference between `OLS` and `White`) with the result of calling `vcovHC()` from the `sandwich` package with `type = "HC1"`.[18] For NW, we call `vcovNW()` from the `plm` package.

```
vcovs <- map(fms, vcov)
vcovs[["White"]] <- vcovHC(fms[["White"]], type = "HC1")
vcovs[["NW"]] <- vcovNW(fms[["NW"]])
```

Table 5.8 provides estimates and standard errors calculated using a number of approaches.

```
modelsummary(fms,
             vcov = vcovs,
             estimate = "{estimate}{stars}",
             coef_omit = "^factor",
             gof_map = c("nobs", "r.squared"),
             stars = c('*' = .1, '**' = 0.05, '***' = .01))
```

TABLE 5.8
Standard errors using various methods

	OLS	White	NW	FM-t	CL-i	CL-t	CL-2
(Intercept)	0.021	0.021	0.021	0.098	0.021	0.021	0.021
	(0.026)	(0.026)	(0.027)	(0.506)	(0.023)	(0.492)	(0.492)
x	1.266***	1.266***	1.266***	0.965***	1.266***	1.266***	1.266***
	(0.027)	(0.021)	(0.021)	(0.026)	(0.018)	(0.172)	(0.172)
Num.Obs.	5,000	5,000	5,000	5,000	5,000	5,000	5,000
R2	0.305	0.305	0.305	0.786	0.305	0.305	0.305

With the exception of OLS, all methods depicted in Table 5.8 have asymptotic foundations and can have problems in small-sample settings. For example, FM-t is predicated on $T \to \infty$ and its distribution may not be well approximated by the normal distribution when $T = 10$. Cameron et al. (2008) document that cluster-robust methods will over-reject a true null when the number of clusters is small, an issue that can be attributed to the use of limiting distributions in making inferences in small samples (e.g., using the standard 1.64, 1.96, and 2.58 critical values from the normal distribution).

Accordingly, researchers should exercise caution when applying any asymptotic methods (e.g., FM-t, CL-t, or CL-2) when the number of time periods is small. Nonetheless, Gow et al. (2010) find CL-2 produces unequivocally better inferences in the presence of time-series and cross-section than any of the other methods considered in Table 5.8 even with as few as 10 clusters. Additionally, as there are approaches to address this concern (e.g., Cameron et al., 2008 identify corrections based on bootstrapping methods), Gow et al. (2010) argue

---

[17]The estimates from the `FM-t` model can be viewed as invoking pooled OLS with a different weighting for observations by year.

[18]The default value is `type = "HC3"`, which is actually preferred for many purposes. We use HC1 to line up with the approaches discussed in Gow et al. (2010) and defer discussion of the differences between HC1, HC2, and HC3 to Chapter 23.

that having few clusters does not warrant relying on approaches that are not robust to cross-sectional and time-series dependence.

5.6.7   Exercises

1. Verify either by direct computation using the data in `coefs_df` or using a little algebra that regressing the first-stage coefficients on a constant (as we did above) yields the desired second-stage results for FM-t.

2. Assume that the null hypothesis is $H_0 : \beta = 1$. Using the reported coefficients and standard errors for each of the methods listed in Table 5.8 (i.e., OLS, White, NW, FM-t, CL-i, CL-t, CL-2), for which methods is the null hypothesis rejected?

3. Based on the analysis in Gow et al. (2010), when should you use the FM-NW or Z2 approaches? What factors might have led researchers to use these approaches and make unsubstantiated claims about them (e.g., robustness to cross-sectional and time-series dependence)?

4. If FM-i is, as Gow et al. (2010) show, so inappropriate for the situations in which it was used, why do you think it was used in those situations?

5. Gow et al. (2010) refer to "the standard 1.64, 1.96, and 2.58 critical values." Using the `pnorm()` function, determine the p-value associated with each of these critical values. Are these one-tailed or two-tailed p-values? The values of 1.64, 1.96, and 2.58 are approximations. Which function in R allows you to recover the precise critical values associated with a chosen p-value? (*Hint*: Read the help provided by `? pnorm` in R.) Provide the critical values to four decimal places.

5.7   Further reading

Chapter 26[19] of *R for Data Science* covers the topic of **iteration**, including the functions `across()` and `map()` discussed above. That chapter also discusses anonymous functions and the `unnest_wider()` function.

The discussion above is a very selective introduction to statistical inference and there are a number of basic topics we have not addressed. Chapter 3 of Davidson and MacKinnon (2004) discusses the covariance matrix of OLS estimates along with other statistical properties of OLS. Angrist and Pischke (2008) provide an excellent treatment of issues related to standard errors.

We provided a very intuitive introduction to ideas related to laws of large numbers and central limit theorems. Chapter 4 of Davidson and MacKinnon (2004) provides a more thorough and rigorous introduction to these topics.

We did not discuss joint hypotheses (i.e., those involving more than one coefficient) or hypotheses that impose non-linear restrictions. Chapter 4 of Kennedy (2008) provides an intuitive introduction to these topics.

---

[19] https://r4ds.hadley.nz/iteration

6

Financial statements: A first look

In this chapter, we provide an introduction to data on financial statements, which are the primary focus of financial reporting. While a lot of contemporary accounting research has only a tangential connection to financial reporting, it's not clear there would be a field of accounting research without financial reporting and most accounting research paper use financial reporting data to some degree (e.g., as controls). In this chapter, we will focus on annual financial statement data of North American firms, as provided by Compustat.

> Tip
>
> In this chapter, we will use four R libraries. We have seen `dplyr`, `ggplot2`, and `farr` in earlier chapters, but the `DBI` library is new. The `DBI` library provides a standardized interface to relational database systems, including PostgreSQL (the system used by WRDS), MySQL, and SQLite. For instructions on how to set up your computer to use the code found in this book, see Section 1.2. Quarto templates for the exercises below are available on GitHub.[a]
>
> ---
> [a]https://github.com/iangow/far_templates/blob/main/README.md

```
library(dplyr)
library(ggplot2)
library(DBI)
library(farr)
```

6.1 Setting up WRDS

Academic researchers generally get Compustat data through Wharton Research Data Services, more commonly referred to as WRDS (pronounced "words"). In the following, we assume that you have run the code above, that you have a WRDS account, and that you are connected to the internet.

Now let's connect to the WRDS database. To actually use the code below, you should replace `your_WRDS_ID` with your actual WRDS ID. You may also need to add a line `Sys.setenv(PGPASSWORD = "your_password")` as the third line of code.[1]

```
Sys.setenv(PGHOST = "wrds-pgdata.wharton.upenn.edu",
           PGPORT = 9737L,
```

---
[1]We put our passwords in a special password file, as described at https://go.unimelb.edu.au/5zw8, so we don't need this step. It's obviously not a good idea to put your password in code.

DOI: 10.1201/9781003456230-6

```
            PGUSER = "your_WRDS_ID",
            PGDATABASE = "wrds")

db <- dbConnect(RPostgres::Postgres(), bigint = "integer")
```

The first line of code above sets up **environment variables** containing information that R can use as defaults when connecting to a PostgreSQL database. The second line of code creates the actual connection to the database and assigns it to the variable db. We set bigint = "integer" here because some packages cannot handle 64-bit integers of PostgreSQL's bigint type, so we ask these to be mapped to the 32-bit integers used in R.[2] We can also set check_interrupts = TRUE if we are running queries that take a long time so that we can interrupt them if we think something has gone wrong. Most queries in this book execute quickly, so this is generally not necessary.[3]

For users of the PostgreSQL database, an alternative to the code above would pass connection information to the dbConnect() function directly, as shown in the code below.[4]

```
db <- dbConnect(RPostgres::Postgres(),
                host = "wrds-pgdata.wharton.upenn.edu",
                port = 9737L,
                # user = "your_WRDS_ID",
                # password = "a_really_good_password",
                dbname ="wrds",
                bigint = "integer")
```

However, we recommend using the environment variable-based approach above. We also recommend using the approach based on a .pgpass file discussed in detail in the WRDS instructions.[5] Putting the connection details outside your code (e.g., in environment variables) makes it possible to write code that works both for you and for others, including your co-authors.

Now that we have established a connection to the WRDS data, we can get some data. Data sources on WRDS are organized into **libraries** according to their source (in PostgreSQL, these libraries are called **schemas**, but we will stick with the WRDS terminology for now).

In this chapter we focus on Compustat data, which is found in the comp library. While there are other sources of financial statement data—especially for firms outside North America—in accounting research the pre-eminent data source is Compustat, which is typically obtained via WRDS.

We first look at the table, comp.company, which contains information about the companies in Compustat's North American database (the equivalent table in Compustat's "Global" database is comp.g_company). By running the following code, we create R objects that point to the data on the WRDS database and behave a lot like the data frames you have seen in earlier chapters.

---

[2]The data we use in this book will not exceed the limits of 32-bit integers, so this involves no loss. See the help for RPostgres for more information. You can access this help by typing ? RPostgres::Postgres in the R console.

[3]We flag longer running queries when they arise.

[4]Remove the # from the lines beginning with # and replace the password and user name if you try this.

[5]See https://go.unimelb.edu.au/kv68. Note that we do *not* recommend setting up a connection to WRDS in your .Rprofile as suggested by WRDS, as this yields code that is less transparent and usable by others. Additionally, it means setting up a connection even when you may not need one.

```
company <- tbl(db, Id(schema = "comp", table = "company"))
```

What have we done here? In the first line, we have created an object company that "points to" the table company in schema comp in the data source pg (which represents the connection to the database). From the help for the tbl() function, we learn that in tbl(src, ...), src refers to the data source. Here pg is an object representing the connection to the WRDS PostgreSQL database. The ... allows us to pass other arguments to the function, and we pass Id(schema = "comp", table = "company") to indicate the underlying database object we are looking for.

But what exactly *is* company? One way to get information about company is to use the class() function.[6]

```
class(company)
```

```
[1] "tbl_PqConnection" "tbl_dbi"          "tbl_sql"          "tbl_lazy"
[5] "tbl"
```

The output here is a little complicated for a new user, but the basic idea is that company is a tbl_PqConnection object, and also a tbl_sql object and a tbl object, which means it a tibble of some kind. The critical idea here is that funda is a tbl_sql object, which means it behaves like a tibble, but is actually effectively a reference to a table or query in a relational database. For want of a better term, company is a **remote data frame**.

While there are some technical details here that are unimportant for our discussion, the important points are that, as a tbl_sql, funda has three important properties:

1. It's not a data frame in memory in R. This means that we don't need to download all of the data and load it into R.
2. It's not a data frame in memory in the database (in this case, the remote WRDS PostgreSQL server). This means that we don't load the full data set into memory even onto the remote computer.
3. Notwithstanding the previous two points, company behaves in many important ways just like a data frame.

We can use filter() to get only rows meeting certain conditions and we can use select() to indicate which columns we want to retrieve:

```
company |>
  filter(gsubind == "45202030",
         str_sub(conm, 1, 1) == "A") |>
  select(gvkey, conm, state) |>
  head(5)
```

```
# Source:   SQL [4 x 3]
# Database: postgres [iangow@wrds-pgdata.wharton.upenn.edu:9737/wrds]
  gvkey  conm           state
  <chr>  <chr>          <chr>
1 001690 APPLE INC      CA
2 001820 ASTRONOVA INC  RI
```

---

[6]If you are using the parquet file–based approach, your output will differ from what is shown here.

```
3 027925 AVID TECHNOLOGY INC MA
4 324055 ADVANCETC LTD         <NA>
```

When we are dealing with a remote data frame, the `dplyr` code that we write is automatically translated into SQL, a specialized language for manipulating tabular data "used by pretty much every database in existence"[7] (Appendix B has more on SQL and `dplyr`).

We can inspect the translated SQL using `show_query()`:

```
company |>
  filter(gsubind == "45202030",
         str_sub(conm, 1, 1) == "A") |>
  select(gvkey, conm, state) |>
  head(5) |>
  show_query()
```

```
<SQL>
SELECT "gvkey", "conm", "state"
FROM "comp"."company"
WHERE ("gsubind" = '45202030') AND (SUBSTR("conm", 1, 1) = 'A')
LIMIT 5
```

Here we see that `filter()` is translated into equivalent SQL using `WHERE`.

While remote data frames behave a lot like local data frames, they are not quite the same thing and, for many purposes, we need to use local data frames. For example, the `lm()` function for fitting regressions assumes that the object provided as its `data` argument is a data frame and it will not work if we give it a remote data frame instead. Conversion of a table from a remote data frame to a local one is achieved using the `collect()` verb. Here we use `collect()` and store the result in `df_company`.

```
df_company <-
  company |>
  filter(gsubind == "45202030",
         str_sub(conm, 1, 1) == "A") |>
  select(gvkey, conm, state) |>
  collect()
```

Looking at `class(df_company)`, we see that it is not only a "tibble" (`tbl_df`), but also `data.frame`, meaning that we can pass it to functions (such as `lm()`) that expect `data.frame` objects.

```
class(df_company)
```

```
[1] "tbl_df"      "tbl"         "data.frame"
```

```
df_company
```

```
# A tibble: 4 x 3
  gvkey   conm              state
  <chr>   <chr>             <chr>
1 001690  APPLE INC         CA
2 001820  ASTRONOVA INC     RI
```

---

[7]https://solutions.posit.co/connections/db/r-packages/dplyr/

```
3 027925 AVID TECHNOLOGY INC MA
4 324055 ADVANCETC LTD          <NA>
```

6.2 Financial statement data

Two core tables for the North American Compustat data are `comp.funda`, which contains annual financial statement data, and `comp.fundq`, which contains quarterly data.

```
fundq <- tbl(db, Id(schema = "comp", table = "fundq"))
funda <- tbl(db, Id(schema = "comp", table = "funda"))
```

We can examine the first object in R by typing its name in the R console:

```
funda
```

```
# Source:   table<"comp"."funda"> [?? x 948]
# Database: postgres [iangow@wrds-pgdata.wharton.upenn.edu:9737/wrds]
   gvkey  datadate   fyear indfmt consol popsrc datafmt tic    cusip   conm
   <chr>  <date>     <int> <chr>  <chr>  <chr>  <chr>   <chr>  <chr>   <chr>
 1 001000 1961-12-31  1961 INDL   C      D      STD     AE.2   000032~ A & ~
 2 001000 1962-12-31  1962 INDL   C      D      STD     AE.2   000032~ A & ~
 3 001000 1963-12-31  1963 INDL   C      D      STD     AE.2   000032~ A & ~
 4 001000 1964-12-31  1964 INDL   C      D      STD     AE.2   000032~ A & ~
 5 001000 1965-12-31  1965 INDL   C      D      STD     AE.2   000032~ A & ~
 6 001000 1966-12-31  1966 INDL   C      D      STD     AE.2   000032~ A & ~
 7 001000 1967-12-31  1967 INDL   C      D      STD     AE.2   000032~ A & ~
 8 001000 1968-12-31  1968 INDL   C      D      STD     AE.2   000032~ A & ~
 9 001000 1969-12-31  1969 INDL   C      D      STD     AE.2   000032~ A & ~
10 001000 1970-12-31  1970 INDL   C      D      STD     AE.2   000032~ A & ~
# i more rows
# i 938 more variables: acctchg <chr>, acctstd <chr>, acqmeth <chr>,
#   adrr <dbl>, ajex <dbl>, ajp <dbl>, bspr <chr>, compst <chr>,
#   curcd <chr>, curncd <chr>, currtr <dbl>, curuscn <dbl>, final <chr>,
#   fyr <int>, ismod <int>, ltcm <chr>, ogm <chr>, pddur <int>, scf <int>,
#   src <int>, stalt <chr>, udpl <chr>, upd <int>, apdedate <date>,
#   fdate <date>, pdate <date>, acchg <dbl>, acco <dbl>, accrt <dbl>, ...
```

From this code, we can see that `comp.funda` is a very *wide* table, with 948 columns. With so many columns, it's a moderately large table (at the time of writing, 1212 MB in PostgreSQL), but by focusing on certain columns, the size of the data can be dramatically reduced. Let's learn more about this table.

First, how many rows does it have?

```
funda |>
  count() |>
  pull()
```

```
[1] 901636
```

From this snippet, we can see that `comp.funda` has 901,636 rows. When given an *ungrouped* data frame as the first argument, the `count()` function simply counts the number of rows in the data frame.[8] In this case, `funda |> count()` returns a server-side data frame with a single column (n). The command `pull()`[9] extracts that single column as a vector in R.[10]

While not evident from casual inspection of the table, the **primary key** of `comp.funda` is (`gvkey, datadate, indfmt, consol, popsrc, datafmt`). One requirement for a valid primary key is that each value of the primary key is associated with only one row of the data set. (Chapter 19[11] of *R for Data Science* discusses this point.) That this requirement is met can be seen with output from the following code, which counts the number of rows associated with each set of values for (`gvkey, datadate, indfmt, consol, popsrc, datafmt`) and stores that number in `num_rows` and then displays the various values of `num_rows`.

```
funda |>
  count(gvkey, datadate, indfmt, consol, popsrc, datafmt,
        name = "num_rows") |>
  collect()
```

```
# A tibble: 901,636 x 7
   gvkey  datadate   indfmt consol popsrc datafmt  num_rows
   <chr>  <date>     <chr>  <chr>  <chr>  <chr>       <int>
 1 028485 2000-03-31 INDL   C      D      STD             1
 2 004607 2011-06-30 INDL   C      D      SUMM_STD        1
 3 028420 2021-12-31 INDL   C      D      STD             1
 4 013148 1990-12-31 INDL   C      D      STD             1
 5 030501 2019-12-31 INDL   C      D      STD             1
 6 113491 2002-12-31 INDL   C      D      STD             1
 7 016721 2008-03-31 INDL   C      D      SUMM_STD        1
 8 031392 2014-09-30 INDL   C      D      STD             1
 9 026748 2009-09-30 INDL   C      D      SUMM_STD        1
10 043667 2023-12-31 INDL   C      D      STD             1
# i 901,626 more rows
```

Because `num_rows` is equal to 1 in 901,636 cases and this is the number of rows in the data set, we have (`gvkey, datadate, indfmt, consol, popsrc, datafmt`) as a valid key.

The second requirement of a primary key is that none of the columns contains null values (in SQL, `NULL`; in R, `NA`). The code below checks this.

```
funda |>
  filter(is.na(gvkey) | is.na(datadate) | is.na(indfmt) |
         is.na(consol) | is.na(popsrc) | is.na(datafmt)) |>
  count() |>
  pull()
```

```
[1] 0
```

Teasing apart the primary key, we have a firm identifier (`gvkey`) and financial period identifier (`datadate`), along with four variables that are more technical in nature: `indfmt`, `consol`, `popsrc`, and `datafmt`. Table 6.1 provides the number of observations for each set of values of these four variables.

---

[8]Many `dplyr` functions take data frames as (first) arguments and return data frames as values.

[9]https://dplyr.tidyverse.org/reference/pull.html

[10]As that vector has a single value, it is displayed without any information about the database connection that we used to get the information. This looks cleaner in this setting, as you can see yourself if you run the command yourself without `pull()` at the end.

[11]https://r4ds.hadley.nz/joins.html

```
funda |>
  count(indfmt, consol, popsrc, datafmt) |>
  arrange(desc(n)) |>
  collect()
```

TABLE 6.1
Observations by (`indfmt`, `consol`, `popsrc`, `datafmt`) values

indfmt	consol	popsrc	datafmt	n
INDL	C	D	STD	573,430
INDL	C	D	SUMM_STD	279,283
FS	C	D	STD	46,814
INDL	P	D	STD	1,171
INDL	R	D	STD	480
INDL	P	D	SUMM_STD	300
FS	R	D	STD	51
INDL	R	D	SUMM_STD	28
INDL	D	D	SUMM_STD	25
INDL	D	D	STD	25
INDL	C	D	PRE_AMENDS	19
INDL	C	D	PRE_AMENDSS	10

As discussed on the CRSP website,[12] the first set of values, which covers 573,430 observations, represents the standard (`STD`) secondary keyset, as it is what is used by most researchers when using Compustat. We can create a version of `funda` that limits data to this `STD` secondary keyset as follows:

```
funda_mod <-
  funda |>
  filter(indfmt == "INDL", datafmt == "STD",
         consol == "C", popsrc == "D")
```

For `funda_mod`, (`gvkey`, `datadate`) should represent a primary key. We can check the first requirement using code like that we used above (we already confirmed the second requirement for the larger data set).

```
funda_mod |>
  group_by(gvkey, datadate) |>
  summarize(num_rows = n(), .groups = "drop") |>
  count(num_rows) |>
  collect()
```

```
# A tibble: 1 x 2
  num_rows        n
     <int>    <int>
1        1   573430
```

Now, let's look at the table `comp.company`, which the following code confirms has 39 columns and for which `gvkey` is a primary key.

---

[12]http://www.crsp.org/products/documentation/keysets

```
company
```

```
# Source:    table<"comp"."company"> [?? x 39]
# Database: postgres  [iangow@wrds-pgdata.wharton.upenn.edu:9737/wrds]
     conm      gvkey add1  add2  add3  add4  addzip busdesc cik    city  conml
     <chr>     <chr> <chr> <chr> <chr> <chr> <chr>  <chr>   <chr>  <chr> <chr>
 1 A & E PL~ 0010~ <NA>  <NA>  <NA>  <NA>  <NA>   A & E ~ <NA>   <NA>  A & ~
 2 A & M FO~ 0010~ 1924~ <NA>  <NA>  <NA>  94104  <NA>    0000~  Tulsa A & ~
 3 AAI CORP  0010~ 124 ~ <NA>  <NA>  <NA>  21030~ Textro~ 0001~  Hunt~ AAI ~
 4 A.A. IMP~ 0010~ 7700~ <NA>  <NA>  <NA>  63125  A.A. I~ 0000~  St. ~ A.A.~
 5 AAR CORP  0010~ One ~ <NA>  <NA>  <NA>  60191  AAR Co~ 0000~  Wood~ AAR ~
 6 A.B.A. I~ 0010~ 1026~ <NA>  <NA>  <NA>  33782  A.B.A.~ <NA>   Pine~ A.B.~
 7 ABC INDS~ 0010~ 301 ~ <NA>  <NA>  <NA>  46590  ABC In~ <NA>   Wino~ ABC ~
 8 ABKCO IN~ 0010~ 1700~ <NA>  <NA>  <NA>  10019  ABKCO ~ 0000~  New ~ ABKC~
 9 ABM COMP~ 0010~ 3 Wh~ <NA>  <NA>  <NA>  92714  ABM Co~ <NA>   Irvi~ ABM ~
10 ABS INDU~ 0010~ Inte~ <NA>  <NA>  <NA>  44904  ABS In~ 0000~  Will~ ABS ~
# i more rows
# i 28 more variables: costat <chr>, county <chr>, dlrsn <chr>, ein <chr>,
#   fax <chr>, fic <chr>, fyrc <int>, ggroup <chr>, gind <chr>,
#   gsector <chr>, gsubind <chr>, idbflag <chr>, incorp <chr>, loc <chr>,
#   naics <chr>, phone <chr>, prican <chr>, prirow <chr>, priusa <chr>,
#   sic <chr>, spcindcd <int>, spcseccd <int>, spcsrc <chr>, state <chr>,
#   stko <int>, weburl <chr>, dldte <date>, ipodate <date>
```

First, each `gvkey` value is associated with just one row.

```
company |>
  group_by(gvkey) |>
  summarize(num_rows = n(), .groups = "drop") |>
  count(num_rows) |>
  collect()
```

```
# A tibble: 1 x 2
  num_rows       n
     <int> <int>
1        1 54397
```

Second, there are no missing values of `gvkey`.

```
company |>
  filter(is.na(gvkey)) |>
  count() |>
  collect()
```

```
# A tibble: 1 x 1
      n
  <int>
1     0
```

6.2.1   Illustration: Microsoft versus Apple

Suppose that we were interested in comparing the profitability of Apple and Microsoft over time. To measure performance, we will calculate a measure of **return on assets**, here measured as the value of "Income Before Extraordinary Items" scaled by "Total Assets" for Microsoft (GVKEY: 012141) and Apple (GVKEY: 001690) over time.[13]

```
sample <-
  company |>
  filter(gvkey %in% c("001690", "012141")) |>
  select(conm, gvkey)
```

```
sample
```

```
# Source:   SQL [2 x 2]
# Database: postgres  [iangow@wrds-pgdata.wharton.upenn.edu:9737/wrds]
  conm          gvkey
  <chr>         <chr>
1 APPLE INC     001690
2 MICROSOFT CORP 012141
```

Here we can see that `sample` is a server-side data frame with the company name (`conm`) and identifier (`gvkey`) for Apple and Microsoft. We can use `show_query()` to see what the server-side data frame represents.

```
sample |>
  show_query()
```

```
<SQL>
SELECT "conm", "gvkey"
FROM "comp"."company"
WHERE ("gvkey" IN ('001690', '012141'))
```

Here `filter()` is translated into a `WHERE` condition, `%in%` is translated to `IN`, and `c("001690", "012141")` is translated to `('001690', '012141')`.

It turns out that "Income Before Extraordinary Items" is represented by `ib` and "Total Assets" is `at` (see the CRSP manual[14] for details). We can further restrict `funda` beyond `funda_mod` so that it contains just the key variables (`gvkey`, `datadate`) and the values for `ib` and `at` as follows. (In the following code, we will gradually refine the data set assigned to `ib_at`.)

```
ib_at <-
  funda_mod |>
  select(gvkey, datadate, ib, at)
```

```
ib_at |> head(5)
```

```
# Source:   SQL [5 x 4]
# Database: postgres  [iangow@wrds-pgdata.wharton.upenn.edu:9737/wrds]
  gvkey   datadate      ib    at
  <chr>   <date>     <dbl> <dbl>
1 001000 1961-12-31  0.05   NA
2 001000 1962-12-31  0.12   NA
3 001000 1963-12-31  0.003  NA
```

---

[13]Financial analysts might object to using an end-of-period value as the denominator; we do so here because it simplifies the coding and our goal here is just illustrative.

[14]http://www.crsp.org/products/documentation/annual-data-industrial

```
4 001000 1964-12-31   0.039   1.42
5 001000 1965-12-31  -0.197   2.31
```

and then restrict this by joining it with `sample` as follows:

```
ib_at <-
  funda_mod |>
  select(gvkey, datadate, ib, at) |>
  inner_join(sample)
```

```
Joining with `by = join_by(gvkey)`
```

```
ib_at |> head(5)
```

```
# Source:   SQL [5 x 5]
# Database: postgres [iangow@wrds-pgdata.wharton.upenn.edu:9737/wrds]
  gvkey  datadate        ib     at conm
  <chr>  <date>       <dbl>  <dbl> <chr>
1 001690 1980-09-30   11.7   65.4 APPLE INC
2 001690 1981-09-30   39.4  255.  APPLE INC
3 001690 1982-09-30   61.3  358.  APPLE INC
4 001690 1983-09-30   76.7  557.  APPLE INC
5 001690 1984-09-30   64.1  789.  APPLE INC
```

In the code above, we used a **natural join**, which joins by variables found in both data sets (in this case, `gvkey`). In general, we want to be explicit about the join variables, which we can do using the `by` argument to the join function:

```
ib_at <-
  funda_mod |>
  select(gvkey, datadate, ib, at) |>
  inner_join(sample, by = "gvkey")
```

```
ib_at |> head(5)
```

```
# Source:   SQL [5 x 5]
# Database: postgres [iangow@wrds-pgdata.wharton.upenn.edu:9737/wrds]
  gvkey  datadate        ib     at conm
  <chr>  <date>       <dbl>  <dbl> <chr>
1 001690 1980-09-30   11.7   65.4 APPLE INC
2 001690 1981-09-30   39.4  255.  APPLE INC
3 001690 1982-09-30   61.3  358.  APPLE INC
4 001690 1983-09-30   76.7  557.  APPLE INC
5 001690 1984-09-30   64.1  789.  APPLE INC
```

We next calculate a value for return on assets (using income before extraordinary items and the ending balance of total assets for simplicity).

```
ib_at <-
  funda_mod |>
  select(gvkey, datadate, ib, at) |>
  inner_join(sample, by = "gvkey") |>
  mutate(roa = ib / at)
```

The final step is to bring these data into R. At this stage, `ib_at` is still a server-side data frame (`tbl_sql`), which we can see using the `class` function.[15]

```
"tbl_sql" %in% class(ib_at)
```

```
[1] TRUE
```

To bring the data into R, we can use the `collect()` function, as follows:

```
ib_at <-
  funda_mod |>
  select(gvkey, datadate, ib, at) |>
  inner_join(sample, by = "gvkey") |>
  mutate(roa = ib / at) |>
  collect()
```

Now we have a local data frame that is no longer an instance of the `tbl_sql` class:

```
class(ib_at)
```

```
[1] "tbl_df"      "tbl"           "data.frame"
```

Note that because we have used `select()` to get to just five fields and `filter()` to focus on two firms, the tibble `ib_at` is just 6 kB in size. Note the placement of `collect()` at the end of this pipeline means that the amount of data we need to retrieve from the database and load into R is quite small. If we had placed `collect()` immediately after `funda_mod`, we would have been retrieving data on thousands of observations and 948 variables. If we had placed `collect()` immediately after the `select()` statement, we would have been retrieving data on thousands of observations, but just 4 variables. But by placing `collect()` after the `inner_join()`, we are only retrieving data for firm-years related to Microsoft and Apple. Retrieving data into our local R instance requires both reading it off disk and, if our database server is remote, transferring it over the internet. Each of these operations is (relatively) slow and thus should be avoided where possible.

Having collected our data, we can make Figure 6.1.[16]

```
ib_at |>
  ggplot(aes(x = datadate, y = roa,
             linetype = conm, color = conm)) +
  geom_line() +
  theme(legend.position = "bottom")
```

---

[15]An alternative to this line would be `inherits(ib_at, "tbl_sql")`. More on classes can be found in "Advanced R." at https://adv-r.hadley.nz.

[16]Note that even if we hadn't called `collect()`, recent versions of `ggplot2` are aware of remote data frames and will take care of this for us. But other functions (e.g., `lm()`, which we saw in Chapter 4) are not aware of remote data frames and thus require us to use `collect()` before passing data to them.

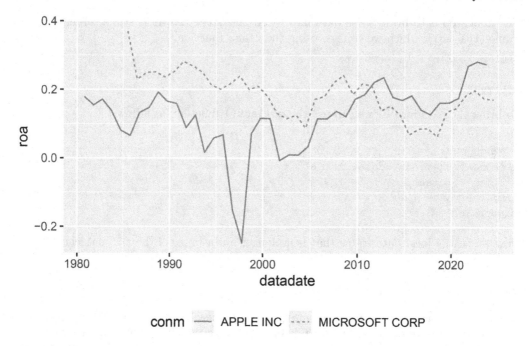

FIGURE 6.1
ROA over time for Microsoft and Apple

6.3 Exercises

1. Suppose we didn't have access to Compustat (or an equivalent database) for the analysis above, describe a process you would use to get the data required to make the plot above comparing performance of Microsoft and Apple.

2. In the following code, how do `funda_mod` and `funda_mod_alt` differ? (For example, where are the data for each table?) What does the statement `collect(n = 100)` at the end of this code do?

```
funda <- tbl(db, Id(schema = "comp", table = "funda"))

funda_mod <-
  funda |>
  filter(indfmt == "INDL", datafmt == "STD",
         consol == "C", popsrc == "D") |>
  filter(fyear >= 1980)

funda_mod_alt <-
  funda_mod |>
  collect(n = 100)
```

3. The table `comp.company` has data on SIC (Standard Industrial Classification) codes in the field `sic`. In words, what is the `case_when()` function doing in the following code? Why do we end up with just two rows?

```
company <- tbl(db, Id(schema = "comp", table = "company"))

sample <-
  company |>
  select(gvkey, sic) |>
  mutate(co_name = case_when(gvkey == "001690" ~ "Apple",
                             gvkey == "012141" ~ "Microsoft")) |>
  filter(!is.na(co_name))
```

4. What does the data frame `another_sample` represent? What happens if we change the `inner_join()` below to simply `inner_join(sample)`? What happens if we change it to `inner_join(sample, by = "sic")` (i.e., omit the `suffix = c("", "_other")` portion)? Why do we use `filter(gvkey != gvkey_other)`?

```
another_sample <-
  company |>
  select(gvkey, sic) |>
  inner_join(sample, by = "sic", suffix = c("", "_other")) |>
  filter(gvkey != gvkey_other) |>
  mutate(group = str_c(co_name, " peer")) |>
  select(gvkey, group)
```

5. What is the following code doing?

```
total_sample <-
  sample |>
  rename(group = co_name) |>
  select(gvkey, group) |>
  union_all(another_sample)
```

6. Suppose that we are interested in firms' R&D activity. One measure of R&D activity is *R&D Intensity*, which can be defined as "R&D expenses" (Compustat item `xrd`) scaled by "Total Assets" (Compustat item `at`). In `xrd_at`, what's the difference between `rd_intensity` and `rd_intensity_alt`? Does `filter(at > 0)` seem like a reasonable filter? What happens if we omit it?

```
xrd_at <-
  funda_mod |>
  select(gvkey, datadate, fyear, conm, xrd, at) |>
  filter(at > 0) |>
  mutate(rd_intensity = xrd / at,
         xrd_alt = coalesce(xrd, 0),
         rd_intensity_alt = xrd_alt / at) |>
  inner_join(total_sample, by = "gvkey")
```

7. Looking at a sample of rows from `xrd_at_sum`, it appears that the three R&D intensity measures are always identical for Apple and Microsoft, but generally different for their peer groups. What explains these differences? Can you say that one measure is "correct"? Or would you say "it depends"?

```
xrd_at_sum <-
  xrd_at |>
  group_by(group, fyear) |>
  summarize(rd_intensity1 = mean(xrd / at),
            rd_intensity2 = mean(xrd_alt / at),
            rd_intensity3 = if_else(sum(at) > 0, sum(xrd) / sum(at), NA),
            .groups = "drop")

xrd_at_sum |>
  arrange(desc(fyear), group) |>
  collect(n = 8)
```

```
# A tibble: 8 x 5
  group          fyear rd_intensity1 rd_intensity2 rd_intensity3
  <chr>          <int>         <dbl>         <dbl>         <dbl>
1 Apple peer      2024        0.105         0.105         0.113
2 Microsoft       2024        0.0576        0.0576        0.0576
3 Microsoft peer  2024        0.256         0.228         0.0672
4 Apple           2023        0.0848        0.0848        0.0848
5 Apple peer      2023        0.151         0.147         0.102
6 Microsoft       2023        0.0660        0.0660        0.0660
7 Microsoft peer  2023        0.224         0.189         0.0796
8 Apple           2022        0.0744        0.0744        0.0744
```

8. Write code to produce Figure 6.2. Also produce plots that use `rd_intensity1`
   and `rd_intensity2` as measures of R&D intensity. Do the plots help you think
   about which of the three measures makes most sense?

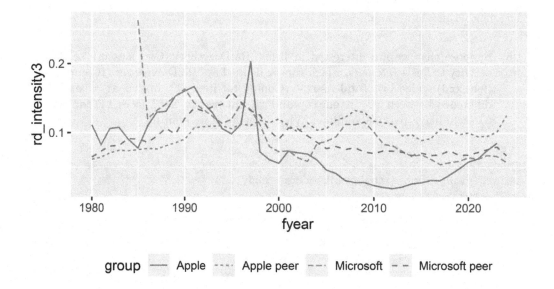

FIGURE 6.2
R&D intensity over time for Microsoft and Apple and their peers

7

Linking databases

Almost any empirical research project in accounting or finance research will involve merging data from multiple sources. When we joined data tables in Chapter 2, we had a common identifier across tables. But this will not always be the case. For example, to evaluate the market reaction to earnings announcements, we might start with data on `comp.fundq` from Compustat, where `gvkey` and `datadate` can be used to identify firm-quarters and associated announcement dates (`rdq`), and then look to merge with daily stock return data from the Center for Research in Security Prices (CRSP, pronounced "crisp") on `crsp.dsf`, which uses `permno` and `date` to identify each firm's trading equity and daily stock returns (`ret`). But this raises the question: Which `permno` (if any) matches a given `gvkey`?

Alternatively, given security price information from CRSP, we might want to know what is the most recent financial statement information for that security. In other words, which `gvkey` (if any) matches a given `permno`? This chapter provides guidance on the standard approaches to answering these questions with a focus on linking CRSP and Compustat.

> 💡 Tip
>
> The code in this chapter uses the packages listed below. For instructions on how to set up your computer to use the code found in this book, see Section 1.2. Quarto templates for the exercises below are available on GitHub.[a]
>
> ---
> [a]https://github.com/iangow/far_templates/blob/main/README.md

```
library(dplyr)
library(DBI)
library(ggplot2)
library(farr)
library(dbplyr)          # window_order()
library(lubridate)       # floor_date()
```

7.1 Firm identifiers

The idea behind a firm **identifier** is that it uniquely identifies a firm for a particular purpose. While Compustat uses GVKEYs, CRSP uses PERMNOs, the SEC uses CIKs, stock exchanges use tickers, and there are also CUSIPs.[1] Table 7.1 lists some common firm (and security) identifiers. Of course, identifiers apply not only to firms, but also people.

---
[1]Of course, PERMNOs, CUSIPs, and tickers (at best) identify *securities*, not firms. More on this below.

Nonetheless, most of this chapter focuses on firm identifiers, in part because of the importance of firms as units of observation in accounting and finance research, and also because identifying firms is much harder than identifying people.[2] While not specific to the platform we describe in this book, the issue of *identifiers* is one that seems less perplexing and better-handled when a relational database provides the backbone of your data store as it does here.

TABLE 7.1
Firm and security identifiers

Data provider	Firm identifiers	Notes
Compustat (`comp`)	`gvkey`	
CRSP (`crsp`)	`permno, permco`	`permno` is a *security* identifier
IBES (`ibes`)	`ticker`	`ticker` is *not* necessarily the ticker assigned to the firm by the exchange on which it trades
SEC EDGAR	CIK	
Audit Analytics (`audit`)	`company_fkey`	`company_fkey` is the same as CIK
Various	CUSIP	CUSIP is a *security* identifier

7.1.1 Firm identifiers: A quiz

Our sense is that firm identifiers is one of those topics that seem easy, but is actually fairly tricky. Here is a quick quiz to test your knowledge of firm identifiers.[3]

- General Motors Corporation declared bankruptcy in June 2009. Does the "successor firm" General Motors *Company* have the same GVKEY as General Motors Corporation? The same PERMNO?
- Can a CUSIP map to more than one PERMNO? To more than one GVKEY?
- Can a PERMNO map to more than one CUSIP?
- Can a GVKEY map to more than one PERMCO?
- Can a PERMCO map to different CIKs?
- If you have two data sets, X and Y and CUSIP is a "firm" identifier on each, can you simply merge using CUSIPs?
- When would a "firm" change CUSIPs?
- When would a "firm" change CIKs?
- If the firm identifier on IBES is `ticker`, should I merge with CRSP using `ticker` from `crsp.stocknames`?

Maybe you know the answers to some questions, but not all. If so, read on; this chapter aims to provide answers to many of these questions.[4]

---

[2]In the United States, a Social Security Number (SSN) is a pretty robust identifier of people, as would be a Tax File Number (TFN) in Australia. Though, as researchers, we generally don't have access to SSNs or TFNs.

[3]Obviously, we are assuming that you recognize the various identifiers. If not, read on.

[4]Coverage of CIKs is deferred to Chapter 23 and we do not use IBES data in this book.

7.2   The CRSP database

According to its website, "the Center for Research in Security Prices, LLC (CRSP) maintains the most comprehensive collection of security price, return, and volume data for the NYSE, AMEX and NASDAQ stock markets. Additional CRSP files provide stock indices, beta-based and cap-based portfolios, treasury bond and risk-free rates, mutual funds, and real estate data. [CRSP] maintains the most comprehensive collection of security price, return, and volume data for the NYSE, AMEX and NASDAQ stock markets. Additional CRSP files provide stock indices, beta-based and cap-based portfolios, treasury bond and risk-free rates, mutual funds, and real estate data." We will discuss the CRSP/COMPUSTAT Merged Database in Section 7.3. CRSP documentation[5] provides details on other CRSP databases, such as CRSP US Treasury and Inflation Series, CRSP Mutual Funds, and CRSP/Ziman Real Estate Data Series.

CRSP provides PERMNO, its own "permanent identifier" for each security in its database. Additionally, CRSP provides a company-level identifier, PERMCO, for each company. WRDS[6] tells us that CRSP's goals in creating these identifiers are to allow "for clean and accurate backtesting, time-series and event studies, measurement of performance, accurate benchmarking, and securities analysis."

According to WRDS,[7] "CRSP contains end-of-day and month-end prices on all listed NYSE, Amex, and NASDAQ common stocks along with basic market indices, and includes the most comprehensive distribution information available, with the most accurate total return calculations." We create remote data frames for `crsp.dsf` (end-of-day prices) and `crsp.msf` (month-end prices).

```
db <- dbConnect(RPostgres::Postgres(), bigint = "integer")

dsf <- tbl(db, Id(schema = "crsp", table = "dsf"))
msf <- tbl(db, Id(schema = "crsp", table = "msf"))
```

Let's look at a few rows from `crsp.dsf`.

```
dsf |> collect(n = 5)
```

```
# A tibble: 5 x 20
  cusip    permno permco issuno hexcd hsiccd date        bidlo askhi   prc
  <chr>     <int>  <int>  <int> <int>  <int> <date>       <dbl> <dbl> <dbl>
1 68391610  10000   7952  10396     3   3990 1986-01-07    2.38  2.75 -2.56
2 68391610  10000   7952  10396     3   3990 1986-01-08    2.38  2.62 -2.5
3 68391610  10000   7952  10396     3   3990 1986-01-09    2.38  2.62 -2.5
4 68391610  10000   7952  10396     3   3990 1986-01-10    2.38  2.62 -2.5
5 68391610  10000   7952  10396     3   3990 1986-01-13    2.5   2.75 -2.62
# i 10 more variables: vol <dbl>, ret <dbl>, bid <dbl>, ask <dbl>,
#   shrout <dbl>, cfacpr <dbl>, cfacshr <dbl>, openprc <dbl>,
#   numtrd <int>, retx <dbl>
```

---

[5]https://www.crsp.org/products/research-products/
[6]https://go.unimelb.edu.au/7cd8
[7]https://go.unimelb.edu.au/7cd8

The CRSP Indices database contains a number of CRSP indices. Here we focus on two index tables, `crsp.dsi` and `crsp.msi`, which can be viewed as complementing `crsp.dsf` and `crsp.msf`, respectively.

```
dsi <- tbl(db, Id(schema = "crsp", table = "dsi"))
msi <- tbl(db, Id(schema = "crsp", table = "msi"))
```

7.2.1 Exercises

1. Looking at `crsp.dsf` and `crsp.msf`, we see that `prc` can be negative. Do negative stock prices make sense economically? What do negative stock prices on CRSP mean?[8] What would be an alternative approach to encode this information? (Write code to recast the data using this approach.) Why do you think that CRSP chose the approach it uses?

2. How do `ret` and `retx` differ? Which variable are you more likely to use in research?

3. Is the `date` variable on `crsp.msf` always the last day of the month? If not, why not?

4. Suggest the "natural" primary key for `crsp.dsf` and `crsp.msf`. Check that this is a valid primary key for `crsp.msf`.

5. In the code below, we are using `collect()` followed by `mutate(month = floor_date(date, "month"))` to calculate `month`. What changes occur in terms of where the processing happens if we replace these two lines with `mutate(month = as.Date(floor_date("month", date))) |> collect()`? Do we get different results? What effect does `as.Date()` have?

```
plot_data <-
  dsf |>
  select(date) |>
  filter(between(date, "2017-12-31", "2022-12-31")) |>
  collect() |>
  mutate(month = floor_date(date, "month"))
```

6. What is being depicted in Figures 7.1 and 7.2? What are the sources of variation across months in Figure 7.1? Can you guess what is the main driver of variation in Figure 7.2? Create an additional plot to visualize the source of variation in Figure 7.1 not depicted in Figure 7.2.

```
plot_data |>
  count(month) |>
  ggplot(aes(x = month, y = n)) +
  geom_bar(stat = "identity") +
  scale_x_date(date_breaks = "2 months", expand = expansion()) +
  theme(axis.text.x = element_text(angle = 90))
```

---

[8]CRSP documentation can be found at https://go.unimelb.edu.au/gcd8.

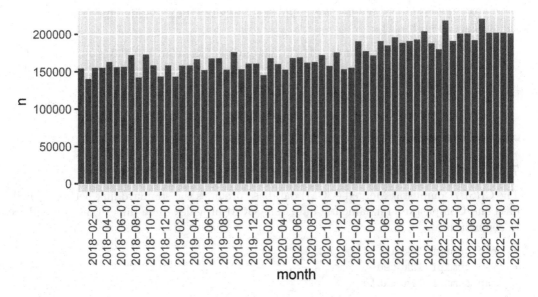

FIGURE 7.1
Number of observations by month (#1)

```
plot_data |>
  distinct() |>
  count(month) |>
  ggplot(aes(x = month, y = n)) +
  geom_bar(stat = "identity") +
  scale_x_date(date_breaks = "2 months", expand = expansion()) +
  theme(axis.text.x = element_text(angle = 90))
```

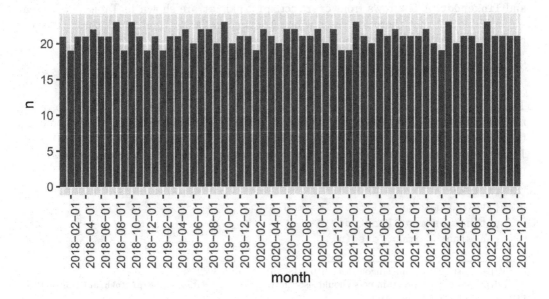

FIGURE 7.2
Number of observations by month (#2)

7. What is the primary key for `crsp.dsi` and `crsp.msi`? Verify that it is a valid key for both tables.

8. Using the `dplyr` verb `anti_join()`, determine if there are any dates on `crsp.dsf` that do not appear on `crsp.dsi` or vice versa. Do the same for `crsp.msi` and `crsp.msf`.

7.3   Linking CRSP and Compustat

The CRSP/Compustat Merged (CCM) database provides the standard link between CRSP data and Compustat's fundamental data. The CCM provides three tables that you will see in common use:

- `crsp.ccmxpf_lnkhist`
- `crsp.ccmxpf_lnkused`
- `crsp.ccmxpf_linktable`

The reality is that the only table we need to worry about is `crsp.ccmxpf_lnkhist`, as the other two tables can be (and likely are) constructed from it (see here[9] for details).[10]

```
ccmxpf_lnkhist <- tbl(db, Id(schema = "crsp", table = "ccmxpf_lnkhist"))
```

7.3.1   Building a link table

In this section, we work towards creating a link table that can be used in subsequent chapters and that represents what appears to the standard approach in research.

The basic idea of `crsp.ccmxpf_lnkhist` one can match a `gvkey`-date combination to a PERMNO (here called `lpermno`) by merging on `gvkey` where the date is between `linkdt` and `linkenddt`. A few rows from `crsp.ccmxpf_lnkhist` are shown in Table 7.2. We can see in Table 7.2 that there are cases where `lpermno` is NA, so "matching" these rows will result in non-matches, which would seem to have no real value.

```
ccmxpf_lnkhist |>
  collect(n = 10)
```

The only value of these non-matches might be data on their `linktype`. Table 7.3 provides the distribution of `linktype` in `crsp.ccmxpf_lnkhist`. If a non-match has `linktype` of NR, then the lack of a link has been "confirmed by research" (presumably by CRSP). If it has `linktype` of NU, the lack of link is "not yet confirmed" by research, so a diligent researcher might want to investigate.[11]

```
ccmxpf_lnkhist |>
  mutate(has_permno = !is.na(lpermno)) |>
  count(has_permno, linktype) |>
  arrange(has_permno, desc(n))
```

---

[9]https://gist.github.com/iangow/fca4cb10b048f5c798113da7039c2688

[10]WRDS says "SAS programmers should use the Link History dataset (`ccmxpf_lnkhist`) from CRSP": https://go.unimelb.edu.au/akw8.

[11]The meaning of a `linktype` of NP is unclear, as no documentation of this code seems to exist.

TABLE 7.2

A sample of observations on `crsp.ccmxpf_lnkhist`

gvkey	linkprim	liid	linktype	lpermno	lpermco	linkdt	linkenddt
001000	C	00X	NU	NA	NA	1961-01-01	1970-09-29
001000	P	01	NU	NA	NA	1970-09-30	1970-11-12
001000	P	01	LU	25881	23369	1970-11-13	1978-06-30
001001	C	00X	NU	NA	NA	1978-01-01	1983-09-19
001001	P	01	LU	10015	6398	1983-09-20	1986-07-31
001002	C	00X	NR	NA	NA	1960-01-01	1970-09-29
001002	C	01	NR	NA	NA	1970-09-30	1972-12-13
001002	C	01	NR	NA	NA	1973-06-06	1973-08-31
001002	C	01	LC	10023	22159	1972-12-14	1973-06-05
001003	C	00X	NU	NA	NA	1980-01-01	1983-12-06

TABLE 7.3

Distribution of `linktype` in `crsp.ccmxpf_lnkhist`

has_permno	linktype	n
FALSE	NR	43,255
FALSE	NU	34,125
FALSE	NP	4
TRUE	LC	17,013
TRUE	LU	15,993
TRUE	LS	5,158

So, in practice, we would likely ignore all matches with `linktype %in% c("NU", "NR", "NP")` or (equivalently) `is.na(lpermno)`.

The cases where `linktype` is LD represent cases where two GVKEYs map to a single PERMNO at the same time and, according to WRDS, "this link should not be used." Table 7.4 provides an example.

```
ccmxpf_lnkhist |> filter(lpermno == 23536)
```

TABLE 7.4

Example of GVKEY mapping to multiple PERMNOs

gvkey	linkprim	liid	linktype	lpermno	lpermco	linkdt	linkenddt
011550	P	01	LC	23536	21931	1962-01-31	NA
013353	P	01	LD	23536	21931	1962-01-31	1986-12-31
013353	C	99X	LD	23536	21931	1987-01-01	2020-12-31

Here we'll take the WRDS's advice and omit these.

The cases where `linktype` is LX are ones where the Compustat security trades on a foreign exchange (so is not covered by CRSP) and CRSP is merely "helpfully" linking to a different security that *is* found on CRSP. Table 7.5 provides an example.

```
ccmxpf_lnkhist |>
  filter(gvkey == "001186", !is.na(lpermno))
```

TABLE 7.5

Example where `linktype` is LX

gvkey	linkprim	liid	linktype	lpermno	lpermco	linkdt	linkenddt
001186	P	01	LC	78223	26174	1982-11-01	NA
001186	N	01C	LX	78223	26174	1982-11-01	NA

These matches are duplicates and we don't want them.

The remaining category for discussion is where `linktype` is LN. These are cases where a link exists, but Compustat does not have price data to allow CRSP to check the quality of the link. While researcher discretion might be used to include these, most researchers appear to exclude these cases and we will do likewise. Given the above, we only include cases where `linktype` is in LC (valid, researched link), LU (unresearched link), or LS (link valid for this `lpermno` only).

```
ccm_link <-
  ccmxpf_lnkhist |>
  filter(linktype %in% c("LC", "LU", "LS"))
```

We now consider `linkprim`, which WRDS explains[12] as follows:

> `linkprim` clarifies the link's relationship to Compustat's marked primary security within the related range. "P" indicates a primary link marker, as identified by Compustat in monthly security data. "C" indicates a primary link marker, as identified by CRSP to resolve ranges of overlapping or missing primary markers from Compustat in order to produce one primary security throughout the company history. "J" indicates a joiner secondary issue of a company, identified by Compustat in monthly security data.

This suggests we should omit cases where `linkprim` equals J. Given that cases where `linkprim` equals N are duplicated links due to the existence of Canadian securities for a US-traded firm, we will exclude these too.

Table 7.6 provides data on `linkprim` for valid links.

```
ccm_link |>
  count(linkprim) |>
  arrange(desc(n))
```

TABLE 7.6

Distribution of `linkprim` for valid links

linkprim	n
P	29,689
C	7,961
J	420
N	94

Based on this, we focus on cases where `linkprim` is either C or P.

---

[12]https://go.unimelb.edu.au/ozw8

A final consideration is the presence of missing values on `linkenddt`. In general, these missing values indicate that the link remains valid at the time the data set was constructed. One approach would be to filter using code something like `datadate <= linkenddt | is.na(linkenddt)`. Another approach would be to fill in missing values with a date that represents the plausible latest date for the data set. We take the latter approach in the code we use to create our final iteration of `ccm_link`.

```
ccm_link <-
  ccmxpf_lnkhist |>
  filter(linktype %in% c("LC", "LU", "LS"),
         linkprim %in% c("C", "P")) |>
  mutate(linkenddt = coalesce(linkenddt,
                              max(linkenddt, na.rm = TRUE)))
```

A natural question is whether, for any given `gvkey`, there is only one PERMNO that is matched with `linkprim IN ('P', 'C')` on any given date. We can examine this by looking for overlapping date ranges between different rows of `ccm_link` for a given `gvkey`. The date range for each row starts with `linkdt` and ends with `linkenddt`. Ordering the rows by `linkdt`, there is overlap between the current row and the next row if the current row's date range ends after the start of the next row's date range.

For this purpose, we use two **window functions**: `lag()` and `lead()`. These are called window functions because each operates on a "window" of data.[13] Here each window comprises data for a single `gvkey` ordered by `linkdt`. For each row within a window, we look at the previous row (if any) to get `lag(linkenddt)` and the following row (if any) to get `lead(linkdt)`. The following code counts the number of overlapping rows.

```
ccm_link |>
  group_by(gvkey) |>
  window_order(linkdt) |>
  mutate(lag_linkenddt = lag(linkenddt),
         lead_linkdt = lead(linkdt),) |>
  filter(lag_linkenddt >= linkdt | linkenddt >= lead_linkdt) |>
  ungroup() |>
  count() |>
  pull()
```

```
[1] 0
```

So, there are no cases of overlapping dates, which confirms that only one `lpermno` is linked to a given `gvkey` for a given date. As can be seen in Table 7.7, the vast majority of GVKEYs map to just one PERMNO even *without* regard to date.

```
ccm_link |>
  group_by(gvkey) |>
  summarize(num_permnos = n_distinct(lpermno)) |>
  count(num_permnos, sort = TRUE)
```

---

[13]A careful reader might have noticed that we actually already use a window function in Chapter 2, namely `fill()`. However, there we used `arrange()` instead of `window_order()`. The `window_order()` function is only available for remote data frames because it provides functionality not available with local data frames. Readers coming from an SQL background might observe that `dplyr`'s `group_by()` is "overloaded" in the sense that it does the work of both the `GROUP BY` statement and the `PARTITION BY` clause in SQL. A short discussion of window functions is found in Chapter 21 of *R for Data Science*.

TABLE 7.7
Distribution of number of PERMNOs
matching each GVKEY

num_permnos	n
1	32,198
2	1,056
3	76
4	3
5	1

The case with 5 PERMNOs is a complicated one involving tracking stock, spin-offs, etc.[14] But one observation doesn't matter much.

Our sense is that the last iteration of `ccm_link` above is more or less the standard approach used by researchers in practice and it's the one we use in the rest of this book. While you may occasionally see code that filters on `usedflag == 1`, which is a variable found on `crsp.ccmxpf_lnkused`, not on `crsp.ccmxpf_lnkhist`, it can be shown[15] that this yields exactly the same result as `ccm_link` above.

7.3.2   Applying the link table

This section provides an analysis of our link table `ccm_link` with a focus on cases where the linked PERMNO changes between the `datadate` associated with the financial statements and the date when earnings were announced (`rdq`). We get `datadate` for fiscal years from `comp.funda` and `rdq` comes from `comp.fundq` and we start by creating remote data tables for these sources.

```
funda <- tbl(db, Id(schema = "comp", table = "funda"))
fundq <- tbl(db, Id(schema = "comp", table = "fundq"))
```

The following code generates from Compustat the table of events that we can link to CRSP. We start with the fourth-quarter data for "report date of quarterly earnings", which is the variable `rdq` on `comp.fundq`.

```
rdqs <-
  fundq |>
  filter(indfmt == "INDL", datafmt == "STD",
         consol == "C", popsrc == "D") |>
  filter(!is.na(rdq), fqtr == 4) |>
  select(gvkey, datadate, rdq)
```

We then link with (`gvkey`, `datadate`) combinations from `comp.funda` and focus on fiscal years ending on or after 1 January 2000.

```
funda_mod <-
  funda |>
  filter(indfmt == "INDL", datafmt == "STD",
         consol == "C", popsrc == "D")
```

---

[14]See https://go.unimelb.edu.au/z9w8 for details.
[15]https://go.unimelb.edu.au/nkw8

```
annc_dates <-
  funda_mod |>
  select(gvkey, datadate) |>
  inner_join(rdqs, by = c("gvkey", "datadate")) |>
  filter(datadate >= "2000-01-01")
```

The resulting table has **gvkey** and **datadate**, as well as **rdq**. But in linking to CRSP, a question arises as to which date to use to check that a link is valid at the relevant time. This question seems difficult to answer in the abstract, so let's try both dates (**datadate** and **rdq**) and drill deeper into cases where the linked **permno** values differ (when the linked **permno** is the same, we are probably safe).

We first merge with **ccm_link** using **datadate** to create **datadate_permnos**.

```
datadate_permnos <-
  annc_dates |>
  inner_join(ccm_link,
             by = join_by(gvkey,
                          between(datadate, linkdt, linkenddt))) |>
  select(gvkey, datadate, rdq, lpermno, lpermco) |>
  rename(permno = lpermno,
         permco = lpermco)
```

We then merge with **ccm_link** using **rdq** to create **rdq_permnos**.

```
rdq_permnos <-
  annc_dates |>
  inner_join(ccm_link,
             join_by(gvkey, between(rdq, linkdt, linkenddt))) |>
  select(gvkey, datadate, rdq, lpermno, lpermco) |>
  rename(permno = lpermno,
         permco = lpermco)
```

The table **link_table_combined** computes a **full_join()** of **rdq_permnos** and **datadate_permnos** and then compares matches across the alternative test dates.

```
link_table_combined <-
  rdq_permnos |>
  full_join(datadate_permnos,
            by = c("gvkey", "datadate", "rdq"),
            suffix = c("_rdq", "_ddate")) |>
  filter(!is.na(permno_rdq) | !is.na(permno_ddate)) |>
  mutate(same_permno = coalesce(permno_rdq == permno_ddate, FALSE),
         same_permco = coalesce(permco_rdq == permco_ddate, FALSE),
         has_permno_rdq = !is.na(permno_rdq),
         has_permno_ddate = !is.na(permno_ddate))
```

Analysis of the data in **link_table_combined** is provided in Table 7.8.

```
link_table_stats <-
  link_table_combined |>
  count(same_permno, same_permco,
        has_permno_rdq, has_permno_ddate) |>
  arrange(same_permno, same_permco,
```

TABLE 7.8

Distribution of CRSP-Compustat links at `datadate` versus `rdq`

same_permno	same_permco	has_permno_rdq	has_permno_ddate	n
TRUE	TRUE	TRUE	TRUE	124,673
FALSE	FALSE	FALSE	TRUE	4,577
FALSE	FALSE	TRUE	FALSE	3,059
FALSE	FALSE	TRUE	TRUE	24
FALSE	TRUE	TRUE	TRUE	22

gvkey	linkprim	liid	linktype	lpermno	lpermco	linkdt	linkenddt
021998	P	01	LC	15075	55117	2014-11-20	2018-01-17
032961	P	01	LC	15075	55117	2018-01-18	2022-08-10

```
          has_permno_rdq, has_permno_ddate) |>
  ungroup() |>
  arrange(desc(n))

link_table_stats
```

So in 124,673 (94.2%) cases, we match to the same `permno` using either date. In many cases, we match to a `permno` using one date but not the other. In 3059 cases, we have a `permno` link on `rdq`, but not on `datadate`. Inspecting some of these cases, they appear to be cases where the stock was not traded on `datadate`, but was trading by `rdq`, perhaps due to an IPO in the meantime.

In 4577 cases, we have a `permno` link on `datadate`, but not on `rdq`. One example is the `gvkey` of 021998, which on 2017-12-31 relates to the firm Neothetics Inc. According to a 10-K filing[16], "on January 17, 2018, Neothetics, Inc., or Neothetics, and privately-held Evofem Biosciences Operations, Inc., or Private Evofem, completed ... the Merger. ... The Merger was structured as a reverse capitalization and Private Evofem was determined to be the accounting acquirer based on the terms of the Merger and other factors." So the continuing entity with respect to financial reporting is Evofem, which has a GVKEY of 032961.

The twist in this case is that "the financial information included in the first 10-K filed after the merger[17] is that of Neothetics prior to the Merger." Thus, the representation of `ni` and `at` as continuous series up to 2017-12-31 provided by Compustat for `gvkey` of 021998 (see below) and a series starting from 2018-12-31 for `gvkey` of 032961 seems correct. Inspection of the financial statements for the period ending 2018-12-31 suggests that there are values that could be supplied for `gvkey` of 032961 and `datadate` of 2017-12-31 (Compustat leaves these missing). However, it seems inappropriate to view the market reaction on `rdq` (2018-02-26) as being to the financial statement information associated with *either* `gvkey` (021998 or 032961) alone and thus a non-match on `ccm_link` seems appropriate.

```
ccm_link |>
  filter(lpermno == 15075)
```

---

[16]https://www.sec.gov/Archives/edgar/data/1618835/000156459018003259/evfm-10k_20171231.htm

[17]https://www.sec.gov/Archives/edgar/data/1618835/000156459018003259/evfm-10k_20171231.htm

gvkey	linkprim	liid	linktype	lpermno	lpermco	linkdt	linkenddt
017010	P	01	LC	15998	54311	2016-04-18	2017-12-31
017010	J	01	LC	15998	54311	2018-01-01	NA
017010	J	03	LC	16000	54311	2016-04-18	2017-12-31
017010	P	03	LC	16000	54311	2018-01-01	NA

```
ccm_link |>
  filter(lpermno == 15075) |>
  select(gvkey) |>
  inner_join(funda_mod, by = "gvkey") |>
  select(gvkey, datadate, at, ni) |>
  arrange(datadate)
```

gvkey	datadate	at	ni
021998	2012-12-31	12.82	−7.83
021998	2013-12-31	4.53	−15.02
021998	2014-12-31	76.90	−10.82
021998	2015-12-31	40.11	−43.16
021998	2016-12-31	12.82	−13.02
021998	2017-12-31	4.12	−9.99

Only in 46 cases do we find *different* permno values on the two dates. In 22 cases, we link to different permno values, but the *same* permco value. One of these cases is for gvkey of 017010. It's not clear why CRSP switched the primary permno on 2018-01-01.[18]

```
ccmxpf_lnkhist |>
  filter(gvkey == "017010",
         lpermno %in% c(15998, 16000))
```

What about the 24 cases with *different* permco values? Looking at the case where gvkey is 183603, the issue is a merger. According to the 10-K filed by Colony Northstar Inc. on 2017-02-28,[19] "on January 10, 2017, NSAM completed the tri-party merger with Colony Capital, Inc., or Colony, and NorthStar Realty Finance Corp., or NorthStar Realty or NRF, under which the companies combined in an all-stock merger of equals transaction, referred to as the Mergers, to create Colony NorthStar, an internally-managed, diversified real estate and investment management company. ... Although NSAM is the legal acquirer in the Mergers, Colony has been designated as the accounting acquirer, resulting in a reverse acquisition of NSAM for accounting purposes." The issue here is that the gvkey of 183603, which related to Colony prior to the merger, survives on Compustat, while the permno of 14686, which related to NSAM prior to the merger, is the surviving permno. The correct approach here depends to some extent on the research question. As the net income reported on 2017-02-28 relates to an entity (a pre-merger Colony) whose securities ceased to trade

---

[18]The same seems to be true for the case with gvkey of 003581. Again, it's not clear why CRSP switched the primary permno on 2018-01-01.

[19]https://go.unimelb.edu.au/23w8

on 2017-01-10, the correct answer seems to be that there is no appropriate `permno` to use for measuring the market reaction to earnings on 2017-02-28.

In summary, it appears that observations that have a `permno` link on `datadate`, but not on `rdq`, or whose matched `permno` changes between these two dates are problematic and should not be used for our hypothetical event study. Observations with a `permno` link on `rdq` but not on `datadate` seem more appropriate, but are likely to be new firms in some way.

7.4   All about CUSIPs

According to CUSIP Global Services,[20] "CUSIP identifiers are the universally accepted standard for classifying financial instruments across institutions and exchanges worldwide. Derived from the Committee on Uniform Security Identification Procedures, CUSIPs are nine-character identifiers that capture an issue's important differentiating characteristics for issuers and their financial instruments in the U.S. and Canada."

CUSIP Global Services[21] uses the CUSIP of Amazon.com's common stock, 023135106, as an example of the components of a nine-character CUSIP. The first six characters—023135—represent the issuer, which is Amazon.com. While in this case the issuer is a company, an issuer could be a municipality or a government agency. The next two characters (10) indicate the type of instrument (e.g., debt or equity) and also uniquely identifies the issue among the issuer's securities. The final character (6) is a check digit created by a mathematical formula. This last character will indicate any corruption of the preceding eight characters. Note that the characters need not be digits. For example, according to an SEC filing,[22] the Class C Common Stock of Dell Technologies Inc. has a CUSIP of 24703L202, which contains the letter L.

While a full CUSIP always comprises nine characters, many data services abbreviate the CUSIP by omitting the check digit (to create an "eight-digit" CUSIP) or both the check digit and the issue identifier (to create a "six-digit" CUSIP). For example, the CRSP table `crsp.stocknames` uses eight-digit CUSIPs.

Notwithstanding the existence of `crsp.ccmxpf_lnkhist`, some researchers choose to link CRSP and Compustat using CUSIPs. For example, the code[23] supplied with Jame et al. (2016) merges CRSP daily data on returns, prices, volume, and shares outstanding with Compustat data on shareholders' equity using CUSIP and "year".[24]

To evaluate the appropriateness of using CUSIPs to link CRSP and Compustat, we can construct a link table for `comp.funda` using CUSIPs (`funda_cusip_link` below) and compare it with an analogous link table constructed using `ccm_link` (`funda_ccm_link` below). First, let's construct the subset of `comp.funda` of interest.

```
funda_mod <-
  funda |>
  filter(indfmt == "INDL", datafmt == "STD",
```

---

[20]https://www.cusip.com/identifiers.html#/CUSIP

[21]https://www.cusip.com/identifiers.html?section=CUSIP

[22]https://investors.delltechnologies.com/static-files/acf98957-c3f6-4363-bbaa-f0525ce5296d

[23]https://go.unimelb.edu.au/f3w8

[24]Only the last observation for a calendar year is kept by Jame et al. (2016) for CRSP and `year` means `year(datadate)` for Compustat.

```
            consol == "C", popsrc == "D") |>
  mutate(mkt_cap = prcc_f * csho) |>
  select(gvkey, datadate, cusip, at, mkt_cap)
```

Our source for PERMNO-CUSIP links is `crsp.stocknames`. There are some cases where there is no value on `ncusip`, but there is a value on `cusip` and we use `coalesce()` to fill in missing values in such cases.

```
stocknames <- tbl(db, Id(schema = "crsp", table = "stocknames"))
```

```
stocknames_plus <-
  stocknames |>
  mutate(ncusip = coalesce(ncusip, cusip))
```

Now we can construct `funda_cusip_link` containing CUSIP-based matches for each (gkvey, datadate).

```
funda_cusip_link <-
  funda_mod |>
  mutate(ncusip = str_sub(cusip, 1L, 8L)) |>
  inner_join(stocknames_plus,
             join_by(ncusip,
                     between(datadate, namedt, nameenddt))) |>
  select(gvkey, datadate, permno, permco)
```

Similarly, we can construct our matches using `ccm_link`.

```
funda_ccm_link <-
  funda_mod |>
  select(gvkey, datadate) |>
  inner_join(ccm_link,
             join_by(gvkey,
                     between(datadate, linkdt, linkenddt))) |>
  select(gvkey, datadate, lpermno, lpermco) |>
  rename(permno = lpermno, permco = lpermco)
```

Finally, we combine both sets of matches in `funda_link_combined` for comparison.

```
funda_link_combined <-
  funda_mod |>
  select(-cusip) |>
  left_join(funda_ccm_link, by = join_by(gvkey, datadate)) |>
  left_join(funda_cusip_link,
            by = join_by(gvkey, datadate),
            suffix = c("_ccm", "_cusip")) |>
  mutate(same_permno = permno_ccm == permno_cusip,
         same_permco = permco_ccm == permco_cusip,
         has_permno_ccm = !is.na(permno_ccm),
         has_permno_cusip = !is.na(permno_cusip)) |>
  filter(has_permno_ccm | has_permno_cusip) |>
  collect()
```

Regarding Table 7.9, we can probably view the cases with `same_permno` as valid matches, but would probably need to check the cases where `same_permno` is FALSE.

```
funda_link_combined |>
  count(same_permno, same_permco)
```

TABLE 7.9
Comparison of CCM- and CUSIP-based links

same_permno	same_permco	n
FALSE	FALSE	131
FALSE	TRUE	160
TRUE	TRUE	242,840
NA	NA	118,522

The cases where `same_permno` is NA in Table 7.9 are explored in Table 7.10. We would need to investigate the cases where one of `permno_ccm` or `permno_cusip` is NA to understand the source of the non-matches in one table or the other. However, a reasonable view seems to be that `ccm_link` provides many valid matches that are lost when matching using CUSIPs and for this reason `crsp.ccmxpf_lnkhist` should be preferred to CUSIP-based matches.

```
funda_link_combined |>
  count(has_permno_ccm, has_permno_cusip)
```

TABLE 7.10
Differences in coverage of CCM- and CUSIP-based link tables

has_permno_ccm	has_permno_cusip	n
FALSE	TRUE	3,233
TRUE	FALSE	115,289
TRUE	TRUE	243,131

7.4.1 Exercises

1. Is there any evidence of "reuse" of CUSIPs on `crsp.stocknames`? In other words, are there any `ncusip` or `cusip` values associated with more than one `permno`?

2. The CRSP table `crsp.stocknames` includes two CUSIP-related fields, `cusip` and `ncusip`. What are the differences between the two fields? What does it mean when `ncusip` is missing, but `cusip` is present?

3. Like CUSIPs, PERMNOs are security-level identifiers. Can a PERMNO be associated with more than one CUSIP at a given point in time? Can a PERMNO be associated with more than one CUSIP over time?

4. Looking at entries on `crsp.stocknames` where `ticker` is DELL, we see two different `permno` values. What explains this?

```
stocknames |>
  filter(str_detect(comnam, '^DELL ')) |>
  select(permno, cusip, ncusip, comnam, siccd, namedt, nameenddt)
```

permno	cusip	ncusip	comnam	siccd	namedt	nameenddt
11081	24702R10	24702510	DELL COM-PUTER CORP	3570	1988-06-22	2003-07-21
11081	24702R10	24702R10	DELL INC	3570	2003-07-22	2013-10-29
16267	24703L10	24703L10	DELL TECH-NOLOGIES INC	3824	2016-09-07	2018-12-27
18267	24703L20	24703L20	DELL TECH-NOLOGIES INC	3824	2018-12-28	2020-03-22
18267	24703L20	24703L20	DELL TECH-NOLOGIES INC	3571	2020-03-23	2023-12-29

5. Looking at `permno` of 11081 (Dell), we see two different CUSIP values. What change appears to have caused the change in CUSIP for what CRSP regards as the same security?

6. Choose a row from `funda_link_combined` where `same_permco` is FALSE. Can you discern from the underlying tables what issue is causing the difference and which match (if any) is valid? (*Hint:* Do rows where gvkey %in% c("065228", "136265") meet this condition? What appears to be the issue for these GVKEYs?) Can you conclude that the CCM-based match is the preferred one in each case?

7. Choose a row from `funda_link_combined` where `has_permno_cusip` is TRUE and `has_permno_ccm` is FALSE. Can you discern from the underlying tables whether the CUSIP-based match is valid? (*Hint:* Do rows where gvkey %in% c("033728", "346027") meet this condition? What appears to be the issue for these GVKEYs?)

8. Given the results shown in Tables 7.9 and 7.10 and your answer to the previous two questions, can you conclude that the CCM-based match is preferred to the CUSIP-based match in each case?

8

Financial statements: A second look

In this chapter, we will dive a bit more deeply into financial statements than we did in Chapter 6. The focus of this chapter is an exploration of core attributes of financial accounting data (e.g., that balance sheets balance). After exploring how financial statement data are represented in Compustat, we dive into our first detailed exploration of an accounting research paper, Koh and Reeb (2015), which explores issues regarding how R&D spending is reported by firms and coded in Compustat.

This chapter is optional in the sense that subsequent chapters do not depend on an understanding of the content of this chapter. However, the material of this chapter provides a good opportunity for readers to better understand the messiness of representing seemingly tidy relationships in accounting data in a database. Some of the exercises help readers to understand better where data in Compustat come from. Working through this chapter also serves to dispel any notion that commercial databases are error-free, as we find discrepancies that are apparent even from just the information in Compustat.

> 💡 Tip
>
> In this chapter, we will use the R libraries below. For instructions on how to set up
> your computer to use the code found in this book, see Section 1.2. Quarto templates
> for the exercises below are available on GitHub.[a]
>
> ---
> [a]https://github.com/iangow/far_templates/blob/main/README.md

```
library(dplyr)
library(DBI)
library(ggplot2)
library(farr)
library(dbplyr)          # window_order()
```

8.1 Core attributes of financial statements

Three core attributes of financial statements that should hold are the following:

1. **Balanced balance sheets.** Balance sheets should balance (i.e., total assets should equal the sum of total liabilities and shareholders' equity). This is a *sine qua non* of double-entry bookkeeping.

2. **Within-statement articulation.** The various sub-totals *within* a financial statement make sense with regard to the items they comprise. For example, the amount in total assets should equal the sum of the component assets.

3. **Across-statement articulation.** Financial statements should **articulate**. For example, the amounts reported for cash on the beginning and ending balance sheets should be explained by the statement of cash flows.

Below we explore how these attributes are reflected in Compustat. As in Chapter 6, we will focus on two tables from Compustat's North American database: `comp.funda` and `comp.company` and we will construct a table `funda_mod` that contains the "standard" set of observations for Compustat.

We begin by setting up the remote data frames we used in Chapter 6.

```
db <- dbConnect(RPostgres::Postgres(), bigint = "integer")


funda <- tbl(db, Id(schema = "comp", table = "funda"))
company <- tbl(db, Id(schema = "comp", table = "company"))
```

```
funda_mod <-
  funda |>
  filter(indfmt == "INDL", datafmt == "STD",
         consol == "C", popsrc == "D")
```

8.2 Balance sheets

An essential feature of statements of financial position is that they balance, hence the more traditional term "balance sheets". Let's check that this holds on Compustat. Total assets is `at` and the sum of liabilities and shareholders' equity is `lse`.

```
funda_mod |>
  filter(at != lse) |>
  select(gvkey, datadate, at, lse) |>
  collect() |>
  nrow()
```

```
[1] 0
```

So everything balances at this level. Note that if either `at` or `lse` is NA, then the filter will not apply. For completeness, the following code examines missingness of `at` and `lse`. Below we see that if `at` is missing, so is `lse`; if `at` is present, so is `lse`.

```
funda_mod |>
  mutate(missing_at = is.na(at), missing_lse = is.na(lse)) |>
  count(missing_at, missing_lse) |>
  ungroup()
```

missing_at	missing_lse	n
True	True	88,653
False	False	484,490

8.3   Within-statement articulation

Having established that balance sheets always balance on Compustat, the next question
we study is whether the various **balancing models** for `funda` hold. Compustat provides
balancing models that explain how various items on financial statements in its database
relate to each other. WRDS provides these balancing models as Excel files on its website[1]
(requires WRDS access).

One example of a balancing model relates to the decomposition of total assets: `at = act +
ppent + ivaeq + ivao + intan + ao`, where the description of each item in the equation
is provided in Table 8.1.

TABLE 8.1
Components of total assets

Item	Description
act	Current Assets—Total
ppent	Property Plant and Equipment—Total (Net)
ivaeq	Investment and Advances—Equity
ivao	Investment and Advances—Other
intan	Intangible Assets—Total
ao	Assets Other—Total
dc	Deferred Charges (component of AO)
aox	Assets Other (Sundry) (component of AO)

The first issue we need to think about is the presence of `NA` values in the components of `at`
even when `at` itself is not `NA`. Let's look at data for one company (GVKEY: 008902) where
this creates issues.

```
na_sample_firm_years <-
  funda_mod |>
  filter(gvkey == "008902", datadate >= "2000-01-01") |>
  select(gvkey, datadate)
```

From the output from the code below, we see that, because `ivaeq` is `NA` for these years,
the sum of the components of `at` is also `NA`, even though the value of `at` is not `NA`. What's
going on? The answer is that `ivaeq` is `NA` because this firm doesn't report amounts for
"Investment and Advances - Equity", because they're either zero or not material. Thus it
seems reasonable to assume that these should be zero.

```
funda_mod |>
  semi_join(na_sample_firm_years, by = c("gvkey", "datadate")) |>
  mutate(at_calc = act + ppent + ivaeq + ivao + intan + ao) |>
  select(gvkey, datadate, at, at_calc, act,
         ppent, ivaeq, ivao, intan, ao) |>
  arrange(datadate)
```

---

[1] https://go.unimelb.edu.au/tzw8

gvkey	datadate	at	at_calc	act	ppent	ivaeq	ivao	intan	ao
008902	2000-05-31	2099.20	NA	785.09	366.23	NA	0	915.74	32.15
008902	2001-05-31	2078.49	NA	819.42	362.04	NA	0	871.65	25.39
008902	2002-05-31	2036.40	NA	801.31	355.80	NA	0	856.86	22.43
008902	2003-05-31	2247.21	NA	928.09	370.79	NA	0	914.20	34.13
008902	2004-05-31	2353.12	NA	994.62	381.06	NA	0	930.62	46.83

To convert `NA` amounts to zero, we can use the `coalesce()` function, which takes two arguments and returns the first argument if it is not `NA` and returns the second argument otherwise.[2] So, `coalesce(ivaeq, 0)` equals `ivaeq` when `ivaeq` is not `NA`, and 0 otherwise.

```
tol <- 1e-3

funda_na_fixed <-
  funda_mod |>
  select(gvkey, datadate, act, ppent, ivaeq, ivao, intan, ao, at,
         dc, aox) |>
  mutate(at_calc = coalesce(act, 0) + coalesce(ppent, 0) +
           coalesce(ivaeq, 0) + coalesce(ivao, 0) +
           coalesce(intan, 0) + coalesce(ao, 0),
         at_diff = at_calc - at,
         balance = abs(at_diff) < tol)
```

Let's see if applying `coalesce(x, 0)` helps. Below we see that, *in this case*, we can tie the value for `at` with its components (`act`, `ppent`, `ivaeq`, `ivao`, `intan`, and `ao`).

```
funda_na_fixed |>
  semi_join(na_sample_firm_years,
            by = c("gvkey", "datadate")) |>
  select(gvkey, datadate, at, at_calc, balance) |>
  arrange(datadate)
```

gvkey	datadate	at	at_calc	balance
008902	2000-05-31	2099.20	2099.20	True
008902	2001-05-31	2078.49	2078.49	True
008902	2002-05-31	2036.40	2036.40	True
008902	2003-05-31	2247.21	2247.21	True
008902	2004-05-31	2353.12	2353.12	True

Note that we specified a "tolerance" (`tol`) equal to 0.001 because testing for equality of **floating-point** calculations is not always exact.[3]

But, does this approach resolve all issues with articulation of `at` with its components? Alas, the output from the following code suggests the answer is "no". We use the exercises to look more closely at a few of the problem cases.

---

[2]As discussed in its documentation, the `coalesce()` function from `dplyr` is "inspired by the SQL `COALESCE` function which does the same thing for NULLs".

[3]https://stackoverflow.com/questions/51134021/floating-point-equality

```
funda_na_fixed |>
  count(balance)
```

balance	n
True	403,315
NA	88,653
False	81,175

8.3.1 Exercises

1. What is the value in `funda_na_fixed` of `at` when `balance` is `NA`? Is this surprising?

2. Write code to calculate `decade`, the decade in which `datadate` for each observation falls. (Hint: The functions `floor()` and `year()` and the number `10` may be helpful.)

3. Are there any cases in `funda_na_fixed` where `at_calc` is greater than zero and `at` is `NA`? Which decades are these mostly found in? If you were doing research with these data, how might you handle these cases?

4. Consider the firm with `gvkey` equal to `016476`. Write code to obtain the company name and CIK from the `company` table defined above. Using that CIK, find the 10-K filing for the year ending November 2003 on the SEC's website.[4] (*Note*: The 10-K will be released some time after `2003-11-30`.)

5. Looking at the 10-K you just found, what seems to be going on with the observation below?

```
funda_na_fixed |>
  filter(gvkey == "016476", datadate == "2003-11-30") |>
  select(datadate, at, at_calc, at_diff, act, ppent, intan, dc, aox)
```

datadate	at	at_calc	at_diff	act	ppent	intan	dc	aox
2003-11-30	2983.76	3038.56	54.80	1675.12	486.71	244.63	54.80	577.30

6. Using the approach above, we can find the 10-K for the observation examined in the following code on the SEC's EDGAR site.[5] What's going on with this case? What's the most significant difference between this case and the one above? (*Hint*: The following additional Compustat balancing model may help: `act = ppent + intan + ivao + ao`.)

```
funda_na_fixed |>
  filter(gvkey == "145003", datadate == "2001-12-31") |>
  select(datadate, at, at_calc, at_diff, act,
         ppent, intan, ivao, ao)
```

---

[4] https://www.sec.gov/edgar/searchedgar/companysearch.html
[5] https://go.unimelb.edu.au/fzw8

datadate	at	at_calc	at_diff	act	ppent	intan	ivao	ao
2001-12-31	6.636	7.252	0.616	4.451	1.062	0.761	0	0.978

8.4 Across-statement articulation

Under IAS 1 *Presentation of Financial Statements*, a complete set of financial statements includes the following four statements:

- A statement of financial position (i.e., a balance sheet)
- A statement of profit or loss and other comprehensive income (i.e., an income statement)
- A statement of changes in equity
- A statement of cash flows

Of these four statements, the balance sheet represents a statement of **stocks**, or balances at particular points of time, while the other three represent statements of **flows**, or changes in balances over time. Of the three flow statements, only the last two explicitly reconcile beginning and ending balance sheet items.[6] The statement of changes in equity "provides a reconciliation of the opening and closing amounts of each component of equity for the period" and the statement of cash flows performs an analogous function with respect to the opening and closing amounts of cash and cash equivalents.

Compustat's balancing models only address the balance sheet, income statement, and statement of cash flows. There is no balancing model for the statement of changes in equity on Compustat. So we only really have the statement of cash flows available to test the across-statement articulation within Compustat and therefore focus on that statement in the discussion below.

Before moving on to the cash flow statement, we examine the balancing model for cash on the balance sheet: ch + ivst = che, where the description of each item in the equation is given in Table 8.2.

TABLE 8.2
Components of cash balance

Item	Description
ch	Cash
ivst	Short-Term Investments
che	Cash and Short-Term Investments—Total

The wrinkle here can be seen in the output below. There are a few cases where we are missing the total (che), but we have values for one of its components (ivst or ch).

```
funda_mod |>
  select(gvkey, datadate, che, ch, ivst) |>
  count(missing_che = is.na(che),
        missing_ch = is.na(ch),
```

---

[6]While the income statement needs to articulate with retained earnings, it does not provide a complete account of changes in that balance sheet account.

```
    missing_ivst = is.na(ivst)) |>
  arrange(missing_che, missing_ch, missing_ivst)
```

missing_che	missing_ch	missing_ivst	n
False	False	False	415,745
False	False	True	1,208
False	True	False	4
False	True	True	62,333
True	False	True	74
True	True	False	61
True	True	True	93,718

While more digging might be appropriate to work out how best to handle these cases, below we take the simple expedient of using `ch` or `ivst` as the value for `che` when `che` is missing, but `ch` is not. The following suggests that this *modified* balancing model holds in *almost* every case where at least one of the components is not missing (though there is a non-trivial number of cases where `che`, `ch`, and `ivst` are *all* missing).

```
funda_mod |>
  select(gvkey, datadate, che, ch, ivst) |>
  mutate(missing_che = is.na(che),
         missing_che_comps = is.na(ch) & is.na(ivst),
         che = case_when(!is.na(che) ~ che,
                         !is.na(ivst) ~ ivst,
                         !is.na(ch) ~ ch),
         che_calc = coalesce(ch, 0) + coalesce(ivst, 0),
         che_diff = che_calc - che,
         che_balance = abs(che_diff) < tol) |>
  count(che_balance, missing_che, missing_che_comps) |>
  arrange(desc(n))
```

che_balance	missing_che	missing_che_comps	n
True	False	False	416,956
NA	True	True	93,718
False	False	True	62,075
True	False	True	258
True	True	False	135
False	False	False	1

Having explored the Compustat balancing model for cash on the balance sheet, we now consider the Compustat balancing model for the *annual* cash flow statement, which is expressed as follows: `oancf + ivncf + fincf + exre = chech`, where the description of each item in the equation is given in Table 8.3.

Let's collect some data to check whether the cash flow statement balancing model holds. The first step is to calculate its left-hand side, `chech_calc`, converting missing values for each of the four elements to zero. Here we use `across()` to modify multiple columns in

TABLE 8.3
Components of cash flow statement

Item	Description
oancf	Operating Activities—Net Cash Flow
ivncf	Investing Activities—Net Cash Flow
fincf	Financing Activities—Net Cash Flow
exre	Exchange Rate Effect
chech	Cash and Cash Equivalents—Increase (Decrease)

one step. (The `across()` function is covered in some detail in Chapter 26[7] of *R for Data Science*.) We then compare `chech_calc` and `chech` and, if these two values are essentially equal, we set `chech_balance` to `TRUE`.

```
funda_mod |>
  select(gvkey, datadate, oancf, ivncf, fincf, exre, chech) |>
  mutate(across(oancf:exre, \(x) coalesce(x, 0)),
         chech_calc = oancf + ivncf + fincf + exre,
         chech_diff = chech_calc - chech,
         chech_balance = abs(chech_diff) < tol) |>
  filter(!is.na(chech), chech_calc != 0) |>
  count(chech_balance)
```

chech_balance	n
True	307,401
False	525

So, there are relatively few cases where the first equation doesn't hold.

Note that `chech` is likely collected from the statement of cash flows itself, so this is really a within-statement relationship. This leaves the final question: Do the amounts provided in Compustat items related to the balance sheet tie to amounts provided in corresponding items related to the statement of cash flows? Here things get a little more complicated because we need to compare changes in cash implied by two balance sheets with `chech`. In practice, it turns out that, while in many cases `ch` ("cash") equals `che` ("cash and cash equivalents"), in cases where `ch` does *not* equal `che`, there are some firms whose statements of cash flows reconcile with changes in `ch` and other firms whose statements of cash flows reconcile with changes in `che`. Note that we loosen the tolerance a bit (`tol <- 0.1`) in this test to allow for rounding errors in the presentation of multiple financial statements.

Note that we use the **window function `lag()`** in this query. With a window function, the data are organized into **partitions** and the function operates on each partition independently and possibly uses data from multiple rows of data within that window. Here we construct partitions based on `gvkey` using `group_by(gvkey)` because we only want to consider values for a given firm. The data within each partition are ordered by `datadate` using `window_order()`. As its name suggests, the `lag()` function for each value simply returns the value for the previous element in the window.[8] We will see several window functions

---

[7]https://r4ds.hadley.nz/iteration#sec-across

[8]A careful reader might have noticed that we actually used the window function `fill()` in Chapter 2. However, there we used `arrange()` instead of `window_order()`. The `window_order()` function is only available

in this book, including `lead()` (the opposite of `lag()`), `row_number()` (the position of the
row in the window), and `cumsum()` (the cumulative sum of the window's values). A short
discussion of window functions is found in Chapter 21[9] of *R for Data Science.*

```
tol <- 0.1

funda_cf_balance <-
  funda_mod |>
  select(gvkey, datadate, oancf, ivncf, fincf,
         exre, chech, che, ch, ivst) |>
  mutate(across(oancf:exre, \(x) coalesce(x, 0)),
         chech_calc = oancf + ivncf + fincf + exre,
         chech_balance = abs(chech_calc - chech) < tol) |>
  filter(chech_calc != 0, chech_balance) |>
  group_by(gvkey) |>
  window_order(datadate) |>
  mutate(lag_datadate = lag(datadate),
         d_che = che - lag(che),
         d_ch = ch - lag(ch)) |>
  ungroup() |>
  filter(!is.na(d_che) | !is.na(d_ch)) |>
  mutate(artic_desc =
           case_when(abs(d_ch - d_che) < tol & abs(chech - d_che) < tol
                       ~ "Articulates (CHE == CH)",
                     abs(chech - d_che) < tol  ~ "Articulates using CHE",
                     abs(chech - d_ch) < tol   ~ "Articulates using CH",
                     abs(chech - d_che) >= tol ~ "Does not articulate",
                     abs(chech - d_ch) >= tol  ~ "Does not articulate",
                     .default = "Other"))
```

Below we see that there is a non-trivial number of cases where the cash flow statement
appears not to articulate with changes in cash balances on the balance sheet. We explore
some cases of this failure to articulate in the exercises.

```
funda_cf_balance |>
  count(artic_desc) |>
  rename(`Description` = artic_desc,
         `Number of cases` = n)
```

Description	Number of cases
Articulates using CHE	16,823
Articulates using CH	80,708
Articulates (CHE == CH)	138,373
Does not articulate	42,478

---

for remote data frames because it relies on the SQL back-end to provide functionality not available with
local data frames. Readers coming from an SQL background might observe that `dplyr`'s `group_by()` is
"overloaded" in the sense that it does the work of both the `GROUP BY` statement and the `PARTITION BY`
clause in SQL.

[9]https://r4ds.hadley.nz/databases.html

8.4.1 Exercises

1. In checking cash flow statement articulation, we used the `across()` function (the documentation `? dplyr::across` provides more detail). Rewrite the code to create `funda_na_fixed` above to use a similar approach. Check that you get the same results with respect to `count(balance)` as you get using the original code.

2. Consider the case of American Airlines (GVKEY: 001045) for the year ended 31 December 2020. Look up the relevant American Airlines 10-K on `sec.gov`. (*Hint*: You can get the CIK from `comp.company`.) What explains the gaps seen in the output below?

```
funda_cf_balance |>
  filter(datadate == "2020-12-31", gvkey == "001045") |>
  select(gvkey, datadate, chech, d_che, d_ch)
```

gvkey	datadate	chech	d_che	d_ch
001045	2020-12-31	109	3489	−35

8.5 Missing R&D

In this section, we discuss Koh and Reeb (2015), who "investigate whether missing R&D expenditures in financial statements indicates a lack of innovation activity. Patent records reveal that 10.5% of missing R&D firms file and receive patents, which is 14 times greater than zero R&D firms."

8.5.1 Discussion questions

1. What is the following code doing? Koh and Reeb (2015) use a similar filter. What is their rationale for this?

```
included_firms <-
  company |>
  mutate(sic = as.integer(sic)) |>
  filter(!between(sic, 4900, 4999), !between(sic, 6000, 6999)) |>
  select(gvkey)

rd_data <-
  funda_mod |>
  semi_join(included_firms, by = "gvkey") |>
  mutate(missing_rd = is.na(xrd),
         zero_rd = xrd == 0,
         year = year(datadate)) |>
  select(gvkey, datadate, year, xrd, missing_rd, zero_rd, at) |>
  collect()
```

2. Is there a balancing model that includes R&D spending?

3. Focusing on years 1980–2019, what have been the trends in the proportion of firms not reporting R&D and the "average" amount of R&D (scaled by assets)?[10] (In calculating the "average" does it make sense to use the `mean()` or `median()`?) Speculate as to what might explain these trends.

4. Koh and Reeb (2015) find that many "studies in *The Accounting Review* use R&D in their analysis and code the missing values as zero, implicitly assuming that blank R&D is equal to zero R&D, [while many] articles in the *Journal of Finance* [code] ... the blank values as zero and including a dummy variable to indicate blank R&D firms. ... In contrast, 42% of the studies in the *Strategic Management Journal* use R&D and they take a very different approach, often replacing the missing R&D values with either the industry average R&D, or a historical value from prior years." Why do scholars from different fields make such different choices? In light of Koh and Reeb (2015), do you think that one approach is more correct than the others?

5. Consider the 10-K filing[11] made by IHS Markit Ltd on 2021-01-22. Based on the information in the filing, do you think that IHS Markit engages in research and development activity? Does IHS Markit generate patents? Are there other forms of intellectual property protection that IHS Markit relies on? Does IHS Markit report an amount for research and development expenditure? Who is IHS Markit's external auditor? Do you think that IHS Markit is in violation of GAAP? Or is its reporting choice with respect to R&D a within-GAAP use of reporting discretion?

6. Koh and Reeb (2015) state that "our first set of tests compare patent activity between non-reporting R&D firms and firms that report zero R&D. ... We use both full sample and propensity score matched samples. ... While full sample tests allow for greater external validity, the matched sample tests potentially improve the local treatment effect. ... Based on the propensity score matched sample, our multivariate tests indicate that, on average, non-reporting R&D firms file about 14 times more patent applications than the matched zero R&D firms." In speaking of "treatment effects", Koh and Reeb (2015) implicitly view disclosure of R&D as a treatment variable and patent applications as an outcome. Does this make sense to you? Do you think that this is really what Koh and Reeb (2015) want to do? If not, what is the inference they are trying to draw?

7. What exactly is the take-away from the analysis in section 4.4 of Koh and Reeb (2015), which uses "the rapid demise of Arthur Andersen (AA) as a quasi-natural experiment"? For example, what is the treatment? What assumptions are needed to generalize from the specific treatment to a treatment of wider applicability? What additional analyses can you suggest that might provide additional assurance that the results reflect a causal effect of interest?

8. What are the suggested implications of the simulation analysis of section 4.5 of Koh and Reeb (2015)? What are the inherent limitations in a simulation analysis like this one?

---

[10]At the time of initial writing, 2020 was too incomplete to be meaningful.
[11]https://go.unimelb.edu.au/3cd8

9

Importing data

Thus far, our data have come from either the `farr` package or WRDS. In most cases, the WRDS data are nicely **rectangular**, meaning that we have N observations of K variables, and they already have assigned data types, etc.[1] In practice, researchers often obtain data from other sources, especially the internet, and such data are often rather messy. This chapter provides an introduction to importing data from such sources.

While this chapter is fairly task-oriented, we think that it serves to reinforce two deeper ideas.

- **Reproducibility.** This is a big idea—certainly bigger than we can cover in a chapter. But what we will do here will emphasize approaches that are completely reproducible, whether by people you don't know, co-authors, or even yourself at some subsequent date.
- **Tidy data.**[2] A lot of data we see out there—including the data we work with in this chapter—is essentially tabular and relational in nature. Once you recognize this, many data problems become much more straightforward.

Even dedicated users of SAS or Stata may find this chapter useful, as R handles tricky data-munging tasks quite easily and the end results are easily exported in formats that other software can read (e.g., `write_dta()` from the `haven` package creates Stata data files).

We also introduce **regular expressions**. Briefly speaking, a regular expression is a sequence of characters that define a pattern which can be used in a kind of search (or search-and-replace) on steroids. Regular expressions are very useful when working with data in many contexts.[3] The chapters on strings[4] and regular expressions[5] in *R for Data Science* provide excellent introductions that complement material in this chapter. We recommend that you refer to those chapters as you work through this chapter.

 Tip

The code in this chapter uses the packages listed below. For instructions on how to set up your computer to use the code found in this book, see Section 1.2. Quarto templates for the exercises below are available on GitHub.[a]

[a]https://github.com/iangow/far_templates/blob/main/README.md

---

[1]Some WRDS data sets have "incorrect" data types and additional work is needed to address these cases.
[2]https://tidyr.tidyverse.org/articles/tidy-data.html
[3]Regular expressions are available in pretty much every package, include Python, R, PostgreSQL, and SAS. Learning on one platform largely carries over to any other. Stata's support for regular expressions is much weaker than the other platforms'.
[4]https://r4ds.hadley.nz/strings.html
[5]https://r4ds.hadley.nz/regexps.html

DOI: 10.1201/9781003456230-9

```
library(dplyr)
library(tidyr)
library(readr)      # read_lines(), read_fwf(), etc.
library(stringr)    # str_c(), str_detect()
library(pdftools)   # pdf_text()
library(lubridate)  # ymd()
```

9.1 Reading (seemingly) non-tabular data

9.1.1 Fama-French industry definitions

The first task that we will explore is collecting data on Fama-French industry definitions, which are widely used in finance and accounting research to map SIC codes, of which there are hundreds, into a smaller number of industry groups for analysis.[6] For example, we might want to group firms into 48, 12, or even 5 industry groups.

The basic data on Fama-French industry definitions are available from Ken French's website[7] at Tuck School of Business.

There are multiple classifications, starting with 5 industries, then 10, 12, 17, 30, 38, 48, and finally 49 industries. The data are supplied as zipped text files. For example, the 48-industry data can be found on this page,[8] by clicking the link displayed as Download industry definitions.

If we download that linked file[9] and unzip it, we can open it in a text editor or even Excel. The first ten lines of the file are as follows:

```
 1 Agric  Agriculture
        0100-0199 Agricultural production - crops
        0200-0299 Agricultural production - livestock
        0700-0799 Agricultural services
        0910-0919 Commercial fishing
        2048-2048 Prepared feeds for animals

 2 Food   Food Products
        2000-2009 Food and kindred products
        2010-2019 Meat products
```

Looking at the second row, we interpret this as saying that firms with SIC codes between 0100 and 0199 are assigned to industry group 1 (let's call this field ff_ind), which has a label or short description (ff_ind_short_desc) Agric and a full industry description (ff_ind_desc) of Agriculture.

One approach to this task might be to write a function like the following (this one is woefully incomplete, as it only covers the first two lines of data above):

---

[6]According to https://siccode.com, "Standard Industrial Classification Codes (SIC Codes) identify the primary line of business of a company. It is the most widely used system by the US Government, public, and private organizations."

[7]http://mba.tuck.dartmouth.edu/pages/faculty/ken.french/data_library.html

[8]http://mba.tuck.dartmouth.edu/pages/faculty/ken.french/Data_Library/det_48_ind_port.html

[9]http://mba.tuck.dartmouth.edu/pages/faculty/ken.french/ftp/Siccodes12.zip

```
get_ff_ind_48 <- function(sic) {
  case_when(sic >= 100 & sic <= 199 ~ 1,
            sic >= 200 & sic <= 299 ~ 1)
}
```

While tedious and time-consuming, this is perfectly feasible. In fact, this is essentially the approach taken in code you can find on the internet (e.g., SAS code here[10] or here[11] or Stata code here[12]).

However, doing this would only solve the problem for the 48-industry grouping. And it certainly could not be described as particularly robust to, for example, changes in Fama-French industry definitions.[13]

Part of the solution that we use below recognizes that the data are really tabular in nature. A relational database purist would likely look at the data above as representing two tables. One table relates Fama-French industries to short and full industry descriptions. The first two rows in this table would look something like this:

ff_ind	ff_ind_short_desc	ff_ind_desc
1	Agric	Agriculture
2	Food	Food Products

The second table would relate Fama-French industries to ranges of SIC codes, and the first few rows of this table would look something like this:

ff_ind	sic_min	sic_max	sic_desc
1	0100	0199	Agricultural production - crops
1	0200	0299	Agricultural production - livestock
1	0700	0799	Agricultural services
1	0910	0919	Commercial fishing

To keep things simple for this exercise, we will disappoint the purists and make a single table with all six fields: `ff_ind`, `ff_ind_short_desc`, `ff_ind_desc`, `sic_min`, `sic_max`, `sic_desc`.[14]

So how do we make this table? One approach to this task might be to download the linked file,[15] unzip it, open it up in some program (e.g., Excel), and then massage the data manually into the desired form. But this would have the same issues as the approach above.

We can do better by using R and tools from the Tidyverse package. The first step is to download the data. While one can easily do this manually, but we want to automate this

---

[10]http://faculty.washington.edu/edehaan/pages/Programming/industries_ff48

[11]https://github.com/JoostImpink/fama-french-industry/blob/master/SAS/Siccodes48.sas

[12]http://fmwww.bc.edu/repec/bocode/s/sicff.ado

[13]While Fama-French industry definitions might not change very often, we will see other benefits from a more robust and general approach below.

[14]We justify this approach using the fact that these data sets are not large by any stretch and the assumption that, in general, only one of `ff_ind`, `ff_ind_short_desc` and `ff_ind_desc` actually gets used in practice.

[15]http://mba.tuck.dartmouth.edu/pages/faculty/ken.french/ftp/Siccodes48.zip

process as much as possible. And we probably don't have any reason to keep the `.zip` file once we have used it. R provides two functions that we can use here: `download.file()` and `tempfile()`. The `tempfile()` function creates a random file name in a location that will be cleaned up by our system automatically once we're no longer using it.

```
t <- tempfile(fileext = ".zip")
url <- str_c("http://mba.tuck.dartmouth.edu/pages/faculty/",
             "ken.french/ftp/Siccodes48.zip")
```

Here `t` is *file13b985739c5ec.zip*, which is random except for the `.zip` extension, something we need for our code to recognize the supplied file as a zipped file. The `download.file()` function downloads the file at `url` and saves it as `t`.

```
download.file(url, t)
```

If you look at Ken French's website, you will see that all the industry-definition files have URLs that follow a certain pattern, with just the number of industry groups (in this case, 48) changing. Recognizing this, we can rewrite the code above as follows:

```
ind <- 48
t <- tempfile(fileext = ".zip")
url <- str_c("http://mba.tuck.dartmouth.edu",
             "/pages/faculty/ken.french/ftp/Siccodes", ind, ".zip")
download.file(url, t)
```

Here the `str_c()` function from the `stringr` package pastes all the pieces back together.[16]

From visual inspection, we can see that our text file is a **fixed-width format** text file. So to read the data, we will use the function `read_fwf()` from the `readr` package.

There are two required arguments to `read_fwf()`: `file` and `col_positions`. From the help for `read_fwf()` (type `? readr::read_fwf` in the R console to see this), we see that `col_positions` refers to "Column positions, as created by `fwf_empty()`, `fwf_widths()` or `fwf_positions()`. If the width of the last column is variable (i.e., we have a **ragged** fixed-width format file), we can supply the last end position as `NA`." We can also see that `fwf_widths()` is itself a function: `fwf_widths(widths, col_names = NULL)`.

Given that we have a very simple file, we can identify the column widths pretty easily. Manually adding a "ruler" of sorts at the top of the file, we can see below that the first column covers columns 1-3, the second column covers 4-10, and the third column starts at 11.

```
123456789-123456789-123456789-123456789-123456789-...
 1 Agric   Agriculture
         0100-0199 Agricultural production - crops
         0200-0299 Agricultural production - livestock
         0700-0799 Agricultural services
```

So we have widths of 3, 7, and we can use `NA` to have R figure out the width of the last column. The first two columns should be named `ff_ind` and `ff_ind_short_desc`, but the third column is problematic, as some rows provide information on `ff_ind_short_desc` and some rows provide data that we will eventually put into `sic_min` and `sic_max`; so let's call that column `temp` for now. Finally, as the first column contains integer values, while the

---

[16]Note that an alternative to `str_c()` would be `paste0()` from base R; we use `str_c()` here because the `stringr` functions seem easier to learn as a group than their base R equivalents.

other two are text columns, we can supply a string to the option `col_types` argument of
`read_fwf()` to ensure that the columns are read as those types.

```
t |>
  read_fwf(col_positions = fwf_widths(c(3, 7, NA),
                                      c("ff_ind", "ff_ind_short_desc",
                                        "temp")),
           col_types = "icc")
```

```
# A tibble: 693 x 3
   ff_ind ff_ind_short_desc temp
    <int> <chr>             <chr>
 1      1 Agric             Agriculture
 2     NA <NA>              0100-0199 Agricultural production - crops
 3     NA <NA>              0200-0299 Agricultural production - livestock
 4     NA <NA>              0700-0799 Agricultural services
 5     NA <NA>              0910-0919 Commercial fishing
 6     NA <NA>              2048-2048 Prepared feeds for animals
 7     NA <NA>              <NA>
 8      2 Food              Food Products
 9     NA <NA>              2000-2009 Food and kindred products
10     NA <NA>              2010-2019 Meat products
# i 683 more rows
```

The next step is to handle the problematic `temp` column. We can see that if `ff_ind` is NA,
then `temp` contains (if it contains anything) a range of SIC codes, but if `ff_ind` is *not*
NA, then `temp` contains the value we want to store in `ff_ind_desc`. We can use `mutate`
along with `if_else` statements to extract the data into the appropriate columns (we'll call
the range of SIC codes `sic_range` for now) and, having done so, we can drop the column
`temp`.[17]

```
t |>
  read_fwf(col_positions = fwf_widths(c(3, 7, NA),
                                      c("ff_ind", "ff_ind_short_desc",
                                        "temp")),
           col_types = "icc") |>
  mutate(ff_ind_desc = if_else(!is.na(ff_ind), temp, NA),
         sic_range = if_else(is.na(ff_ind), temp, NA)) |>
  select(-temp)
```

```
# A tibble: 693 x 4
   ff_ind ff_ind_short_desc ff_ind_desc sic_range
    <int> <chr>             <chr>       <chr>
 1      1 Agric             Agriculture <NA>
 2     NA <NA>              <NA>        0100-0199 Agricultural productio~
 3     NA <NA>              <NA>        0200-0299 Agricultural productio~
 4     NA <NA>              <NA>        0700-0799 Agricultural services
 5     NA <NA>              <NA>        0910-0919 Commercial fishing
 6     NA <NA>              <NA>        2048-2048 Prepared feeds for ani~
```

---

[17]Throughout this section, we will repeat code, but in practice, we would simply add additional lines as we
work through the code. This is simply a way to represent the idea of working through the data interactively,
something that the Tidyverse makes very easy.

```
7      NA <NA>          <NA>           <NA>
8       2 Food          Food Products <NA>
9      NA <NA>          <NA>           2000-2009 Food and kindred produ~
10     NA <NA>          <NA>           2010-2019 Meat products
# i 683 more rows
```

We are getting closer. Now, we see that the issue is that our `sic_range` column does not line up with the other three columns. To solve this, we can use `fill()` from the `tidyr` package. The fill function accepts arguments for the columns to "fill". In this case, we can fill missing values with the previous non-missing value.

```
t |>
  read_fwf(col_positions = fwf_widths(c(3, 7, NA),
                                      c("ff_ind", "ff_ind_short_desc",
                                        "temp")),
           col_types = "icc") |>
  mutate(ff_ind_desc = if_else(!is.na(ff_ind), temp, NA),
         sic_range = if_else(is.na(ff_ind), temp, NA)) |>
  select(-temp) |>
  fill(ff_ind, ff_ind_short_desc, ff_ind_desc)
```

```
# A tibble: 693 x 4
   ff_ind ff_ind_short_desc ff_ind_desc    sic_range
    <int> <chr>             <chr>          <chr>
1       1 Agric             Agriculture    <NA>
2       1 Agric             Agriculture    0100-0199 Agricultural productio~
3       1 Agric             Agriculture    0200-0299 Agricultural productio~
4       1 Agric             Agriculture    0700-0799 Agricultural services
5       1 Agric             Agriculture    0910-0919 Commercial fishing
6       1 Agric             Agriculture    2048-2048 Prepared feeds for ani~
7       1 Agric             Agriculture    <NA>
8       2 Food              Food Products <NA>
9       2 Food              Food Products 2000-2009 Food and kindred produ~
10      2 Food              Food Products 2010-2019 Meat products
# i 683 more rows
```

At this point, we have no further use for the rows where `sic_range` is NA, so we can filter them out.

```
t |>
  read_fwf(col_positions = fwf_widths(c(3, 7, NA),
                                      c("ff_ind", "ff_ind_short_desc",
                                        "temp")),
           col_types = "icc") |>
  mutate(ff_ind_desc = if_else(!is.na(ff_ind), temp, NA),
         sic_range = if_else(is.na(ff_ind), temp, NA)) |>
  select(-temp) |>
  fill(ff_ind, ff_ind_short_desc, ff_ind_desc) |>
  filter(!is.na(sic_range))
```

```
# A tibble: 598 x 4
   ff_ind ff_ind_short_desc ff_ind_desc    sic_range
    <int> <chr>             <chr>          <chr>
```

```
1       1 Agric        Agriculture    0100-0199 Agricultural productio~
2       1 Agric        Agriculture    0200-0299 Agricultural productio~
3       1 Agric        Agriculture    0700-0799 Agricultural services
4       1 Agric        Agriculture    0910-0919 Commercial fishing
5       1 Agric        Agriculture    2048-2048 Prepared feeds for ani~
6       2 Food         Food Products  2000-2009 Food and kindred produ~
7       2 Food         Food Products  2010-2019 Meat products
8       2 Food         Food Products  2020-2029 Dairy products
9       2 Food         Food Products  2030-2039 Canned & preserved fru~
10      2 Food         Food Products  2040-2046 Flour and other grain ~
# i 588 more rows
```

The last issue to address is the column `sic_range`. We want to split that into three target columns (`sic_min`, `sic_max`, and `sic_desc`). To do this, we can use `extract()` from the `tidyr` package. The two required arguments for `extract()` are `col`, the column from which data are being extracted, and `into`, the columns that will get the data.

The `regex` argument to `extract()` allows us to specify a regular expression that is used to split the data. We will discuss regular expressions more carefully later in the chapter, so it is fine if you don't fully follow what's going on here. For now, we just need to know that the portions contained in each pair of parentheses (`(` and `)`) are what is **captured** for each field. The first pair of parentheses surround `[0-9]+`, which we can read as "one or more digit characters". So, for the case of `0100-0199 Agricultural production - crops`, this would match `0100`. This is followed by a `-` which is a literal dash that matches the `-` after `0100`. The next pair of parentheses is again "one or more digit characters" and captures `0199`. This is followed by `\\s*`. The `\\s` represents "any space" and the `*` means "zero or more", so `\\s*` means "zero or more spaces", which are matched, but not captured. Finally we have `(.*)$`. The `.` represents "any character", so `.*` means "zero or more of any character", which are captured as the third variable. The `$` just means "the end of the string"; this mirrors the `^` at the start of the regular expression, which means "the start of the string". Together, `^` and `$` ensure that we are considering the whole string in our analysis.[18]

The argument `convert = TRUE` asks `extract()` to convert the data types of the extracted fields to types that seem appropriate (e.g., integers for `sic_min` and `sic_max`).

```
t |>
  read_fwf(col_positions = fwf_widths(c(3, 7, NA),
                                      c("ff_ind", "ff_ind_short_desc",
                                        "temp")),
           col_types = "icc") |>
  mutate(ff_ind_desc = if_else(!is.na(ff_ind), temp, NA),
         sic_range = if_else(is.na(ff_ind), temp, NA)) |>
  select(-temp) |>
  fill(ff_ind, ff_ind_short_desc, ff_ind_desc) |>
  filter(!is.na(sic_range)) |>
  extract(sic_range,
          into = c("sic_min", "sic_max", "sic_desc"),
          regex = "^([0-9]+)-([0-9]+)\\s*(.*)$",
          convert = TRUE)
```

```
# A tibble: 598 x 6
```

---

[18]It seems that omitting `^` and `$` has no effect in this case.

ff_ind	ff_ind_short_desc	ff_ind_desc	sic_min	sic_max	sic_desc
<int>	<chr>	<chr>	<int>	<int>	<chr>
1	1 Agric	Agriculture	100	199	Agricultural pro~
2	1 Agric	Agriculture	200	299	Agricultural pro~
3	1 Agric	Agriculture	700	799	Agricultural ser~
4	1 Agric	Agriculture	910	919	Commercial fishi~
5	1 Agric	Agriculture	2048	2048	Prepared feeds f~
6	2 Food	Food Products	2000	2009	Food and kindred~
7	2 Food	Food Products	2010	2019	Meat products
8	2 Food	Food Products	2020	2029	Dairy products
9	2 Food	Food Products	2030	2039	Canned & preserv~
10	2 Food	Food Products	2040	2046	Flour and other ~

```
# i 588 more rows
```

Lastly, we can put all of the above into a function. But, as we do so, let's take a gamble that the same code will work for any of the Fama-French industry classifications if we only change the URL. To do this, we use an argument `ind` that reflects the industry grouping of interest and inserts that in the URL. Here we use the `str_c()` function from the `stringr` package to create the URL using the value supplied as `ind`.

```
get_ff_ind <- function(ind) {
  t <- tempfile(fileext = ".zip")
  url <- str_c("https://mba.tuck.dartmouth.edu/pages/",
               "faculty/ken.french/ftp/Siccodes", ind, ".zip")
  download.file(url, t)

  t |>
    read_fwf(col_positions = fwf_widths(c(3, 7, NA),
                                        c("ff_ind", "ff_ind_short_desc",
                                          "temp")),
             col_types = "icc") |>
    mutate(ff_ind_desc = if_else(!is.na(ff_ind), temp, NA),
           sic_range = if_else(is.na(ff_ind), temp, NA)) |>
    select(-temp) |>
    fill(ff_ind, ff_ind_short_desc, ff_ind_desc) |>
    filter(!is.na(sic_range)) |>
    extract(sic_range,
            into = c("sic_min", "sic_max", "sic_desc"),
            regex = "^([0-9]+)-([0-9]+)\\s*(.*)$",
            convert = TRUE)
}
```

We can test out for 48-industry classification above (just to make sure we didn't mess up what we already had working):

```
get_ff_ind(48)
```

```
# A tibble: 598 x 6
```

ff_ind	ff_ind_short_desc	ff_ind_desc	sic_min	sic_max	sic_desc
<int>	<chr>	<chr>	<int>	<int>	<chr>
1	1 Agric	Agriculture	100	199	Agricultural pro~
2	1 Agric	Agriculture	200	299	Agricultural pro~
3	1 Agric	Agriculture	700	799	Agricultural ser~

```
4      1 Agric          Agriculture         910     919 Commercial fishi~
5      1 Agric          Agriculture        2048    2048 Prepared feeds f~
6      2 Food           Food Products      2000    2009 Food and kindred~
7      2 Food           Food Products      2010    2019 Meat products
8      2 Food           Food Products      2020    2029 Dairy products
9      2 Food           Food Products      2030    2039 Canned & preserv~
10     2 Food           Food Products      2040    2046 Flour and other ~
# i 588 more rows
```

And now let's try it for 5- and 12-industry classifications:

```
get_ff_ind(5)
```

```
# A tibble: 58 x 6
   ff_ind ff_ind_short_desc ff_ind_desc                sic_min sic_max sic_desc
    <int> <chr>             <chr>                        <int>   <int> <chr>
 1      1 Cnsmr             Consumer Durables, No~         100     999 ""
 2      1 Cnsmr             Consumer Durables, No~        2000    2399 ""
 3      1 Cnsmr             Consumer Durables, No~        2700    2749 ""
 4      1 Cnsmr             Consumer Durables, No~        2770    2799 ""
 5      1 Cnsmr             Consumer Durables, No~        3100    3199 ""
 6      1 Cnsmr             Consumer Durables, No~        3940    3989 ""
 7      1 Cnsmr             Consumer Durables, No~        2500    2519 ""
 8      1 Cnsmr             Consumer Durables, No~        2590    2599 ""
 9      1 Cnsmr             Consumer Durables, No~        3630    3659 ""
10      1 Cnsmr             Consumer Durables, No~        3710    3711 ""
# i 48 more rows
```

```
get_ff_ind(12)
```

```
# A tibble: 49 x 6
   ff_ind ff_ind_short_desc ff_ind_desc                sic_min sic_max sic_desc
    <int> <chr>             <chr>                        <int>   <int> <lgl>
 1      1 NoDur             Consumer Nondurables ~         100     999 NA
 2      1 NoDur             Consumer Nondurables ~        2000    2399 NA
 3      1 NoDur             Consumer Nondurables ~        2700    2749 NA
 4      1 NoDur             Consumer Nondurables ~        2770    2799 NA
 5      1 NoDur             Consumer Nondurables ~        3100    3199 NA
 6      1 NoDur             Consumer Nondurables ~        3940    3989 NA
 7      2 Durbl             Consumer Durables -- ~        2500    2519 NA
 8      2 Durbl             Consumer Durables -- ~        2590    2599 NA
 9      2 Durbl             Consumer Durables -- ~        3630    3659 NA
10      2 Durbl             Consumer Durables -- ~        3710    3711 NA
# i 39 more rows
```

9.1.2 Exercises

1. Follow the steps below to produce a data set where each column has the appropriate data type.

    a. Go to the MSCI GICS website.[19]

---

[19]https://www.msci.com/gics

    b. Get the link to the file under *historical GICS structures* that is "Effective until Sep 28, 2018".

    c. Use this link and the `tempfile()` and `download.file()` functions to download the linked file.

    d. Use `read_excel()` from the `readxl` package to read the downloaded file.

    e. Identify any variables that need to be handled like `temp` in the Fama-French data set above and process accordingly.

    f. Use the `fill()` function from the `tidyr` package to fill in rows as necessary.

    g. Make sure that each column has the appropriate data type.

Hints:

    a. You may find it helpful to look at the Excel file so you can see how the data are structured.

    b. The `read_excel()` function has `skip` and `col_names` arguments that you will probably want to use.

    c. Your *final* column names should be `sector`, `sector_desc`, `ind_group`, `ind_group_desc`, `industry`, `industry_desc`, `sub_ind`, `sub_ind_desc`, and `sub_ind_details`.

    d. The following code snippets *might* be useful:

- `filter(!is.na(sub_ind_details))`
- `fill(sector:sub_ind_desc, .direction = "down")`
- `mutate(across(where(is.numeric), as.integer))`

9.2 Extracting data from messy formats

Sometimes data are provided in formats even messier than fixed-width text files. For example, we may want to extract data (perhaps tabular data) from a PDF. While the data may appear tabular to our eyes, the reality is that PDFs retain very little information about the structure of data, as the PDF format is a lightweight way of making a computer or printer present text for consumption by *humans*, not statistical software packages.

In this section, we will examine a case study in extracting tabular data from a PDF. Our initial focus will be on extracting data about money "left on the table" by firms in initial public offerings (IPOs). These data are provided by Jay Ritter at the University of Florida in a PDF found here.[20]

In this case, we will use `pdf_text()` from the `pdftools` package. From the help for `pdf_text()`, we learn that "the `pdf_text` function renders all text boxes on a text canvas and returns a character vector of equal length to the number of pages in the PDF file." We also learn that the function accepts an argument for the "pdf file path or raw vector with pdf data". Since the above URL actually provides an absolute file path for the PDF, we can simply call `pdf_text(url)` to download the PDF from the url and convert it to a text representation in R.

```
url <- str_c("https://site.warrington.ufl.edu/ritter/files/",
             "money-left-on-the-table.pdf")
```

---

[20]https://site.warrington.ufl.edu/ritter/files/money-left-on-the-table.pdf

Looking at the PDF, we see that the first page is text and the table starts on page 2. We can omit the first page returned by `pdf_text(url)` below by appending `[-1]`.

```
output <- pdf_text(url)[-1]
```

While the output from `pdf_text(url)` or even `pdf_text(url)[-1]` uses up too much space to be shown here, if you run the code immediately above, then call each of these alternatives, you can see that the second function returns data that *appears* quite regular. In this case, it makes sense to run the text through `read_lines()` from `readr`. We will do this and store the result in `temp` so that we can examine it here more closely.[21]

```
temp <-
  output |>
  read_lines()
```

Let's look at a few selected lines. Lines `1:5` represent the top of the table, with the column names unfortunately split into two rows.[22]

```
print_width <- 70
str_sub(temp[1:5], 1, print_width)
```

```
[1] " Dollar amount                                IPO     Offer   F"
[2] "left on the table    Company                  date    Price   m"
[3] ""
[4] "$5,075,000,000       Visa                     080319  $44.00  $"
[5] "$3,937,028,063       Airbnb                   201210  $68.00  $"
```

Rather than trying to deal with the column-names-spread-over-two-rows issue with code, we can just manually specify the column names and skip the first two rows when we import the data. For now, we merely create variables to reflect those choices; we will use these variables later.

```
col_names <-  c("amount_left_on_table", "company", "ipo_date", "offer_price",
                "first_close_price", "shares_offered", "ticker")
skip_rows <- 2
```

While the top portion of the table looks like it might be able to work with `read_fwf()`, once we look at the lines 56 through 62 at the bottom of the first page and top of the second pages (pp. 2–3 of the original PDF), we can see that they are not aligned.

```
str_sub(temp[56:62], 1, print_width)
```

```
[1] " $541,875,000        Charter Communications*   991109  $19.00  $"
[2] " $541,328,968        Blackstone Group LP        070621  $31.00  $"
[3] ""
[4] "                                               2"
[5] "$540,929,382  XP Inc            191211  $27.00  $34.46   "
[6] "$538,630,000  Gigamedia         000218  $27.00  $88.00   "
[7] "$531,999,999  CureVac B.V.      200814  $16.00  $55.90   "
```

The same is true of lines 114 through 120 at the bottom of the second page and top of the third pages (pp. 3–4 of the original PDF).

---

[21]Again, you may find it useful to inspect the full contents of `temp` yourself. We don't do that here due to space constraints.

[22]We only print the first 70 characters of each row to keep the output on the page.

```
str_sub(temp[114:120], 1, print_width)
```

```
[1] ""
[2] "$360,000,000    Niku                          000229    $24.00    $69.00    "
[3] ""
[4] "                                                  3"
[5] "$355,384,125    Lucent Technologies*         960403    $27.00    $30.625   "
[6] "$354,046,000    Fitbit                       150618    $20.00    $29.68    "
[7] "$352,400,000    Tradeweb Markets             190404    $27.00    $35.81    "
```

Finally, rows 461 through 463 contain a footer that is not part of the table.

```
str_sub(temp[458:464], 1, print_width)
```

```
[1] " $128,724,000    HCA Holdings                      110310    $30.00    $"
[2] " $128,562,500    Netscape Communications*          950808    $28.00    $"
[3] ""
[4] "Source: Prof. Jay R. Ritter, Warrington College of Business, Universit"
[5] "jay.ritter@warrington.ufl.edu). Data come from IPOScoop.com, Bloomberg"
[6] "WSJ.com, and the S.E.C.'s Edgar electronic database of 424B forms (fin"
[7] ""
```

We can flag that by looking for `Source:` at the start of a row. To match at the start of a string, we need to **anchor** the regular expression using ^ to match the start.[23] We want to read only up to one row before 461, and we will be skipping `skip_rows` at the start. So we can calculate the maximum number of rows we want to read as follows:

```
max_rows <- str_which(temp, "^Source:") - 1 - skip_rows
```

Also, as we will use Tidyverse tools to munge the data, we will find it useful to put the data in a tibble, albeit one with just a single column, which we call `temp`.

```
ritter_data_raw <-
  output |>
  read_lines(skip = skip_rows, n_max = max_rows) |>
  tibble(temp = _)
```

Coming as it does from a PDF, there are some messy elements of the data.

First, there are empty rows, which we can detect by matching on ^$ (i.e., the start and end with nothing in between) and we will want to filter out these rows.

```
ritter_data_raw |>
  filter(str_detect(temp, "^$"))
```

```
# A tibble: 49 x 1
   temp
   <chr>
 1 ""
 2 ""
 3 ""
 4 ""
 5 ""
```

---

[23]Note that in other contexts, such as inside [and], ^ will act as a kind of "not" operator.

```
 6 ""
 7 ""
 8 ""
 9 ""
10 ""
# i 39 more rows
```

Second, there are rows with just spaces and page numbers, which we can match with:

```
ritter_data_raw |>
   filter(str_detect(temp, "^\\s+\\d+$"))
```

```
# A tibble: 7 x 1
  temp
  <chr>
1 "                                    2"
2 "                          3"
3 "                          4"
4 "                          5"
5 "                          6"
6 "                              7"
7 "                          8"
```

So we want to filter out these rows too:

```
ritter_data_raw <-
  output |>
    read_lines(skip = 2, n_max = max_rows) |>
    tibble(temp = _) |>
    filter(!str_detect(temp, "^$"),
           !str_detect(temp, "^\\s+\\d+$"))
```

The next step will be, as before, to use **extract()** from **tidyr** and a regular expression to arrange the data into columns. However, the regular expression that we will need to use will be a bit more complicated than the one above. If it weren't for the second column, we could use a function from the **readr** package such as **read_delim()** that is designed to read **delimited** text data, such as comma-separated values (CSVs), tab-separated values, or (most relevant here) values separated by spaces.

Unfortunately, the second column (company) contains spaces (e.g., United Parcel Service*), so this would be read as three columns (for this row, at least).[24]

Fortunately, company is the only column with embedded spaces and it is followed by a column (ipo_date) that is strictly six digits (it has the form yymmdd, where yy is the last two digits of the year, mm are the month digits, and dd represents the date of the month). So we can use this to effectively "delimit" the company column from the rest of the data.

The first column (amount_left_on_table) contains non-space characters, which we can represent as [^\\s]. Here the ^ functions to negate the expression following it \\s, which is the general way of representing spaces in regular expressions. So [^\\s]+ denotes "one or more non-space characters" and embedding this in parentheses (i.e., as ([^\\s]+)) allows us to **capture** the matching characters. The second column could contain pretty much

---

[24]If the values in company had been **quoted**, e.g., as "United Parcel Service*", then reading the data as space-delimited would work. But these are "wild" data from a PDF and such niceties cannot be expected.

anything (that is . in regex terms), followed by one or more spaces, which we could represent as (.+)\\s+. The third column (ipo_date) can be captured by a regular expression for six digits (again followed by one or more spaces that we don't capture): ([0-9]{6})\\s+. The next four columns are like the first column, so we can use ([^\\s]+) to capture these, with all but the last column being followed by one or more spaces (\\s+). One issue that is not evident initially is that some, but not all, rows begin with spaces. We don't want to capture these spaces, but we want to allow for them to be there, so we can use ^\\s* to represent "the start of the line followed by zero or more spaces" to do this.

Now we can put all of this together as follows:

```
regex <- str_c("^\\s*",            # Start string (perhaps followed by spaces)
               "([^\\s]+)\\s+",    # Non-space characters (followed by spaces)
               "(.+)\\s+",         # Any characters, which may include spaces
                                   #  (followed by spaces)
               "([0-9]{6})\\s+",   # Six digits (followed by spaces)
               "([^\\s]+)\\s+",    # Non-space characters (followed by spaces)
               "([^\\s]+)\\s+",    # Non-space characters (followed by spaces)
               "([^\\s]+)\\s+",    # Non-space characters (followed by spaces)
               "([^\\s]+)",        # Non-space characters
               "$")                # End of string
```

We can then run this through extract().

```
ritter_data_raw |>
    # Here we use the regular expression to split the data into columns
    extract(temp, col_names, regex)
```

```
# A tibble: 402 x 7
   amount_left_on_table company     ipo_date offer_price first_close_price
   <chr>                <chr>       <chr>    <chr>       <chr>
 1 $5,075,000,000       "Visa     ~ 080319   $44.00      $56.50
 2 $3,937,028,063       "Airbnb    ~ 201210   $68.00      $144.71
 3 $3,750,040,000       "Snowflake ~ 200916   $120        $253.93
 4 $3,477,690,000       "Rivian Aut~ 211110  $78.00      $100.73
 5 $2,887,830,000       "DoorDash  ~ 201209   $102        $189.51
 6 $1,852,500,000       "Coupang   ~ 210311   $35.00      $49.25
 7 $1,586,300,000       "United Par~ 991110   $50.00      $68.125
 8 $1,540,730,469       "Corvis    ~ 000728   $36.00      $84.71875
 9 $1,496,000,000       "Snap      ~ 170302   $17.00      $24.48
10 $1,365,500,000       "Bumble    ~ 210211   $43.00      $70.31
# i 392 more rows
# i 2 more variables: shares_offered <chr>, ticker <chr>
```

We are getting close. Now we want to convert the fields amount_left_on_table, offer_price, first_close_price, and shares_offered to numerical values. For this task, the parse_number() function from readr is ideal. Note that we can use across() to apply a single function parse_number() to multiple columns (in this case, all_of() these four columns) and, by default, the results will replace the values previously found in those columns. And we also want to convert ipo_date to an actual date, which we can do using ymd() from the lubridate package. At this stage, we store the result in a data frame ritter_data.

```
ritter_data <-
  ritter_data_raw |>
  # Here we use the regular expression to split the data into columns
  extract(temp, col_names, regex) |>
  # Finally, fix up the data types of the columns
  mutate(across(all_of(c("amount_left_on_table", "first_close_price",
                         "offer_price", "shares_offered")),
             parse_number),
       ipo_date = ymd(ipo_date),
       company = str_trim(company))

ritter_data
```

```
# A tibble: 402 x 7
   amount_left_on_table company   ipo_date   offer_price first_close_price
                  <dbl> <chr>     <date>           <dbl>             <dbl>
1            5075000000 Visa      2008-03-19          44              56.5
2            3937028063 Airbnb    2020-12-10          68             145.
3            3750040000 Snowflake 2020-09-16         120             254.
4            3477690000 Rivian Au~ 2021-11-10          78             101.
5            2887830000 DoorDash  2020-12-09         102             190.
6            1852500000 Coupang   2021-03-11          35              49.2
7            1586300000 United Pa~ 1999-11-10          50              68.1
8            1540730469 Corvis    2000-07-28          36              84.7
9            1496000000 Snap      2017-03-02          17              24.5
10           1365500000 Bumble    2021-02-11          43              70.3
# i 392 more rows
# i 2 more variables: shares_offered <dbl>, ticker <chr>
```

At this stage, we should check if we have any rows with NA values, as the existence of such rows would suggest possible parsing issues.

```
ritter_data |>
  filter(if_any(.cols = everything(), .fns = is.na))
```

```
# A tibble: 0 x 7
# i 7 variables: amount_left_on_table <dbl>, company <chr>,
#   ipo_date <date>, offer_price <dbl>, first_close_price <dbl>,
#   shares_offered <dbl>, ticker <chr>
```

We have none!

9.2.1 Exercises

1. In the PDF, Ritter states "IPOs marked with an * also had international tranches (and probably some others, too)." Write code to remove this "*" when present and create an additional column `intl_tranche` that indicates an international tranche. (*Hint*: The `extract()` function can be used here. In a first pass, you might use `into = c("company", "intl_tranche")` and `regex = "^(.*?)(\\*?)$"` with this function. Can you see what the ? in .*? is doing?

There is some explanation here.[25] Does the regex work without this ?? What does *? match?)

2. Ritter defines *money left on the table* as "the difference between the closing price on the first day of trading and the offer price, multiplied by the number of shares sold." Can you calculate this from the data provided? Is the calculated amount (`amount`) equal to the amount in `amount_left_on_table` in each case? What explains the differences? (*Hints*: There will be more than one reason. You may find it helpful to calculate `ratio = amount/amount_left_on_table` and to focus on differences of more than 1% with `filter(abs(ratio - 1) > 0.01)`.)

3. In words, what do each of the following regular expressions *match*? What do they *capture*?[26]

   - `"^\\s*"`
   - `"(.+)\\s+"`
   - `"([^\\s]+)\\s+"`
   - `"([0-9]{1,2}/[0-9]{4})\\s+"`
   - `"([0-9,]+)\\s+"`

4. The online appendix[27] for "The Customer Knows Best: The Investment Value of Consumer Opinions" contains Table OA.1 entitled *List of firms with customer reviews on Amazon.com*. Using an approach similar to that we used for `ritter_data` above, create a data frame `huang_data` with columns `company_name`, `industry`, `start`, `end`, `months`, `reviews`.

   - `months` and `reviews` should be numerical values.
   - `start` and `end` should be dates (use the first day of the month if only the month and year is specified).
   - You can solve this by combining the partial regular expressions above into a single regular expression (in the way we combined partial regular expressions above for `ritter_data` using `str_c()`).

5. Using the following code, where the URL provides a sample of observations on mergers and acquisitions from SDC, create a first-pass import of the data.[28] What is the first issue you see when you look at the data in `ma_sdc`? (*Hint*: Look at the first five rows.) Adapt the code to address this issue. (*Hint*: You may have to experiment with different values for the `skip` argument to get the right setting.)

```
col_names <- c("date_announced", "date_effective", "tgt_name",
               "tgt_nation", "acq_name", "acq_nation",
               "status", "pct_of_shares_acq", "pct_owned_after_transaction",
               "acq_cusip", "tgt_cusip", "value_of_transaction_mil",
               "acq_prior_mktval", "tgt_prior_mktval",
               "acq_nation_code", "tgt_nation_code")

url <- str_c("https://gist.githubusercontent.com/iangow/",
             "eb7dfe1cd0913821429bdf0566465d41/raw/",
             "358d60a4429f5747abc61f8acc026d335fc165f3/sap_sample.txt")
```

---

[25]https://stackoverflow.com/questions/3075130/what-is-the-difference-between-and-regular-expressions
[26]https://javascript.info/regexp-groups
[27]https://www.dropbox.com/s/58ba7d3wg9q2ijw/Huang_app.pdf?dl=1
[28]We messed with these data, so these cannot be used for research! But they are a realistic representation of an actual data set.

```
ma_sdc_file <- tempfile()
download.file(url, ma_sdc_file)
```

```
ma_sdc_cols <- fwf_empty(ma_sdc_file, col_names = col_names)
ma_sdc <- read_fwf(ma_sdc_file, col_positions = ma_sdc_cols)
```

6. Open the file found at `url` in your browser (`browseURL(url)` will help here) and locate the row containing the word `Coffey`. What do you see there as relates to the variable `status`? How does this compare with what you see in `status` if you filter using `tgt_name == "Coffey International Ltd"`)? What do you think has happened here? How can setting a value for `n` in `fwf_empty()` help here? (*Hint*: Using `which(ma_sdc$status == "Unconditi")` might help here.)

7. Using an appropriate function from the `lubridate` package, fix the variables `date_announced` and `date_effective` so that they have type `Date`.

8. What are the minimum and maximum values of `date_announced` and `date_effective`? What explains missing values (if any) here?

9. What do you observe about `acq_cusip` and `tgt_cusip`? Can you write some code to check that these variables have been read in correctly? (*Hint*: The function `str_length()` might be useful here.)

9.3 Further reading

Chapter 7[29] of *R for Data Science* provides an introduction to importing data, including functions such as `read_csv()` and `read_fwf()`. Chapter 14[30] in *R for Data Science* covers strings and Chapter 15[31] covers regular expressions.

The topic of regular expressions is surprisingly deep and regular expressions are useful in more contexts than might be apparent when first learning about them. Friedl (2006) provides a deep treatment of the topic. Goyvaerts and Levithan (2009) is full of examples of solutions to common cases.

---

[29] https://r4ds.hadley.nz/data-import.html
[30] https://r4ds.hadley.nz/strings.html
[31] https://r4ds.hadley.nz/regexps.html

Part II

Capital Markets Research

10

FFJR

The (brief!) introduction in Fama et al. (1969) ("FFJR") confirms that the goal of the paper is to shed light on market efficiency. Prior empirical work had inferred "market efficiency from the observed independence of successive price changes." In contrast, FFJR focuses on the "speed of adjustment of prices to *specific kinds* of new information ... [specifically] the information (if any) that is implicit in a stock split." We start this chapter with a brief introduction to the **efficient markets hypothesis**.

To examine their research question, FFJR conduct one of the earliest **event studies**. Event studies have a tight connection with notions of efficient markets, especially the "semi-strong" form of the hypothesis offered by Fama (1970), which in turn has important implications for the study of accounting information in capital markets. Thus we use FFJR to provide an initial introduction to event studies, which we revisit in more depth in Chapter 13.

We also use this chapter to introduce the reader to important additional data sets on CRSP, such as those related to dividends and stock splits, and to some R functions for efficiently manipulating data and models (e.g., the `unnest()` function from the `tidyr` package). To better understand FFJR, we also provide some background on stock splits and dividends. We then conduct a replication of FFJR and provide exercises for the reader. We conclude the chapter with a guide to this part of the book.

> 💡 Tip
>
> The code in this chapter uses the packages listed below. For instructions on how to set up your computer to use the code found in this book, see Section 1.2. Quarto templates for the exercises below are available on GitHub.[a]
>
> ---
> [a]https://github.com/iangow/far_templates/blob/main/README.md

```r
library(dplyr)
library(DBI)
library(ggplot2)
library(farr)
library(dbplyr)      # window_order()
library(tidyr)       # nest(), unnest()
library(purrr)       # map(), map2()
```

10.1   Efficient capital markets

One of the core ideas in capital market research is the **efficient markets hypothesis** (EMH). Fama (1991) defines the EMH as "the simple statement that security prices fully reflect all available information." The EMH is perhaps the most empirically tested proposition in all of social sciences.

In Fama's formulation, the terms *fully reflect* and *all available information* are doing a lot of work. One widely understood implication of the notion that security prices fully reflect a piece of information is that there are no opportunities to generate risk-adjusted profits by trading on that information.[1]

The EMH is particularly important for accounting research and practice for at least two reasons. First, accounting information is often a component of "all available information" against which the EMH is tested. In particular, Beaver (1998, p. 136) points out that accounting earnings "are widely analyzed by the investment community. No other firm-specific variable receives more attention by the analysts and other capital market participants than earnings." Second, whether the EMH holds or not has significant implications for preparers, users, and regulators of accounting information.

Richard Thaler[2] identifies two notions of the EMH, which he labels the "price is right" and "no free lunch" principles. The "price is right" principle says asset prices will "fully reflect" available information and thus "provide accurate signals for resource allocation". The "no free lunch" principle holds that market prices are impossible to predict making it very hard for an investor to beat the market after taking account of risk. The "no free lunch" principle is the variant that is both less demanding of market foresight and more empirically testable.

Most empirical studies of the EMH test the "no free lunch" principle. However, Shiller (1984) laments the existence of "claims that because real returns are nearly unforecastable, the real price of stocks is close to the intrinsic value" and suggest that "this argument for the efficient markets hypothesis represents one of the most remarkable errors in the history of economic thought" (1984, p. 459). In other words, a common fallacy is to conflate the two variants so that evidence for the "no free lunch" variant is adduced as supporting the "price is right" theory.

The "price is right" theory often underlies papers that use event studies to evaluate the merits of corporate policies or regulation. We will discuss some issues with conflation of these two principles in Chapter 13.

10.2   Stock splits

Because much about FFJR and its setting is implicit rather than explicit and might not be clear to a reader entering the field more than fifty years later, we first elucidate some elements of the paper.

---

[1]Another deeper implication of the EMH is that prices should be "correct" in some sense.

[2]https://www.newyorker.com/news/john-cassidy/interview-with-richard-thaler

Walker (2021, p. 1) describes a **stock split** as "the issuance of an additional number of shares, at no cost to the shareholder, in proportion to the number of shares already owned. For example, a 2-for-1 split is implemented by a firm issuing one new share for each existing share, thereby doubling the total number of shares outstanding."

From a fundamental perspective, a stock split might appear to have no economic consequences. Suppose we hold 1,000 shares in a company with an intrinsic value of $500 million and 10 million shares outstanding. Our shares would be worth $50,000 ($1000 \times 500 \div 10$). If the company then does a 2-for-1 split so that it now has 20 million shares outstanding, we would have 2,000 shares worth $50,000 ($2000 \times 500 \div 20$).

As such, it seems natural to ask why firms engage in stock splits. One explanation posits that, even though there is no underlying effect of a stock splits on the intrinsic value of the firm, markets are not efficient and the stock does not properly adjust for the change in the number of shares, leading to our hypothetical stock holding being worth (say) more than $50,000 after the split. If managers would prefer the firm to have a higher stock market value, this would be a reason to do a stock split.[3]

But if we assume market efficiency (and being able to calculate the stock-price impact of a stock split is surely a modest level of efficiency), we need to find alternative explanations for stock splits. Walker (2021, p. 2) identifies "two leading explanations".

The first explanation is the *information-signalling hypothesis*, which posits that "management uses the split as a channel to signal their private information about the firm's positive outlook" (Walker, 2021, p. 2). But this is a non-explanation unless there is some underlying effect of a split that is differentially beneficial or costly for firms with private information. For example, a **signalling equilibrium** might be sustained if a stock split is less costly for firms with better prospects.[4] But we need to account for such differences in costs.

The second explanation is the *liquidity hypothesis*, which "posits that management use a split as a way to improve their stock's liquidity" (Walker, 2021, p. 2). Delving into the detailed explanations of a liquidity effect of splits would take us too far away from the objectives of this chapter, so we offer a more stylized account that is adequate for our purposes here. We can label this account as a kind of Goldilocks stock price theory whereby a firm's liquidity is maximized when its stock price trades in a certain range—neither too high nor too low—and that firms tend split their shares if doing so gets their stock prices into this range. For example, if a company's liquidity is maximized with a stock price around $40, but it is trading at around $80, then a 2-for-1 split would enhance liquidity.

The liquidity hypothesis *might* fill the gap we identified in the information-signalling hypothesis. If a firm expects to trade at $100 in the future (due to inside information that will come to light later), it might want to do a 2-for-1 split, as a stock at $40 rising to $50 will enjoy better liquidity. But a firm that has inside information and expects to trade closer to $40 in the future (without a split) would be reluctant to do a 2-for-1 split, as it would then expect to trade closer to $20 in the future, and thus would have worse liquidity.

An important element of an event study is the identification of the period during which the market is expected to react to the event information. Like dividends, stock splits will generally be announced prior to their effective date. Fama et al. (1969, p. 7) "define month 0 as the month in which the effective date of a split occurs". Fama et al. (1969, p. 9) mention that for "a random sample of fifty-two splits" from their sample "the median time

---

[3]In some cases, managers might benefit from a *lower* post-split stock market value, say, if they want to purchase shares themselves after the split.

[4]For more on signalling theory, see Chapter 17 of Kreps (1990).

between the announcement date and the effective date of the split was 44.5 days." If there is information content in stock splits, we would expect an efficient market to react when the split is announced, not when it is effective. The analyses in FFJR use monthly data, so a median announcement date for splits of 44.5 days before the effective date implies that the median split is announced in month -2. FFJR do not provide details on the distribution of split announcement dates, but some splits may be announced in month -1, while others could occur in month -3 or even earlier.

10.2.1 Discussion question

Consider the following alternative theories.

- **Theory A**: Firms like to keep their stock price within certain bounds. When the stock price rises above a certain threshold, a firm may initiate a split, announcing it several weeks in advance of its effective date. Firms do not use splits to signal private information about firm prospects.
- **Theory B**: Firms use splits to signal private information about firm prospects. A firm will announce a split several weeks in advance of its effective date.
- **Theory C**: Capital market participants don't fully adjust for the effect of splits, tending to anchor on the pre-split price to some degree. Firms do not use splits to signal private information about firm prospects.

Produce a set of indicative plots (e.g., drawn by hand) for the predicted behaviour of cumulative abnormal returns for assuming that the split is announced in month -2. What impact would variation in the announcement dates relative to the split effective date have on the plots?

10.3 Dividend policy

Black (1976, p. 9) points out that the "Miller-Modigliani theory ... says that the dividends a corporation pays do not affect the value of its shares or the returns to investors." Of course, assumptions underlying the Miller-Modigliani theory include the absence of tax effects or transaction costs. While these assumptions are violated in reality, the violations generally do not provide an explanation for the existence of dividends (e.g., dividends are generally treated unfavourably for tax purposes).

Black (1976, p. 10) summarizes the common understanding of dividends: "For one reason or another, managers and directors do not like to cut the dividend. So they will raise the dividend only if they feel the company's prospects are good enough to support the higher dividend for some time. And they cut the dividend only if they think the prospects for a quick recovery are poor." While this theory is "behavioural" (see Thaler, 2015, p. 166, for discussion of this), it has real teeth because managers and directors have inside information and therefore, if they act in accordance with this theory, dividend policy will convey some of this information.

Fama et al. (1969, pp. 2–3) endorse this account of dividends: "Studies ... have demonstrated that, once dividends have been increased, large firms show great reluctance to reduce them, except under the most extreme conditions. Directors have appeared to hedge against such dividend cuts by increasing dividends only when they are quite sure of their ability to maintain them in the future, i.e., only when they feel strongly that future earnings will be sufficient to maintain the dividends at their new higher rate. Thus dividend changes may be assumed to convey important information to the market concerning management's assessment of the firm's long-run earning and dividend paying potential."

The connection between stock splits and dividends is not spelt out clearly in FFJR. Fama et al. (1969, p. 2) do say that "in the past a large fraction of stock splits have been followed closely by dividend increases." Fama et al. (1969, pp. 12–16) offer the following explanation: "When a split is announced or anticipated, the market interprets this (and correctly so) as greatly improving the probability that dividends will soon be substantially increased. (In fact, ... in many cases the split and dividend increase will be announced at the same time.)"

Several assumptions are buried in that explanation, which appears to be based on some kind of signalling story.[5] For example, why would firms signal future dividend increases using stock splits? And, what are firms signalling with splits when they announce splits and dividend increases at the same time?

10.3.1 Discussion questions

1. Does the research design of FFJR include the use of a control group? If so, how? What alternative methods could have been used to introduce a control group?

2. Fama et al. (1969, p. 9) state "the most important empirical results of this study are summarized in Tables 2 and 3 and Figures 2 and 3." What does Table 3 of FFJR tell us? (*Hint*: Read p. 11.) Do you find the presentation of Table 3 to be effective?

3. Consider Table 2 of FFJR. Is it more or less important than Table 3? What is the relationship between Table 2 and Figures 2 and 3?

4. What statistical tests are used to test the hypotheses of the paper?

10.4 Replication of FFJR

We will now conduct a rough replication of FFJR.[6]

All the data we will use come from the `crsp` schema. The new table here is `dsedist`, which contains information on distributions, including dividends and stock splits.

```
db <- dbConnect(RPostgres::Postgres(), bigint = "integer")

msf <- tbl(db, Id(schema = "crsp", table = "msf"))
msi <- tbl(db, Id(schema = "crsp", table = "msi"))
stocknames <- tbl(db, Id(schema = "crsp", table = "stocknames"))
dsedist <- tbl(db, Id(schema = "crsp", table = "dsedist"))
```

Fama et al. (1969, p. 3) "define a 'stock split' as an exchange of shares in which at least five shares are distributed for every four formerly outstanding. Thus this definition of splits includes all stock dividends of 25 per cent or greater."

---

[5]Evidence that FFJR have a signalling story in mind comes from subsequent sentences in the paper.
[6]We thank James Kavourakis for an earlier replication effort that helped us in preparing this code.

```
splits <-
  dsedist |>
  filter(between(exdt, "1927-01-01", "1959-12-31"),
         distcd %in% c(5523L, 5533L),
         facshr >= 0.25) |>
  group_by(permno, exdt) |>
  summarize(facshr = sum(facshr, na.rm = TRUE), .groups = "drop") |>
  mutate(split = TRUE)
```

Fama et al. (1969, p. 3) continue: "Since the data cover only common stocks listed on the New York Stock Exchange, our rules require that to qualify for inclusion in the tests a split security must be listed on the Exchange for at least twelve months before and twelve months after the split. From January, 1927, through December, 1959, 940 splits meeting these criteria occurred on the New York Stock Exchange." NYSE-listed stocks have `exchcd == 1` and ordinary common shares are those for which the first character of `shrcd` is `1`. Data on `exchcd` and `shrcd` on various dates is found on `crsp.stocknames`. The following code creates a table with the `permno` values and date ranges for the securities meeting these criteria.

```
nyse_stocks <-
  stocknames |>
  filter(exchcd == 1,
         str_sub(as.character(shrcd), 1L, 1L) == "1") |>
  select(permno, namedt, nameenddt)
```

We can then use this table to focus `splits` on these securities.

```
nyse_splits_raw <-
  splits |>
  inner_join(nyse_stocks,
             join_by(permno, between(exdt, namedt, nameenddt)))
```

We will need to combine these data with data on returns from `crsp.msf`. The table `crsp.msf` contains monthly data indexed by `permno` and `date`, where `date` is the last trading day of the month. The splits in `nyse_splits_raw` will generally not occur on dates found on `crsp.msf`, so we need to create a "month" variable so that we can line up observations on `crsp.msf` and `nyse_splits_raw`.

To this end we create a table `month_indexes` that includes the "index" of the month, which will also be used in doing the arithmetic to go backwards and forwards in months. In performing this arithmetic, we use **window functions**, which are discussed in the documentation for `dplyr`[7] and in the "Databases" chapter[8] of *R for Data Science*.[9] Here the "index" of a month refers to its placement in the sequence of months found on `crsp.msf` and `crsp.msi`.[10]

---

[7]https://dplyr.tidyverse.org/articles/window-functions.html

[8]https://r4ds.hadley.nz/databases.html

[9]Also see https://www.postgresql.org/docs/current/tutorial-window.html.

[10]Note that we are operating on remote data here, so we use `window_order()` instead of `arrange()` when using window functions. In fact, `window_order()` offers power that is not available with local data frames. In Chapters 6 and 8, we used `dbplyr` implicitly when connecting to databases through the `dplyr` package, but `window_order()` is not "re-exported" by the `dplyr` package and thus we needed to invoke `library(dbplyr)` above.

```
month_indexes <-
  msi |>
  mutate(month = as.Date(floor_date(date, 'month'))) |>
  window_order(month) |>
  mutate(month_index = row_number()) |>
  select(date, month, month_index) |>
  collect()
```

We actually used window functions in Chapter 2 when we used fill() to complete data for calculating market-to-book ratios. It turns out that the window-function functionality made available via the **dbplyr** package and PostgreSQL (using window_order()) are more powerful in this case than those available via the base **dplyr** package and **tibble** data frames (using arrange()), as the **dbplyr** version allows us to specify a window_frame() over which window functions are applied, as we will see below.

We next bring our splits data (nyse_splits_raw) into R and merge it with month_indexes.

```
nyse_splits <-
  nyse_splits_raw |>
  mutate(month = as.Date(floor_date(exdt, 'month'))) |>
  collect() |>
  inner_join(month_indexes, by = "month") |>
  rename(ex_month_index = month_index) |>
  select(-namedt, -nameenddt, -date)
```

We construct nyse_msf, which is essentially crsp.msf restricted to NYSE stocks and months with non-missing returns, and with the added variable month_index.

```
nyse_msf <-
  msf |>
  filter(!is.na(ret)) |>
  inner_join(nyse_stocks,
             join_by(permno, between(date, namedt, nameenddt))) |>
  collect() |>
  inner_join(month_indexes, by = "date") |>
  select(permno, month_index, date, ret)
```

The following code merges data on splits from nyse_splits with data on returns from nyse_msf. We create a variable (month_rel_ex) that measures the number of months between the split and the return.

Because we may have multiple splits for a given permno in nyse_splits (each with a different exdt) and each will be joined with multiple return months in nyse_msf, we will have a many-to-many relationship between rows in nyse_splits and rows in nyse_msf. To suppress a warning that dplyr issues in such cases, we indicate that we anticipate such matches by specifying relationship = "many-to-many" as an argument to left_join().

```
split_return_data <-
  nyse_splits |>
  left_join(nyse_msf, by = "permno", relationship = "many-to-many") |>
  mutate(month_rel_ex = month_index - ex_month_index) |>
  select(permno, exdt, month_rel_ex, date, ret)
```

However, we want to ensure that we have just one row for each (`permno`, `exdt`, `month_rel_ex`) combination:

```
split_return_data |>
  count(permno, exdt, month_rel_ex) |>
  count(n_rows = n)
```

```
# A tibble: 1 x 2
  n_rows        n
   <int>    <int>
1       1  640200
```

Fama et al. (1969, p. 3) "decided, arbitrarily, that in order to get reliable estimates of the parameters that will be used in the analysis, it is necessary to have at least twenty-four successive months of price-dividend data around the split date." Note that Fama et al. (1969) actually look at 12 months before and 12 months after the split, so including the month of the split, we have 25 months. The table `split_sample` imposes this restriction and, by doing a `semi_join()` of this with `split_return_data`, we create `split_returns`, which contains the data from `crsp.msf` for our sample of splits.

```
split_sample <-
  split_return_data |>
  filter(between(month_rel_ex, -12, 12)) |>
  group_by(permno, exdt) |>
  summarize(n_obs = n(), .groups = "drop") |>
  filter(n_obs == 25L) |>
  select(permno, exdt)

split_returns <-
  split_return_data |>
  semi_join(split_sample, by = "permno")
```

The sample of Fama et al. (1969, p. 4) comprises 940 splits for 622 securities. The output from the code below suggests that our sample is slightly larger, perhaps due to changes in the underlying data since 1966.

```
split_sample |> distinct(permno) |> count() |> pull()
```

```
[1] 626
```

```
split_sample |> count() |> pull()
```

```
[1] 948
```

We follow the basic approach of FFJR in estimating excess returns. FFJR estimate a market model by regressing the log of gross returns of each stock on the log of gross returns of a market index and taking the residual.

The market index used in Fama et al. (1969, p. 4) is Fisher's Combination Investment Performance Index, which is made available as Table A1 in Fisher (1966). To use this table, which is not available in a machine-readable form, we would have to type in the hundreds of numbers contained therein. Instead of doing that, we will use one of the standard indexes provided by CRSP in `crsp.msi`. There are two widely used indexes on `crsp.msi`: `vwretd` provides the value-weighted return including distributions and `ewretd` provides the equal-weighted return including distributions. A value-weighted index weights each security in the portfolio based on its market value at the time of portfolio formation, while an equal-weighted index puts equal weight on each security in the portfolio regardless of its market value. For our base analysis, we will focus on `vwretd`.

```
index_returns <-
  msi |>
  select(date, vwretd, ewretd) |>
  collect()
```

Fama et al. (1969, pp. 4–5) express concern about non-zero excess returns causing "specification error" and exclude "fifteen months before the split for all securities and fifteen months after the split for splits followed by dividend decreases." We *partially* implement this by excluding fifteen months before the split for all securities, but not excluding any months after the split, as this requires data on dividends that would not have been available at the time of the split—which we do not compile until later (see below)—and thus induces additional look-ahead issues.

Note that because a (`permno`, `date`) combination may appear for more than one split, and be within fifteen months before the split for one split, and not so for another split, we need to aggregate data across such observations. The following code sets `exclude` to true if the (`permno`, `date`) combination is within fifteen months before *any* split for that stock.

```
omit_returns <-
  split_returns |>
  mutate(exclude = between(month_rel_ex, -15, 0)) |>
  group_by(permno, date) |>
  summarize(exclude = any(exclude), .groups = "drop")
```

The following code adds `exclude` to the data from `split_returns`. Note that we need to select `distinct` values for return data from `split_returns`, as a (`permno`, `date`) combination may appear more than once, but with different values of `month_rel_ex` due to it relating to more than one split. Note also that we do not simply drop rows from the data set where `exclude` is `TRUE`, as we want to calculate abnormal returns for these months. Instead, we merely exclude these observations from the regression analysis by specifying `subset = !exclude`.

```
split_returns_reg <-
  split_returns |>
  inner_join(omit_returns, by = c("permno", "date")) |>
  select(permno, date, ret, exclude) |>
  distinct()
```

The next step is to estimate regressions by `permno`. Like FFJR, we estimate a market model by regressing the log of gross returns of each stock on the log of gross returns of a market index. In the code below, we make extensive use of ideas from the "many models" chapter[11] of the first edition of *R for Data Science* and you may find it helpful to read that chapter.[12]

Because we want to fit one model for each `permno`, we facilitate this by using `nest()` to put all the data for each value of `permno` (other than the value of `permno`) into a single column. We can then use `map()` from the `purrr` library to create a new column `fit` that contains a fitted linear (OLS) model for each `permno` value. If we apply `predict()` to the model,

---

[11]https://r4ds.had.co.nz/many-models.html
[12]There is no close equivalent in the second edition of *R for Data Science*. Instead readers of that book are directed to *Tidy Modeling with R* https://www.tmwr.org (Kuhn and Silge, 2022).

we will get fitted values only for those observations used to estimate the model, which will mean we will not have predicted values for months within fifteen months before the split. Instead, we need to supply data for all observations as the second argument (`newdata`) of the `predict` function. To do this, we use `map2()` from the `purrr` library and store the result in the variable `predicted`.

Note that `predicted` will be a list-column in which each element is a vector of predicted values using the market model for the applicable `permno`. We can use `unnest()` from the `tidyr` package to "expand" the data frame so that each row relates to a given (`permno`, `date`) pair. Finally, we calculate excess returns by subtracting `predicted` from the actual value of `lpr` and store the result in the column named `resid`.

```
abnormal_returns <-
  split_returns_reg |>
  left_join(index_returns, by = "date") |>
  mutate(lpr = log(1 + ret),
         lm = log(1 + vwretd)) |>
  select(permno, date, lpr, lm, exclude) |>
  nest(data = !permno) |>
  mutate(fit = map(data, \(x) lm(lpr ~ lm, data = x, subset = !exclude,
                                 na.action = "na.exclude"))) |>
  mutate(predicted = map2(fit, data, \(x, y) predict(x, newdata = y))) |>
  unnest(cols = c(predicted, data)) |>
  mutate(resid = lpr - predicted) |>
  select(permno, date, resid)
```

To facilitate joining `abnormal_returns` with `nyse_splits`, we create a version of `nyse_splits` with `start` and `end` columns that indicate the `month_index` of the first and last months that we want abnormal returns for (i.e., within 30 months of the split).

```
nyse_splits_join <-
  nyse_splits |>
  mutate(start = ex_month_index - 30,
         end = ex_month_index + 30)
```

We can combine `abnormal_returns` with `nyse_splits`, but before doing so, we need to pull in `month_index` for each `date` from `month_indexes`. We join `abnormal_returns` with `nyse_splits` using `permno` and `month_index` between `start` and `end`.

```
table2_data <-
  abnormal_returns |>
  inner_join(month_indexes, by = "date") |>
  inner_join(nyse_splits_join,
             join_by(permno, between(month_index, start, end))) |>
  mutate(month_gap = month_index - ex_month_index) |>
  select(permno, exdt, month_gap, resid)
```

Before proceeding, we check that we have just one row for each split (`permno` and `exdt`) and relative month (`month_gap`).[13]

---

[13]It seems that by using an inequality join (i.e., `join_by(between())`), a "many-to-many" check is not applied automatically.

```
table2_data |>
  count(permno, exdt, month_gap) |>
  count(n_rows = n)
```

```
# A tibble: 1 x 2
  n_rows       n
   <int> <int>
1      1 57986
```

We now have the data we need to produce Figure 10.1, our analogue of Figure 2b from Fama et al. (1969, p. 13).

```
table2_data |>
  group_by(month_gap) |>
  summarize(all_u = mean(resid), .groups = "drop") |>
  arrange(month_gap) |>
  mutate(all_U = cumsum(all_u)) |>
  ggplot(aes(x = month_gap, y = all_U)) +
  geom_point()
```

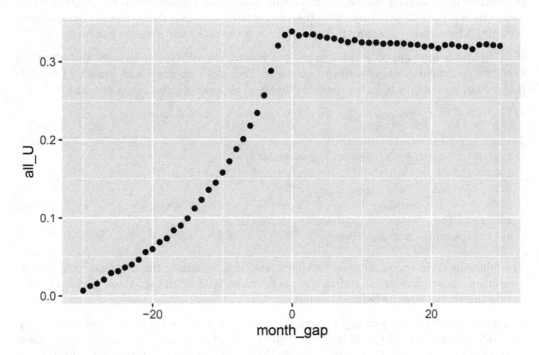

FIGURE 10.1
Cumulative average residuals—All splits

10.4.1 Data on dividends

Most columns of Table 2 of FFJR and all plots in Figure 3 of FFJR require data on dividends. We collect data on dividends from `dsedist`.[14] Ordinary dividends are distinguished from

---

[14]The table `crsp.dsedist` is a subset of the table `crsp.dse` focused on distributions.

other distributions (e.g., liquidating dividends or exchanges and reorganizations) by the first digit of `distcd`, which is 1 for ordinary dividends.

It turns out that there can be more than one dividend paid in a given month for a given security, so we aggregate dividends by (`permno, month`).

```
div_months <-
  dsedist |>
  filter(str_sub(as.character(distcd), 1L, 1L) == "1") |>
  mutate(month = as.Date(floor_date(exdt, 'month'))) |>
  group_by(permno, month) |>
  summarize(divamt = sum(divamt, na.rm = TRUE), .groups = "drop")
```

For each month, FFJR define "the *dividend change ratio* as total dividends (per equivalent unsplit share) paid in the twelve months after the split, divided by total dividends paid during the twelve months before the split" Fama et al. (1969, p. 8). So, for each month t, we want to sum up dividends from month $t - 11$ to month t and also dividends for months $t + 1$ through $t + 12$.

Care is needed to calculate dividends "per equivalent unsplit share" as a firm paying 80 cents per share each quarter prior to a 2-for-1 split should be considered to have increased its dividends if it paid 45 cents per share (i.e., 90 cents per *original* share) in each quarter after the split. The variable `cfacshr` from `msf` allows us to make the necessary adjustment.[15]

If no dividends are paid on a stock in a given month, we set `divamt` to zero; this allows us to distinguish months where a stock is not an NYSE stock on `crsp.msf` (missing from this table) from ones where it is, but pays no dividend that month (i.e., zero `divamt`).[16]

```
nyse_divs_raw <-
  msf |>
  inner_join(nyse_stocks, by = "permno") |>
  filter(between(date, namedt, nameenddt)) |>
  mutate(month = as.Date(floor_date(date, 'month'))) |>
  select(permno, date, month, cfacshr) |>
  left_join(div_months, by = c("permno", "month")) |>
  mutate(divamt = coalesce(divamt / cfacshr, 0)) |>
  select(permno, month, divamt)
```

The following code aggregates the variables needed to calculate the dividend change ratio mentioned above: dividends paid in the twelve months after the split and dividends paid during the twelve months before the split.

In this case, we specify `group_by(permno)` as we want windows to be applied on a by-PERMNO basis (i.e., when calculating trailing and forward dividends, we are only interested in dividends related to a single PERMNO) and then `window_order(month)`, as we want to arrange the data within each window by month so that "go back 11 months" is

---

[15]Note that, as `cfacshr` comes from `crsp.msf`, it is presumably the value applicable at the end of the month. It seems possible that a firm could pay a dividend and then split its stock in the same month, in which case applying the end-of-month `cfacshr` to the dividend would be incorrect. For the purposes of this chapter, we ignore this issue, but a careful researcher would investigate further when conducting a proper research study.

[16]We want to do this for every month that a security was an NYSE security, so we start from `msf` and join with `nyse_stocks`. Note that we cannot use the `nyse_msf` data frame created above because this is a local data frame and we are using remote data from `div_months` in this analysis. This may be a case where a little duplicate code is preferred to shifting data from a local machine to a remote one or vice versa.

meaningfully defined. In the next step, we specify `window_frame(from = -11, to = 0)` to indicate that in the subsequent calculation, we want to include values from $t-11$ to t in the window.[17] This drives the values considered in the subsequent `mutate()` step, which calculates `div_trailing` and `mths_trailing`. Note that `na.rm = TRUE` is always the case for SQL (which this code is ultimately translated into) and that `n()` counts the rows included in the window.

We then respecify the window using `window_frame(from = 1, to = 12)` so that we can calculate `div_forward` and `mths_forward`.

Finally, to exclude cases where, say, a stock has listed within the last twelve months or delists in the subsequent twelve months and thereby perhaps makes the calculation of the dividend change ratio less meaningful, we use `filter(mths_trailing == 12, mths_forward == 12)`. Finally, we `ungroup` to remove the `group_by(permno)` grouping that we no longer need, select the variables of interest, and then collect the data.

```
nyse_divs <-
  nyse_divs_raw |>
  group_by(permno) |>
  window_order(month) |>
  window_frame(from = -11, to = 0) |>
  mutate(div_trailing = sum(divamt, na.rm = TRUE),
         mths_trailing = n()) |>
  window_frame(from = 1, to = 12) |>
  mutate(div_forward = sum(divamt, na.rm = TRUE),
         mths_forward = n()) |>
  filter(mths_trailing == 12, mths_forward == 12) |>
  ungroup() |>
  select(permno, month, div_trailing, div_forward) |>
  collect()
```

Now that we have data on `div_trailing` and `div_forward` for every `(permno, month)` combination on NYSE, we can use these data to calculate the dividend change ratio (`div_ratio`) for each of the split months on `nyse_splits`.

```
split_firm_dividends <-
  nyse_splits |>
  left_join(nyse_divs, by = c("permno", "month")) |>
  mutate(div_ratio = if_else(div_trailing > 0,
                             div_forward / div_trailing, NA)) |>
  filter(!is.na(div_ratio)) |>
  select(permno, month, exdt, div_ratio)
```

Fama et al. (1969, p. 8) measure dividend changes "relative to the average dividends paid by all securities on the New York Stock Exchange during the relevant time periods. ... Dividend 'increases' are then defined as cases where the dividend change ratio of the split stock is greater than the ratio for the Exchange as a whole." We calculate the "market" dividend change ratio (`mkt_div_ratio`) by averaging the `div_trailing` and `div_forward` values for each `month` across all stocks.

---

[17]Like `window_order()`, `window_frame()` is only available for remote data frames because it relies on the SQL backed to provide functionality not available with local data frames.

```
div_mkt <-
  nyse_divs |>
  group_by(month) |>
  summarize(div_trailing = mean(div_trailing, na.rm = TRUE),
            div_forward = mean(div_forward, na.rm = TRUE),
            .groups = "drop") |>
  mutate(mkt_div_ratio = if_else(div_trailing > 0,
                                 div_forward / div_trailing, NA))
```

We can then combine data on split- and market-level dividend change ratios for each month in which there is a split to calculate `up_div`, which is an indicator for the dividend change ratios for a stock undergoing a split being greater than that for the market in that month.

Fama et al. (1969, p. 9) do not pretend that their measure of dividend increases is perfect and merely use it as "a simple and convenient way of ... classifying year-to-year dividend changes for individual securities."[18]

```
dividends_file <-
  split_firm_dividends |>
  inner_join(div_mkt, by = "month") |>
  select(permno, exdt, div_ratio, mkt_div_ratio) |>
  mutate(up_div = div_ratio >= mkt_div_ratio)
```

Finally, we can combine data on relative dividend change ratios with abnormal return data.

```
table2_w_divs <-
  table2_data |>
  left_join(dividends_file, by = c("permno", "exdt"))
```

We can now produce Figure 10.2, our analogue of Figure 3c from Fama et al. (1969, p. 15).

```
table2_w_divs |>
  filter(up_div) |>
  group_by(month_gap) |>
  summarize(u = mean(resid), .groups = "drop") |>
  arrange(month_gap) |>
  mutate(U = cumsum(u)) |>
  ggplot(aes(x = month_gap, y = U)) +
  geom_point()
```

---

[18]See Fama et al. (1969, p. 9) for discussion of some limitations of the measure.

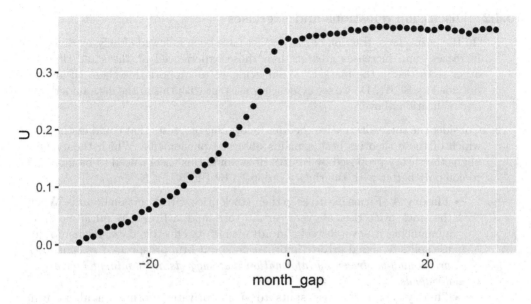

FIGURE 10.2
Cumulative average residuals—Dividend "increases"

Figure 10.3 provides our equivalent of Figure 3d from Fama et al. (1969, p. 15).

```
table2_w_divs |>
  filter(!up_div) |>
  group_by(month_gap) |>
  summarize(u = mean(resid), .groups = "drop") |>
  arrange(month_gap) |>
  mutate(U = cumsum(u)) |>
  ggplot(aes(x = month_gap, y = U)) +
  geom_point()
```

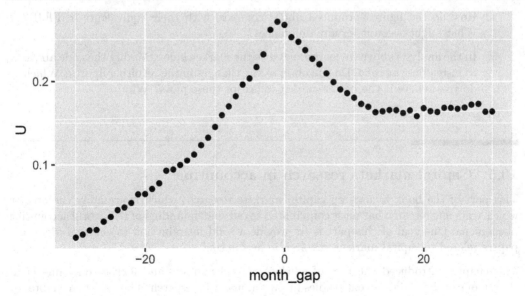

FIGURE 10.3
Cumulative average residuals—Dividend "decreases"

10.4.2   Discussion questions and exercises

1. "In the past a large fraction of stock splits have been followed closely by dividend increases—and increases greater than those experienced at the same time by other securities in the market." What evidence to support this claim (if any) is provided by FFJR? Do we see evidence consistent with this in the data underlying our replication above?

2. Consider the three alternative theories below. Suggest tests that could distinguish which of these theories best explains observed phenomena. Which theory best aligns the theory proposed by FFJR? How would this theory need to be modified to comport better with the theory proposed by FFJR?

    *   **Theory A':** Firms like to keep their stock price within certain bounds. When the stock price rises above a certain threshold, a firm may initiate a split, announcing it several weeks in advance of its effective date. Firms do not use splits to signal private information about firm prospects. *Pre-split stock prices may be driven by information that suggests an imminent increase in dividends.*
    *   **Theory B':** Firms use splits to signal private information about firm prospects. A firm will announce a split several weeks in advance of its effective date. *Firms may also use dividend changes to signal private information about firm prospects.*
    *   **Theory B'':** *Firms use splits to signal private information about future dividend increases, which in turn signal private information about firm prospects.* A firm will announce a split several weeks in advance of its effective date.

3. On p. 17 of Fama et al. (1969), it is argued that "our data suggest that once the information effects of associated dividends are properly considered, a split *per se* has no net effect on common stock returns." Is it clear what meaning the words "per se" have in this sentence? Does FFJR provide persuasive evidence in support of this claim? Describe how you might test this claim using the richer data available today. What data would you use beyond that used in FFJR?

4. How do the figures produced above compare with their equivalents in FFJR? What might account for any differences?

5. In the analysis above we used `vwretd` as the market index. Modify the code above to instead use `ewretd`. Do you observe any changes in the resulting figures? Which do you believe is the better market index for these plots? Why?

10.5   Capital markets research in accounting

This part of the book focuses on capital markets research, which is arguably the area in which academic research has most contributed to our understanding of real-world accounting phenomena. One goal of this part is to provide a solid introduction to classical ideas and papers related to capital markets research in accounting.

This chapter introduced the idea of efficient capital markets and focuses on Fama et al. (1969), one of the earliest event studies in capital markets research. The next two chapters—Chapters 11 and 12—cover two seminal papers, Ball and Brown (1968) and Beaver (1968), respectively. While the first three chapters of this part study papers from more than 50 years ago, we believe that they provide an excellent introduction to the foundations of **research design** that remain relevant to this day.

Chapter 13 builds on the three previous chapters in providing more depth on event studies, including coverage of how event studies have evolved since Fama et al. (1969). Chapter 14 examines **post-earnings announcement drift**, a much-studied **anomaly** in the pricing of accounting information in capital markets. Chapter 15 goes deeper into the measurement of accruals, a critical element of financial accounting, and also provides an opportunity to explore simulation analysis in more detail. Chapter 16 explores **earnings management**, which has been the focus of significant body of accounting research over several decades, and also provides an opportunity to understand issues related to measurement in research and the power of statistical tests.

11

Ball and Brown (1968)

In this and the following chapter, we cover the first two winners of the Seminal Contributions to Accounting Literature Award: Ball and Brown (1968) and Beaver (1968).[1]

Ball and Brown (1968) won the inaugural Seminal Contribution to the Accounting Literature Award with a citation: "No other paper has been cited as often or has played so important a role in the development of accounting research during the past thirty years." However, Philip Brown (Brown, 1989) recalled in a presentation to the 1989 JAR conference that the paper was rejected by *The Accounting Review* with the editor indicating a willingness to "reconsider the manuscript if Ray and I wished to cut down the empirical stuff and expand the 'bridge' we had tried to build between our paper and the accounting literature." (Brown, 1989, p. 205)

According to Kothari (2001, p. 113), "Ball and Brown (1968) and Beaver (1968) heralded empirical capital markets research as it is now known." Prior to that period, accounting research was a largely theoretical discipline focused on **normative research**, that is, research concerned with the "right" or "best" way to account for various events and transactions. In addition to being normative, accounting theory was largely *deductive*, meaning that detailed theories were derived from general principles.

Beaver (1998) identifies one approach as asking, say, "what properties should the 'ideal' net income have?" One answer to this question is that accounting income a period should reflect the change in the net present value of cash flows (plus cash distributions) to shareholders during the period. But other answers existed. Accounting researchers would start with a set of desired properties and use these to derive the "best" approach to accounting for depreciation of long-lived assets, inventory, or lease assets. Kothari (2001) points out that there was "little emphasis on the empirical validity" of theory.

Similar ideas still permeate the thinking of standard-setters, who purport to derive detailed accounting standards from their "conceptual frameworks", which outline broad definitions of things such as *assets* and *liabilities* that standard-setters can supposedly use to derive the correct accounting approach in any given setting.

However, in the period since Ball and Brown (1968), these approaches have been largely discarded in academic research. A largely normative, theoretical emphasis has been replaced by a positive, empirical one.

This chapter uses Ball and Brown (2019) as a kind of reading guide to Ball and Brown (1968) before considering a replication of Ball and Brown (1968) styled on that provided by Nichols and Wahlen (2004).

---

[1]The list of winners can be found at https://go.unimelb.edu.au/yzw8.

DOI: 10.1201/9781003456230-11

> **Tip**
>
> The code in this chapter uses the packages listed below. Because we use several packages from the Tidyverse, we save time by using `library(tidyverse)` to load them in one go. For instructions on how to set up your computer to use the code found in this book, see Section 1.2. Quarto templates for the exercises below are available on GitHub[a].
>
> ---
>
> [a]https://github.com/iangow/far_templates/blob/main/README.md

```
library(tidyverse)
library(DBI)
library(farr)
library(dbplyr)        # window_order()
```

11.1 Principal results of Ball and Brown (1968)

The first two pages of Ball and Brown (1968) address the (then) extant accounting literature. This discussion gives the impression that the (academic) conventional wisdom at that time was that accounting numbers were (more or less) meaningless ("the difference between twenty-seven tables and eight chairs") and further research was needed to devise more meaningful accounting systems.

Arguably this notion informs the null hypothesis of Ball and Brown (1968). If accounting numbers are meaningless, then they should bear no relationship to economic decisions made by rational actors. Ball and Brown (1968) seek to test this (null) hypothesis by examining the relationship between security returns and unexpected income changes. Ball and Brown (1968) argue that "recent developments in capital theory ... [justify] selecting the behavior of security prices as an operational test of usefulness."

The evidence provided by Ball and Brown (1968) might not convince critics of the usefulness of adding chairs and tables unless the market is a rational, efficient user of a broader set of information. There are accounts for the results of Ball and Brown (1968) that do not rely on such rationality and efficiency. First, the market might react to "twenty-seven tables less eight chairs" because it does not know what it is doing. Second, the market might know that "twenty-seven tables less eight chairs" is meaningless, but has no better information to rely upon. Given the arguments they seek to address, assumptions that the market is (a) efficient and (b) has access to a rich information set beyond earnings seem implicit in the use of security returns in a test of usefulness.

Ball and Brown (2019, p. 414) identify three main results of Ball and Brown (1968): "The most fundamental result was that accounting earnings and stock returns were correlated. ... Nevertheless, annual accounting earnings lacked timeliness. ... After the earnings announcement month, the API (which cumulated abnormal returns) continued to drift in the same direction."

Figure 1 of Ball and Brown (1968) depicts the data provided in Table 5 and comprises all three of the principal results flagged by Ball and Brown (2019).

The returns reported in Figure 1 of Ball and Brown (1968) are not feasible portfolios because the earnings variables used to form each portfolio at month -12 are not reliably available until month 0 or later. Yet one can posit the existence of a mechanism (e.g., a time machine or a magical genie) that would suffice to make the portfolios notionally feasible. For example, if a genie could tell us whether earnings for the upcoming year for each firm will increase or decrease at month -12, we could go long on firms expecting positive news, and go short on firms expecting negative news.[2] Note that this hypothetical genie is sparing with the information she provides. For example, we might want further details on how much earnings increased or decreased, but our genie gives us just the sign of the earnings news.

Additionally, we have implicit constraints on the way we can use this information in forming portfolios. We might do better to adjust the portfolio weights according to other factors, such as size or liquidity, but the portfolios implicit in Figure 1 of Ball and Brown (1968) do not do this. Nor do the portfolios represented in Figure 1 involve any opportunity to adjust portfolio weights during the year.

In assessing the relative value of various sources of information, Ball and Brown (1968) consider three approaches to constructing portfolios, which they denote as *TI*, *NI*, and *II* (Ball and Brown (2019) denote *II* as *AI* and we follow this notation below). Using these metrics, Ball and Brown (1968, p. 176) conclude that "of all the information about an individual firm which becomes available during a year, one-half or more is captured in that year's income number. ... However, the annual income report does not rate highly as a timely medium, since most of its content (about 85 to 90 per cent) is captured by more prompt media which perhaps include interim reports."[3]

The third principal result is shown in Table 5, where we see that the income surprise is correlated with the API to a statistically significant extent in each month up to two months after the earnings announcement.[4]

This result, which later become known as **post-earnings announcement drift** (or simply **PEAD**), was troubling to Ball and Brown (1968), who argue that some of it may be explained by "peak-ahead" in the measure of market income and transaction costs causing delays in trade. We study PEAD more closely in Chapter 14.

11.1.1   Discussion questions

1. What is the research question of Ball and Brown (1968)? Do you find the paper to be persuasive?

2. What do you notice about the references in Ball and Brown (1968, pp. 177–178)?

3. Given that "the most fundamental result" of Ball and Brown (1968) relates to an association or correlation, is it correct to say that the paper provides no evidence on causal linkages? Does this also mean that Ball and Brown (1968) is a "merely" descriptive paper according to the taxonomy of research papers outlined in Chapter 4. How might the results of Ball and Brown (1968) be represented in a causal diagram assuming that accounting information is meaningful and markets are efficient? Would an alternative causal diagram be assumed by a critic who viewed accounting information as meaningless?

---

[2]Actually, because the lines of Figure 1 represent *abnormal* returns, the associated portfolios involve going long or short in each of a group of stocks and short or long in a broader market index at the same time.

[3]For more on the "apparent paradox" discussed on p. 176, see Leftwich and Zmijewski (1994).

[4]There are some months where this does not hold, but the statement is broadly true.

4. Describe how Figure 1 of Ball and Brown (1968) supports each of principal results identified by Ball and Brown (2019).

5. Consider the causal diagrams you created above. Do the results of Ball and Brown (1968) provide more support for one causal diagram than the other.

6. Compare Figure 1 of Ball and Brown (2019) with Figure 1 of BB68. What is common between the two figures? What is different?

7. What does "less their average" mean in the title of Figure 1 of Ball and Brown (2019)? What effect does this have on the plot? (Does it make this plot different from Figure 1 of BB68? Is information lost in the process?)

8. Ball and Brown (2019, p. 418) say "in this replication we address two issues with the BB68 significance tests." Do you understand the points being made here?

9. Ball and Brown (2019, p. 418) also say "the persistence of PEAD over time is evidence it does not constitute market inefficiency." What do you make of this argument?

10. What is the minimum amount of information that our hypothetical genie needs to provide to enable formation of the portfolios underlying *TI*, *NI*, and *II*? What are the rules for construction of each of these portfolios?

11. Ball and Brown (1968) observe a ratio of *NI* to *TI* of about 0.23. What do we expect this ratio to be? Does this ratio depend on the information content of accounting information?

12. Consider the paragraph in Ball and Brown (2019, p. 418) beginning "an innovation in BB68 was to estimate …". How do the discussions of these results differ between Ball and Brown (1968) and Ball and Brown (2019)?

13. Consider column (4) of Table 2 of Ball and Brown (2019). Is an equivalent set of numbers reported in BB68? What is the underlying investment strategy associated with this column (this need not be feasible in practice)?

14. Heading 6.3 of Ball and Brown (2019) is "Does 'useful' disprove 'meaningless'?" Do you think that "not meaningless" implies "not useless"? Which questions (or facts) does BB68 address in these terms?

11.2 Replicating Ball and Brown (1968)

In this section, we follow Nichols and Wahlen (2004) in conducting an updated replication of Ball and Brown (1968).

We get earnings and returns data from Compustat and CRSP, respectively.

```
db <- dbConnect(RPostgres::Postgres(),
                bigint = "integer",
                check_interrupts = TRUE)


msf <- tbl(db, Id(schema = "crsp", table = "msf"))
msi <- tbl(db, Id(schema = "crsp", table = "msi"))
ccmxpf_lnkhist <- tbl(db, Id(schema = "crsp",
```

```
                               table = "ccmxpf_lnkhist"))
stocknames <- tbl(db, Id(schema = "crsp",
                         table = "stocknames"))


funda <- tbl(db, Id(schema = "comp", table = "funda"))
fundq <- tbl(db, Id(schema = "comp", table = "fundq"))
```

11.2.1 Announcement dates and returns data

Getting earnings announcement dates involved significant data-collection effort for Ball and
Brown (1968). Fortunately, as discussed in Ball and Brown (2019), quarterly Compustat
(`comp.fundq`) has data on earnings announcement dates from roughly 1971 onwards. Like
Ball and Brown (1968), we are only interested in fourth quarters and firms with 31 December
year-ends. Because we will need to line up these dates with data from monthly CRSP
(`crsp.msf`), we create an `annc_month` variable.

```
annc_events <-
  fundq |>
  filter(indfmt == "INDL", datafmt == "STD",
         consol == "C", popsrc == "D") |>
  filter(fqtr == 4, fyr == 12, !is.na(rdq)) |>
  select(gvkey, datadate, rdq) |>
  mutate(annc_month = as.Date(floor_date(rdq, unit = "month")))
```

To compile returns for months $t-11$ through $t+6$ for each earnings announcement date
(t) (as Ball and Brown (1968) and Nichols and Wahlen (2004) do), we will need the `date`
values on CRSP associated with each of those months. We will create a table `td_link` that
will provide the link between announcement events in `annc_events` and dates on CRSP's
monthly stock file (`crsp.msf`).

The first step is to create a table (`crsp_dates`) that orders the dates on monthly CRSP
and assigns each month a corresponding "trading date" value (`td`), which is 1 for the first
month, 2 for the second month, and so on. Because the `date` values on `crsp.msf` line up
with the `date` values on `crsp.msi`, we can use the latter (much smaller) table.

```
crsp_dates <-
  msi |>
  select(date) |>
  window_order(date) |>
  mutate(td = row_number()) |>
  mutate(month = as.Date(floor_date(date, unit = "month")))


crsp_dates |> collect(n = 10)
```

```
# A tibble: 10 x 3
   date           td month
   <date>      <dbl> <date>
 1 1925-12-31      1 1925-12-01
 2 1926-01-30      2 1926-01-01
 3 1926-02-27      3 1926-02-01
 4 1926-03-31      4 1926-03-01
 5 1926-04-30      5 1926-04-01
```

```
 6 1926-05-28     6 1926-05-01
 7 1926-06-30     7 1926-06-01
 8 1926-07-31     8 1926-07-01
 9 1926-08-31     9 1926-08-01
10 1926-09-30    10 1926-09-01
```

We want to construct a table that allows us to link earnings announcements (`annc_events`) with returns from `crsp.msf` Because we are only interested in months where returns are available, we can obtain the set of potential announcement months from `crsp_dates`. The table `annc_months` has each value of `annc_month` and its corresponding `annc_td` from `crsp_dates`, along with the boundaries of the window that contains all values of `td` within the range $(t - 11, t + 6)$, where t is the announcement month.

```
annc_months <-
  crsp_dates |>
  select(month, td) |>
  rename(annc_month = month, annc_td = td) |>
  mutate(start_td = annc_td - 11L,
         end_td = annc_td + 6L)

annc_months |> collect(n = 10)
```

```
# A tibble: 10 x 4
   annc_month annc_td start_td end_td
   <date>       <dbl>    <dbl>  <dbl>
 1 1925-12-01       1      -10      7
 2 1926-01-01       2       -9      8
 3 1926-02-01       3       -8      9
 4 1926-03-01       4       -7     10
 5 1926-04-01       5       -6     11
 6 1926-05-01       6       -5     12
 7 1926-06-01       7       -4     13
 8 1926-07-01       8       -3     14
 9 1926-08-01       9       -2     15
10 1926-09-01      10       -1     16
```

We can then join `annc_months` with `crsp_dates` to create the table `td_link`.

```
td_link <-
  crsp_dates |>
  inner_join(annc_months, join_by(between(td, start_td, end_td))) |>
  mutate(rel_td = td - annc_td) |>
  select(annc_month, rel_td, date)
```

Here are the data for one `annc_month`:

```
td_link |>
  filter(annc_month == "2001-04-01") |>
  collect() |>
  print(n = Inf)
```

```
# A tibble: 18 x 3
   annc_month rel_td date
   <date>      <dbl> <date>
```

```
 1 2001-04-01      6 2001-10-31
 2 2001-04-01      5 2001-09-28
 3 2001-04-01      4 2001-08-31
 4 2001-04-01      3 2001-07-31
 5 2001-04-01      2 2001-06-29
 6 2001-04-01      1 2001-05-31
 7 2001-04-01      0 2001-04-30
 8 2001-04-01     -1 2001-03-30
 9 2001-04-01     -2 2001-02-28
10 2001-04-01     -3 2001-01-31
11 2001-04-01     -4 2000-12-29
12 2001-04-01     -5 2000-11-30
13 2001-04-01     -6 2000-10-31
14 2001-04-01     -7 2000-09-29
15 2001-04-01     -8 2000-08-31
16 2001-04-01     -9 2000-07-31
17 2001-04-01    -10 2000-06-30
18 2001-04-01    -11 2000-05-31
```

We use `ccm_link` (as used in Chapter 7) to connect earnings announcement dates on Compustat with returns from CRSP.

```
ccm_link <-
  ccmxpf_lnkhist |>
  filter(linktype %in% c("LC", "LU", "LS"),
         linkprim %in% c("C", "P")) |>
  rename(permno = lpermno) |>
  mutate(linkenddt = coalesce(linkenddt, max(linkenddt, na.rm = TRUE))) |>
  select(gvkey, permno, linkdt, linkenddt)
```

Nichols and Wahlen (2004) focus on firms listed on NYSE, AMEX, and NASDAQ, which correspond to firms with `exchcd` values of 1, 2, and 3, respectively. The value of `exchcd` for each firm at each point in time is found on `crsp.stocknames`. Following Nichols and Wahlen (2004), we get data on fiscal years from 1988 to 2002 (2004, p. 270).

```
rets_all <-
  annc_events |>
  inner_join(td_link, by = "annc_month") |>
  inner_join(ccm_link, by = "gvkey") |>
  filter(annc_month >= linkdt,
         annc_month <= linkenddt | is.na(linkenddt)) |>
  inner_join(msf, by = c("permno", "date")) |>
  inner_join(stocknames, by = "permno") |>
  filter(between(date, namedt, nameenddt),
         exchcd %in% c(1, 2, 3)) |>
  select(gvkey, datadate, rel_td, permno, date, ret) |>
  filter(between(year(datadate), 1987L, 2002L)) |>
  collect()
```

To keep things straightforward, we focus on firms that have returns for each month in the $(t-11, t+6)$ window and the table `full_panel` identifies these firms.

```
full_panel <-
  rets_all |>
  group_by(gvkey, datadate) |>
  mutate(n_obs = n()) |>
  ungroup() |>
  filter(n_obs == max(n_obs)) |>
  select(gvkey, datadate)

rets <-
  rets_all |>
  semi_join(full_panel, by = c("gvkey", "datadate"))
```

Note that, unlike other early papers (e.g., Beaver, 1968; Fama et al., 1969), Ball and Brown (1968) do not exclude observations due to known confounding events.[5]

11.2.2 Data on size-portfolio returns

Ball and Brown (1968) focus on **abnormal returns** and estimate a market model with firm-specific coefficients as the basis for estimating residual returns, which they denote API. The use of residuals from a market model addresses a concern about cross-sectional correlation that would arise if raw returns were used. Ball and Brown (1968) note that about 10% of returns are due to industry factors, but conclude that the likely impact of this on inference is likely to be small.

In contrast, Nichols and Wahlen (2004) use **size-adjusted returns** as their measure of abnormal returns. To calculate size-adjusted returns, we get two kinds of data from the website of Ken French[6] (as seen in Chapter 9).

First, we get data on size-decile returns. Ken French's website supplies a comma-delimited text file containing monthly and annual data for value-weighted and equal-weighted portfolio returns.

```
t <- "Portfolios_Formed_on_ME_CSV.zip"
url <- str_c("http://mba.tuck.dartmouth.edu",
             "/pages/faculty/ken.french/ftp/",
             "Portfolios_Formed_on_ME_CSV.zip")
if (!file.exists(t)) download.file(url, t)
```

From inspection of the downloaded text file, we observe that there are several data sets in this file. We want monthly returns and will extract both value-weighted and equal-weighted data. We see that the equal-weighted returns begin with a row starting with text `Equal Weight Returns -- Monthly` and end a few rows before a row starting with text `Value Weight Returns -- Annual`.

---

[5]This issue seems related to that discussed on p.164, where Ball and Brown (1968) state "our prediction [is] that, or certain months around the report dates, the expected values of the v_j's are nonzero." They defend the absence of an exclusion period on the basis that there is a low, observed autocorrelation in the v_j's, and in no case was the stock return regression fitted over less than 100 observations." (Ball and Brown, 1968, p. 164). But note that the basis for assuming that the expected value of v_j is nonzero in any given month is much less clear in this setting than it was in Fama et al. (1969). Given that a company will announce earnings in a particular month, this announcement could be either good or bad news, so the expected abnormal return seems likely to be zero.

[6]http://mba.tuck.dartmouth.edu/pages/faculty/ken.french/data_library.html

```
# Determine breakpoints (lines) for different tables
temp <- read_lines(t)
vw_start <- str_which(temp, "^\\s+Value Weight Returns -- Monthly")
vw_end <- str_which(temp, "^\\s+Equal Weight Returns -- Monthly") - 4

ew_start <- str_which(temp, "^\\s+Equal Weight Returns -- Monthly")
ew_end <- str_which(temp, "^\\s+Value Weight Returns -- Annual") - 4
```

Having identified these separating rows, we can use the following function to read in the data set and organize the associated data tables appropriately. Note that `NA` values are represented as `-99.99` in the text files and that the dates have a form `yyyymm` that we convert to dates of form `yyyy-mm-01`, which we call `month`.

While the original data come in a "wide" format with returns at every fifth percentile, we rearrange the data into a "long" format, retain only the deciles (i.e., every tenth percentile), and rename the decile labels from `Lo 10`, `Dec 2`, ..., `Dec 9`, and `Hi 10` to 1, 2, ..., 9, and 10.

```
read_data <- function(start, end) {

  Sys.setenv(VROOM_CONNECTION_SIZE = 500000)

  fix_names <- function(names) {
    str_replace_all(names, "^$", "date")
  }

  read_csv(t, skip = start, n_max = end - start,
           na = "-99.99",
           name_repair = fix_names,
           show_col_types = FALSE) |>
    mutate(month = ymd(str_c(date, "01"))) |>
    select(-date) |>
    pivot_longer(names_to = "quantile",
                 values_to = "ret",
                 cols = -month) |>
    mutate(ret = ret / 100,
           decile = case_when(quantile == "Hi 10" ~ "10",
                              quantile == "Lo 10" ~ "1",
                              str_detect(quantile, "^Dec ") ~
                                sub("^Dec ", "", quantile)),
           decile = as.integer(decile)) |>
    filter(!is.na(decile)) |>
    select(-quantile)
}
```

Now we can apply this function to extract the relevant data, which we combine into a single data frame `size_rets`.

```
vw_rets <-
  read_data(vw_start, vw_end) |>
  rename(vw_ret = ret)

ew_rets <-
```

```
  read_data(ew_start, ew_end) |>
  rename(ew_ret = ret)

size_rets <-
  ew_rets |>
  inner_join(vw_rets, by = c("month", "decile")) |>
  select(month, decile, everything())

size_rets
```

```
# A tibble: 11,770 x 4
   month      decile  ew_ret   vw_ret
   <date>     <int>   <dbl>    <dbl>
 1 1926-07-01      1 -0.0142 -0.0012
 2 1926-07-01      2  0.0029  0.0052
 3 1926-07-01      3 -0.0015 -0.0005
 4 1926-07-01      4  0.0088  0.0082
 5 1926-07-01      5  0.0145  0.0139
 6 1926-07-01      6  0.0185  0.0189
 7 1926-07-01      7  0.0163  0.0162
 8 1926-07-01      8  0.0138  0.0129
 9 1926-07-01      9  0.0338  0.0353
10 1926-07-01     10  0.0329  0.0371
# i 11,760 more rows
```

The second set of data we need to get from Ken French's website is data on the cut-offs we will use in assigning firms to decile portfolios in calculating size-adjusted returns.

Again the original data come in a "wide" format with cut-offs at every fifth percentile, so again we rearrange the data into a "long" format, retain only the deciles (i.e., every tenth percentile), and rename the decile labels from p10, p20, ..., p90, and p100, to 1, 2, ..., 9, and 10.[7] Also, we are only interested in the cut-offs for December in each year and use `filter()` to retain only these.

```
me_breakpoints_raw <-
  read_csv(t, skip = 1,
           col_names = c("month", "n",
                         str_c("p", seq(from = 5, to = 100, by = 5))),
           col_types = "c",
           n_max = str_which(temp, "^Copyright") - 3) |>
  mutate(month = ymd(str_c(month, "01"))) |>
  select(-ends_with("5"), -n) |>
  pivot_longer(cols = - month,
               names_to = "decile",
               values_to = "cutoff") |>
  mutate(decile = str_replace(decile, "^p(.*)0$", "\\1")) |>
  mutate(decile = as.integer(decile))
```

Finally, we organize the data to facilitate their use in joining with other data. Specifically, we create variables for the range of values covered by each decile (from `me_min` to `me_max`).

---

[7]Unlike the data set on returns above, there are no column labels in this data set and we are making those ourselves here.

We specify the minimum value for the first decile as zero and the maximum value for the tenth decile to infinity (`Inf`).

```
me_breakpoints <-
  me_breakpoints_raw |>
  group_by(month) |>
  arrange(decile) |>
  mutate(me_min = coalesce(lag(cutoff), 0), me_max = cutoff) |>
  mutate(me_max = if_else(decile == 10, Inf, me_max)) |>
  select(-cutoff) |>
  ungroup() |>
  arrange(month, decile)
```

To assign stocks to size deciles, we collect data on market capitalization from `crsp.msf`

```
me_values <-
  msf |>
  mutate(mktcap = abs(prc) * shrout / 1000) |>
  select(permno, date, mktcap) |>
  mutate(month = as.Date(floor_date(date, unit = "month"))) |>
  filter(month(month) == 12) |>
  collect()
```

We can compare market capitalization for each firm-year with the cut-offs in `me_breakpoints` to obtain its decile assignment.

```
me_decile_assignments <-
  me_values |>
  inner_join(me_breakpoints,
             join_by(month, mktcap >= me_min, mktcap < me_max)) |>
  mutate(year = as.integer(year(date)) + 1L) |>
  select(permno, year, decile)
```

11.2.3 Earnings news variables

The main issue that Ball and Brown (1968) need to tackle regarding earnings news is one that persists in accounting research today: how does one measure the unanticipated component of income? Alternatively, how does one estimate earnings expectations of the "market"?

Ball and Brown (1968) use two measures of expected earnings. The first is a naive model that simply "predicts that income will be the same for this year as for the last" (1968, p. 163). The second uses a model that estimates a relationship between the changes in income for a firm and for the market and then applies that relationship to the contemporaneous observation of the market's income. Note that this is an interesting variable: the equity market does not know the income for all firms at the start of the year. So the expectation is conditional with respect to a rather peculiar information set. In effect, the question is whether, given information about the market's earnings and market returns, information about accounting earnings helps predict the unexpected portion of earnings. In any case, the main results (see the famous Figure 1) are robust to the choice of the expectations model.

```
news <-
  funda |>
```

```
filter(indfmt == "INDL", datafmt == "STD",
       consol == "C", popsrc == "D") |>
filter(fyr == 12) |>
group_by(gvkey) |>
window_order(datadate) |>
mutate(lag_ibc = lag(ibc),
       lag_oancf = lag(oancf),
       lag_at = lag(at),
       lag_fyear = lag(fyear)) |>
ungroup() |>
filter(between(fyear, 1987, 2002),
       lag_fyear + 1 == fyear) |>
mutate(earn_chg = if_else(lag_at > 0, (ibc - lag_ibc) / lag_at, NA),
       cfo_chg = if_else(lag_at > 0, (oancf - lag_oancf) / lag_at, NA),
       earn_gn = earn_chg > 0,
       cfo_gn = cfo_chg > 0) |>
filter(!is.na(cfo_gn), !is.na(earn_gn)) |>
select(gvkey, datadate, earn_chg, cfo_chg, earn_gn, cfo_gn) |>
group_by(datadate) |>
mutate(earn_decile = ntile(earn_chg, 10),
       cfo_decile = ntile(cfo_chg, 10)) |>
ungroup() |>
collect()
```

11.2.4 Figure 1 of Ball and Brown (1968)

We can now merge our data tables to create the data set we can use to make variants of
Figure 1 of Ball and Brown (1968).

```
merged <-
  news |>
  mutate(year = year(datadate)) |>
  inner_join(rets, by = c("gvkey", "datadate")) |>
  inner_join(me_decile_assignments, by = c("permno", "year")) |>
  mutate(month = floor_date(date, unit = "month")) |>
  inner_join(size_rets, by = c("decile", "month")) |>
  select(-permno, -month)
```

To prepare the data for our plot, we first need to accumulate returns over time for each
firm. We then need to aggregate these returns by portfolio (here `earn_gn`) and relative
trading date (`rel_td`). Following Ball and Brown (1968), we calculate abnormal returns by
subtracting market returns from the portfolio returns. Here we calculate measures using
both equal-weighted (`ew_ret`) and value-weighted (`vw_ret`) market returns.

```
plot_data <-
  merged |>
  filter(!is.na(ret)) |>
  group_by(gvkey, datadate) |>
  arrange(rel_td) |>
  mutate(across(ends_with("ret"),
                \(x) cumprod(1 + x))) |>
```

```
group_by(rel_td, earn_gn) |>
summarize(across(ends_with("ret"),
                 \(x) mean(x, na.rm = TRUE)),
          .groups = "drop") |>
mutate(aret_ew = ret - ew_ret, aret_vw = ret - vw_ret)
```

Figure 1 of Ball and Brown (1968) confirms that a picture is worth a thousand words. We produce our analogue of Figure 1 in Figure 11.1.

```
plot_data |>
  mutate(`Earnings news` = if_else(earn_gn, "Good", "Bad")) |>
  ggplot(aes(x = rel_td, y = aret_ew, linetype = `Earnings news`)) +
  geom_line() +
  theme(legend.position = "bottom")
```

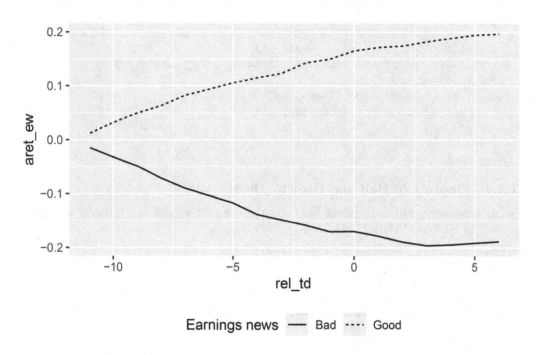

FIGURE 11.1
Replication of Figure 1 of Ball and Brown (1968)

11.2.5   Exercises

1. From the data below, we see that the upper bound for the tenth decile is about US$544 billion. How can we reconcile this with the existence of firms with market capitalizations over US$1 trillion? *Bonus*: Using data from `crsp.msf`, identify the firm whose market capitalization was US$544 billion in December 2020? (*Hint*: For the bonus question, you can add a `filter()` to code in the template to obtain the answer. Why do we need to group by `permco`, not `permno`, to find the answer?)

```
me_breakpoints_raw |>
  filter(month == "2020-12-01")
```

```
# A tibble: 10 x 3
     month      decile  cutoff
     <date>     <int>   <dbl>
 1 2020-12-01      1     327.
 2 2020-12-01      2     756.
 3 2020-12-01      3    1439.
 4 2020-12-01      4    2447.
 5 2020-12-01      5    3655.
 6 2020-12-01      6    5544
 7 2020-12-01      7    9656.
 8 2020-12-01      8   17056.
 9 2020-12-01      9   37006.
10 2020-12-01     10  543615.
```

2. To keep things straightforward, we focused on firms that have returns for each month in the $(t-11, t+6)$ window. Can you tell what approach Nichols and Wahlen (2004) took with regard to this issue?

3. Table 2 of Nichols and Wahlen (2004) measures cumulative abnormal returns as the "cumulative raw return minus cumulative size decile portfolio to which the firm begins." Apart from the use of a size-decile portfolio rather than some other market index, how does this measure differ from the Abnormal Performance Index (API) defined on p.168 of Ball and Brown (1968)? Adjust the measure depicted in the replication of Figure 1 to more closely reflect the API definition used in Ball and Brown (1968) (but retaining the size-decile as the benchmark). Does this tweak significantly affect the results? Which approach seems most appropriate? That of Nichols and Wahlen (2004) or that of Ball and Brown (1968)?

4. Create an alternative version of Figure 11.1 using the sign of "news" about cash flows in the place of income news. Do your results broadly line up with those in Panel A of Figure 2 of Nichols and Wahlen (2004)? Do these results imply that accounting income is inherently more informative than cash flows from operations? Why or why not?

5. Create an alternative version of the figure above focused on the extreme earnings deciles in place of the good-bad news dichotomy. Do your results broadly line up with those in Panel B of Figure 2 of Nichols and Wahlen (2004)?

6. Calculate *AI* by year following the formula on p. 175 of Ball and Brown (1968) (there denoted as II_0). You may find it helpful to start with the code producing `plot_data` above. You may also find it helpful to use the function `pivot_wider` to get information about portfolios in each year into a single row. Note that you will only be interested in rows at $t = 0$ (e.g., `filter(rel_td == 0)`).

7. Calculate *NI* by year following the formula on p. 175 of Ball and Brown (1968) (there denoted as NI_0). Note that you will only be interested in rows at $t = 0$ (e.g., `filter(rel_td == 0)`).

8. Using the data on *NI* and *AI* from above, create a plot of *AI/NI* like that in Figure 2 of Ball and Brown (2019). Do you observe similar results to those shown in Figure 2 of Ball and Brown (2019)?

12

Beaver (1968)

In this chapter, we cover Beaver (1968), the second winner of the Seminal Contributions to Accounting Literature Award.[1] Beaver (1968) followed Ball and Brown (1968) and examines whether investors appear to react to earnings announcements.

Beaver (1968) uses two approaches to measure investor reaction. The first approach measures market reaction using trading volume and the second uses the squared return residuals, where return residuals are the difference between observed returns and fitted returns using a market model (1968, p. 78).

In this chapter, we start with a close study of Beaver (1968) itself before introducing Bamber et al. (2000), which provides a critique of Beaver (1968). We then conduct a "replication" of Beaver (1968) using a more recent sample period and use that replication to think about Beaver (1968) and the issues raised by Bamber et al. (2000).

> 💡 Tip
>
> The code in this chapter uses the packages listed below. For instructions on how to set up your computer to use the code found in this book, see Section 1.2. Quarto templates for the exercises below are available on GitHub[a].
>
> ---
> [a]https://github.com/iangow/far_templates/blob/main/README.md

```
library(dplyr)
library(DBI)
library(ggplot2)
library(farr)
library(lubridate)    # For year()
library(tidyr)        # For fill()
library(dbplyr)       # For copy_inline()
```

12.1 Market reactions to earnings announcements

Results for the volume analysis are provided in Figures 1 and 3 of Beaver (1968). Figure 1 provides analysis of unadjusted volume and Figure 3 provides analysis of residual volume, which is calculated as the residual from a kind of "volume market model" (see p. 76 of Beaver, 1968). Figure 6 contains the results of the analysis of return residuals.

---
[1]The list of winners can be found at https://go.unimelb.edu.au/yzw8.

DOI: 10.1201/9781003456230-12

Some aspects of the Beaver (1968) are confusing unless read closely, so we provide some explanation here. The sample in Beaver (1968) comprises 506 earnings announcements for 143 firms. Each earnings announcement is associated with a 17-week announcement window ($t = -8, \dots, +8$), which Beaver (1968) calls the "report period". Observations on returns and volumes within the sample period for a given firm, but outside report periods for that firm, make up the non-report period for that firm. There are 261 periods in the sample as there are 261 weeks between the start of 1961 and the end of 1965.

Figure 2 of Beaver (1968) provides information about the distribution of residual volume measured for each of the 261 weeks. The formula in the top-left of this figure is not strictly correct, as the number of firms for which residual volume would be available would generally be less than 143 due to the exclusion of values for firms during their report periods.

Figures 4 and 7 of Beaver (1968) are used to support inferences discussed in the text of the paper. The dashed line in Figure 4 is presumably around the value of 179 ($179 \approx 506 \times 0.354$, as 35.4% of residual volume values in the non-report period are positive [see discussion on p. 76]), even though the mean of residual volume is (by definition) equal to zero during the non-report period. The dashed line in Figure 7 is presumably around the value of 132 ($132 \approx 506 \times 0.26$, as 26% of return residuals in the non-report period exceed one [see footnote 26 on p. 80]), even though the mean value is (by definition) equal to one during the non-report period. Note that the figures in Beaver (1968) would have been created using tools such as rulers, pens, and scissors rather than anything like `ggplot2`.

12.1.1 Discussion questions

1. How do the research questions of Beaver (1968) and Ball and Brown (1968) differ? If there is overlap, do you think that one paper provides superior evidence to the other? Or are they just different?

2. What differences are there in the data (e.g., sample period) used in Beaver (1968) from the data used in Ball and Brown (1968)? Why do these differences exist?

3. Do the reasons given by Beaver (1968) for his sample selection criteria seem reasonable? Do you think these were made prior to looking at results? (Why or why not?) Does it matter at what stage in the research process these decisions were made?

4. Which do you think is the better variable—price or volume—for addressing the research questions of Beaver (1968)? Do you think it is helpful to have both variables?

5. Beaver (1968) compares event-week volume and volatility with their average equivalents in non-event periods. Ball and Shivakumar (2008) "quantify the relative importance of earnings announcements in providing new information to the share market" and argue that earnings announcements provide relatively little of this information. If volume is a measure of information content, what does Figure 1 of Beaver (1968) imply regarding the proportion of information conveyed during earnings announcement weeks? Can this be reconciled with reported results in Beaver (1968) and the arguments in Ball and Shivakumar (2008)?

6. Beaver (1968) discusses the statistical significance of his results on p. 77 (residual volume analysis) and pp. 81–82 (return residual analysis). Why do think Beaver (1968) uses the approaches discussed there? How might you evaluate statistical significance more formally if you were writing the paper today?

7. The primary analyses in Beaver (1968) are provided in plots. While the cost of producing plots has surely plummeted since 1968, we generally do not see primary analyses presented as plots today. Why do you think this is the case?

12.2   A re-evaluation of Beaver (1968)

Bamber et al. (2000) analyse the setting of Beaver (1968) using the same sample—announcement of annual earnings between 1961 and 1965—but consider modifications to sample selection and empirical tests. Bamber et al. (2000) replicate the results of Beaver (1968), but argue that these are not robust to two alternative research design choices. First, they find that Beaver's "focus on mean effects obscures the fact that most individual earnings announcements are not associated with unusual price reactions" (2000, p. 105). Second, rather than apply Beaver's sample selection criteria, Bamber et al. (2000, p. 105) "choose an alternative set of firms that would have been at least equally interesting at the time when there was no empirical evidence on the information content of any firms' [sic] earnings announcements—the era's Fortune 200 firms."

Bamber et al. (2000) suggest that "had the initial information content studies made different research design choices ... the apparent information content would have been much less dramatic, and perhaps even non-existent. [After early studies] the conclusion that earnings announcements convey new information to the market soon became the received wisdom."

In this chapter, we will revisit the question of whether "earnings announcements convey new information to the market" using the basic research design of Beaver (1968) and the same focus on NYSE firms, but using all of them (subject to data availability), focusing on *recent* earnings announcements—specifically those related to fiscal periods ending between 1 January 2010 and 31 December 2019—and making a few simplifications to the empirical analysis of Beaver (1968).

12.2.1   Core set of events from Compustat

We begin by connecting to the several tables we will use in our analysis.

```
db <- dbConnect(RPostgres::Postgres(), bigint = "integer")

dsf <- tbl(db, Id(schema = "crsp", table = "dsf"))
dsi <- tbl(db, Id(schema = "crsp", table = "dsi"))
msf <- tbl(db, Id(schema = "crsp", table = "msf"))
msi <- tbl(db, Id(schema = "crsp", table = "msi"))
ccmxpf_lnkhist <- tbl(db, Id(schema = "crsp", table = "ccmxpf_lnkhist"))
stocknames <- tbl(db, Id(schema = "crsp", table = "stocknames"))

funda <- tbl(db, Id(schema = "comp", table = "funda"))
fundq <- tbl(db, Id(schema = "comp", table = "fundq"))
```

Our next task is to collect the set of earnings announcements that we will use in our analysis. Beaver (1968) obtained earnings announcement dates from the *Wall Street Journal Index* (1968, p. 72), but for our newer sample period, we can use the field `rdq` found on the

Compustat quarterly data file (`comp.fundq`). Following Beaver (1968), we focus on annual earnings announcements.[2]

```
first_date <- "2010-01-01"
last_date <- "2019-12-31"

earn_annc_dates <-
  fundq |>
  filter(indfmt == "INDL", datafmt == "STD",
         consol == "C", popsrc == "D") |>
  filter(!is.na(rdq), fqtr == 4L) |>
  select(gvkey, datadate, rdq) |>
  filter(between(datadate, first_date, last_date)) |>
  collect()
```

The data frame `earn_annc_dates` provides our set of event dates. Like Beaver (1968), we also need a set of non-event dates for comparison. Implicitly, Beaver (1968) used weeks $(-8, \dots, -1)$ and $(+1, \dots, +8)$ as non-event weeks. In our mini-study, we will use daily data and look at dates between 20 **trading days** before and 20 trading days after the announcement of earnings.

For our purposes, a trading day (or **trading date**) will be a date on which CRSP stocks traded. In other words, dates found on `crsp.dsf` will be considered trading days. In the last chapter, we confirmed that the same dates are found on `crsp.dsf` and `crsp.dsi`, so we can use either. Because each date is only found once on `crsp.dsi`, we will use that smaller table.

12.2.2 Replication for a single event

For concreteness, we begin with a focus on a single earnings announcement: CTI BioPharma Corporation, a biopharmaceutical company with GVKEY of `064515` that announced earnings for the year ending 31 December 2017 on 2018-03-04.

```
single_event <-
  earn_annc_dates |>
  filter(gvkey == "064515", rdq == "2018-03-04")
```

First, we need to map our GVKEY to a PERMNO and we use `ccm_link` as seen in Chapter 7 for this purpose:

```
ccm_link <-
  ccmxpf_lnkhist |>
  filter(linktype %in% c("LC", "LU", "LS"),
         linkprim %in% c("C", "P")) |>
  rename(permno = lpermno) |>
  mutate(linkenddt = coalesce(linkenddt, max(linkenddt, na.rm = TRUE))) |>
  collect()
```

Now we can make our link table, which here covers just one firm.

```
single_event_link <-
  single_event |>
```

---

[2]As pointed out by Dechow et al. (2014), quarterly earnings announcements were not required during the sample period of Beaver (1968).

```
inner_join(ccm_link, join_by(gvkey, rdq >= linkdt, rdq <= linkenddt)) |>
select(gvkey, datadate, permno)
```

Next, we need to identify the dates for which we need return and volume data, namely the period extending from twenty trading days before the earnings announcement to twenty days after.

One issue we might consider is: when (at what time) did the firm announce earnings? If the firm announced *after* trading hours, then the first opportunity that the market would have to react to that information will be the next day. But, for our purposes, we don't need to be this precise and we will assume that, if a firm announces earnings on a trading date, then that date will be **trading day zero**.

A second issue to consider is: What do we do with announcements that occur on non-trading dates? For example, CTI BioPharma Corporation's earnings announcement occurred on Sunday (a firm might also announce earnings on a public holiday). In this case, it seems reasonable to assume that trading day zero will be the next available trading day (e.g., if a firm announces earnings on Sunday and the Monday is a trading date, then that Monday would be the relevant day zero).

The final issue relates to the arithmetic of counting forwards and backwards. Suppose a firm announced earnings on 2018-03-05, a Monday. Because that date is a trading date, day zero would be 2018-03-05. But day -1 would be 2018-03-02, as 2018-03-03 and 2018-03-04 are non-trading dates (a Saturday and a Sunday, respectively).

Let's address the last issue first. If we take the dates on `crsp.dsi` and order the data by `date`, then we can create a variable `td` using the `row_number()` function where `td` stands for "trading day" and represents, for each date, the number of trading days since the start of `crsp.dsi`.

```
trading_dates <-
  dsi |>
  select(date) |>
  collect() |>
  arrange(date) |>
  mutate(td = row_number())
```

In our example, 2018-03-05 has `td` equal to 24333 and subtracting 1 from that `td` gives 24332, which is associated with 2018-03-02.

```
trading_dates |>
  filter(date >= "2018-02-27")
```

```
# A tibble: 1,471 x 2
   date            td
   <date>       <int>
 1 2018-02-27   24329
 2 2018-02-28   24330
 3 2018-03-01   24331
 4 2018-03-02   24332
 5 2018-03-05   24333
 6 2018-03-06   24334
 7 2018-03-07   24335
 8 2018-03-08   24336
```

```
 9 2018-03-09 24337
10 2018-03-12 24338
# i 1,461 more rows
```

Similar calculations can be done with any number of trading days.[3] This solves the final issue, but what about the second issue of announcements on non-trading dates? To address this, we make a table `annc_dates` as follows:

1. Create a table with all possible announcement dates. (In this case, we use the `seq` function to create a list of *all* dates over the period represented on `crsp.dsi`.)
2. Where possible, line these dates up with dates (and their associated `td` values) on `trading_dates`. (The "where possible" translates into using a `left_join()` in this case.)
3. Fill in the missing `td` entries by grabbing the next available `td`. (For this, we use the `fill()` function from the `tidyr` package.)

```
min_date <-
  trading_dates |>
  summarize(min(date, na.rm = TRUE)) |>
  pull()

max_date <-
  trading_dates |>
  summarize(max(date, na.rm = TRUE)) |>
  pull()

annc_dates <-
  tibble(annc_date = seq(min_date, max_date, 1)) |>
  left_join(trading_dates, by = join_by(annc_date == date)) |>
  fill(td, .direction = "up")
```

To illustrate how we can use this table, let's return to our earnings announcement on 2018-03-04 where we're interested in returns running over the window from $t - 20$ to $t + 20$. We can see below that 2018-03-04 has `td` equal to 24333 (this is because the relevant day zero is 2018-03-05, which has `td` of 24333). So, we identify $t - 20$ as the date with `td` equal to 24313 and $t + 20$ as the date with `td` equal to 24353.

```
trading_dates |>
  filter(td %in% c(24313, 24353))
```

```
# A tibble: 2 x 2
  date          td
  <date>     <int>
1 2018-02-02 24313
2 2018-04-03 24353
```

```
days_before <- 20L
days_after <- 20L

single_event_window <-
  single_event |>
```

---

[3]We will demonstrate shortly how we actually use this table.

```
  left_join(annc_dates, by = join_by(rdq == annc_date)) |>
  mutate(start_td = td - days_before,
         end_td = td + days_after) |>
  inner_join(trading_dates, by = join_by(start_td == td)) |>
  rename(start_date = date) |>
  inner_join(trading_dates, by = join_by(end_td == td)) |>
  rename(end_date = date,
         event_td = td) |>
  select(-start_td, -end_td)

single_event_window
```

```
# A tibble: 1 x 6
  gvkey   datadate    rdq         event_td start_date end_date
  <chr>   <date>      <date>         <int> <date>     <date>
1 064515  2017-12-31  2018-03-04     24333 2018-02-02 2018-04-03
```

We merge this data set with our link table to make a table that we can link with CRSP.

```
single_event_window_permno <-
  single_event_window |>
  inner_join(single_event_link, by = c("gvkey", "datadate"))
```

Now we have the dates and PERMNO for which we want CRSP data; we now need to go grab those data. Here, we are interested in stock and market returns (`ret` and `ret_mkt`, respectively) and volume (`vol`) for each date.

```
mkt_rets <-
  dsf |>
  inner_join(dsi, by = "date") |>
  mutate(ret_mkt = ret - vwretd) |>
  select(permno, date, ret, ret_mkt, vol)
```

We next merge `mkt_rets` with our event data to create `single_event_crsp`. Note that we have a slight problem here because we want to merge data on events, which is found in the *local* data frame `single_event_window_permno`, with data from `crsp.dsf`, which is found in the WRDS database, which we have as a *remote* data frame. One approach would run `collect()` on `crsp.dsf` to turn it into a local data frame, but this would be very cumbersome (it's about 20 GB in size). Another approach would be to copy the local data frame to the WRDS PostgreSQL server, but WRDS does not allow us to write data to its server. Fortunately, we can use `copy_inline()` from the `dbplyr` package to create something like a temporary table on the server using local data.[4]

```
single_event_crsp <-
  copy_inline(db, single_event_window_permno) |>
  inner_join(mkt_rets, by = "permno") |>
  filter(date >= start_date, date <= end_date) |>
  select(gvkey, datadate, event_td, date, ret, ret_mkt, vol)
```

The final step is to convert certain date variables into event time. For example, 2018-02-02 has `td` of 21313, but this isn't particularly meaningful here. What *is* important is that

---

[4]In fact, `copy_inline()` was added to `dbplyr` after a request related to this book: https://github.com/t idyverse/dbplyr/issues/628.

2018-02-02 represents $t - 20$ for the earnings announcement on 2018-03-04. To convert each date to a "relative trading day", we just need to join our `single_event_crsp` table with `trading_dates` to get the relevant `td` data, then subtract the `event_td` from the retrieved `td`. The following code does this and puts the results in a new data frame.

```
single_event_tds <-
  single_event_crsp |>
  collect() |>
  inner_join(trading_dates, by = "date") |>
  mutate(relative_td = td - event_td) |>
  select(-event_td, -td)
```

Now we can do a simple plot (Figure 12.1) of the trading volume (`vol`) against the number of trading days from the earnings announcement (`relative_td`).

```
single_event_tds |>
  ggplot(aes(x = relative_td, y = vol)) +
  geom_line()
```

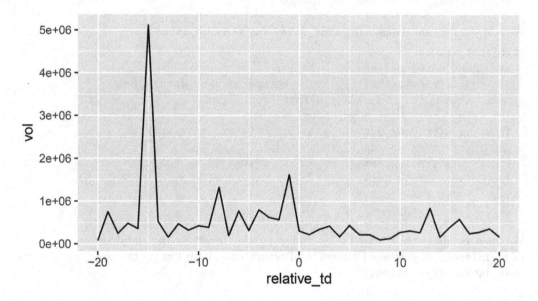

FIGURE 12.1
Return volatility around earnings announcement: Single event

Interestingly there is no volume spike on the earnings announcement date, but there is a small spike just before it and a larger one in February, when CTI BioPharma executed a stock offering. Note that there was actually no Sunday earnings announcement in this case; CTI BioPharma actually announced earnings at 4:01 pm EST on 7 March 2018[5] suggesting that the `rdq` on Compustat is simply incorrect.

---

[5]https://go.unimelb.edu.au/bzw8

12.2.3   Replication for the full sample

We now retrace the steps we took above, but using the full data set `earn_annc_dates` in place of the `single_event` data frame. While our goal is not to replicate Beaver (1968) precisely, we follow Beaver (1968) in focusing on NYSE-listed firms. Data on the exchanges that firms are listed on is found in `exchcd`, which is found on `crsp.stocknames`.[6]

```
nyse <-
  stocknames |>
  filter(exchcd == 1) |>
  select(permno, namedt, nameenddt) |>
  collect()
```

We add PERMNOs from `ccm_link` and limit the data to NYSE firms.

```
earn_annc_links <-
  earn_annc_dates |>
  inner_join(ccm_link, join_by(gvkey, rdq >= linkdt, rdq <= linkenddt)) |>
  semi_join(nyse, join_by(permno, rdq >= namedt, rdq <= nameenddt)) |>
  select(gvkey, datadate, rdq, permno)
```

We add the relevant dates using `annc_dates` and `trading_dates`.

```
earn_annc_windows <-
  earn_annc_dates |>
  inner_join(annc_dates, by = join_by(rdq == annc_date)) |>
  mutate(start_td = td - days_before,
         end_td = td + days_after) |>
  inner_join(trading_dates, by = join_by(start_td == td)) |>
  rename(start_date = date) |>
  inner_join(trading_dates, by = join_by(end_td == td)) |>
  rename(end_date = date,
         event_td = td) |>
  select(-start_td, -end_td)
```

We then combine these two tables and copy the result to the database server (using `copy_inline()` as discussed above), add return data from `mkt_rets`, and then pull the result into R using `collect()`.

```
earn_annc_window_permnos <-
  earn_annc_windows |>
  inner_join(earn_annc_links, by = c("gvkey", "datadate", "rdq"))

earn_annc_crsp <-
  mkt_rets |>
  inner_join(copy_inline(db, earn_annc_window_permnos),
             join_by(permno,
                     date >= start_date, date <= end_date)) |>
  select(gvkey, datadate, rdq, event_td, date, ret, ret_mkt, vol) |>
  collect()
```

---

[6]The variable `exchcd` is a **header variable**, meaning that there is one value for each PERMNO, even if a stock has traded on different exchanges over time. In Section 19.2.1, we are more careful and obtain the exchange on which a stock was traded on the relevant dates, but for our simple replication analysis, we ignore this detail.

The data frame `earn_annc_crsp` contains all the data on (raw and market-adjusted) returns and trading volumes that we need. The next step is to calculate relative trading dates, which are trading dates expressed in **event time**.

```
earn_annc_rets <-
  earn_annc_crsp |>
  inner_join(trading_dates, by = "date") |>
  mutate(relative_td = td - event_td)
```

We now calculate a measure of relative volume using the average volume for each stock over the window around each earnings announcement.

```
earn_annc_vols <-
  earn_annc_rets |>
  group_by(gvkey, datadate) |>
  mutate(avg_vol = mean(vol, na.rm = TRUE)) |>
  mutate(rel_vol = vol / avg_vol,
         year = year(datadate)) |>
  ungroup()
```

Finally, we calculate summary statistics for each year by trading date relative to the earnings announcement date (`relative_td`).

```
earn_annc_summ <-
  earn_annc_vols |>
  group_by(relative_td, year) |>
  summarize(obs = n(),
            mean_ret = mean(ret, na.rm = TRUE),
            mean_ret_mkt = mean(ret_mkt, na.rm = TRUE),
            mean_rel_vol = mean(rel_vol, na.rm = TRUE),
            sd_ret = sd(ret, na.rm = TRUE),
            sd_ret_mkt = sd(ret_mkt, na.rm = TRUE),
            mad_ret = mean(abs(ret), na.rm = TRUE),
            mad_ret_mkt = mean(abs(ret_mkt), na.rm = TRUE),
            .groups = "drop")
```

Now we can produce two plots. First, in Figure 12.2, we plot a measure of the standard deviation of market-adjusted returns. Each line represents a different year. The lines are largely indistinguishable from each other with the exception of 2015, which has high volatility throughout.

```
earn_annc_summ |>
  filter(year > 2010, year < 2019) |>
  mutate(year = as.factor(year)) |>
  ggplot(aes(x = relative_td, y = sd_ret_mkt,
             group = year, colour = year,
             linetype = year == 2015)) +
  geom_line()
```

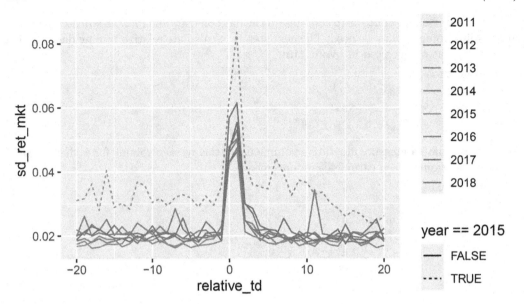

FIGURE 12.2
Residual return volatility around earnings announcements

Second, in Figure 12.3, we plot a measure of the relative trading volumes.

```
earn_annc_summ |>
  filter(year > 2010, year < 2019) |>
  mutate(year = as.factor(year)) |>
  ggplot(aes(x = relative_td, y = mean_rel_vol,
             group = year, color = year)) +
  geom_line()
```

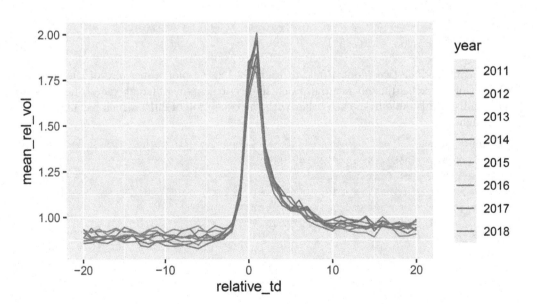

FIGURE 12.3
Relative trading volume around earnings announcements

12.3 Discussion questions

1. After reading Bamber et al. (2000), do the reasons given by Beaver (1968) for his sample selection criteria still seem reasonable? Why or why not?

2. Bamber et al. (2000) do a replication and an extension of Beaver (1968). Why was it important to include the replication? What alternative approaches to extension could Bamber et al. (2000) have considered? Does it make sense to limit extensions to the sample period, available data, and methods of Beaver (1968)? What do you think of the claim that "the first research bricks (i.e., Beaver, 1968) affect the whole wall [of accounting research]"?

3. What's the basic conclusion from Figures 12.2 and 12.3 in terms of whether "earnings announcements convey new information to the market"? Do the results support the conclusions made by subsequent researchers based on Beaver (1968)? Or do the concerns of Bamber et al. (2000) remain applicable?

4. In our replication analysis above, we made a number of measurement and design choices that differed from those made by Beaver (1968). What are those differences? Do you expect these to materially affect the tenor of the results? Do the choices we made seem appropriate if we were writing a research paper?

5. Figures 12.2 and 12.3 use *daily* data (Beaver, 1968 used weekly data). Apart from more **statistical power**, do the daily plots provide novel insights in this case?

6. What does the variable `mad_ret_mkt` on the data frame `earn_annc_summ` represent? Do results look different if this variable is used? Does this address the first concern about research design that Bamber et al. (2000) raise? If not, can you suggest (and apply) an alternative measure?

7. In Figures 12.2 and 12.3, two filters have been applied: `year > 2010` and `year < 2019`. Does it make sense to remove one or the other of these filters? What do you observe if you remove one or both of these? Can you explain these observations?

13

Event studies

In previous chapters, we studied papers that have examined the reaction of financial markets to information. Chapter 10 covered Fama et al. (1969), which studied the adjustment of prices to the information in stock splits. Chapter 11 examined Ball and Brown (1968), which shows that market returns over a year are correlated with earnings news for the period. Chapter 12 studied Beaver (1968), which shows that volume and price volatility increase around earnings announcements.

The decades since those early papers have seen significant evolution in how researchers study the market reaction to information and this chapter provides an introduction to current event-study methods.

> **Tip**
>
> The code in this chapter uses the packages listed below. For instructions on how to set up your computer to use the code found in this book, see Section 1.2. Quarto templates for the exercises below are available on GitHub.[a]
>
> ---
> [a]https://github.com/iangow/far_templates/blob/main/README.md

```
library(dplyr)
library(DBI)
library(ggplot2)
library(farr)
library(modelsummary)
library(dbplyr)          # window_order()
```

13.1 Overview

MacKinlay (1997, p. 13) defines an **event study** as one that, "using financial market data ... measures the impact of a specific event on the value of a firm. The usefulness of such a study comes from the fact that, assuming highly efficient markets, the effects of an event will be reflected immediately in security prices.[1] Thus a measure of the event's economic impact can be constructed using security prices observed over a relatively short time period."

MacKinlay (1997, p. 14) continues: "In the late 1960s seminal studies by Ray Ball and Philip Brown (1968) and Eugene Fama et al. (1969) introduced the methodology that is essentially the same as that which is in use today. Ball and Brown considered the information content

---
[1]In terms of Section 10.1, it's the "price is right" notion of market efficiency that applies here.

DOI: 10.1201/9781003456230-13

of earnings, and Fama et al. studied the effects of stock splits after removing the effects of simultaneous dividend increases."

Event studies examine the impact of a class of identifiable events on one or more variables of economic interest. In capital markets research, the variable of economic interest is typically the returns of a firm's shares around the event.

The basic ingredients of an event study are:

- A class of events: earnings announcements, merger announcements, stock splits, earnings forecast changes. In many research papers, these events represent the treatment of interest.
- An outcome variable of interest: accounting policy, market returns, trading volume.
- Control observations: observations for which the event did not occur.
- Control variables: additional variables that may be correlated with both the event—such as dividend announcements with earnings announcements or forecast revisions with earnings announcements—and also returns (see Chapter 4 for more on control variables).

Additionally, as we saw in Chapter 12, lining up observations in **event time** is a critical feature of most event studies.

13.1.1 Discussion questions

1. Does Ball and Brown (1968) meet the MacKinlay (1997) definition of an event study? What features are present and what, if any, are missing?

2. Does Beaver (1968) meet the MacKinlay (1997) definition of an event study? What features are present and what, if any, are missing?

13.2 The modern event study

The event study has evolved over time. While event studies today generally meet the definition in MacKinlay (1997), the event-study approach has changed as it has been adapted to a wider variety of situations.

One change is that researchers have become more interested in using event studies to understand the economic effects of regulation, rather than the market reaction to firm-specific announcements. Fama et al. (1969) studied stock-splits and Beaver (1968) studied earnings announcements. In each case, the events are firm-level events that are largely independent of each other (e.g., they are not excessively clustered in time). In contrast, each of the three more recent event studies we study below uses regulatory events such as events affecting the probability of legislation (Larcker et al., 2011; Zhang, 2007) or accounting standards going into effect (Khan et al., 2017).

A related change is an increased reliance on market efficiency, as the typical modern event study uses (in the words of MacKinlay, 1997) "security prices observed over a relatively short time period" as a "measure of the event's economic impact". Neither Fama et al. (1969) nor Beaver (1968) relies heavily on market efficiency in establishing that markets appear to react to stock-splits and earnings announcements, respectively, and neither study seeks to show whether stock-splits or earnings announcements create (or destroy) value. In contrast, the modern event study is often leaning on market efficiency to evaluate regulation. For example, using an event study to ask "do the FASB's standards add shareholder value?" (Khan et al., 2017) relies heavily on the market having an informed view of the kind implied by stronger forms of market efficiency.

A consequence of this changed emphasis is that often there are fewer independent observations for the researcher to work with. For example, the primary analysis of Zhang (2007) focuses on four event windows, far fewer than the 506 earnings announcements in Beaver (1968). As we will see, researchers often use supplementary analyses to address the relative paucity of data.

13.2.1   A small event study

To better understand the modern event study, we conduct a mini-study of our own. Suppose that we want to understand better the value-creation process at Apple with a particular emphasis on Apple's product development process. At the time of writing (mid-2022), Apple is the most valuable company in the world, with a market capitalization over US$2 trillion; so understanding how it creates value for shareholders may be of interest to researchers.

As Apple is notoriously secretive about its product pipeline, the media events at which its products are launched are closely watched affairs. For example, at the Macworld Conference & Expo San Francisco 2007 (9 January 2007), Apple announced the iPhone, which would go on to become Apple's primary revenue source and one of the largest phone products in the world. At an Apple Special Event on 27 January 2010, Apple announced the iPad, Apple's tablet computer.

So, to understand whether Apple's products create value for Apple shareholders, we might run an event study using Apple's media events as the events of interest.

The `farr` package includes the data frame `apple_events`, which is derived from data found on Wikipedia.[2] Let's look at the last few rows of this table:

```
tail(apple_events)
```

```
# A tibble: 6 x 3
  event               event_date  end_event_date
  <chr>               <date>      <date>
1 Apple Special Event 2019-09-10  2019-09-10
2 Apple Special Event 2019-12-02  2019-12-02
3 WWDC 2020           2020-06-22  2020-06-26
4 Apple Special Event 2020-09-15  2020-09-15
5 Apple Special Event 2020-10-13  2020-10-13
6 Apple Special Event 2020-11-10  2020-11-10
```

We will need return data from CRSP to conduct our event study. We first need to get Apple's PERMNO so we can look up returns on CRSP. Knowing that Apple's ticker is `AAPL` helps.

```
db <- dbConnect(RPostgres::Postgres())

stocknames <- tbl(db, Id(schema = "crsp", table = "stocknames"))
dsf <- tbl(db, Id(schema = "crsp", table = "dsf"))
dsi <- tbl(db, Id(schema = "crsp", table = "dsi"))
```

[2]https://en.wikipedia.org/wiki/List_of_Apple_Inc._media_events

```
idx_daily <- tbl(db, Id(schema = "comp", table = "idx_daily"))
```

```
apple_permno <-
  stocknames |>
  filter(ticker == "AAPL") |>
  select(permno) |>
  distinct() |>
  pull()
```

We then use Apple's PERMNO (`apple_permno` equals 14593) to get return data from CRSP. In this case, we will get daily returns for Apple (`ret`) from `crsp.dsf` and value-weighted "market" returns (`vwretd`) from `crsp.dsi` and calculate **market-adjusted returns** as ret - vwretd.[3] In this case, we will grab all returns since the start of 2005, which covers all the events on `apple_events`.

```
apple_rets <-
  dsf |>
  inner_join(dsi, by = "date") |>
  mutate(ret_mkt = ret - vwretd) |>
  select(permno, date, ret, ret_mkt, vol) |>
  filter(permno == apple_permno,
         date >= "2005-01-01") |>
  collect()
```

Unlike the earnings announcements that we studied in Chapter 12, Apple's media events extend over multiple days; so our event windows also need to extend over multiple days. To allow for some leakage of information in the day before the start of the media events and to allow the market some time to process the value implications of the media event, we will set our **event window** from one trading day before the start of each media event through to one day after the end of the media event. We will use the `get_event_dates()` function from the `farr` package to this end; behind the scenes,[4] `get_event_dates()` uses `get_trading_dates()` and `get_annc_dates()` to get tables like `trading_dates` and `annc_dates`, respectively, which we studied in Chapter 12.[5]

```
apple_event_dates <-
  apple_events |>
  mutate(permno = apple_permno) |>
  get_event_dates(db,
                  end_event_date = "end_event_date",
                  win_start = -1, win_end = +1)
```

```
tail(apple_event_dates)
```

```
# A tibble: 6 x 5
  permno event_date end_event_date start_date end_date
   <int> <date>     <date>         <date>     <date>
1  14593 2019-09-10 2019-09-10     2019-09-09 2019-09-11
2  14593 2019-12-02 2019-12-02     2019-11-29 2019-12-03
3  14593 2020-06-22 2020-06-26     2020-06-19 2020-06-29
```

---

[3]MacKinlay (1997, p. 18) points out that "the market-adjusted return model can be viewed as a restricted market model with α_i constrained to be zero and β_i constrained to be one."

[4]https://github.com/iangow/farr/blob/main/R/get_event_dates.R

[5]For more on `get_event_dates()`, type `? get_event_dates` in the console of RStudio.

```
4   14593 2020-09-15 2020-09-15      2020-09-14 2020-09-16
5   14593 2020-10-13 2020-10-13      2020-10-12 2020-10-14
6   14593 2020-11-10 2020-11-10      2020-11-09 2020-11-11
```

We now organize the data in a way that allows us to depict Apple's returns graphically over time including information about media events.

```
apple_data <-
  apple_rets |>
  left_join(apple_event_dates,
            join_by(permno, date >= start_date, date <= end_date)) |>
  mutate(is_event = !is.na(start_date)) |>
  select(permno, date, ret, ret_mkt, vol, is_event)
```

Now we have the data we need, we can calculate cumulative returns using the `cumprod()` function and then plot these returns over time.[6]

```
apple_data |>
  arrange(date) |>
  mutate(cumret = cumprod(1 + coalesce(ret, 0)),
         switch = coalesce(is_event != lead(is_event), FALSE)) |>
  ggplot(aes(x = date, y = cumret)) +
  geom_line() +
  geom_ribbon(aes(ymax = if_else(!is_event | switch, cumret, NA),
                  ymin = 0, fill = "Non-event")) +
  geom_ribbon(aes(ymax = if_else(is_event | switch, cumret, NA),
                  ymin = 0, fill = "Event")) +
  theme(legend.position = "inside", legend.position.inside = c(0.2, 0.8))
```

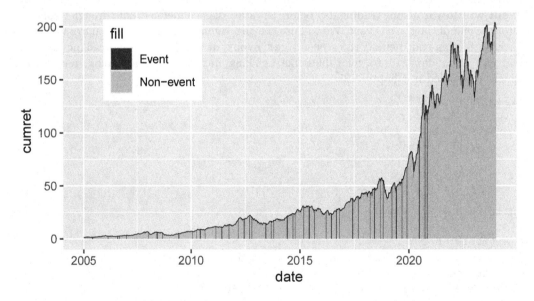

FIGURE 13.1
Apple returns

---

[6]The purpose of the variable `switch` is to "fill" gaps in the plot that would arise in its absence. To see what this variable is doing, replace `switch = coalesce(is_event != lead(is_event), FALSE)` with `switch = FALSE`. These gaps are more apparent in Figure 13.2 than they are in Figure 13.1.

The line in Figure 13.1 represents cumulative returns since the start of the window. Two "area" plots are added to this line: one for the non-event windows and one for the event windows. The vast majority of dates are non-event dates, making the event windows difficult to discern in the plot. But "zooming in" makes the event windows easier to discern, as seen in Figure 13.2, which focuses on the last quarter of 2020.

```
apple_data |>
  arrange(date) |>
  mutate(cumret = cumprod(1 + coalesce(ret, 0)),
         switch = coalesce(is_event != lead(is_event), FALSE)) |>
  filter(date >= "2020-09-01", date <= "2020-12-31") |>
  ggplot(aes(x = date, y = cumret)) +
  geom_line() +
  geom_ribbon(aes(ymax = if_else(!is_event | switch, cumret, NA),
                  ymin = 0, fill = "Non-event")) +
  geom_ribbon(aes(ymax = if_else(is_event | switch, cumret, NA),
                  ymin = 0, fill = "Event")) +
  theme(legend.position = "bottom")
```

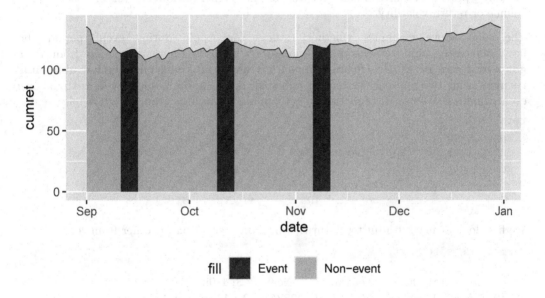

FIGURE 13.2
Apple returns: "Zoomed in"

There is little in the plots above to suggest that Apple media events are associated with unusual returns, but we will use regression analysis to test this more formally. We consider whether returns are different when the indicator variable `is_event` is `TRUE`. Inspired by , see 12, we also consider the absolute value of returns (similar to squared return residuals used in Beaver, 1968) and relative trading volume.

```
fms <- list("ret_mkt" = lm(ret_mkt ~ is_event, data = apple_data),
            "abs(ret)" = lm(abs(ret) ~ is_event, data = apple_data),
            "Volume" = lm(vol / mean(vol) ~ is_event, data = apple_data))
```

TABLE 13.1

Regression analysis of Apple media events

	ret_mkt	abs(ret)	Volume
(Intercept)	0.001***	0.014***	0.997***
is_eventTRUE	−0.002**	0.001	0.060
Num.Obs.	4,781	4,781	4,781
R2	0.001	0.000	0.000

```
modelsummary(fms,
             estimate = "{estimate}{stars}",
             statistic = NULL,
             gof_map = c("nobs", "r.squared"),
             stars = c('*' = .1, '**' = 0.05, '***' = .01))
```

Results from these regressions are reported in Table 13.1. These results—interpreted fairly casually—provide evidence of *lower* returns, but not higher (or lower) levels of either trading volume or return volatility.

The code above examines whether returns for Apple during event periods behave differently from returns during non-event periods. Another function from `farr`, `get_event_cum_rets()`, calculates cumulative raw returns and **cumulative abnormal returns** using two approaches: market-adjusted returns and **size-adjusted returns** over event windows. We use this function to get cumulative returns around each Apple event.

```
rets <-
  apple_events |>
  mutate(permno = apple_permno) |>
  get_event_cum_rets(db,
                     win_start = -1, win_end = +1,
                     end_event_date = "end_event_date")
```

We first look at market-adjusted returns, which on average barely differ from zero.

```
summary(rets$ret_mkt)
```

```
    Min.   1st Qu.    Median      Mean   3rd Qu.      Max.
-0.103895 -0.026442 -0.007217 -0.005840  0.014285  0.135833
```

How many media events are positive-return events? (Answer: Fewer than half!)

```
summary(rets$ret_mkt > 0)
```

```
   Mode   FALSE    TRUE
logical      26      21
```

Finally, Figure 13.3 depicts market-adjusted returns for Apple media events by event date.

```
rets |>
  ggplot(aes(x = event_date, y = ret_mkt)) +
  geom_point() +
  geom_smooth(method = "lm", formula = "y ~ 1")
```

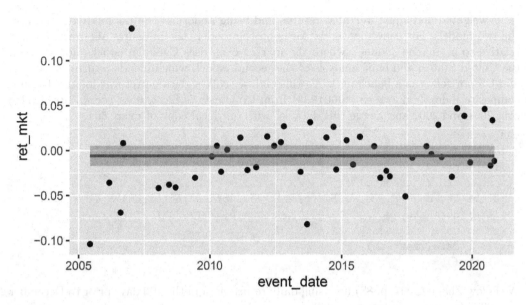

FIGURE 13.3
Apple market-adjusted returns and media events

13.2.2 Exercises

1. How would you expect the plot to change if we used `cumret = exp(cumsum(log(1 + coalesce(ret, 0))))` in place of `cumret = cumprod(1 + coalesce(ret, 0))` in creating the plot above? Is there any reason to prefer one calculation over time other?

2. Do we get different results in this case if we use `cumret = cumprod(1 + ret)` (i.e., remove the `coalesce()` function)? If so, why? If not, would we always expect this to be case (e.g., for stocks other than Apple)?

13.3 Event studies and regulation

Zhang (2007, p. 74) "investigates the economic consequences of the Sarbanes–Oxley Act (SOX) by examining market reactions to related legislative events." Zhang (2007, p. 75) finds that "the cumulative value-weighted (equal-weighted) raw return of the U.S. market amounts to −15.35% (−12.53%) around the key SOX events." As Zhang (2007) uses CRSP returns for the US market, we collect a local copy of the relevant data.

```
dsi_local <-
  dsi |>
  select(date, vwretd, ewretd) |>
  collect()
```

Zhang (2007, p. 76) focuses some analyses on "key SOX events" (defined below) and finds that "the estimated U.S. cumulative abnormal returns range from −3.76% and −8.21% under alternative specifications and are all statistically significant." Here "all" means for each of value-weighted and equal-weighted returns and using abnormal returns relative to each of the two models. For convenience, we focus on the model that measures abnormal returns relative to a "market model", where the market comprises Canadian stocks not listed in the United States, and omit analysis of the second model, which blends returns on several non-US portfolios as a benchmark. To this end, we collect data on returns for the Toronto composite index (`gvkeyx == "000193"`) from Compustat's index data (`comp.idx_daily`) over 2001 and 2002 and merge this data set with our local copy of `crsp.dsi`.

```
can_rets <-
  idx_daily |>
  filter(gvkeyx == "000193") |>
  window_order(datadate) |>
  mutate(ret_can = if_else(lag(prccd) > 0, prccd/lag(prccd) - 1, NA)) |>
  filter(datadate >= "2000-01-01", datadate <= "2002-12-31") |>
  rename(date = datadate) |>
  select(date, ret_can) |>
  collect()
```

We follow Zhang (2007, p. 88) in "using daily return data in the 100 days prior to December 28, 2001" for expected return models.

```
reg_data <-
  dsi_local |>
  inner_join(can_rets, by = "date") |>
  filter(date < "2001-12-28") |>
  top_n(100, wt = date)
```

We then fit models for market returns for both equal-weighted and value-weighted portfolios against Canadian returns.

```
fm_vw <- lm(vwretd ~ ret_can, data = reg_data)
fm_ew <- lm(ewretd ~ ret_can, data = reg_data)
```

We then use these models to calculate excess returns for all observations. In the following code, we use `pick(everything())`, which is an idiom that we use in a few places in this book. In the function following the native pipe (`|>`), we often access the data frame supplied by the item preceding `|>` using the pipe placeholder `_`. However, this placeholder is not accessible to functions inside `mutate()`. Fortunately, we can use `pick()` instead. The `pick()` function provides a way to easily select columns from data inside a function like `mutate()`. In the current setting, we can use `everything()` to indicate that we want to select all variables in the source data. As such, `pick(everything())` is a handy workaround to the limitations of the pipe placeholder `_`.

```
dsi_merged <-
  dsi_local |>
  inner_join(can_rets, by = "date") |>
  mutate(abret_vw = vwretd - predict(fm_vw, pick(everything())),
         abret_ew = ewretd - predict(fm_ew, pick(everything()))) |>
  select(-ret_can)
```

From Table 2, Zhang (2007) appears to calculate the daily standard deviation of returns at about 1.2%. The exact basis for this calculation is unclear, but similar analyses are "estimated using daily return data in the 100 days prior to December 28, 2001" (Zhang, 2007, p. 88). So, we calculate the daily volatility on this basis using the following calculation, which yields the value 1.28%.

```
sd_ret <-
  dsi_local |>
  filter(date < "2001-12-28") |>
  top_n(100, wt = date) |>
  summarize(sd(vwretd)) |>
  pull()
```

The `farr` package contains the data frame `zhang_2007_windows` containing the dates of the event windows found in Table 2 of Zhang (2007). We can combine these data with return data from `dsi_local` to calculate cumulative returns for each event window. Following Zhang (2007), we can estimate the standard error by scaling the daily return volatility by the square-root of the number of trading days in each window to calculate a t-statistic for each event. We use the standard deviation of residuals to estimate the daily volatility of the abnormal-return models.

```
zhang_2007_rets <-
  zhang_2007_windows |>
  inner_join(dsi_merged, join_by(beg_date <= date, end_date >= date)) |>
  group_by(event) |>
  summarize(n_days = n(),
            vwret = sum(vwretd),
            ewret = sum(ewretd),
            abret_vw = sum(abret_vw),
            abret_ew = sum(abret_ew),
            vw_t = vwret / (sqrt(n_days) * sd_ret),
            ew_t = ewret / (sqrt(n_days) * sd_ret),
            abret_vw_t = abret_vw / (sqrt(n_days) * sd(fm_vw$residuals)),
            abret_ew_t = abret_ew / (sqrt(n_days) * sd(fm_ew$residuals)))
```

In subsequent analyses, Zhang (2007) focuses on "key SOX events", which seem to be those events with a "statistically significant" return at the 10% level in a two-tailed test, and reports results in Panel D of Table 1 (2007, pp. 91–92). We replicate the key elements of this procedure and our results correspond roughly with those reported in Zhang (2007) as "CAR2".

```
zhang_2007_res <-
  zhang_2007_rets |>
  filter(abs(vw_t) > abs(qnorm(.05))) |>
  summarize(vwret = sum(vwret),
            ewret = sum(ewret),
            abret_vw = sum(abret_vw),
            abret_ew = sum(abret_ew),
            n_days = sum(n_days),
            vw_t = vwret / (sqrt(n_days) * sd_ret),
            ew_t = ewret / (sqrt(n_days) * sd_ret),
            abret_vw_t = abret_vw / (sqrt(n_days) * sd(fm_vw$residuals)),
            abret_ew_t = abret_ew / (sqrt(n_days) * sd(fm_ew$residuals)))
```

We estimate cumulative raw value-weighted returns for the four "key SOX events" at -15.2% (t-statistic -3.18), quite close to the value reported in Zhang (2007) (-15.35% with a t-statistic of -3.49). However, our estimate of cumulative abnormal value-weighted returns for the four "key SOX events" is -3.18% (t-statistic -1.02), which is closer to zero than

the value reported in Zhang (2007) (−8.21% with a t-statistic of −2.99), which is the only
value of eight reported in Panel D of Table 1 that is statistically significant at conventional
levels (5% in two-tailed tests).

13.3.1 Discussion questions

13.3.1.1 Zhang (2007)

1. What are the relative merits of raw and abnormal returns in evaluating the effect
   of SOX on market values of US firms? What do you observe in the raw returns
   for Canada, Europe, and Asia for the four events that are the focus of Panel B
   of Table 2 of Zhang (2007)? Does this raise concerns about the results of Zhang
   (2007)?

2. Describe the process for constructing the test statistics reported in Panel D of
   Table 2. How compelling are these results? Do you agree with the assessment by
   Leuz (2007, p. 150) that Zhang (2007) is "very careful in assessing the significance
   of the event returns"?

3. Describe in detail how you might conduct statistical inference using **randomization inference** in the setting of Zhang (2007) (see Section 19.7 for more on this
   approach)? What are the challenges faced and design choices you need to make
   in applying this approach? Does your approach differ from the bootstrapping
   approach used in Zhang (2007)?

4. Leuz (2007) identifies studies other than Zhang (2007) that find evidence that
   SOX was beneficial to firms? How can these sets of results be reconciled? What
   steps would you look to undertake to evaluate the conflicting claims of the two
   papers?

13.3.1.2 Khan et al. (2017)

1. What is the research question examined in Khan et al. (2017)? (Hint: Read the
   title.)

2. Khan et al. (2017, p. 210) argue that "an ideal research design to evaluate the
   benefits of accounting standards is to compare a voluntary disclosure regime,
   in which firms disclose information required by a particular standard, with a
   mandatory disclosure regime, in which firms are required to disclose that same
   information." Do you agree that this research design would be "ideal" to address
   the question? What is the implied treatment in this ideal design?

3. Compare the Apple event study above with Khan et al. (2017). What are the
   relative strengths and weaknesses of the two studies? Do you think an event-
   study approach is appropriate for addressing the question "do Apple products
   add value?" Do you think an event-study approach is appropriate for addressing
   the research question of Khan et al. (2017)? Why or why not?

4. Do you think that standard-setters would view "reduction in estimation risk"
   as a goal of accounting standards? Evaluate the quality of the arguments linking
   improved standards to reduced estimation risk. The null hypothesis for Panel A is
   that the CAR of affected firms is not different from CAR of unaffected firms. How
   appropriate is it to report "most negative" and "most positive" CAR differences
   only? (Hint: If the null hypothesis is true, how many standards might you expect
   to have "statistically significant" coefficients?)

5. Interpret the results of Table 5, Panel B of Khan et al. (2017).

13.3.1.3 Larcker et al. (2011) "LOT"

1. How do LOT and FFJR differ in terms of the role of market efficiency in their research designs?

2. Consider Table 1 of LOT. What are the differences between the event study design in LOT from that in FFJR? What are implications of these differences?

3. How do you think Table 1 was developed? Do you see potential problems in the process underlying Table 1? Can you suggest alternative approaches to developing Table 1?

4. Consider proxy access, as some of the core results of the paper relate to proxy access. If you were a shareholder in a company, what concerns might you have about proxy access? Why might this decrease the value of your shares? Think about this is concrete terms; be specific about the kinds of circumstances where value will be reduced. How well do the variables *NLargeBlock* and *NSmallCoalitions* measure the exposure of firms to the issues you identified in the previous question? (As part of this, consider the timing of variable measurement relative to the timing of possible value-reducing outcomes.)

5. LOT makes use of a number of **Monte Carlo simulations**. How do these compare with the bootstrapping analyses conducted by Zhang (2007)? Are the simulations addressing the same underlying issues as Zhang (2007) bootstrapping approach?

14

Post-earnings announcement drift

In Chapter 11, we saw that the cumulative returns of "good news" firms and "bad news" firms continued to drift apart even *after* earnings announcements. This result was considered an **anomaly** as, once earnings are announced, an efficient market should quickly impound the implications of those earnings and there should be no association with subsequent excess returns. Later research expanded on Ball and Brown (1968), finding that the **post-earnings announcement drift** ("PEAD") existed with more refined measures and approaches.

In this chapter, we will build some foundation concepts before more closely studying a seminal paper in the PEAD literature (Bernard and Thomas, 1989). As most of the research on PEAD has focused on quarterly earnings, we will spend some time understanding Compustat's database of quarterly financial statement information, `comp.fundq`.

A core concept in PEAD research is **earnings surprise**, which can be defined quite generally as actual earnings minus expected earnings. Thus to measure earnings surprise, we need a measure of expected earnings. Early research used **time-series models** of earnings to develop earnings expectation models. We will look closely at Foster (1977), which is an early study of the behaviour of quarterly accounting data.

> 💡 Tip
>
> The code in this chapter uses the packages listed below. For instructions on how to set up your computer to use the code found in this book, see Section 1.2. Quarto templates for the exercises below are available on GitHub.[a]
>
> ----
>
> [a]https://github.com/iangow/far_templates/blob/main/README.md

```
library(tidyverse)
library(DBI)
library(farr)
library(modelsummary)
library(furrr)
```

14.1 Fiscal years

A concept frequently encountered in accounting and finance research is the **fiscal year**. Most US firms have financial reporting periods ending on 31 December of each year. Most Australian firms have financial reporting periods ending on 30 June of each year, perhaps because accountants don't want to be preparing financial statements during the summer month of January.

DOI: 10.1201/9781003456230-14

TABLE 14.1

Date variables on `comp.funda`

Variable	Description
`datadate`	Period-end date
`apdedate`	Actual period-end date
`pdate`	Date the data are updated on a preliminary basis
`fdate`	Date the data are finalized

Some US firms (often retailers) have fiscal year-ends in January. For example, Autodesk—"a global leader in 3D design, engineering, and entertainment software and services"—had a fiscal year-end on 31 January 2021.[1] In contrast, Akamai Technologies had a fiscal year-end ending 31 December 2020. An analyst wishing to compare financial performance of Autodesk and Akamai is likely to line up both of these periods as "fiscal 2020".

Note that there is no standard definition of "fiscal year" in practice. Fedex describes the year ending 31 May 2020 as "fiscal 2019" and General Mills describes the same period as "fiscal 2020".

Compustat has the variable `fyear` (called `fyearq` on `comp.fundq`), which is described as "fiscal year". Before zooming in on `fyear`, note that there are four variables of type `Date` on `comp.funda` and these are listed in Table 14.1.

Before proceeding further, we set up connections to the database tables that we will use in this chapter. Note that we set `check_interrupts = TRUE` in connecting to PostgreSQL so we can cancel a long-running query if necessary.

```
db <- dbConnect(RPostgres::Postgres(),
                bigint = "integer",
                check_interrupts = TRUE)


funda <- tbl(db, Id(schema = "comp", table = "funda"))
fundq <- tbl(db, Id(schema = "comp", table = "fundq"))
company <- tbl(db, Id(schema = "comp", table = "company"))


ccmxpf_lnkhist <- tbl(db, Id(schema = "crsp", table = "ccmxpf_lnkhist"))
```

We construct the same `funda_mod` remote data frame we saw in earlier chapters.

```
funda_mod <-
  funda |>
  filter(indfmt == "INDL", datafmt == "STD",
         consol == "C", popsrc == "D")
```

To understand the date variables on `comp.funda`, we focus on Apple (`gvkey == "001690"`), whose "fiscal year is the 52- or 53-week period that ends on the last Saturday of September".

```
apple_dates <-
  funda_mod |>
  filter(gvkey == "001690") |>
  collect() |>
  select_if(is.Date)
```

---

[1] https://www.sec.gov/ix?doc=/Archives/edgar/data/769397/000076939721000014/adsk-20210131.htm

While there are several date variables, it seems that data on fields other than `datadate` are only available for more recent periods.

```
apple_dates |>
  filter(!is.na(apdedate)) |>
  head()
```

```
# A tibble: 6 x 4
  datadate    apdedate    fdate       pdate
  <date>      <date>      <date>      <date>
1 2003-09-30  2003-09-27  NA          NA
2 2004-09-30  2004-09-25  NA          NA
3 2005-09-30  2005-09-24  NA          NA
4 2006-09-30  2006-09-30  2007-01-01  2006-10-18
5 2007-09-30  2007-09-29  2007-11-16  2007-10-22
6 2008-09-30  2008-09-27  2008-11-05  2008-10-21
```

In its 10-K filing[2] for 2019, Apple says "the Company's fiscal years 2019 and 2018 spanned 52 weeks each, whereas fiscal year 2017 included 53 weeks. A 14th week was included in the first quarter of 2017, as is done every five or six years, to realign the Company's fiscal quarters with calendar quarters." We also see that `datadate` is the last day of the month in which the period ended (`apdedate`).

These facts are evident in the following:

```
apple_dates |>
  arrange(datadate) |>
  mutate(fyear_length = apdedate - lag(apdedate)) |>
  tail()
```

```
# A tibble: 6 x 5
  datadate    apdedate    fdate       pdate       fyear_length
  <date>      <date>      <date>      <date>      <drtn>
1 2018-09-30  2018-09-29  2018-11-05  2018-11-01  364 days
2 2019-09-30  2019-09-28  2019-11-04  2019-10-30  364 days
3 2020-09-30  2020-09-26  2020-11-02  2020-10-29  364 days
4 2021-09-30  2021-09-25  2021-11-01  2021-10-28  364 days
5 2022-09-30  2022-09-24  2022-10-31  2022-10-27  364 days
6 2023-09-30  2023-09-30  2023-11-04  2023-11-02  371 days
```

(The output above notes that `fyear_length` has type `drtn`, which is an abbreviation of "duration". The type of variable that results from subtracting one `Date` object from another is a `difftime`, which is a class of durations. See here[3] for more details.)

While Compustat's documentation[4] explains how `fyear` is determined, many database providers do not adequately explain variables, so being able to *infer* what an item describes is a useful skill. To help build this skill, we will try to deduce what `fyear` means here *without* reading the manual. We will figure out `fyear` by collecting data from Compustat on `fyear` and `datadate`, and then doing some statistical and graphical analysis of these data. As a side effect, we will also reinforce some aspects of regression analysis using R that we learnt about in Chapter 3.

---

[2]https://www.sec.gov/Archives/edgar/data/320193/000032019319000119/a10-k20199282019.htm
[3]https://tibble.tidyverse.org/articles/types.html
[4]https://go.unimelb.edu.au/r7d8

TABLE 14.2

Regression of `fyear` on month indicators

	(1)	(2)
factor(month)1	−1.000	
factor(month)2	−1.000	
factor(month)3	−1.000	
factor(month)4	−1.000	
factor(month)5	−1.000	
factor(month)6	0.000	
factor(month)7	0.000	
factor(month)8	0.000	
factor(month)9	0.000	
factor(month)10	0.000	
factor(month)11	0.000	
factor(month)12	0.000	
year	1.000	1.000
(Intercept)		−1.124
month		0.122
R2	1.000	1.000

We begin by collecting all combinations of `fyear` and `datadate` found on `comp.funda` and extracting the month and year applicable to each of these.

```
fyear_data <-
  funda_mod |>
  distinct(fyear, datadate) |>
  mutate(month = month(datadate),
         year = year(datadate)) |>
  filter(!is.na(fyear)) |>
  collect()
```

We then run a couple of regression models on the data (the discussion questions below will explore these in more detail).

```
fms <- list(lm(fyear ~ factor(month) + year - 1, data = fyear_data),
            lm(fyear ~ month + year, data = fyear_data))
```

The results of these regressions are shown in Table 14.2.

```
modelsummary(fms,
             estimate = "{estimate}",
             statistic = NULL,
             align = "ldd",
             gof_map = "r.squared")
```

Having run the regressions, we can add the fitted values for each model to the data set.

```
fyear_plot_data <-
  fyear_data |>
  mutate(pred_1 = predict(fms[[1]]),
         pred_2 = predict(fms[[2]]),
```

```
            resid_1 = pred_1 - fyear,
            resid_2 = pred_2 - fyear)
```

Finally we take a sample from `fyear_plot_data` to make Figure 14.1.

```
fyear_plot_data |>
  filter(year %in% c(2001, 2002)) |>
  select(datadate, fyear, pred_1, pred_2) |>
  distinct() |>
  arrange(datadate) |>
  ggplot(aes(x = datadate)) +
  geom_point(aes(y = fyear, color = "fyear")) +
  geom_line(aes(y = pred_1, color = "pred_1")) +
  geom_line(aes(y = pred_2, color = "pred_2")) +
  scale_x_date(date_breaks = "1 month") +
  theme(axis.text.x = element_text(angle = 90),
        legend.position = "inside", legend.position = c(.85, .35))
```

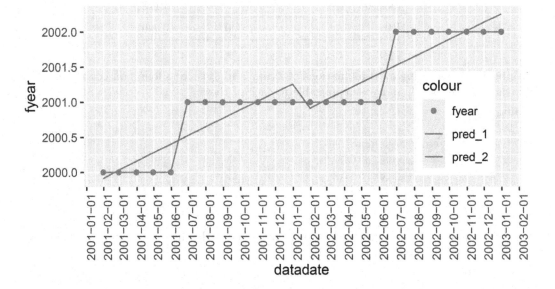

FIGURE 14.1
Plot of fyear against datadate for sample values

14.1.1 Exercises

1. What is different between the first and second models in `fms`? What is `factor()` doing here?

2. What does the inclusion of `- 1` do in the first model in `fms`? Would the omission of `- 1` affect the fit of that model? Would it affect the interpretability of results? Would the inclusion of `- 1` affect the fit of the second model in `fms`? Would it affect the interpretability of results?

3. Does Figure 14.1 help you understand what's going on? Why did we focus on a relatively short period in Figure 14.1? (*Hint*: What happens if you remove the line `filter(year %in% c(2001, 2002))` |> from the code?)

4. Using `year` and `month`, add some code along the lines of `mutate(fyear_calc = ...)` to *calculate* `fyear`. Use this code to create a function `fiscal_year(datadate)`. Check that you match `fyear` in each case.

14.2 Quarterly data

In Chapters 6 and 8, we focused on *annual* data from Compustat. But for many purposes, we will want to use *quarterly* data. While Ball and Brown (1968) used annual data, the PEAD literature has generally focused on quarterly data.

We begin by creating `fundq_mod`, the quarterly analogue of `funda_mod`.

```
fundq_mod <-
  fundq |>
  filter(indfmt == "INDL", datafmt == "STD", consol == "C", popsrc == "D")
```

In Chapter 6, we saw that, by focusing on the "standard" subset of observations, we have (`gvkey`, `datadate`) forming a primary key. Alas, the same is not true for quarterly data, as we can see in the output below:

```
fundq_mod |>
  group_by(gvkey, datadate) |>
  summarize(num_rows = n(), .groups = "drop") |>
  count(num_rows) |>
  collect()
```

```
# A tibble: 2 x 2
  num_rows       n
     <dbl>   <dbl>
1        2    1544
2        1 2017936
```

To dig deeper, we create the data frame `fundq_probs` that captures the problem cases implied by the output above.

```
fundq_probs <-
  fundq_mod |>
  group_by(gvkey, datadate) |>
  filter(n() > 1) |>
  ungroup() |>
  select(gvkey, datadate, fyearq, fqtr, fyr, rdq) |>
  arrange(gvkey, datadate) |>
  collect()
```

It turns out that the cause of the problems can be found by looking more closely at `funda_mod` for firms in `fundq_probs` around the problematic periods. One such firm is `gvkey == "001224"`, which has observations in `fundq_probs` for fiscal years 2013 and 2014.

```
funda_mod |>
  filter(gvkey == "001224", between(fyear, 2012, 2015)) |>
  select(gvkey, datadate) |>
  arrange(gvkey, datadate) |>
  collect()
```

```
# A tibble: 4 x 2
  gvkey  datadate
  <chr>  <date>
1 001224 2012-12-31
2 001224 2013-12-31
3 001224 2014-09-30
4 001224 2015-09-30
```

In this case, the firm changed its year-end from December to September during 2014. Thus, Q4 of the year ending 2013-12-31 became Q1 of the year ending 2014-09-30. Compustat retains the data for the quarter ending 2013-12-31 twice: once as Q4 and once as Q1.

Meanwhile, Q1 of *what would have been* the year ending 2014-12-31 became Q2 of the year ending 2014-09-30.[5] But there is no year ending 2014-12-31 on `comp.funda` for this firm, so Compustat sets `fqtr` to `NA` for the second row of data for the quarter ending 2014-03-31, as can be seen in the following:

```
fundq_probs |>
  filter(gvkey == "001224") |>
  arrange(gvkey, datadate, fqtr)
```

```
# A tibble: 6 x 6
  gvkey  datadate   fyearq fqtr  fyr rdq
  <chr>  <date>     <int> <int> <int> <date>
1 001224 2013-12-31  2014     1    9 2014-02-11
2 001224 2013-12-31  2013     4   12 2014-02-11
3 001224 2014-03-31  2014     2    9 2014-04-30
4 001224 2014-03-31  2014    NA   12 2014-04-30
5 001224 2014-06-30  2014     3    9 2014-08-11
6 001224 2014-06-30  2014    NA   12 2014-08-11
```

So, the variable `fyr` allows us to distinguish rows and, from the following analysis, we see that (`gvkey`, `datadate`, `fyr`) is a valid primary key for the "standard" subset of `comp.fundq`.

```
fundq_mod |>
  count(gvkey, datadate, fyr, name = "num_rows") |>
  count(num_rows) |>
  collect()
```

---

[5]From the firm's transitional 10-K filing at https://go.unimelb.edu.au/j7d8: "Effective September 2, 2014, the Company amended its bylaws to change the Company's fiscal year from beginning January 1st and ending on December 31st, to beginning October 1st and ending on September 30th. As a result, this Form 10-K is a transition report and includes financial information for the period from January 1, 2014 through September 30, 2014. Subsequent to this report, the Company's annual reports on Form 10-K will cover the fiscal year October 1st to September 30th. The period beginning January 1, 2014 through September 30, 2014 is referred to as the 'current period' or 'transition period' and the period beginning October 1, 2014 through September 30, 2015 as 'fiscal 2015'."

```
# A tibble: 1 x 2
  num_rows         n
     <dbl>     <dbl>
1        1 2021024
```

```
fundq_mod |>
  select(gvkey, datadate, fyr) |>
  mutate(across(everything(), is.na)) |>
  filter(if_any(everything())) |>
  count() |>
  pull()
```

```
[1] 0
```

Presumably the idea is to allow researchers to link data from `comp.funda` with data from `comp.funda`. Using `fyr` and `fyear`, we can recover the relevant *annual* `datadate`.

```
link_table <-
  fundq_mod |>
  rename(datadateq = datadate) |>
  select(gvkey:fyr) |>
  mutate(year = if_else(fyr <= 5L, fyearq + 1L, fyearq)) |>
  mutate(month = lpad(as.character(fyr), 2L, "0")) |>
  mutate(datadate = as.Date(str_c(year, month, '01', sep = "-"))) |>
  mutate(datadate = as.Date(datadate + months(1) - days(1))) |>
  select(-month, -year, -fqtr) |>
  collect()

link_table |>
  arrange(gvkey, datadateq)
```

```
# A tibble: 2,021,024 x 5
   gvkey  datadateq  fyearq   fyr datadate
   <chr>  <date>      <int> <int> <date>
 1 001000 1966-03-31   1966    12 1966-12-31
 2 001000 1966-06-30   1966    12 1966-12-31
 3 001000 1966-09-30   1966    12 1966-12-31
 4 001000 1966-12-31   1966    12 1966-12-31
 5 001000 1967-03-31   1967    12 1967-12-31
 6 001000 1967-06-30   1967    12 1967-12-31
 7 001000 1967-09-30   1967    12 1967-12-31
 8 001000 1967-12-31   1967    12 1967-12-31
 9 001000 1968-03-31   1968    12 1968-12-31
10 001000 1968-06-30   1968    12 1968-12-31
# i 2,021,014 more rows
```

We could then link the table above with (gvkey, datadate) combinations from `comp.funda` and (gvkey, datadateq, fyr) combinations from `comp.fundq` (we rename `datadate` on `comp.fundq` to "datadateq" to avoid a clash between annual and quarter period-ends for any given quarter).

14.2.1 Exercises

1. Pick a couple of `gvkey` values from `fundq_probs`. Is it possible to construct a "clean" sequence of consecutive quarterly earnings announcements for each of these firms? (Here "clean" means that, at the very least, each quarter shows up just once in the series.) What challenges does one face in this task?

2. The code below produces Figure 14.2. From Figure 14.2, it seems that Q2 has been the most profitable on average over the last three decades, while in all decades, Q4 has seen the most sales. Can you speculate as to what might explain these facts?

```
ni_annual <-
  funda_mod |>
  select(gvkey, datadate, fyr, sale, ni) |>
  collect()

ni_qtrly <-
  fundq_mod |>
  select(gvkey, datadate, fyr, fqtr, saleq, niq, ibq) |>
  rename(datadateq = datadate) |>
  collect()

ni_merged <-
  ni_annual |>
  inner_join(link_table, by = c("gvkey", "datadate", "fyr")) |>
  inner_join(ni_qtrly, by = c("gvkey", "fyr", "datadateq"))

plot_data <-
  ni_merged |>
  mutate(decade = str_c(floor(fyearq / 10) * 10, "s")) |>
  filter(!is.na(fqtr), fyearq < 2020) |>
  group_by(decade, fqtr) |>
  summarize(prop_ni = sum(niq, na.rm = TRUE)/
              sum(ni, na.rm = TRUE),
            prop_sale = sum(saleq, na.rm = TRUE)/
              sum(sale, na.rm = TRUE),
            .groups = "drop") |>
  mutate(fqtr = factor(fqtr)) |>
  pivot_longer(cols = c(prop_ni, prop_sale),
               names_to = "metric",
               values_to = "value")
```

```
plot_data |>
  ggplot(aes(x = fqtr, y = value, fill = fqtr)) +
  geom_bar(stat = "identity") +
  facet_grid(metric ~ decade) +
  theme(legend.position = "none")
```

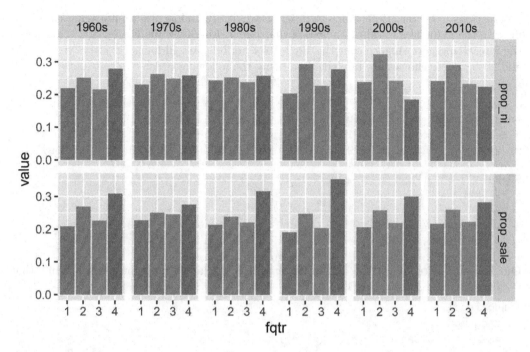

FIGURE 14.2
Sales and net income by quarter over decades

3. Create another plot using data in `ni_merged` that you think might be interesting. (Feel free to add variables to `ni_annual` or `ni_qtrly` before merging.)

14.3 Time-series properties of earnings

Foster (1977) studies the properties of quarterly accounting variables (sales, net income, and expenses) and considers the predictive ability of six models. Models 1 and 2 relate values in quarter Q_t to values in quarter Q_{t-4} and is therefore a **seasonal** model. Models 3 and 4 relate values in quarter Q_t to values in the adjacent quarter (Q_{t-1}). Model 5 builds on Model 2 by including a component related to $(Q_{t-1} - Q_{t-5})$. Finally, Foster (1977) considers the "Box-Jenkins time-series methodology" as Model 6.

In this section, we will do a *loose* replication of Foster (1977). We say "loose" because we will not try to match the sample composition or even all the details of the approach, but the basic ideas and some of the details will be the same. We replace the Box-Jenkins approach with an alternative model, as Box-Jenkins forecasting methods are more complex and not often used in accounting research today.

Foster (1977) focuses on 69 firms meeting sample-selection criteria, including the availability of quarter sales and earnings information over the period of 1946–1974. To keep our analysis to a comparable number of firms, we will choose 70 firms based on criteria detailed below. While Foster (1977) uses what he calls an "adaptive forecasting" approach whereby "*all* data available at the time the forecast was made were used to forecast", we will limit ourselves to 20 quarters of data available at the time the model is developed. As a substantial majority of observations in Foster (1977) come from 12 two-digit SIC industries, we limit our sample to firms in the industries listed in Table 1 of that paper. Finally, while Foster (1977) evaluates the forecasting performance over the 1962–1974 period, we will focus on just 2 years: 1974 and 2019. These choices are implemented using the following code:

```
n_qtrs <- 20
n_firms <- 70
focus_years <- c(1974L, 2019L)
# See Table 1 of Foster (1977) for SICs
sic2s <- as.character(c(29, 49, 28, 35, 32, 33, 37, 20, 26, 10, 36, 59))
```

```
companies <-
  company |>
  mutate(sic2 = str_sub(sic, 1L, 2L)) |>
  filter(sic2 %in% sic2s) |>
  select(gvkey, sic2)
```

Our first step is to collect the data from `comp.fundq` that we need into a data frame that we give the name `fundq_local`.

```
fundq_local <-
  fundq_mod |>
  semi_join(companies, by = "gvkey") |>
  filter(saleq > 0 & !is.na(saleq)) |>
  select(gvkey, datadate, fyr, fqtr, fyearq, rdq, niq, saleq, ibq) |>
  rename(datadateq = datadate) |>
  collect()
```

Next we select firm-years available on `comp.funda`.

```
firm_years <-
  funda_mod |>
  select(gvkey, datadate) |>
  collect()
```

We then link `firm_years` to `fundq_local` using the `link_table` we created earlier.

```
merged_data <-
  firm_years |>
  inner_join(link_table, by = c("gvkey", "datadate")) |>
  inner_join(fundq_local,
             by = c("gvkey", "datadateq", "fyearq", "fyr"))
```

We identify "regular" fiscal years, which we define as fiscal years comprising four quarters and extending over exactly a year (i.e., either 365 or 366 days).

```
qtr_num <-
  merged_data |>
  group_by(gvkey, datadate) |>
  count(name = "num_quarters") |>
  ungroup()
```

```
regular_fyears <-
  firm_years |>
  inner_join(qtr_num, by = c("gvkey", "datadate")) |>
```

```
  group_by(gvkey) |>
  arrange(gvkey, datadate) |>
  mutate(fyear_length = datadate - lag(datadate)) |>
  ungroup() |>
  mutate(regular_year = num_quarters == 4 &
           (is.na(fyear_length) | fyear_length %in% c(365, 366))) |>
  filter(regular_year) |>
  select(gvkey, datadate)
```

We next limit the sample to the regular fiscal years that we defined above and calculate the variables we need to estimate and apply the models.

```
reg_data <-
  merged_data |>
  semi_join(companies, copy = TRUE, by = "gvkey") |>
  semi_join(regular_fyears, by = c("gvkey", "datadate")) |>
  select(gvkey, datadateq, fyearq, rdq, niq, saleq) |>
  group_by(gvkey) |>
  arrange(datadateq) |>
  mutate(sale_lag_1 = lag(saleq, 1L),
         sale_lag_4 = lag(saleq, 4L),
         sale_lag_5 = lag(saleq, 5L),
         sale_diff = saleq - sale_lag_1,
         sale_seas_diff = saleq - sale_lag_4,
         lag_sale_seas_diff = lag(sale_seas_diff, 1L),
         ni_lag_1 = lag(niq, 1L),
         ni_lag_4 = lag(niq, 4L),
         ni_lag_5 = lag(niq, 5L),
         ni_diff = niq - ni_lag_1,
         ni_seas_diff = niq - ni_lag_4,
         lag_ni_seas_diff = lag(ni_seas_diff, 1L)) |>
  ungroup()
```

We next create a `fit_model()` function that estimates all six models using data from the given `gvkey` and the 20 quarters prior to `datadateq` and predicts values for `datadateq` for all models.

The first thing done in the function is a filter for observations that relate to the firm with the `gvkey` supplied as an argument to the function. Note that `gvkey` has two meanings in the context of the code at the point where `filter()` is called. First, it refers to the column of `reg_data`. Second, it refers to the value of the `gvkey` supplied as an argument to the function. It is for this reason that we use `gvkey` to get the first meaning and use `!!gvkey` as a way to get the second meaning.[6] (We do a similar thing for the same reason when we filter on dates.)

Having obtained data related to the firm, we split the data into `train_data`, the data from previous periods that we will use to estimate the models, and `test_data`, the data for the period that we will forecast. (Note that we exit the function and return `NULL` if we don't have 20 quarters of data to train the model.)

---

[6]Chapter 19 of Wickham (2019) has more details on the "unquote" operator `!!`.

We first estimate models 2 and 4, which require no more than estimation of a drift term (δ in the notation of Foster, 1977), which we estimate as the mean of the respective changes over the training period.

We then create predicted values for models 1–4.

The next thing we do is fit models 5 and 6. Model 6 in Foster (1977) was based on the Box-Jenkins method. Our model 6 can be viewed as an "unconstrained" version of model 5. In model 6, we regress (in the notation of Foster, 1977) Q_t on Q_{t-1}, Q_{t-4}, and Q_{t-5}.

In the final step, we add predicted values for models 5 and 6 to the same for models 1–4, then calculate the prediction errors and return the results of our analysis.

```r
fit_model <- function(gvkey, datadateq) {

  firm_data <-
    reg_data |>
    filter(gvkey == !!gvkey)

  train_data <-
    firm_data |>
    filter(datadateq < !!datadateq) |>
    top_n(n_qtrs, datadateq)

  if (nrow(train_data) < n_qtrs) return(NULL)

  test_data <-
    firm_data |>
    filter(datadateq == !!datadateq)

  # Estimate models 2 & 4
  model_24 <-
    train_data |>
    group_by(gvkey) |>
    summarize(sale_diff = mean(sale_diff, na.rm = TRUE),
              ni_diff = mean(ni_diff, na.rm = TRUE),
              sale_seas_diff = mean(sale_seas_diff, na.rm = TRUE),
              ni_seas_diff = mean(ni_seas_diff, na.rm = TRUE))

  # Fit models 1, 2, 3 & 4
  df_model_1234 <-
    test_data |>
    # We drop these variables because we will replace them with
    # their means from model_24
    select(-sale_diff, -ni_diff, -sale_seas_diff, -ni_seas_diff) |>
    inner_join(model_24, by = "gvkey") |>
    mutate(ni_m1 = ni_lag_4,
           sale_m1 = sale_lag_4,
           ni_m2 = ni_lag_4 + ni_seas_diff,
           sale_m2 = sale_lag_4 + sale_seas_diff,
           ni_m3 = ni_lag_1,
           sale_m3 = sale_lag_1,
```

```
                ni_m4 = ni_lag_1 + ni_diff,
             sale_m4 = sale_lag_1 + sale_diff)

# Fit model 5
sale_fm5 <- tryCatch(lm(sale_seas_diff ~ lag_sale_seas_diff,
                        data = train_data, model = FALSE),
                     error = function(e) NULL)


ni_fm5 <- tryCatch(lm(ni_seas_diff ~ lag_ni_seas_diff,
                      data = train_data, model = FALSE),
                   error = function(e) NULL)

# Fit model 6
sale_fm6 <- tryCatch(lm(saleq ~ sale_lag_1 + sale_lag_4 + sale_lag_5,
                        data = train_data, model = FALSE),
                     error = function(e) NULL)


ni_fm6 <- tryCatch(lm(niq ~ ni_lag_1 + ni_lag_4 + ni_lag_5,
                      data = train_data, model = FALSE),
                   error = function(e) NULL)

if (!is.null(sale_fm5) & !is.null(ni_fm5)) {
  results <-
    df_model_1234 |>
    mutate(ni_m5 = ni_lag_4 + predict(ni_fm5,
                                      newdata = test_data)) |>
    mutate(sale_m5 = sale_lag_4 + predict(sale_fm5,
                                          newdata = test_data)) |>
    mutate(ni_m6 = predict(ni_fm6, newdata = test_data)) |>
    mutate(sale_m6 = predict(sale_fm6, newdata = test_data))|>
    select(gvkey, datadateq, fyearq, niq, saleq,
           matches("(ni|sale)_m[0-9]")) |>
    pivot_longer(cols = ni_m1:sale_m6,
                 names_to = "item", values_to = "predicted") |>
    mutate(actual = if_else(str_detect(item, "^ni"), niq, saleq),
           abe = abs(predicted - actual) / predicted,
           se = abe^2 * sign(predicted)) |>
    separate(item, into = c("item", "model"), sep = "_m") |>
    select(-niq, -saleq)

  results
}
}
```

Note that the calculations of `abe` and `se` above ensure that the sign of each result is the same as the sign of the denominator. This will prove useful in the approach to addressing outliers that we discuss below.

In each test period, we choose the 70 firms with the largest sales within our larger sample.

```
top_firms <-
  reg_data |>
  filter(fyearq %in% focus_years) |>
  group_by(gvkey, fyearq) |>
  summarize(total_sales = sum(saleq),
            .groups = "drop") |>
  group_by(fyearq) |>
  arrange(desc(total_sales)) |>
  mutate(rank = row_number()) |>
  filter(rank <= n_firms)

test_years <-
  reg_data |>
  semi_join(top_firms, by = c("gvkey", "fyearq")) |>
  select(gvkey, datadateq)
```

We then use `pmap()` to apply `fit_model()` from above to each firm-year in `test_years` and store the results in `results`.

```
results <-
  pmap(test_years, fit_model) |>
  list_rbind()
```

Before graphing results comparable to Table 3 in Foster (1977), let's look at the results in a less processed form. We address outliers with a `fix_outliers()` function.

```
fix_outliers <- function(x) {
  if_else(x < 0 | x > 1, 1, x)
}
```

Figure 14.3 provides histograms of `abe` based on net income for both years for all six models.

```
results |>
  filter(item == "ni") |>
  filter(!is.na(abe)) |>
  mutate(abe = fix_outliers(abe)) |>
  ggplot(aes(x = abe)) +
  geom_histogram(bins = 40) +
  facet_grid(model ~ fyearq)
```

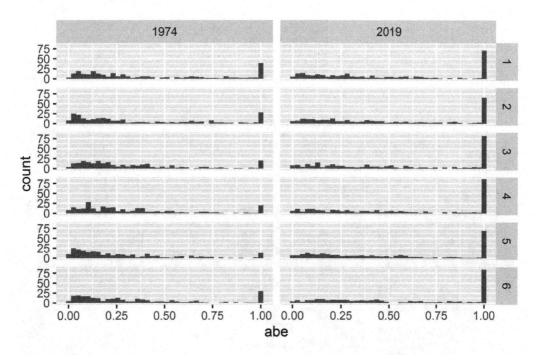

FIGURE 14.3
Histograms of abe

Next, we rank the models for each observation.

```
model_ranks <-
  results |>
  group_by(gvkey, datadateq, item) |>
  arrange(gvkey, datadateq, item, abe) |>
  mutate(rank = row_number()) |>
  group_by(fyearq, item, model) |>
  summarize(avg_rank = mean(rank, na.rm = TRUE),
            .groups = "drop") |>
  pivot_wider(names_from = c("model"), values_from = "avg_rank")
```

Finally, we produce results analogous to Table 3 of Foster (1977).

```
results_summ <-
  results |>
  mutate(abe = fix_outliers(abe),
         se = fix_outliers(se)) |>
  group_by(fyearq, item, model) |>
  summarize(mabe = mean(abe, na.rm=TRUE),
            mse = mean(se, na.rm=TRUE),
            .groups = "drop")
```

We plot these results in Figure 14.4.

```
results_summ |>
  pivot_longer(cols = mabe:mse, names_to = "metric", values_to = "value") |>
```

```
ggplot(aes(x = model, y = value, fill = model)) +
  geom_bar(stat = "identity") +
  facet_grid(fyearq ~ item + metric) +
  theme(legend.position = "none")
```

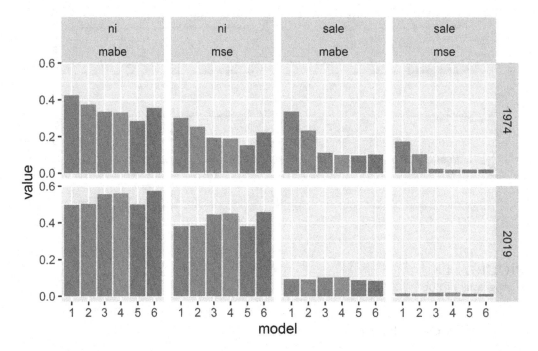

FIGURE 14.4
Plot of results like Table 3 of Foster (1977)

14.3.1 Exercises

1. How do the denominators in our calculations of `abe` and `se` differ from those in the analogous calculations in Foster (1977)? Is one approach to the denominator more correct than the other?

2. What does `fix_outliers()` do? Does Foster (1977) do anything to address outliers? If so, how does the approach in Foster (1977) compare to that of `fix_outliers()`? Do you agree with the approach taken in `fix_outliers()`? What would you do differently?

3. How do the results in Figure 14.4 compare with those in Foster (1977)?

4. What do you make of the significantly "worse" performance of models predicting `ni` than those predicting `sale`? Does this imply that `ni` is simply more difficult to forecast? Can you suggest an alternative approach to measuring performance that might place these models on a more "level playing field"?

14.4 Post-earnings announcement drift

The paper by Bernard and Thomas (1989) is a careful and persuasive one that rewards close reading when you have time. Bernard and Thomas (1989) build on the work of Foster et al. (1984) in measuring earnings surprise relative to expectations from Model 5 of Foster (1977); Bernard and Thomas (1989) also consider two alternative denominators.

$$FE_i^1 = \frac{Q_{i,t} - E[Q_{i,t}]}{|Q_{i,t}|}$$

$$FE_i^2 = \frac{Q_{i,t} - E[Q_{i,t}]}{\sigma\left(Q_{i,t} - E[Q_{i,t}]\right)}$$

In the first model, FE_i^1 is calculated with the absolute value of the forecast item as the denominator. In the second model, FE_i^2 is calculated with the standard deviation of the forecast error as the denominator.

Rather than net income, Bernard and Thomas (1989) use "Income Before Extraordinary Items" (`ibq`) and we follow them in this regard. While Bernard and Thomas (1989) focus on NYSE/AMEX-listed firms, we use all firms that we can match from Compustat to CRSP.

```
reg_data_fos <-
  merged_data |>
  semi_join(regular_fyears, by = c("gvkey", "datadate")) |>
  select(gvkey, datadateq, fyearq, rdq, ibq) |>
  group_by(gvkey) |>
  arrange(datadateq) |>
  mutate(ib_lag_4 = lag(ibq, 4L),
         ib_seas_diff = ibq - ib_lag_4,
         lag_ib_seas_diff  = lag(ib_seas_diff, 1L),
         qtr = quarter(datadateq, with_year = TRUE)) |>
  ungroup()
```

Bernard and Thomas (1989) require at least 10 quarters of data to produce a forecast and use up to 24 quarters of data. If there are fewer than 16 observations, then Bernard and Thomas (1989) use the simpler Model 1 of Foster (1977). The `fit_model_fos()` function below is adapted from `fit_model()` above to reflect these features.

As before, we split the data into `train_data`, which we use to fit the data, and `test_data`, which is the **holdout period** for the forecast. The calculation `denom_m2` provides the denominator for FE_i^2.

```
fit_model_fos <- function(gvkey, quarter) {

  n_qtrs <- 24
  min_qtrs_fos <- 16
  min_qtrs <- 10

  firm_data <-
    reg_data_fos |>
    filter(gvkey == !!gvkey)
```

```r
train_data <-
  firm_data |>
  filter(qtr < !!quarter) |>
  top_n(n_qtrs, datadateq)

test_data <-
  firm_data |>
  filter(qtr == !!quarter)

if (nrow(train_data) < min_qtrs) return(NULL)
if (nrow(train_data) >= min_qtrs_fos) {
  # Fit model 5
  ib_fm <- tryCatch(lm(ib_seas_diff ~ lag_ib_seas_diff,
                       data = train_data, na.action = na.exclude,
                       model = FALSE),
                    error = function(e) NULL)
} else {
  ib_fm <- NULL
}

if (!is.null(ib_fm)) {
  train_results <-
    train_data |>
    mutate(fib = ib_lag_4 + predict(ib_fm))
} else {
  train_results <-
    train_data |>
    mutate(fib = ib_lag_4)
}

denom_m2 <-
  train_results |>
    mutate(fe = ibq - fib) |>
    pull() |>
    sd()

if (is.null(ib_fm)) {
  results <-
    test_data |>
    mutate(fib = ib_lag_4)
} else {
  results <-
    test_data |>
    mutate(fib = ib_lag_4 + predict(ib_fm, newdata = test_data))
}

results |>
  mutate(fe1 = (ibq - fib) / abs(ibq),
         fe2 = (ibq - fib) / denom_m2)
}
```

So far, we have not worried about calendar time very much. But in this analysis, we are going to form portfolios each quarter based on the earnings surprise of each firm in that quarter. For this purpose, we want to measure earnings surprise in each *calendar* quarter. In constructing `reg_data_fos` above, we calculated `qtr` as `quarter(datadateq, with_year = TRUE)`. The `quarter()` function comes from the `lubridate` package and, with `with_year = TRUE`, will return (say) the number `2014.2` if given the data 2014-06-30.

We follow Bernard and Thomas (1989) in focusing on quarters from 1974 through 1986. The following code gets a list of all quarters on the data set in that range:

```
quarters <-
  reg_data_fos |>
  filter(qtr >= 1974, qtr < 1987) |>
  select(qtr) |>
  distinct() |>
  arrange(qtr) |>
  pull()
```

The `get_results()` function calls `fit_model_fos()`, which produces data for a given gvkey-quarter combination, then assembles the results into a data frame and returns them.

```
get_results <- function(quarter) {

  gvkeys <-
    reg_data_fos |>
    filter(qtr == quarter) |>
    select(gvkey) |>
    distinct() |>
    pull()

  map2(gvkeys, quarter, fit_model_fos) |>
    list_rbind()
}
```

> ⚠ **Warning**
>
> After setting `plan(multisession)`, we use `future_map()` from the `furrr` package to use parallel processing. This would take much more time if we instead used `map()` from the `purrr` package. Only run this code if you want to play around with the output yourself.

The following code calls `get_results()` for each quarter:

```
plan(multisession)

results <-
  quarters |>
  future_map(get_results) |>
  list_rbind() |>
  system_time()
```

```
  user  system elapsed
 6.849   0.530 142.278
```

Bernard and Thomas (1989) form portfolios based on deciles of earnings surprise. Decile 1 will have the 10% of firms with the lowest (most negative) earnings surprise. Decile 10 will have the 10% of firms with the highest (most positive) earnings surprise. But to avoid **lookahead bias**, the cut-offs for assigning firms to deciles will be based on the distribution of earnings surprises for the *previous* quarter. The `get_deciles()` function takes a vector of data and returns the decile cut-offs for that vector. Note that we set the highest limit to `Inf`, as we want to put any firms whose earnings surprise is greater than the maximum in Decile 10. Note also that we return the `breaks` inside a list so that the returned value can be stored in a column of a data frame, which we do in creating `decile_cuts`. The code creating `decile_cuts` is fairly self-explanatory, but note that we will be most interested in `fe1_deciles_lag` and `fe2_deciles_lag`, which are the decile cut-offs based on data from the previous quarter.

```
get_deciles <- function(x) {
  breaks <- quantile(x, probs = seq(from = 0, to = 1, by = 0.1),
                     na.rm = TRUE)
  breaks[length(breaks)] <- Inf
  list(breaks)
}


decile_cuts <-
  results |>
  group_by(qtr) |>
  summarize(fe1_deciles = get_deciles(fe1),
            fe2_deciles = get_deciles(fe2),
            .groups = "drop") |>
  arrange(qtr) |>
  mutate(fe1_deciles_lag = lag(fe1_deciles),
         fe2_deciles_lag = lag(fe2_deciles))
```

We add the decile cut-offs to the `results` data frame in the following code. Note that we need to use `rowwise()` here to calculate `fe1_decile` and `fe2_decile` row by row. While `cut()` is a **vectorized function**, it accepts just one value for the second argument `breaks`. However, we want to apply different breaks to each row (based on which quarter it is in). Using `rowwise()` allows us to do this. Once we have applied the breaks, we no longer need them, so we drop them in the last line below.

```
results_deciles <-
  results |>
  inner_join(decile_cuts, by = "qtr") |>
  rowwise() |>
  mutate(fe1_decile = cut(fe1, fe1_deciles_lag, labels = FALSE),
         fe2_decile = cut(fe2, fe2_deciles_lag, labels = FALSE)) |>
  filter(!is.na(fe1_decile) | !is.na(fe2_decile)) |>
  ungroup() |>
  select(-matches("^fe[12]_deciles"))
```

We will need to grab stock returns for our earnings announcements. The following code is more or less copy-pasted from earlier chapters.

```
ccm_link <-
  ccmxpf_lnkhist |>
  filter(linktype %in% c("LC", "LU", "LS"),
```

```
            linkprim %in% c("C", "P")) |>
  mutate(linkenddt = coalesce(linkenddt,
                                    max(linkenddt, na.rm = TRUE))) |>
  rename(permno = lpermno) |>
  collect()

link_table <-
  results_deciles |>
  select(gvkey, rdq) |>
  inner_join(ccm_link,
            join_by(gvkey, between(rdq, linkdt, linkenddt))) |>
  select(gvkey, rdq, permno)
```

> ⚠ Warning
>
> Note that the following code chunk can take a few minutes to run. Parallel processing
> does not help much in this case, as the code is not written in a way to facilitate this.

```
rets <-
  link_table |>
  get_event_rets(db, event_date = "rdq",
                    win_start = -60, win_end = 60) |>
  nest_by(rdq, permno) |>
  system_time()
```

```
   user  system elapsed
 25.420   4.299   7.587
```

Finally, we calculate mean **size-adjusted returns** for each `decile` at each trading day
relative to the earnings announcement. Here we basically follow the approach described in
Bernard and Thomas (1989, p. 7). There are some issues implicit in these calculations that
are discussed in Bernard and Thomas (1989) and covered in the discussion questions below.

Note that there are some duplicates in terms of (`gvkey`, `rdq`) on `results_deciles`. Many
of these cases appear to be ones in which announcement of earnings from a prior quarter
has been delayed, causing two quarters of earnings to be announced at the same time. It
seems reasonable to attribute market reactions on those dates to the latest quarter, so we
limit our data to the latest values of `datadateq` for each (`gvkey`, `rdq`).

```
plot_data <-
  results_deciles |>
  filter(!is.na(rdq)) |>
  group_by(gvkey, rdq) |>
  filter(datadateq == max(datadateq)) |>
  ungroup() |>
  mutate(decile = fe2_decile) |>
  inner_join(link_table, by = c("gvkey", "rdq")) |>
  inner_join(rets, by = c("rdq", "permno")) |>
  unnest(cols = c(data)) |>
  group_by(decile, relative_td) |>
  summarize(ar = mean(ret - decret, na.rm = TRUE), .groups = "drop")
```

In Figures 14.5 and 14.6, we set abnormal returns for the first trading day to zero. Thus, Figure 14.5 depicts returns starting from day $t-59$ and Figure 14.6 depicts returns starting from day $t+1$. Adding a label to the last day of the series means we don't need a legend.

```
plot_data |>
  filter(relative_td <= 0) |>
  filter(!is.na(decile)) |>
  mutate(decile = as.factor(decile)) |>
  mutate(first_day = relative_td == min(relative_td),
         last_day = relative_td == max(relative_td),
         ar = if_else(first_day, 0, ar),
         label = if_else(last_day, as.character(decile), NA)) |>
  select(-first_day) |>
  group_by(decile) |>
  arrange(relative_td) |>
  mutate(car = cumsum(ar)) |>
  ggplot(aes(x = relative_td, y = car,
             group = decile, color = decile)) +
  geom_line() +
  geom_label(aes(label = label), na.rm = TRUE) +
  theme(legend.position = "none")
```

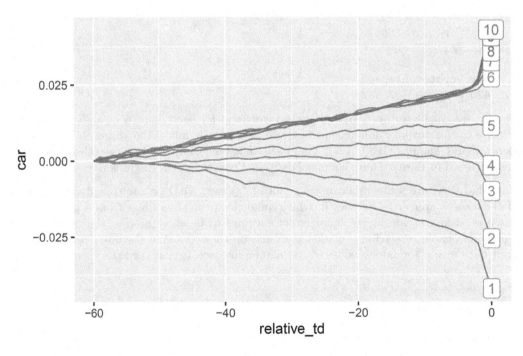

FIGURE 14.5
Pre-announcement returns

Figure 14.6 shows post-announcement returns for each of the ten portfolios.

```
plot_data |>
  filter(relative_td >= 0) |>
```

```
filter(!is.na(decile)) |>
mutate(decile = as.factor(decile)) |>
mutate(first_day = relative_td == min(relative_td),
       last_day = relative_td == max(relative_td),
       ar = if_else(first_day, 0, ar),
       label = if_else(last_day, as.character(decile), NA)) |>
group_by(decile) |>
arrange(relative_td) |>
mutate(car = cumsum(ar)) |>
ggplot(aes(x = relative_td, y = car, group = decile, color = decile)) +
geom_line() +
geom_label(aes(label = label), na.rm = TRUE) +
theme(legend.position = "none")
```

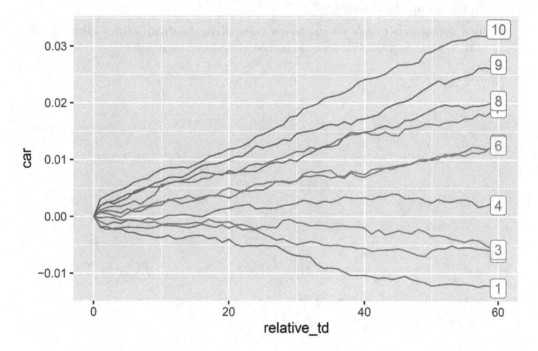

FIGURE 14.6
Post-announcement returns

14.4.1 Discussion questions

1. Both Bernard and Thomas (1989) and Ball and Brown (1968) were addressing issues with "conventional wisdom" at their respective times. How had conventional wisdom changed in the years between 1968 and 1989?

2. Evaluate the introduction of Bernard and Thomas (1989). How clear is the research question to you from reading this? How would you compare this introduction with other papers you have read in this course or elsewhere?

3. How persuasive do you find Bernard and Thomas (1989) to be? What evidence and arguments do you find persuasive or not persuasive? (Answering this question requires you to read the paper fairly closely.)

4. The analysis underlying Figure 14.6 considers the 13-year period from 1974 to 1986. What changes would you need to make to the code to run the analysis for the 13-year period from 2007 to 2019? (If you choose to make this change and run the code, what do you notice about the profile of returns in the post-announcement period? Does it seem necessary to make an additional tweak to the code to address this?)

5. Considering a single stock, what trading strategy is implicit in calculating `ar` as `ret - decret`?

6. In calculating mean returns by `decile` and `relative_td` (i.e., first using `group_by(decile, relative_td)` and then calculating `ar` by aggregating `mean(ret - decret, na.rm = TRUE)`), are we making assumptions about the trading strategy? What issues are created by this trading strategy? Can you suggest an alternative trading strategy? What changes to the code would be needed to implement this alternative?

7. Is it appropriate to *add* returns to get cumulative abnormal returns as is done in `car = cumsum(ar)`? What would be one alternative approach?

15

Accruals

In this chapter, we use Sloan (1996) to provide a focus for a study of accrual processes. We use simulation analysis to understand better accounting processes, with a particular focus on **accruals**, which for this chapter we define as the portion of earnings in excess of operating cash flows.[1] We finish up the chapter with an examination of the so-called **accrual anomaly**.

> Tip
>
> The code in this chapter uses the packages listed below. For instructions on how to set up your computer to use the code found in this book, see Section 1.2. Quarto templates for the exercises below are available on GitHub.[a]
>
> ---
> [a]https://github.com/iangow/far_templates/blob/main/README.md

```
library(tidyverse)
library(DBI)
library(farr)
library(modelsummary)
library(dbplyr)        # window_order()
library(car)           # linearHypothesis()
library(furrr)         # future_map()
```

15.1   Sloan (1996)

While we saw evidence in Chapter 14 that capital markets do not fully price earnings surprises, Sloan (1996) goes further and examines how capital markets price *components* of earnings. Sloan (1996) points out that a number of practitioners provide investment advice predicated on identifying firms whose earnings depend on accruals rather than cash flows. Such investment advice is based on a claimed tendency for capital markets to "fixate" on earnings and to fail to recognize differences in the properties of cash flow and accrual components of earnings. In some respects, Sloan (1996) provides a rigorous evaluation of such investment advice and the premises underlying it.

---
[1]This is one definition that can be tightened and vary by context.

DOI: 10.1201/9781003456230-15

15.1.1 Discussion questions

The following discussion questions provide an approach to reading Sloan (1996). While one approach to reading a paper involves a careful reading from start to finish, a useful skill is being able to read a paper quickly with a focus on the empirical results and the hypotheses that these relate to.

1. Read the material preceding the formal statement of H1. What reasons for differential persistence of earnings components does Sloan (1996) offer? How important is it for these reasons to be correct in light of the empirical support for H1 provided in Table 3? How important is the empirical support for H1 to H2(i)?

2. Which hypothesis (if any) does Table 4 test? How would you interpret the results of Table 4 in words?

3. Which hypothesis (if any) does Table 5 test? How would you interpret the results of Table 5 in words?

4. Which hypothesis (if any) does Table 6 test? How would you interpret the results of Table 6 in words? There are similarities between the results of Table 6 of Sloan (1996) and the results in Bernard and Thomas (1989). Both involve forming portfolios of firms based on deciles of some variable (accruals in Sloan, 1996; earnings surprise in Bernard and Thomas, 1989) and examining how those portfolios perform subsequently. Apart from the measure used to form portfolios, what are the significant differences between the analyses in the two papers that you can think of looking at Table 6?

5. With which hypothesis (if any) is Figure 2 related? What does Figure 2 show according to Sloan (1996)?

6. With which hypothesis (if any) is Figure 3 related? What does Figure 3 show according to Sloan (1996)?

15.2 Measuring accruals

Hribar and Collins (2002) include a definition of accruals similar to that used in Sloan (1996). Referring to prior research, they state (2002, p. 10):

Specifically, accruals (ACC_{bs}) are typically calculated (firm and time subscripts omitted for convenience):

$$ACC_{bs} = (\Delta CA - \Delta CL - \Delta Cash + \Delta STDEBT - DEP)$$

where

- ΔCA = the change in current assets during period t (Compustat #4)
- ΔCL = the change in current liabilities during period t (Compustat #5)
- $\Delta Cash$ = the change in cash and cash equivalents during period t (Compustat #1);
- $\Delta STDEBT$ = the [change in] current maturities of long-term debt and other short-term debt included in current liabilities during period t (Compustat #34);
- and ΔDEP = depreciation and amortization expense during period t (Compustat #14).

All variables are deflated by lagged total assets (TA_{t-1}) to control for scale differences.

The first thing you may ask is "what does (say) 'Compustat #4' mean?". Prior to 2006, Compustat data items were referred to using numbers such as Compustat #4 or `data4`. So older papers may refer to such items. Fortunately, Wharton Research Data Services (WRDS) provides translation tables[2] from these items to the current variables and relevant translations are provided in Table 15.1.

TABLE 15.1

Translation of key pre-2006 Compustat items

Old item	Current item	Item description
#1	che	Cash and Short-Term Investments
#4	act	Current Assets—Total
#5	lct	Current Liabilities—Total
#14	dp	Depreciation and Amortization
#34	dlc	Debt in Current Liabilities—Total

Hribar and Collins (2002) point out that calculating current accruals by subtracting the change in current liabilities from the change in noncash current assets is incorrect "because other non-operating events (e.g., mergers, divestitures) impact the current asset and liability accounts with no earnings impact."

15.2.1 Discussion questions

1. In the equation above, why is $\Delta Cash$ subtracted?

2. In the equation above, why is $\Delta STDEBT$ added?

3. Is it true that mergers and divestitures have "no earnings impact"? Is the absence of earnings impact important to the estimation issue? Are there transactions that have no earnings impact, but do affect cash flow from operations?

4. Are there any differences between Hribar and Collins (2002) (ACC_{bs} above) and Sloan (1996) in their definitions of accruals? Which definition makes more sense to you? Why?

15.3 Simulation analysis

We now consider some **simulation analysis**. One reason for this analysis is to better understand the basis for H1 of Sloan (1996).

A second reason for conducting simulation analysis here is to illustrate the power it offers. In many contexts, derivation of the properties of estimators or understanding how phenomena interact is very complex. While many researchers rely on intuition to guide their analyses, such intuition can be unreliable. As an example, the idea that the "FM-NW" method provides standard errors robust to both time-series and cross-sectional dependence has strong intuitive appeal, but we saw in Chapter 5 that this intuition is simply wrong.

---

[2]https://go.unimelb.edu.au/i7d8

15.3.1   Vectors

In our simulation analysis, we make more extensive use of base R functionality than we have in prior chapters. Chapter 28[3] of *R for Data Science*—"A field guide to base R"—provides material that might be helpful if code in the next section is unclear. Here we are simulating the cash flows and accounting for a simple firm that buys goods for cash and sells them on account after adding a mark-up.

15.3.2   Simulation function

As we have seen before, it is often a good coding practice to use functions liberally in analysis.[4] In this case, we embed the core of the simulation in a function.

The simulation function `get_data()` below generates a time-series of data for a single "firm" and accepts two arguments. The first argument to `get_data()` is `add_perc`, which has a default value of 0.03. The value of `add_perc` drives the amount of *allowance for doubtful debts*. The second argument is `n_years`, which has a default value of 20. The value of `n_years` drives the number of years of data generated by the simulation.

The simulation generates various cash flows and the financial statements to represent them. The main driver of the model is sales, which follows an autoregressive process. Denoting sales in period t as S_t, we have

$$S_t - \overline{S} = \rho(S_{t-1} - \overline{S}) + \epsilon_t$$

where $\rho \in (0,1)$ and $\epsilon_t \sim N(0, \sigma^2)$.

Sales then drives both cost of goods sold, which are assumed to require cash outlays in the period of sale, and accounts receivable, as all sales are assumed to be on account. The model also addresses collections, write-offs, and dividends. There are no inventories in our model.

In the simulation function, we use "base R" functionality to a fair degree. Rather than using `mutate()` to generate variables, we refer to variables using $ notation, which returns the variable as a vector. For example `df |> select(ni)` returns a data frame with a single column. In contrast, `df$ni` gets the same underlying data, but as a vector. To calculate shareholders' equity (`se`), we set the initial ($t = 0$) value to `beg_se`. Then we calculate the ending balance of shareholders' equity as beginning shareholders' equity plus net income minus dividends.

```
get_data <- function(add_perc = 0.03, n_years = 20) {

  # Parameters
  add_true <- 0.03
  gross_margin <- 0.8
  beg_cash <- beg_se <- 1500
  div_payout <- 1
  mean_sale <- 1000
  sd_sale <- 100
  rho <- 0.9


  # Generate sales as an AR(1) process around mean_sale
```

---

[3]https://r4ds.hadley.nz/base-r
[4]Wickham et al. (2023, p. 441) suggest that "a good rule of thumb is to consider writing a function whenever you've copied and pasted a block of code more than twice."

```
    sale_err <- rnorm(n_years, sd = sd_sale)
    sales <- vector("double", n_years)
    sales[1] <- mean_sale + sale_err[1]
    for (i in 2:n_years) {
      sales[i] = mean_sale +  rho * (sales[i-1] - mean_sale) + sale_err[i]
    }


    # Combine data so far into a data frame;
    # add slots for variables to come
    df <- tibble(year = 1:n_years,
                 add_perc = add_perc,
                 sales,
                 writeoffs = NA, collect = NA,
                 div = NA, se = NA, ni = NA,
                 bde = NA, cash = NA)


    # All sales at the same margin
    df$cogs <- (1 - gross_margin) * df$sales


    # All sales are on credit;
    # collections/write-offs occur in next period
    df$ar <- df$sales


    # Allowance for doubtful debts
    df$add <- add_perc * df$sales


    # Calculate year-1 values
    df$writeoffs[1] <- 0
    df$collect[1] <- 0
    df$bde[1] <- df$add[1]
    df$ni[1] <- df$sales[1] - df$cogs[1] - df$bde[1]
    df$div[1] <- df$ni[1] * div_payout
    df$cash[1] <- beg_cash + df$collect[1] - df$cogs[1] - df$div[1]
    df$se[1] <- beg_se + df$ni[1] - df$div[1]


    # Loop through years from 2 to n_years
    for (i in 2:n_years) {
      df$writeoffs[i] <- add_true * df$ar[i-1]
      df$collect[i] <- (1 - add_true) * df$ar[i-1]
      df$bde[i] = df$add[i] - df$add[i-1] + df$writeoffs[i]
      df$ni[i] <- df$sales[i] - df$cogs[i] - df$bde[i]
      df$div[i] <- df$ni[i] * div_payout
      df$cash[i] <- df$cash[i-1] + df$collect[i] - df$cogs[i] - df$div[i]
      df$se[i] <- df$se[i-1] + df$ni[i] - df$ni[i]
    }


    df
}
```

To understand a function like this, it can be helpful to set values for the arguments (e.g., `add_perc <- 0.03; n_years <- 20`) and step through the lines of code one by one, intermittently inspecting the content of variables such as `df` as you do so.[5]

Let's generate 1000 years of data.

```
set.seed(2021)
df_1000 <- get_data(n_years = 1000)
```

The first 20 years are shown in Figure 15.1.

```
df_1000 |>
  filter(year <= 20) |>
  ggplot(aes(x = year)) +
  geom_line(aes(y = sales), colour = "red") +
  geom_line(aes(y = mean(sales)), colour = "blue")
```

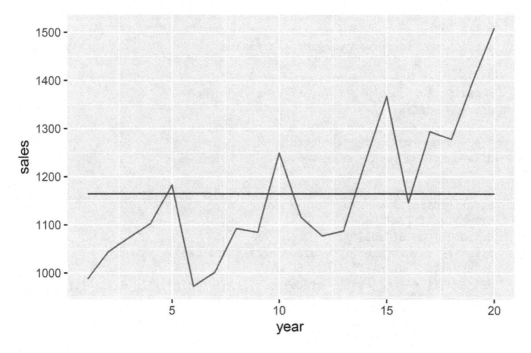

FIGURE 15.1
Sales over time for a simulated firm

Now, let's generate 5,000 random values for the `add_perc` parameter that we can use to generate simulated data.

```
add_percs <- runif(n = 5000, min = 0.01, max = 0.05)
```

We will generate simulated data for each value `add_percs` and store these data in a list called `res_list`.

---

[5]Indeed, this is the process often used to create a function like this in the first place.

```
set.seed(2021)

res_list <-
  map(add_percs, get_data) |>
  system_time()
```

```
   user  system elapsed
 16.354   0.036  16.405
```

While production of `res_list` takes just a few seconds, because each iteration of `get_data()` is independent of the others, we could use the `future` package and `plan(multisession)` to do it even more quickly.

```
plan(multisession)

res_list <-
  future_map(add_percs, get_data,
             .options = furrr_options(seed = 2021)) |>
  system_time()
```

```
   user  system elapsed
  0.541   0.027   4.305
```

We then make two data frames. The first data frame (`res_df`) stores all the data in a single data frame using the field `id` to distinguish one simulation run from another. These runs might be considered as "firms" with each run being independent of the other.

```
res_df <- list_rbind(res_list, names_to = "id")
```

To make it easier to compile results, we create `get_coefs()`, which calculates *persistence* as the coefficient in a regression of income on its lagged value—a specification similar to that in Sloan (1996)—and returns that value.

```
get_coefs <- function(df) {
  fm <-
    df |>
    arrange(year) |>
    mutate(lag_ni = lag(ni)) |>
    lm(ni ~ lag_ni, data = _)

  tibble(add_perc = mean(df$add_perc),
         persistence = fm$coefficients[2])
}
```

We apply `get_coefs()` to `res_list` and store the results in the second data frame, `results`.

```
results <-
  res_list |>
  map(get_coefs) |>
  list_rbind(names_to = "id")
```

We plot the estimated *persistence* value against the assumed value for `add_perc` in Figure 15.2.

```
results |>
  ggplot(aes(x = add_perc, y = persistence)) +
  geom_point()
```

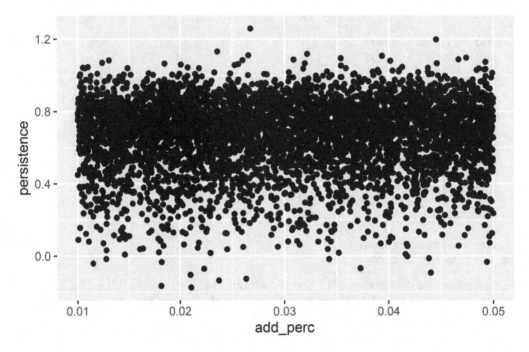

FIGURE 15.2
Relationship between persistence and `add_perc`

15.3.3 Exercises

1. When generating simulated financial statement data, it is generally important
   to ensure that the generated data meet some basic requirements. What is one
   fundamental relation that we expect to hold for these data? Does it hold for the
   data in `df_1000`?

2. Calculate values for cash flows from operating activities and cash flows from
   financing activities. (Treat payment of dividends as a financing activity. *Hint*:
   You may find it easier to use the **direct method** to calculate cash flows from
   operating activities.) Does the cash flow statement articulate as it should?

3. How evident are the details of the underlying process generating sales from Figure 15.1? Does looking at more data help? (Obviously, having a thousand years
   of data on a firm with a stationary process is not common.)

4. What is the "correct" value of `add_perc` that should be used? Using the plot from
   `results` above, what is the relation between values of departing from that value
   and persistence? Does this agree with your intuition? What's going on? What
   aspects of the `add_perc`-related accounting seem unrealistic? (*Hint*: It may help to
   use variant of the following code `set.seed(2021); get_data(0.03)` for various
   values in place of `0.03` and to examine how the earnings process is affected.)

5. Does the simulation analysis speak to the underlying rationale for H1 of Sloan (1996)? If so, why? If not, what might be missing from the analysis? How might we modify the simulation to incorporate the missing elements?

15.4   Replicating Sloan (1996)

To better understand some elements of the empirical analysis of Sloan (1996), we conduct a replication analysis.

We start by collating the data. We first connect to the tables we will use in our analysis.

```
db <- dbConnect(RPostgres::Postgres(), bigint = "integer")

funda <- tbl(db, Id(schema = "comp", table = "funda"))
company <- tbl(db, Id(schema = "comp", table = "company"))
ccmxpf_lnkhist <- tbl(db, Id(schema = "crsp", table = "ccmxpf_lnkhist"))
msf <- tbl(db, Id(schema = "crsp", table = "msf"))
```

We then make a subset of `comp.funda`, add SIC data from `comp.company`, and call the result `funda_mod`. We construct SIC codes using `sich`, which provides the "historical" SIC code (`sich`), where available, but we use the "header" SIC code (`sic`) found on `comp.company` when `sich` is unavailable.[6] For some reason, `sic` is a character variable on `comp.company` and `sich` is an integer on `comp.funda`. So, we convert `sic` to an integer before merging the two tables here.

```
sics <-
  company |>
  select(gvkey, sic) |>
  mutate(sic = as.integer(sic))

funda_mod <-
  funda |>
  filter(indfmt == "INDL", datafmt == "STD",
         consol == "C", popsrc == "D") |>
  left_join(sics, by = "gvkey") |>
  mutate(sic = coalesce(sich, sic))
```

We next apply the same sample selection criteria as Sloan (1996). We focus on NYSE and AMEX firm-years (i.e., ones with `exchg` equal to 11 and 12, respectively) and years between 1962 and 1991.

Sloan (1996, p. 293) suggests that "the financial statement data required to compute operating accruals are not available ... on Compustat for banks, life insurance or property and casualty companies." However, it is not clear if these firms are explicitly excluded (e.g., by filtering on SIC codes) or implicitly excluded by simply requiring that data for calculating accruals be available. As such, we retain these firms and merely create a related indicator variable (`finance`).

---

[6]Recall from Chapter 12 that a header variable is one where only the most recent value is retained in the database.

The next step is to create variables to reflect changes in key variables. We can use `lag()` to do this.[7]

```
acc_data_raw <-
  funda_mod |>
  filter(!is.na(at),
         pddur == 12,
         exchg %in% c(11L, 12L)) |>
  mutate(finance = between(sic, 6000, 6999),
         across(c(che, dlc, txp), \(x) coalesce(x, 0))) |>
  group_by(gvkey) |>
  window_order(datadate) |>
  mutate(avg_at = (at + lag(at)) / 2,
         d_ca = act - lag(act),
         d_cash = che - lag(che),
         d_cl = lct - lag(lct),
         d_std = dlc - lag(dlc),
         d_tp = txp - lag(txp)) |>
  select(gvkey, datadate, fyear, avg_at, at, oiadp, dp, finance,
         starts_with("d_"), sic, pddur) |>
  mutate(acc_raw =  (d_ca - d_cash) - (d_cl - d_std - d_tp) - dp) |>
  ungroup() |>
  filter(between(fyear, 1962, 1991),
         avg_at > 0)
```

The final step in our data preparation calculates the core variables `earn`, `acc`, and `cfo` according to the definitions found in Sloan (1996), creates a variable to store the leading value of `earn` (using the `lead()` function, which is a window function that complements the `lag()` function we used above), creates deciles for `acc`, `earn`, `cfo`, and `lead_earn`, creates a two-digit SIC code, and finally filters out `finance` firms and observations without values for `acc`.

```
acc_data <-
  acc_data_raw |>
  mutate(earn = oiadp / avg_at,
         acc = acc_raw / avg_at,
         cfo = earn - acc) |>
  group_by(gvkey) |>
  window_order(datadate) |>
  mutate(lead_earn = lead(earn)) |>
  ungroup() |>
  collect() |>
  mutate(acc_decile = ntile(acc, 10),
         earn_decile = ntile(earn, 10),
         cfo_decile = ntile(cfo, 10),
         lead_earn_decile = ntile(lead_earn, 10),
         sic2 = str_sub(as.character(sic), 1, 2)) |>
  filter(!finance, !is.na(acc))
```

The next step is to collect data on stock returns for each firm-year. We use `ccm_link`, which we saw in Chapter 7, to link GVKEYs (Compustat) to PERMNOs (CRSP).

---

[7]We introduced the `lag()` function in Chapters 7 and 8.

```
ccm_link <-
  ccmxpf_lnkhist |>
    filter(linktype %in% c("LC", "LU", "LS"),
           linkprim %in% c("C", "P")) |>
    rename(permno = lpermno) |>
    mutate(linkenddt = coalesce(linkenddt, max(linkenddt, na.rm = TRUE))) |>
  select(gvkey, permno, linkdt, linkenddt)
```

Following Sloan (1996), we link (gvkey, datadate) combinations with permnos and a date range for the twelve-month period beginning four months after the end of the fiscal period.

```
crsp_link <-
  acc_data_raw |>
  select(gvkey, datadate) |>
  inner_join(ccm_link,
             join_by(gvkey, between(datadate, linkdt, linkenddt))) |>
  select(gvkey, datadate, permno) |>
  mutate(start_month = as.Date(floor_date(datadate + months(4L), "month")),
         end_month = as.Date(floor_date(datadate + months(16L) - days(1L),
                                        "month")),
         month = floor_date(datadate, 'month'))
```

We then calculate compounded returns over this window for each (gvkey, datadate, permno) combination.

```
crsp_data <-
  msf |>
  inner_join(crsp_link,
             by = join_by(permno, between(date, start_month, end_month))) |>
  group_by(gvkey, permno, datadate) |>
  summarize(ret = exp(sum(log(1 + ret), na.rm = TRUE)) - 1,
            n_months = n(),
            .groups = "drop") |>
  collect()
```

Table 4 of Sloan (1996, p. 304) uses abnormal returns, which are "computed by taking the raw buy-hold return ... and subtracting the buy-hold return of a size-matched, value-weighted portfolio of firms. The size portfolios are based on market value of equity deciles of NYSE and AMEX firms." We obtained the returns of individual firms above, but need to collect data on size portfolios, both the returns for each portfolio and the market capitalization cut-offs.

Data for size portfolios come from Ken French's website, as we saw in Chapter 11. Code like that used in Chapter 11 is included in the farr package in two functions: get_size_rets_monthly() and get_me_breakpoints().

```
size_rets <- get_size_rets_monthly()
size_rets
```

```
# A tibble: 11,770 x 4
   month      decile  ew_ret  vw_ret
   <date>      <int>   <dbl>   <dbl>
 1 1926-07-01      1 -0.0142 -0.0012
 2 1926-07-01      2  0.0029  0.0052
```

```
 3 1926-07-01       3 -0.0015 -0.0005
 4 1926-07-01       4  0.0088  0.0082
 5 1926-07-01       5  0.0145  0.0139
 6 1926-07-01       6  0.0185  0.0189
 7 1926-07-01       7  0.0163  0.0162
 8 1926-07-01       8  0.0138  0.0129
 9 1926-07-01       9  0.0338  0.0353
10 1926-07-01      10  0.0329  0.0371
# i 11,760 more rows
```

The table returned by `get_size_rets_monthly()` has four columns, including two measures of returns: one based on equal-weighted portfolios (`ew_ret`) and one based on value-weighted portfolios (`vw_ret`). Like Sloan (1996), we use `vw_ret`.

```
me_breakpoints <- get_me_breakpoints()
me_breakpoints
```

```
# A tibble: 11,840 x 4
   month      decile me_min me_max
   <date>      <int>  <dbl>  <dbl>
 1 1925-12-01      1    0      2.38
 2 1925-12-01      2    2.38   4.95
 3 1925-12-01      3    4.95   7.4
 4 1925-12-01      4    7.4   10.8
 5 1925-12-01      5   10.8   15.6
 6 1925-12-01      6   15.6   22.9
 7 1925-12-01      7   22.9   38.4
 8 1925-12-01      8   38.4   65.8
 9 1925-12-01      9   65.8  142.
10 1925-12-01     10  142.    Inf
# i 11,830 more rows
```

The table returned by `get_me_breakpoints()` identifies the size decile (`decile`) to which firms with market capitalization between `me_min` and `me_max` in a given `month` should be assigned.

To join CRSP with `crsp_link`, we construct the variable `month` for each value of `date` on `crsp.msf`. This prevents non-matches due to non-alignment of `datadate` values with `date` values on `crsp.msf`.

```
crsp_dates <-
  msf |>
  distinct(date) |>
  mutate(month = floor_date(date, 'month'))
```

The following code assigns firm-years (i.e., (`permno`, `datadate`) combinations) to size deciles according to market capitalization and size cut-offs applicable during the month of `datadate`.

```
me_values <-
  crsp_link |>
  inner_join(crsp_dates, by = "month") |>
  inner_join(msf, by = c("permno", "date")) |>
  mutate(mktcap = abs(prc) * shrout / 1000) |>
```

```
  select(permno, datadate, month, mktcap) |>
  collect()

me_decile_assignments <-
  me_breakpoints |>
  inner_join(me_values,
             join_by(month, me_min <= mktcap, me_max > mktcap)) |>
  select(permno, datadate, decile)
```

For each `datadate` and size decile, the following code calculates the cumulative returns over the twelve-month period beginning four months after `datadate`.

```
cum_size_rets <-
  me_decile_assignments |>
  select(datadate, decile) |>
  distinct() |>
  mutate(start_month = datadate + months(4),
         end_month =  datadate + months(16)) |>
  inner_join(size_rets,
             join_by(decile, start_month <= month, end_month >= month)) |>
  group_by(datadate, decile) |>
  summarize(ew_ret = exp(sum(log(1 + ew_ret), na.rm = TRUE)) - 1,
            vw_ret = exp(sum(log(1 + vw_ret), na.rm = TRUE)) - 1,
            n_size_months = n(),
            .groups = "drop")
```

Now we have the data we need to calculate size-adjusted returns. We simply combine `crsp_data` with `me_decile_assignments` and then with `cum_size_rets` and calculate `size_adj_ret` as a simple difference.

```
size_adj_rets <-
  crsp_data |>
  inner_join(me_decile_assignments, by = c("permno", "datadate")) |>
  inner_join(cum_size_rets, by = c("datadate", "decile")) |>
  mutate(size_adj_ret = ret - vw_ret) |>
  select(gvkey, datadate, size_adj_ret, n_months, n_size_months)
```

For our regression analysis, we simply join our processed data from Compustat (`acc_data`) with our new data on size-adjusted returns (`size_adj_rets`).

```
reg_data <-
  acc_data |>
  inner_join(size_adj_rets, by = c("gvkey", "datadate"))
```

Before running regression analyses, it is important to examine our data. One useful benchmark is the set of descriptive statistics reported in Table 1 of Sloan (1996). Some degree of assurance is provided by the similar values seen in Table 15.2 with those reported in Sloan (1996).

```
reg_data |>
  group_by(acc_decile) |>
  summarize(across(c(acc, earn, cfo), \(x) mean(x, na.rm = TRUE)))
```

TABLE 15.2
Means by accruals decile

acc_decile	acc	earn	cfo
1	−0.1751	0.0355	0.2106
2	−0.0871	0.0946	0.1817
3	−0.0614	0.1035	0.1649
4	−0.0442	0.1109	0.1551
5	−0.0302	0.1178	0.1481
6	−0.0168	0.1182	0.1351
7	−0.0006	0.1269	0.1275
8	0.0226	0.1353	0.1127
9	0.0599	0.1422	0.0823
10	0.1724	0.1539	−0.0185

15.4.1   Table 2 of Sloan (1996)

Having done a very basic check of our data, we can create analogues of some of the regression analyses found in Sloan (1996).

The output shown in Table 15.3 parallels the "pooled" results in Table 2 of Sloan (1996).

```
fms <- list(lm(lead_earn ~ earn, data = reg_data),
            lm(lead_earn_decile ~ earn_decile, data = reg_data))

modelsummary(fms,
             estimate = "{estimate}{stars}",
             gof_map = c("nobs", "r.squared"),
             stars = c('*' = .1, '**' = 0.05, '***' = .01))
```

TABLE 15.3
Replication of Table 2 of Sloan (1996)

	(1)	(2)
(Intercept)	0.030***	1.396***
	(0.001)	(0.039)
earn	0.704***	
	(0.005)	
earn_decile		0.749***
		(0.006)
Num.Obs.	12,877	12,877
R2	0.607	0.553

To produce "industry level" analysis like that in Table 2 of Sloan (1996), we create `run_table_ind()`, a small function to produce regression coefficients by industry. As seen in Chapter 14, we use `!!` to distinguish the value `sic2` supplied to the function from the variable `sic2` found in `reg_data`.[8]

```
run_table_ind <- function(sic2, lhs = "lead_earn", rhs = "earn") {
  df <-
```

---

[8]Chapter 19 of Wickham (2019) has more details on this "unquote" operator `!!`.

```
    reg_data |>
    filter(sic2 == !!sic2)

  fm <- lm(as.formula(str_c(lhs, " ~ ", rhs)), data = df)

  coefs <- as_tibble(t(fm$coefficients))
  names(coefs) <- colnames(t(fm$coefficients))
  bind_cols(sic2 = sic2, coefs)
}
```

The function `stats_for_table()` compiles descriptive statistics.

```
stats_for_table <- function(x) {
  qs <- quantile(x, probs = c(0.25, 0.50, 0.75), na.rm = TRUE)

  tibble(mean = mean(x, na.rm = TRUE),
         q1 = qs[1], median = qs[2], q3 = qs[3])
}
```

Finally, `summ_for_table()` calls `run_table_ind()` and `stats_for_table()` and produces a summary table.

```
summ_for_table <- function(lhs = "lead_earn", rhs = "earn") {
  reg_data |>
    distinct(sic2) |>
    pull() |>
    map(run_table_ind, lhs = lhs, rhs = rhs) |>
    list_rbind() |>
    select(-sic2) |>
    map(stats_for_table) |>
    list_rbind(names_to = "term")
}
```

Tables 15.4 and 15.5 parallel the "industry level" results reported in Table 2 of Sloan (1996).

```
summ_for_table(lhs = "lead_earn", rhs = "earn")
```

TABLE 15.4
Industry-level persistence with actual earnings

term	mean	q1	median	q3
(Intercept)	0.026	0.012	0.026	0.036
earn	0.664	0.611	0.697	0.815

```
summ_for_table(lhs = "lead_earn_decile", rhs = "earn_decile")
```

TABLE 15.5
Industry-level persistence with earnings deciles

term	mean	q1	median	q3
(Intercept)	1.612	0.937	1.257	1.745
earn_decile	0.670	0.649	0.739	0.819

Our results thus far might be described as "qualitatively similar" to those in Table 2 of Sloan (1996). The main difference may be in the magnitude of the pooled coefficient on `earn` in the regression with `lead_earn` as the dependent variable. Table 2 of Sloan (1996) reports a coefficient of 0.841, notably higher than the mean coefficient from the industry-level regressions (0.773). In contrast, the mean coefficients in our pooled and industry-level analyses are much closer to each other.

15.4.2   Table 3 of Sloan (1996)

In Table 3, Sloan (1996) decomposes the right-hand side variables from Table 15.3 into accrual and cash-flow components. We replicate these analyses in Table 15.6.

```
fms <- list(lm(lead_earn ~ acc + cfo, data = reg_data),
            lm(lead_earn_decile ~ acc_decile + cfo_decile, data = reg_data))
```

```
modelsummary(fms,
             estimate = "{estimate}{stars}",
             gof_map = c("nobs", "r.squared"),
             stars = c('*' = .1, '**' = 0.05, '***' = .01))
```

TABLE 15.6
Replication of Table 3 of Sloan (1996)

	(1)	(2)
(Intercept)	0.028***	−1.312***
	(0.001)	(0.073)
acc	0.640***	
	(0.007)	
cfo	0.723***	
	(0.005)	
acc_decile		0.534***
		(0.007)
cfo_decile		0.760***
		(0.007)
Num.Obs.	12,877	12,877
R2	0.613	0.470

Tables 15.7 and 15.8 parallel the "industry level" results reported in Table 3 of Sloan (1996). Again we have "qualitatively similar" results to those found in Sloan (1996).

```
summ_for_table(lhs = "lead_earn", rhs = "acc + cfo")
```

TABLE 15.7
Industry-level persistence with earnings components

term	mean	q1	median	q3
(Intercept)	0.0247	0.0098	0.0239	0.0315
acc	0.6708	0.5895	0.6669	0.7659
cfo	0.7135	0.6514	0.7228	0.8276

```
summ_for_table(lhs = "lead_earn_decile", rhs = "acc_decile + cfo_decile")
```

TABLE 15.8
Industry-level persistence with earnings component deciles

term	mean	q1	median	q3
(Intercept)	−0.8800	−2.0286	−1.2842	−0.6100
acc_decile	0.4854	0.4376	0.5201	0.5982
cfo_decile	0.6966	0.6724	0.7353	0.8455

15.4.3 Pricing of earnings components

An element of the analysis reported in Table 5 of Sloan (1996) regresses abnormal returns on contemporaneous earnings and components of lagged earnings and we do likewise.

```
mms <- list(lm(size_adj_ret ~ lead_earn + acc + cfo,
            data = reg_data),
        lm(size_adj_ret ~ lead_earn_decile + acc_decile + cfo_decile,
            data = reg_data))
```

Table 15.9 provides results from this regression.

```
modelsummary(mms,
            estimate = "{estimate}{stars}",
            gof_map = c("nobs", "r.squared"),
            stars = c('*' = .1, '**' = 0.05, '***' = .01))
```

TABLE 15.9
Abnormal returns, earnings, and lagged components

	(1)	(2)
(Intercept)	−0.027***	0.186***
	(0.006)	(0.018)
lead_earn	2.538***	
	(0.061)	
acc	−1.961***	
	(0.062)	
cfo	−1.826***	
	(0.057)	
lead_earn_decile		0.082***
		(0.002)
acc_decile		−0.055***
		(0.002)
cfo_decile		−0.056***
		(0.002)
Num.Obs.	12,867	12,867
R2	0.124	0.109

In the notation of Sloan (1996), the coefficient on `acc` can be expressed as $-\beta\gamma_1^*$, which is minus one times the product of β, the coefficient on `lead_earn` (i.e., earnings roughly contemporaneous with `size_adj_ret`), and γ_1^*, the implied market coefficient on accruals.

With estimates of $\hat{\beta} = 2.538$ and $\widehat{\beta\gamma_1^*} = 1.961$, we have an implied estimate of $\hat{\gamma}_1^* = 0.7726$.

This estimate $\hat{\gamma}_1^* = 0.7726$ is higher than the estimate of $\hat{\gamma}_1 = 0.6399$. But can we conclude that the difference between these two coefficients is statistically significant?

One approach to this question would be to evaluate whether $\widehat{\beta\gamma_1^*} = 1.961$ as estimated from the market regression is statistically different from the value implied by $\hat{\beta} \times \hat{\gamma}_1 = 2.538 \times 0.6399 = 1.624$.

However, as pointed out by Mishkin (1983), this procedure "implicitly assumes that there is no uncertainty in the estimate of $\hat{\gamma}_1$. This results in inconsistent estimates of the standard errors of the parameters and hence test statistics that do not have the assumed F distribution. This can lead to inappropriate inference ...."[9]

Given the issue of "inappropriate inference" described above, Mishkin (1983) uses "iterative weighted non-linear least squares" (Sloan, 1996, p. 302) to estimate a system of equations and then calculates an F-statistic based on comparison of goodness-of-fit of an unconstrained system of equations with that of a constrained system of equations (i.e., one in which γ is constrained equal in both equations). While Sloan (1996) uses this "Mishkin (1983)" test in his analysis reported in Tables 4 and 5, this approach involves significant complexity.[10]

Fortunately, Abel and Mishkin (1983) suggest a simpler approach that they show is asymptotically equivalent to the Mishkin test. The intuition for this approach is that if components of lagged earnings (accruals and cash flows) are mispriced in a way that predicts stock returns, then this should be apparent from a regression of stock returns on those lagged earnings components. Kraft et al. (2007) provide additional discussion of the Mishkin test and the approach used by Abel and Mishkin (1983).[11] In effect, this allows us to skip the "middleman" of contemporaneous earnings in the regression analysis.

The regression results in Table 15.10 come from applying the approach suggested by Abel and Mishkin (1983).

```
eff <- list(lm(size_adj_ret ~ acc + cfo, data = reg_data),
            lm(size_adj_ret ~ acc_decile + cfo_decile, data = reg_data))

modelsummary(eff,
             estimate = "{estimate}{stars}",
             gof_map = c("nobs", "r.squared"),
             stars = c('*' = .1, '**' = 0.05, '***' = .01))
```

---

[9]Note that Mishkin (1983) is actually critiquing a different econometric procedure whereby residuals from the first regression are included in a version of the second, but the quoted criticism is equally applicable to the procedure we describe here.

[10]This is apparent from inspection of the Stata `.ado` file provided by Judson Caskey to implement the Mishkin (1983) approach at https://sites.google.com/site/judsoncaskey/data.

[11]Note that Kraft et al. (2007) appear to assume that the OLS test used by Mishkin (1983) is the same as the test proposed by Abel and Mishkin (1983), but differences in these tests do not affect the substance of the discussion of Abel and Mishkin (1983).

TABLE 15.10
Regressions for Abel and Mishkin (1983) test

	(1)	(2)
(Intercept)	0.041***	0.078***
	(0.006)	(0.018)
acc	−0.327***	
	(0.049)	
cfo	0.038	
	(0.037)	
acc_decile		−0.011***
		(0.002)
cfo_decile		0.006***
		(0.002)
Num.Obs.	13,798	13,798
R2	0.005	0.007

From Table 15.10, we see that lagged accruals are negatively associated with abnormal returns. This result is consistent with the market overpricing accruals because it assumes a level of persistence that is too high.

15.4.4 Exercises

1. In creating `acc_data_raw`, we used `coalesce()` to set the value of certain variables to zero when missing on Compustat. Does this seem appropriate here? Are the issues similar to those observed with regard to R&D in Chapter 8? It may be helpful to find some observations from recent years where this use of the `coalesce()` function has an effect and think about the issues in context of financial statements for those firm-years.

2. Can you reconcile the results from the Abel and Mishkin (1983) test with those from the previous regressions? (*Hint*: Pay attention to sample composition; you may need to tweak these regressions.)

3. The equations estimated in Table 5 of Sloan (1996) could be viewed as a structural (causal) model. Can you represent this model using a causal diagram? In light of the apparent econometric equivalence between that structural model and the estimation approach used in Abel and Mishkin (1983), how might the coefficients from the structural model be recovered from the latter approach?

4. A critique of Sloan (1996) made by Kraft et al. (2007) is that the coefficients may be biased due to omitted variables. This critique implies a causal interpretation of the coefficients in Sloan (1996). How might the critique of Kraft et al. (2007) be represented on the causal diagrams above? How persuasive do you find the critique of Kraft et al. (2007) to be?

5. Apart from the different data sources used, another difference between the simulation analysis earlier in this chapter and the regression analysis in Table 3 of Sloan (1996) is the regression model used. Modify the code below to incorporate the appropriate formulas for cash flow from operating activities (`cfo`) and accruals (`acc`). Then replicate the pooled analysis of Panel A of Table 3 of Sloan (1996) using the resulting `sim_reg_data` data frame. What do you observe?

```
sim_reg_data <-
  res_df |>
  mutate(cfo = [PUT CALC HERE], acc = [PUT CALC HERE]) |>
  group_by(id) |>
  arrange(id, year) |>
  mutate(lag_cfo = lag(cfo),
         lag_acc = lag(acc)) |>
  ungroup()
```

6. Which hypothesis does Figure 1 of Sloan (1996) relate to? What aspects of the plot make it easier or more difficult to interpret the results? The following code replicates a version of Figure 1 from Sloan (1996) using our simulated data. On the basis of Figures 15.3–15.5 and the arguments given in Sloan (1996), is H1 true in our simulated data? Given the other analysis above, is H1 true in our simulated data?

```
year_of_event <- 10

decile_data <-
  sim_reg_data |>
  filter(year == year_of_event) |>
  mutate(cfo_decile = ntile(cfo, 10),
         ni_decile = ntile(ni, 10),
         acc_decile = ntile(acc, 10)) |>
  select(id, ends_with("decile"))

reg_data_deciles <-
  sim_reg_data |>
  inner_join(decile_data, by = "id")
```

The following code produces Figure 15.3:

```
reg_data_deciles |>
  filter(ni_decile %in% c(1, 10)) |>
  mutate(ni_decile = as.factor(ni_decile),
         event_year = year - year_of_event) |>
  group_by(ni_decile, year) |>
  summarize(ni = mean(ni, na.rm = TRUE), .groups = "drop") |>
  ggplot(aes(x = year, y = ni, group= ni_decile, color = ni_decile)) +
  geom_line()
```

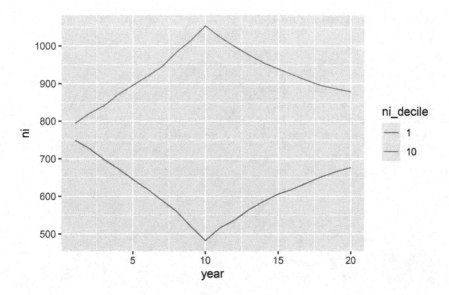

FIGURE 15.3
Persistence of net income by decile of income

The following code produces Figure 15.4:

```
reg_data_deciles |>
  filter(cfo_decile %in% c(1, 10)) |>
  mutate(cfo_decile = as.factor(cfo_decile),
         event_year = year - year_of_event) |>
  group_by(cfo_decile, year) |>
  summarize(ni = mean(ni, na.rm = TRUE), .groups = "drop") |>
  ggplot(aes(x = year, y = ni, group = cfo_decile, color = cfo_decile)) +
  geom_line()
```

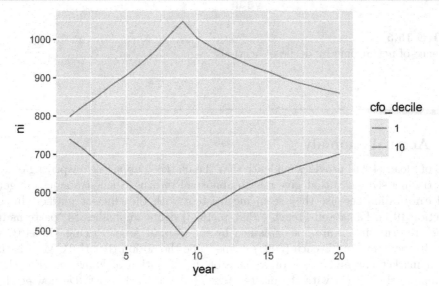

FIGURE 15.4
Persistence of net income by decile of cash flows

The following code produces Figure 15.5:

```
reg_data_deciles |>
  filter(acc_decile %in% c(1, 10)) |>
  mutate(acc_decile = as.factor(acc_decile),
         event_year = year - year_of_event) |>
  group_by(acc_decile, year) |>
  summarize(ni = mean(ni, na.rm = TRUE), .groups = "drop") |>
  ggplot(aes(x = year, y = ni, group = acc_decile, color = acc_decile)) +
  geom_line()
```

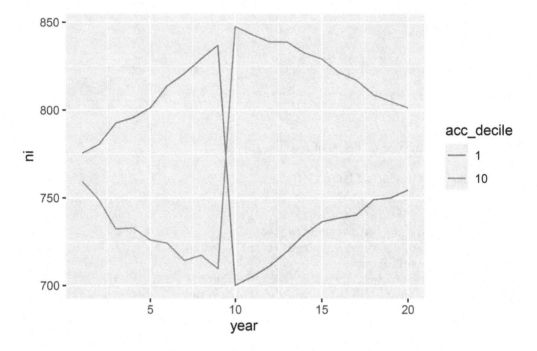

FIGURE 15.5
Persistence of net income by decile of accruals

15.5   Accrual anomaly

Table 6 of Sloan (1996) provides evidence that the market's apparent mispricing of accruals implies trading strategies that give rise to abnormal returns. Such strategies are generally termed **anomalies** because they seem inconsistent with the efficient markets hypothesis (see Section 10.1). Fama and French (2008, p. 1653) define anomalies as "patterns in average stock returns that ... are not explained by the Capital Asset Pricing Model (CAPM)." Implicit in the Fama and French (2008) seems to be the notion that the CAPM is the true model of market risk and a general version of the definition of Fama and French (2008) would replace the CAPM with the posited true model of market risk. Dechow et al. (2011, p. 23) argue that "the accrual anomaly is not really an anomaly at all. In fact, the original research documenting the accrual anomaly predicted that it would be there. The term anomaly is usually reserved for behavior that deviates from existing theories, but when Sloan (1996) first documented the accrual anomaly, he was testing a well-known theory and found that it was supported."

While Table 6 provides portfolio returns for years $t + 1$, $t + 2$, and $t + 3$, we only collected returns for year $t + 1$ in the steps above. Therefore, Table 15.11 only replicates the first column of Table 6 of Sloan (1996).

```
fm <-
  reg_data |>
  group_by(fyear, acc_decile) |>
  summarize(size_adj_ret = mean(size_adj_ret, na.rm = TRUE),
            .groups = "drop") |>
  mutate(acc_decile = as.factor(acc_decile)) |>
  lm(size_adj_ret ~ acc_decile - 1, data = _)

modelsummary(fm,
             estimate = "{estimate}{stars}",
             gof_map = c("nobs", "r.squared"),
             stars = c('*' = .1, '**' = 0.05, '***' = .01))
```

TABLE 15.11
Partial replication of Table 6 of Sloan (1996)

	(1)
acc_decile1	0.120***
	(0.020)
acc_decile2	0.109***
	(0.020)
acc_decile3	0.042**
	(0.020)
acc_decile4	0.058***
	(0.020)
acc_decile5	0.067***
	(0.020)
acc_decile6	0.042**
	(0.020)
acc_decile7	0.025
	(0.020)
acc_decile8	0.034*
	(0.020)
acc_decile9	−0.003
	(0.020)
acc_decile10	−0.012
	(0.020)
Num.Obs.	300
R2	0.255

```
hedge_ret <- fm$coefficients["acc_decile1"] - fm$coefficients["acc_decile10"]
p_val <- linearHypothesis(fm, "acc_decile1 = acc_decile10")$`Pr(>F)`[2]
```

The hedge portfolio return (`hedge_ret`) is 0.1322 with a p-value of 4.4e-06 (`p_val`).

15.5.1 Discussion questions

1. In estimating the hedge portfolio regression, we included a line `summarize(size_adj_ret = mean(size_adj_ret))`. Why is this step important?

2. Green et al. (2011) say "the simplicity of the accruals strategy and the size of the returns it generates have led some scholars to conclude that the anomaly is illusory. For example, Khan (2008) and Wu et al. (2010) argue that the anomaly can be explained by a mis-specified risk model and the q-theory of time-varying discount rates, respectively; Desai et al. (2004) conclude that the anomaly is deceptive because it is subsumed by a different strategy; Kraft et al. (2006) attribute it to outliers and look-ahead biases; Ng (2005) proposes that the anomaly's abnormal returns are compensation for high exposure to bankruptcy risk; and Zach (2006) argues that there are firm characteristics correlated with accruals that cause the return pattern." Looking at Sloan (1996), but without necessarily looking at each of the papers above, what evidence in Sloan (1996) seems inconsistent with the claims made by each paper above? Which do you think you would need to look more closely at the paper to understand? What evidence do you think Zach (2006) would need to provide to support the claim of an alternative "cause"?

3. Do Green et al. (2011) address the alternative explanations advanced in the quote in Q1 above? Do you think that they need to do so?

4. How persuasive do you find the evidence regarding the role of hedge funds provided by Green et al. (2011)?

5. Xie (2001) (p. 360) says that "for firm-years prior to 1988 when Compustat item #308 is unavailable, I estimate CFO_t as follows ...". Why would item #308 be unavailable prior to 1988? What is the equivalent to #308 in Compustat today?

6. Study the empirical model on p. 361 of Xie (2001), which is labelled equation (1). (This is the much-used "Jones model" from Jones (1991), which we examine in Chapter 16.) What are the assumptions implicit in this model and the labelling of the residual as "abnormal accruals"? (Hint: Take each component of the model and identify circumstances where it would be a reasonable model of "normal" accruals.)

7. What is "channel stuffing"? (Hint: Wikipedia has a decent entry on this.) What effect would channel stuffing have on abnormal accruals? (Hint: Think about this conceptually and with regard to equation (1). Do you need more information than is provided in Xie (2001) to answer this?)

16

Earnings management

A significant body of accounting research focuses on **earnings management**. One definition of earnings management might be "intervention by managers in the accounting process with a view to achieving financial reporting outcomes that benefit managers."

A classic form of earnings management is **channel stuffing**, which is a way for a company to report higher sales (and perhaps higher profits) in a period by pushing more products through a distribution channel than is needed to meet underlying demand. A classic case of channel stuffing involved the Contact Lens Division ("CLD") of Bausch and Lomb ("B&L"). According to the SEC[1], "B&L materially overstated its net income for 1993 by improperly recognizing revenue from the sale of contact lenses. These overstatements of revenue ... arose from sales of significant amounts of contact lenses to the CLD's distributors less than two weeks before B&L's 1993 fiscal year-end in connection with a marketing program that effectively resulted in consignment sales." In the case, the sales were not appropriately recognized as revenue during fiscal 1993 because "certain employees of the CLD granted unauthorized rights of return to certain distributors and shipped contact lenses after the fiscal year-end."

While B&L's channel stuffing was clearly earnings management (and in violation of generally accepted accounting principles, or **GAAP**), firms may engage in less extreme practices that are motivated by a desire to deliver higher sales in the current period, but do not involve any violation of GAAP or direct manipulation of the accounting process, yet would be generally regarded as earnings management. In such cases, earnings management is achieved by so-called **real activities** (i.e., those affecting business realities such as when products are delivered) and this form of earnings management is called **real earnings management**. Thus, not all forms of earnings management involve direct manipulation of the accounting process.

But once we allow for real earnings management, it can be difficult to distinguish, even in principle, actions taken to increase firm value that happen to benefit managers because of their financial reporting effects from actions that might fit more conventional notions of earnings management.

Another difficulty discussed by Beaver (1998) is the existence of alternative views of earnings management. While actions by managers to "manipulate the financial reporting system in ways that enhance management's well-being to the [detriment] of capital suppliers and others" (Beaver, 1998, p. 84) clearly meet the definition above, it is possible that earnings management allows managers "to reveal ... private information to investors."

We also note that Beaver (1998, p. 83) views earnings management as just one form a wide class of "discretionary behaviour", which also includes voluntary disclosures, such as earnings forecasts.

---

[1]https://www.sec.gov/litigation/admin/3439329.txt

DOI: 10.1201/9781003456230-16

Tip

The code in this chapter uses the packages listed below. Rather than invoking several Tidyverse packages separately, we load the `tidyverse` package. For instructions on how to set up your computer to use the code found in this book, see Section 1.2. Quarto templates for the exercises below are available on GitHub.[a]

---

[a]https://github.com/iangow/far_templates/blob/main/README.md

```
library(tidyverse)
library(DBI)
library(ggplot2)
library(farr)
library(broom)        # tidy()
library(furrr)        # future_map(), future_map2(), future_map_lgl()
```

16.1 Measuring earnings management

Even putting aside definitional issues, a challenge for researchers seeking to understand earnings management—the prevalence of the mechanisms through which it is achieved, and the effects that it has—is detecting and measuring it. In this section, we use discussion questions to explore two early papers (Jones, 1991; McNichols and Wilson, 1988) that illustrate some key issues and approaches that researchers have used to address these.

16.1.1 Discussion questions

1. Jones (1991) focuses on a small number of firms. Why does Jones (1991) have such a small sample? What are the disadvantages of a small sample? Are there advantages of a smaller sample or narrower focus?

2. What are the primary conclusions of Jones (1991)? Which table presents the main results of Jones (1991)? Describe the empirical test used in that table. Can you suggest an alternative approach? What do you see as the primary challenges to the conclusions of Jones (1991)?

3. Can you think of refinements to the broad research question? What tests might you use to examine these?

4. McNichols and Wilson (1988) state at the outset that their paper "examines whether managers manipulate earnings." Is this a good statement of the main research question of McNichols and Wilson (1988)? If not, suggest an alternative summary of the research questions of McNichols and Wilson (1988).

5. What do McNichols and Wilson (1988) mean by "nondiscretionary accruals"? How "operationalizable" is this concept?[2]

6. McNichols and Wilson (1988) say "if DA were observable, accrual-based tests of earnings management would be expressed in terms of the following regression:

---

[2]See https://en.wiktionary.org/wiki/operationalizable#English.

$$DA = \alpha + \beta PART + \epsilon$$

where $PART$ is a dummy variable that partitions the data into two groups for which earnings management predictions are specified". Healy (1985) points out that bonus plans can give managers incentives to increase earnings or decrease earnings depending on the situation. How is this problematic for the formulation of McNichols and Wilson (1988) above? How might a researcher address this?

7. What are the benefits and costs of focusing on a single item (bad debt expense) in a study of earnings management?

8. The main results of McNichols and Wilson (1988) are in Tables 6 and 7. How persuasive do you find the evidence of earnings management found in the "residual provision" columns of those tables?

9. How well does the $PART$ framework apply to Jones (1991)? Does the framework require modification for this paper? In which periods would $PART$ be set to one in Jones (1991)?

16.2   Evaluating measures of earnings management

A natural question that arises is how well measures of earnings management such as that used in Jones (1991) perform. An ideal measure would detect earnings management when it is present, but not detect earnings management when it is absent. This leads to two questions. First, how well does a given measure detect earnings management when it is present? Second, how does a given measure behave when earnings management is not present?

Dechow et al. (1995) evaluate five earnings management measures from prior research on these terms. Each of these measures uses an estimation period to create a model of non-discretionary accruals which is then applied to measure discretionary accruals for a test period as the difference between total accruals and estimated non-discretionary accruals. Assuming that the estimation period runs from $t = 1$ to $t = T$, these measures are defined, for firm i in year τ, as follows:

- The **Healy Model** (Healy, 1985) measures non-discretionary accruals as mean total accruals during the estimation period

$$NDA_{i,\tau} = \frac{\sum_{t=1}^{T} TA_{i,t}}{T}$$

where $TA_{i,t}$ is (both here and below) total accruals scaled by lagged total assets.

- The **DeAngelo Model** (DeAngelo, 1986) uses last period's total accruals as the measure of nondiscretionary accruals.

$$NDA_{i,\tau} = TA_{i,\tau-1}$$

- The **Jones Model** (Jones, 1991) "attempts to control for the effect of changes in a firm's economic circumstances on nondiscretionary accruals" using the following model:

$$NDA_{i,\tau} = \alpha_1(1/AT_{i,\tau-1}) + \alpha_2 \Delta REV_{i,\tau} + \alpha_3 PPE_{i,\tau}$$

where $AT_{i,\tau-1}$ is total assets for firm i at $\tau - 1$, ΔREV_τ is revenues in year τ less revenues in year $\tau - 1$ scaled by $AT_{i,\tau-1}$, and $PPE_{i,\tau}$ is gross property plant and equipment for firm i in year τ scaled by $AT_{i,\tau-1}$.

- **Modified Jones Model.** Dechow et al. (1995) consider a modified version of the Jones Model "designed to eliminate the conjectured tendency of the Jones Model to measure discretionary accruals with error when discretion is exercised over revenues" (1995, p. 199). In this model, non-discretionary accruals are estimated during the event period as:

$$NDA_{i,\tau} = \alpha_1(1/AT_{i,\tau-1}) + \alpha_2(\Delta REV_{i,\tau} - \Delta REC_{i,\tau}) + \alpha_3 PPE_{i,\tau}$$

where $\Delta REC_{i,\tau}$ is net receivables for firm i in year τ less net receivables in year $\tau-1$ scaled by $AT_{i,\tau-1}$.

- The **Industry Model** "relaxes the assumption that non-discretionary accruals are constant over time. The Industry Model assumes that variation in the determinants of non-discretionary accruals are common across firms in the same industry" (1995, p. 199). In this model, non-discretionary accruals are calculated as:

$$NDA_{i,\tau} = \gamma_1 + \gamma_2 \text{median}(TA_{I,\tau})$$

where $TA_{I,\tau}$ are the values of $TA_{j,\tau}$ for all the firms in industry I ($\forall j \in I$).

In each of the models above, the parameters (i.e., $(\alpha_1, \alpha_2, \alpha_3)$ or (γ_1, γ_2)) are estimated on a firm-specific basis during the estimation period.

Dechow et al. (1995) conduct analyses on four distinct samples, with each designed to test a different question. Drawing on the framework from McNichols and Wilson (1988), an indicator variable $PART$ is set to one for a subset of firm-years in each sample:

1. Randomly selected samples of 1000 firm-years.
2. Samples of 1000 firm-years randomly selected from firm-years experiencing extreme financial performance.
3. Samples of 1000 firm-years randomly selected to which a fixed and known amount of accrual manipulation is introduced.
4. Samples based on SEC enforcement actions.

Here we conduct a replication of sorts of Dechow et al. (1995). We consider the first three samples, but omit the fourth sample. Data for our analysis come from two tables on Compustat: `comp.funda` and `comp.company`.[3]

```
db <- dbConnect(RPostgres::Postgres(), bigint = "integer")


funda <- tbl(db, Id(schema = "comp", table = "funda"))
company <- tbl(db, Id(schema = "comp", table = "company"))
```

For financial statement data, we construct `funda_mod` in exactly the same way as we did in Chapter 15.

```
sics <-
  company |>
  select(gvkey, sic) |>
  mutate(sic = as.integer(sic))


funda_mod <-
  funda |>
  filter(indfmt == "INDL", datafmt == "STD",
         consol == "C", popsrc == "D") |>
```

---

[3]As in Chapter 15, we supplement data on `comp.funda` with SIC codes from `comp.company`.

```
  left_join(sics, by = "gvkey") |>
  mutate(sic = coalesce(sich, sic))
```

Sloan (1996, p. 293) suggests that the data needed to calculate accruals are not available for "banks, life insurance or property and casualty companies", so we exclude these firms (those with SIC codes starting with 6). Following Dechow et al. (1995), we restrict the sample to the years 1950–1991.[4] We also limit our sample to firm-years with non-missing assets and with twelve months in the fiscal period (`pddur == 12`).

```
acc_data_raw <-
  funda_mod |>
  filter(!is.na(at),
         pddur == 12,
         !between(sic, 6000, 6999)) |>
  mutate(across(c(che, dlc, sale, rect), \(x) coalesce(x, 0))) |>
  select(gvkey, datadate, fyear, at, ib, dp, rect, ppegt, ni, sale,
         act, che, lct, dlc, sic) |>
  filter(between(fyear, 1950, 1991)) |>
  arrange(gvkey, fyear) |>
  collect()
```

Like Sloan (1996) and Jones (1991), Dechow et al. (1995) measure accruals using a balance-sheet approach. The following function takes a data frame with the necessary Compustat variables, and calculates accruals for each firm-year, returning the resulting data set.

```
calc_accruals <- function(df) {
  df |>
    group_by(gvkey) |>
    arrange(datadate) |>
    mutate(lag_at = lag(at),
           d_ca = act - lag(act),
           d_cash = che - lag(che),
           d_cl = lct - lag(lct),
           d_std = dlc - lag(dlc),
           d_rev = sale - lag(sale),
           d_rec = rect - lag(rect)) |>
    ungroup() |>
    mutate(acc_raw =  (d_ca - d_cash - d_cl + d_std) - dp)
}
```

Like Jones (1991), Dechow et al. (1995) split firm-level data into an estimation period and a test firm-year and estimate earnings management models on a firm-specific basis. Dechow et al. (1995) require at least 10 years in the estimation period and each sample firm will have one test firm-year by construction. To give effect to this, we construct a sample of candidate firm-years comprising firms with at least 11 years with required data.

```
test_sample <-
  acc_data_raw |>
  calc_accruals() |>
```

---

[4]Because we need lagged values for most analyses, in only collecting data from 1950, we will lose that first year from most analyses. It is unclear whether Dechow et al. (1995) collected data for 1949 to be able to use 1950 firm-years in their analysis, but it is unlikely to have much of an impact (there are few firms in the data for 1950) and it would be easy to tweak if we wanted to include 1950 in our analysis.

```
  filter(lag_at > 0, sale > 0, ppegt > 0, !is.na(acc_raw),
         !is.na(d_rev), !is.na(d_rec), !is.na(ppegt)) |>
  group_by(gvkey) |>
  filter(n() >= 11) |>
  ungroup() |>
  arrange(gvkey, fyear) |>
  select(gvkey, fyear)
```

Most of our analysis will focus on a single random sample of 1000 firms. For each of these 1000 firms, we select a single fiscal year for which we set `part` to `TRUE`. Because we use the lagged value of accruals for the DeAngelo Model, we constrain the random choice to be any year but the first year.

```
set.seed(2022)

sample_1_firm_years <-
  test_sample |>
  mutate(rand = rnorm(n = nrow(pick(everything())))) |>
  group_by(gvkey) |>
  filter(rand == min(rand),
         fyear > min(fyear)) |>
  ungroup() |>
  top_n(1000, wt = rand) |>
  select(gvkey, fyear) |>
  mutate(part = TRUE)
```

To create `sample_1`, we use two joins. The first join—a `semi_join()`—is by `gvkey` and ensures that we draw those firm-years in `test_sample` that have been selected in `sample_1_firm_years`. The second join—a `left_join()`—is by `gvkey` and `fyear` and has the effect of adding the `part` indicator from `sample_1_firm_years` to the data for the applicable firm-years. Most firm-years will not be found in `sample_1_firm_years` and the final step uses `coalesce()` to set `part` to `FALSE` for firm-years missing from `sample_1_firm_years`.

```
sample_1 <-
  test_sample |>
  semi_join(sample_1_firm_years, by = "gvkey") |>
  left_join(sample_1_firm_years, by = c("gvkey", "fyear")) |>
  mutate(part = coalesce(part, FALSE))
```

We combine the data on `part` for our sample firm-years with the Compustat data in `acc_data_raw` to form `merged_sample_1`.[5]

```
merged_sample_1 <-
  sample_1 |>
  inner_join(acc_data_raw, by = c("gvkey", "fyear"))
```

---

[5]Our sampling approach deviates from that in Dechow et al. (1995), where firm-years are selected (without replacement) subject to the constraint that "a firm-year is not selected if its inclusion in the random sample leaves less than ten unselected observations for the estimation period." One important difference is that the approach in Dechow et al. (1995) could lead to a firm having two years of earnings management. There seems to be little upside in this, while our approach is much simpler to code and unlikely to impact results in a significant way.

If we were conducting a simple study of observed earnings management, it would be natural to calculate our measures of earnings management and then proceed to our analyses. However, in our analysis here we will—like Dechow et al. (1995)—be manipulating accounting measures ourselves and doing so will require us to recalculate earnings management measures and inputs to these, such as measures of total accruals. To facilitate this process, we embed the calculations for all five earnings management measures in the function `get_nda()` below.[6] Note that we use `reframe()` in place of `summarize()` because the former does not assume that the result will be a single row for each group.

```
fit_jones <- function(df) {
  fm <- lm(acc_at ~ one_at + d_rev_at + ppe_at - 1,
           data = df, model = FALSE, subset = !part)

  df |>
    mutate(nda_jones = predict(fm, newdata = df),
           da_jones = acc_at - nda_jones) |>
    select(fyear, nda_jones, da_jones)
}

fit_mod_jones <- function(df) {
  fm <- lm(acc_at ~ one_at + d_rev_alt_at + ppe_at - 1,
           data = df, model = FALSE, subset = !part)
  df |>
    mutate(nda_mod_jones = predict(fm, newdata = df),
           da_mod_jones = acc_at - nda_mod_jones) |>
    select(fyear, nda_mod_jones, da_mod_jones)
}

get_nda <- function(df) {

  df_mod <-
    df |>
    calc_accruals() |>
    mutate(sic2 = str_sub(as.character(sic), 1, 2),
           acc_at = acc_raw / lag_at,
           one_at = 1 / lag_at,
           d_rev_at = d_rev / lag_at,
           d_rev_alt_at = (d_rev - d_rec) / lag_at,
           ppe_at = ppegt / lag_at) |>
    group_by(sic2) |>
    mutate(acc_ind = median(if_else(part, NA, acc_at), na.rm = TRUE)) |>
    ungroup()

  da_healy <-
    df_mod |>
    group_by(gvkey) |>
    arrange(fyear) |>
    mutate(nda_healy = mean(if_else(part, NA, acc_at), na.rm = TRUE),
           da_healy = acc_at - nda_healy,
```

---

[6]We put `fit_jones()` and `fit_mod_jones()` outside the function for reasons that will become clear if you attempt the exercises.

```
             nda_deangelo = lag(acc_at),
             da_deangelo = acc_at - nda_deangelo) |>
    ungroup() |>
    select(gvkey, fyear, part, nda_healy, da_healy, nda_deangelo,
           da_deangelo)

  df_jones <-
    df_mod |>
    nest_by(gvkey) |>
    reframe(fit_jones(data))

  df_mod_jones <-
    df_mod |>
    nest_by(gvkey) |>
    reframe(fit_mod_jones(data))

  fit_industry <- function(df) {
    fm <- lm(acc_at ~ acc_ind, data = df, model = FALSE, subset = !part)

    df |>
      mutate(nda_industry = suppressWarnings(predict(fm, newdata = df)),
             da_industry = acc_at - nda_industry) |>
      select(fyear, nda_industry, da_industry)
  }

  df_industry <-
    df_mod |>
    nest_by(gvkey) |>
    reframe(fit_industry(data))

  da_healy |>
    left_join(df_jones, by = c("gvkey", "fyear")) |>
    left_join(df_mod_jones, by = c("gvkey", "fyear")) |>
    left_join(df_industry, by = c("gvkey", "fyear"))
}
```

Applying `get_nda()` to our main sample (`merged_sample_1`) to create `reg_data` for further analysis requires just one line:

```
reg_data <- get_nda(merged_sample_1)
```

16.2.1   Results under the null hypothesis: Random firms

Table 1 of Dechow et al. (1995) presents results from regressions of discretionary accruals on *PART* for each of the five models. For each model, three rows are provided. The first row provides summary statistics for the estimated coefficients on *PART* from firm-specific regressions for the 1000 firms in the sample. The second and third rows provide summary statistics on the estimated standard errors of the coefficients on *PART* and t-statistic testing the null hypothesis that the coefficients on *PART* are equal to zero.

To facilitate creating a similar table, we make two functions. The first function—`fit_model()`—takes a data frame and, for each firm, regresses the measure of discretionary accruals corresponding to `measure` on the `part` variable, returning the fitted models. As we did in Chapter 14, we use `!!` to distinguish the `measure` supplied to the function from `measure` found in `df`.[7]

```
fit_model <- function(df, measure = "healy") {
  df |>
    nest_by(gvkey) |>
    summarize(model = list(lm(as.formula(str_c("da_", measure, " ~ part")),
                              model = FALSE, data = data)),
              .groups = "drop") |>
    mutate(measure = !!measure)
}
```

The second function—`multi_fit()`—runs `fit_model()` for all five models, returning the results as a data frame.

```
multi_fit <- function(df) {
  models <- c("healy", "deangelo", "jones", "mod_jones", "industry")
  models |>
    map(\(x) fit_model(df, x)) |>
    list_rbind()
}
```

With these functions in hand, estimating firm-specific regressions for the five models requires a single line of code.

```
results <- multi_fit(reg_data)
```

The returned results comprise three columns: gvkey, `model`, and `type`, with `model` being the fitted model for the firm and model indicated by gvkey and `type`. Note that `model` is a **list column** and contains the values returned by `lm()`. We can interrogate the values stored in `model` to extract whatever details about the regression we need.

```
head(results)
```

```
# A tibble: 6 x 3
  gvkey  model   measure
  <chr>  <list>  <chr>
1 001021 <lm>    healy
2 001033 <lm>    healy
3 001043 <lm>    healy
4 001058 <lm>    healy
5 001070 <lm>    healy
6 001072 <lm>    healy
```

We will use `tidy()` to extract the coefficients, standard error, t-statistics, and p-values in each fitted model as a data frame. For Table 16.1 (our version of Table 1 of Dechow et al. (1995)), we are only interested in the coefficient on `part` (i.e., the one labelled `partTRUE`) and thus can discard the other row (this will be the constant of each regression) and the column `term` in the function `get_stats()` that will be applied to each model.

---

[7]Chapter 19 of Wickham (2019) has more details on this "unquote" operator `!!`.

TABLE 16.1

Results of tests of earning management: Sample 1

measure	stat	mean	sd	q1	median	q3
deangelo	estimate	0.0124	0.3083	−0.0778	−0.0036	0.0676
deangelo	statistic	−0.0071	1.1151	−0.6277	−0.0297	0.6169
deangelo	std.error	0.1841	0.3048	0.0781	0.1298	0.2030
healy	estimate	0.0104	0.3820	−0.0604	−0.0062	0.0477
healy	statistic	0.0589	1.3072	−0.6509	−0.0869	0.6303
healy	std.error	0.1287	0.2330	0.0556	0.0909	0.1413
industry	estimate	0.0099	0.3780	−0.0601	−0.0061	0.0474
industry	statistic	0.0577	1.3189	−0.6520	−0.0758	0.6252
industry	std.error	0.1275	0.2308	0.0553	0.0903	0.1402
jones	estimate	0.0142	0.3664	−0.0498	−0.0014	0.0550
jones	statistic	0.0421	1.9776	−0.7897	−0.0256	0.7928
jones	std.error	0.0890	0.0942	0.0444	0.0699	0.1039
mod_jones	estimate	0.0108	0.3755	−0.0531	−0.0020	0.0535
mod_jones	statistic	0.0163	1.8795	−0.8040	−0.0307	0.7861
mod_jones	std.error	0.0918	0.0968	0.0456	0.0715	0.1089

```
get_stats <- function(fm) {
  fm |>
    tidy() |>
    filter(term == "partTRUE") |>
    select(-term)
}
```

The function `table_1_stats()` calculates the statistics presented in the columns of Table 1 of Dechow et al. (1995).

```
table_1_stats <- function(x) {
  tibble(mean = mean(x, na.rm = TRUE),
         sd = sd(x, na.rm = TRUE),
         q1 = quantile(x, p = 0.25, na.rm = TRUE),
         median = median(x, na.rm = TRUE),
         q3 = quantile(x, p = 0.75, na.rm = TRUE))
}
```

To produce Table 16.1, our version of Table 1 of Dechow et al. (1995), we use `map()` from the `purrr` library to apply `get_stats()` to each model, then `unnest_wider()` and `pivot_longer()` (both from the `tidyr` package) to arrange the statistics in a way that can be summarized to create a table.

```
results |>
  mutate(stats = map(model, get_stats)) |>
  unnest_wider(stats) |>
  pivot_longer(estimate:statistic, names_to = "stat") |>
  group_by(measure, stat) |>
  summarize(table_1_stats(value), .groups = "drop")
```

TABLE 16.2

Type I error rates

measure	neg_p01	neg_p05	pos_p01	pos_p05
deangelo	0.0141	0.0489	0.0141	0.0522
healy	0.0100	0.0320	0.0290	0.0690
industry	0.0100	0.0320	0.0300	0.0670
jones	0.0320	0.0860	0.0500	0.1000
mod_jones	0.0310	0.0870	0.0490	0.1010

Table 2 of Dechow et al. (1995) presents rejection rates for the null hypothesis of no earnings management in the *PART* year for the five measures. Given that Sample 1 comprises 1000 firms selected at random with the *PART* year in each case also being selected at random, we expect the rejection rates to equal the size of the test being used (i.e., either 5% or 1%). To help produce a version of Table 2 of Dechow et al. (1995), we create `h_test()`, which extracts statistics from fitted models and returns data on rejection rates for different hypotheses and different size tests.

```
h_test <- function(fm) {
  coefs <- coef(summary(fm))

  if (dim(coefs)[1]==2) {
    t_stat <- coefs[2 ,3]
    df <- fm$df.residual

    tibble(neg_p01 = pt(t_stat, df, lower = TRUE) < 0.01,
           neg_p05 = pt(t_stat, df, lower = TRUE) < 0.05,
           pos_p01 = pt(t_stat, df, lower = FALSE) < 0.01,
           pos_p05 = pt(t_stat, df, lower = FALSE) < 0.05)
  } else {
    tibble(neg_p01 = NA, neg_p05 = NA, pos_p01 = NA, pos_p05 = NA)
  }
}
```

We then map this function to the models in `results` and store the results in `test_results`.

```
test_results <-
  results |>
  mutate(map_dfr(model, h_test))
```

Using, `test_results`, Table 16.2 provides our analogue of Table 2 of Dechow et al. (1995).

```
test_results |>
  group_by(measure) |>
  summarize(across(matches("p0"),
                   \(x) mean(x, na.rm = TRUE)))
```

Dechow et al. (1995) indicate cases where the Type I error rate is statistically significantly different from the size of the test using a "two-tailed binomial test". This may be confusing at an initial reading, as the statistics presented in Table 2 of Dechow et al. (1995) are—like those in Table 16.2—based on one-sided tests. But note that whether we are conducting one-sided tests or two-sided tests of the null hypothesis, we should expect rejection rates

TABLE 16.3
p-values for null that Type I error rates equal size of tests

measure	neg_p01	neg_p05	pos_p01	pos_p05
deangelo	0.1860	0.9398	0.1860	0.7619
healy	1.0000	0.0071	0.0000	0.0088
industry	1.0000	0.0071	0.0000	0.0165
jones	0.0000	0.0000	0.0000	0.0000
mod_jones	0.0000	0.0000	0.0000	0.0000

to equal the size of the test (e.g., 5% or 1%) if we have constructed the tests correctly. For example, if we run 1000 tests with a true null and set the size of the test at 5%, then rejecting the null hypothesis 10 times (1%) or 90 times (9%) will lead to rejection of the null (meta-)hypothesis that our test of our null hypothesis is properly sized, as the following p-values confirm. The function `binom.test()` provides the p-value that we need here.

```
binom.test(x = 10, n = 1000, p = 0.05)$p.value
```

```
[1] 6.476681e-12
```

```
binom.test(x = 90, n = 1000, p = 0.05)$p.value
```

```
[1] 1.322284e-07
```

We embed `binom.test()` in a small function (`binom_test()`) that will be convenient in our analysis. Without an *a priori* reason to expect over-rejection or under-rejection, it makes sense to consider two-sided test statistics against a null hypothesis that the rejection rate equals the size of the test. Such statistics are returned by `binom.test()` by default.

```
binom_test <- function(x, p) {
  x <- x[!is.na(x)]
  binom.test(sum(x), length(x), p = p)$p.value
}
```

We apply `binom_test()` to the test results above, adjusting the `p` argument based on the size of the test used in each case and report the results in Table 16.3.

```
test_results |>
  group_by(measure) |>
  summarize(neg_p01 = binom_test(neg_p01, p = 0.01),
            neg_p05 = binom_test(neg_p05, p = 0.05),
            pos_p01 = binom_test(pos_p01, p = 0.01),
            pos_p05 = binom_test(pos_p05, p = 0.05))
```

Turning to the Jones Model and the Modified Jones Model, it is quite clear that we are over-rejecting the (true) null hypothesis. One possible explanation for this over-rejection is provided by footnote 11 of Dechow et al. (1995, p. 204):

> The computation of the standard error of $\hat{b}_j$ requires special attention because the measures of discretionary accruals in the event period (estimation period) are prediction errors (fitted residuals) from a first-pass estimation process. An adjustment must therefore be made to reflect the fact that the standard errors of the prediction errors are greater than the standard errors of the fitted residuals. Likewise, the degrees of freedom in the t-test must reflect the degrees of freedom used up in the first-pass estimation.

This can be' accomplished by ... estimating a single-stage regression that includes both *PART* and the determinants of nondiscretionary accruals.

The invocation of a single-stage regression might remind some readers of the Frisch-Waugh-Lovell theorem, which we discussed in Section 3.3. But an important element of the single-stage regression approach suggested by the Frisch-Waugh-Lovell theorem is that the first- and second-stage regressions that are shown by the theorem to be equivalent have the same observations in both stages. In contrast, the first stages of the Jones Model and Modified Jones Model approaches used by Dechow et al. (1995) use only the estimation sample (i.e., they exclude the test firm-year of primary interest). But this is not an issue here because the single-stage regression invoked by Dechow et al. (1995) is actually that attributed to Salkever (1976).

Salkever (1976) demonstrates that the estimated value of discretionary accruals in the test year can be obtained by running a single regression including both the estimation and test periods and a dummy variable for the test year. The prediction error for the test observation (i.e., the estimated discretionary accruals for the test firm-year) will be equal to the coefficient on the *PART* variable and the *correct* standard error for this prediction will be the standard error of that coefficient.

Because the Salkever (1976) approach is infrequently used in accounting research, but seems quite relevant in a number of settings, we spend some time exploring it in the discussion questions below. (Note that in the following, to keep things manageable, we pull a single GVKEY value at random from our sample. You may need to modify this code to ensure that you are drawing the GVKEY of a firm in your sample, which may differ from ours.)

To keep things simple, we pull one firm from our sample.

```
df_test <-
  merged_sample_1 |>
  filter(gvkey == "001304")
```

We then create the variables needed to run the Jones Model.

```
df_mod <-
  df_test |>
  calc_accruals() |>
  mutate(acc_at = acc_raw / lag_at,
         one_at = 1 / lag_at,
         d_rev_at = d_rev / lag_at,
         d_rev_alt_at = (d_rev - d_rec) / lag_at,
         ppe_at = ppegt / lag_at) |>
  ungroup()
```

We then fit a (differently) modified Jones Model on the estimation sample, which we store in `fm1a`.[8]

```
fm1a <- lm(acc_at ~ one_at + d_rev_at + ppe_at,
           data = df_mod, subset = !part)
```

We can then calculate non-discretionary accruals for the full sample using `predict()`. Again we use `pick(everything())` for the reasons discussed in Chapter 13.

---

[8]We explore the meaning of "differently" in the discussion questions.

```
res1 <-
  df_mod |>
  mutate(nda_jones = predict(fm1a, newdata = pick(everything()))) |>
  select(fyear, part, acc_at, nda_jones) |>
  mutate(da_jones = acc_at - nda_jones)
```

Finally, we can estimate the regression of discretionary accruals on the *PART* variable and store the results in **fm2a**.

```
fm2a <- lm(da_jones ~ part, data = res1)
```

To implement the approach suggested by Salkever (1976)—and used by Dechow et al. (1995)—we run a single regression on the entire sample with the addition of the *PART* indicator and store the result in **fm2**.

```
fm2 <- lm(acc_at ~ one_at + d_rev_at + ppe_at + part,
          data = df_mod)
```

16.2.2   Results under the null hypothesis: Extreme performance

Table 3 of Dechow et al. (1995) presents results from regressions using the second set of samples ("samples of 1000 firm-years randomly selected from firm-years experiencing extreme financial performance"). Table 3 is analogous to Table 2, for which we provided parallel results in Table 16.2.

We leave reproduction of a parallel analysis to that reported in Table 3 of Dechow et al. (1995) as an exercise for the reader and merely provide code producing a sample that can be used for that purpose.

The following code proceeds in four steps. First, we create **earn_deciles**, which contains a variable **earn_dec** that sorts firm-years into earnings deciles for all firms meeting the sample criteria (i.e., those in **test_sample**).

```
earn_deciles <-
  acc_data_raw |>
  semi_join(test_sample, by = c("gvkey", "fyear")) |>
  group_by(gvkey) |>
  arrange(fyear) |>
  mutate(earn = ib / lag(at)) |>
  ungroup() |>
  mutate(earn_dec = ntile(earn, 10)) |>
  select(gvkey, fyear, earn_dec)
```

Second, we create **sample_2_firm_years**, which selects firm-years from the top earnings decile (subject to the constraint that the year is not the first year for the firm, as a prior year is required for the DeAngelo Model). When a firm has more than one firm-year in the top earnings decile, one of those firm-years is selected at random.

```
sample_2_firm_years <-
  earn_deciles |>
  filter(earn_dec == 10) |>
  select(gvkey, fyear) |>
  mutate(rand = rnorm(n = nrow(pick(everything())))) |>
  group_by(gvkey) |>
```

```
filter(rand == min(rand), fyear > min(fyear)) |>
ungroup() |>
top_n(1000, wt = rand) |>
select(gvkey, fyear) |>
mutate(part = TRUE)
```

Third, we create `sample_2` by pulling firm-years from `test_sample` for firms found in `sample_2_firm_years` and then pulling in the firm-years where `part` is `TRUE` based on the value of `part` from `sample_2_firm_years` and then setting the value of `part` to `FALSE` when it is missing (i.e., not found on `sample_2_firm_years`).

```
sample_2 <-
  test_sample |>
  semi_join(sample_2_firm_years, by = "gvkey") |>
  left_join(sample_2_firm_years, by = c("gvkey", "fyear")) |>
  mutate(part = coalesce(part, FALSE))
```

Finally, we create `merged_sample_2`—the analogue of `merged_sample_1`—by merging `sample_2` with the underlying accounting data in `acc_data_raw`.

```
merged_sample_2 <-
  sample_2 |>
  inner_join(acc_data_raw, by = c("gvkey", "fyear"))
```

Table 3 of Dechow et al. (1995) actually involves two samples. One sample is similar to the above and a second sample would be based on the above, but with `filter(earn_dec == 1)` being used in the creation of `sample_2_firm_years`.

Table 4 of Dechow et al. (1995) is similar, but is based on deciles of cash flow from operations, where cash flow from operations is calculated using earnings and accruals, as cash flow statements were not required for most of the sample period in Dechow et al. (1995).

16.2.3   Discussion questions and exercises

1. What interpretation do Dechow et al. (1995) provide for their Table 1 results?

2. Compare the results in Table 16.1 with those in Table 1 of Dechow et al. (1995). What differences appear to be significant?

3. Compare the values in the standard deviation column of Table 1 of Dechow et al. (1995) with other statistics. Do these differences make sense? Or do they suggest anomalies in the underlying data?

4. Compare the values in the standard deviation column of the "earnings management" rows of Table 1 of Dechow et al. (1995) with the values in the mean column of the standard error rows. What is the relationship between these values? What would you expect the relationship between these values to be? Do you observe similar relations in Table 16.1?

5. Focusing on the Healy Model, DeAngelo Model, and the Industry Model, compare the rejection rates in Table 16.2 with those presented in Table 2 of Dechow et al. (1995). What might explain any differences? Could these be attributed to differences between our results in Table 16.1 and those reported in Table 1 of Dechow et al. (1995)? Or do you expect that these differences have another cause?

6. How do you interpret the results from `binom_test()` reported in Table 16.3? Does it make sense to interpret each of the columns independent of the others?

7. Confirm that the coefficient on *PART* from the regression in `fm2a` can be recovered from the regression in `fm2`. Are the standard errors the same?

8. Modify the code above to check that the same holds for the Modified Jones Model.

9. We described the Jones Model above as "a (differently) modified Jones Model". In what way is the model different from the Jones Model estimated in `fit_jones()` above? Does the Salkever (1976) equivalence hold if we use the Jones Model from `fit_jones()`? If so, why? If not, how might this affect how you would use the Jones Model and the Salkever (1976) approach? (For example, do we expect the "(differently) modified Jones Model" to produce materially different results from the Jones Model?)

10. Do the issues related to a first and second stage apply to either the Healy Model or the DeAngelo Model or both? If so, could we apply the Salkever (1976) approach to address these issues? If not, are there "one-stage" equivalents to the Healy Model and DeAngelo Model approaches as implemented above?

11. Produce an equivalent of Table 3 from Dechow et al. (1995) by adapting the code used above to create `merged_sample_2` and Table 16.2. (*Challenge version*: Implement the approach of Salkever (1976) in doing so.)

12. Produce an equivalent of Table 4 from Dechow et al. (1995) by adapting the code used above to create `merged_sample_2` and Table 16.2.

16.3 Power of tests of earnings management

The final analysis of Dechow et al. (1995) that we consider here relates to the third set of samples considered by Dechow et al. (1995, p. 200), namely "samples of 1000 randomly selected firm-years in which a fixed and known amount of accrual manipulation has been artificially introduced."

Figure 4 of Dechow et al. (1995) presents power functions for three different forms of earnings management, the five measures of earnings management, and levels of induced earnings management from zero to 100% of total assets.

To implement a "fixed and known amount of accrual manipulation", we use the function `manipulate()`, which takes a data set with the required variables from Compustat (e.g., `gvkey`, `fyear`, `sale`, `at`), an argument for the `level` of earnings management as a percentage of lagged total assets, and an argument for the `type` of earnings management, which can be `"expense"`, `"revenue"` or `"margin"`, as described in Dechow et al. (1995).

```
manipulate <- function(df, level = 0, type) {
  df <-
    df |>
    group_by(gvkey) |>
    arrange(datadate) |>
    mutate(ni_ratio = median(if_else(part, NA, ni / sale), na.rm = TRUE),
           lag_at = lag(at),
```

```
            manip_amt = lag_at * level,
            manip_amt_gross = manip_amt / ni_ratio)

  if (type == "expense") {
    df |>
      mutate(lct = if_else(part, lct - manip_amt, lct)) |>
      ungroup()
  } else if (type == "revenue") {
    df |>
      mutate(sale = case_when(part ~ sale + manip_amt,
                              lag(part) ~ sale - manip_amt,
                              .default = sale),
             rect = if_else(part, rect + manip_amt, rect),
             act = if_else(part, act + manip_amt, act)) |>
      ungroup()
  } else if (type == "margin") {
    df |>
      mutate(sale = case_when(part & ni_ratio > 0 ~
                                sale + manip_amt_gross,
                              lag(part) & ni_ratio > 0 ~
                                sale - manip_amt_gross,
                              .default = sale),
             rect = if_else(part & ni_ratio > 0,
                              rect + manip_amt_gross, rect),
             act = if_else(part & ni_ratio > 0,
                              act + manip_amt_gross, act),
             lct = if_else(part & ni_ratio > 0,
                              lct + manip_amt_gross - manip_amt, lct)) |>
      ungroup()
  } else {
    df |>
      ungroup()
  }
}
```

We use the `manipulate()` function above and apply it to levels of earning management from 0 to 100% of lagged total assets for each of the three types. The result from the step above is fed to `get_nda()` from above, the results of which are then fed to `multi_fit()` to calculate the results of regressing discretionary accruals on the *PART* variable. The result of these steps is stored in the data frame named `manip_df`.

> ❗ Important
>
> Note that creating `manip_df` takes some time. The exercises below do not require this code to be run. In addition to processing time, this code is quite memory-intensive—requiring more than 10 GB of RAM. So, if you have less than about 16 GB of RAM, this code might require modification to run smoothly on your machine.

```
plan(multisession)
```

```
manip_df <-
  expand_grid(level = seq(from = 0, to = 1, by = 0.1),
              manip_type = c("expense", "revenue", "margin")) |>
  mutate(data = future_map2(level, manip_type,
                            \(x, y) manipulate(merged_sample_1, x, y))) |>
  mutate(accruals = future_map(data, get_nda)) |>
  mutate(results = future_map(accruals, multi_fit)) |>
  select(-data, -accruals) |>
  system_time()
```

```
   user  system elapsed
 29.586   4.686  90.194
```

With results from regressions for various values of `level`, the three values of `manip_type`, and the five models (`manip_type`) stored in `manip_df`, we can create plots like those presented in Figure 4 of Dechow et al. (1995) using the following code. We first create a function (`h_test_5()`) that takes a fitted model and returns a logical value indicating whether the null hypothesis is rejected at the 5% level.

```
h_test_5 <- function(fm) {
  coefs <- coef(summary(fm))

  if (dim(coefs)[1]==2) {
    t_stat <- coefs[2 ,3]
    df <- fm$df.residual
    pt(t_stat, df, lower = FALSE) < 0.05
  } else {
    NA
  }
}
```

The code below applies `h_test_5()` to each row of `manip_df` to calculate the proportion of firms for which the null is rejected for each value of (`level`, `manip_type`, `measure`).

> **! Important**
>
> Note that creating `power_plot_data` takes some time. The exercises below do not require this code to be run. In addition to processing time, this code is quite memory-intensive—requiring more than 10 GB of RAM. So, if you have less than about 16 GB of RAM, this code might require modification to run smoothly on your machine.

```
plan(multisession)
```

```
power_plot_data <-
  manip_df |>
  unnest(results) |>
  group_by(level, manip_type, measure) |>
  mutate(rej_null = future_map_lgl(model, h_test_5)) |>
  summarize(prop_reject = mean(rej_null, na.rm = TRUE), .groups = "drop") |>
  system_time()
```

This data set is easily plotted using `facet_grid()`, with the results shown in Figure 16.1.

```
power_plot_data |>
  ggplot(aes(x = level, y = prop_reject)) +
  geom_line() +
  facet_grid(measure ~ manip_type)
```

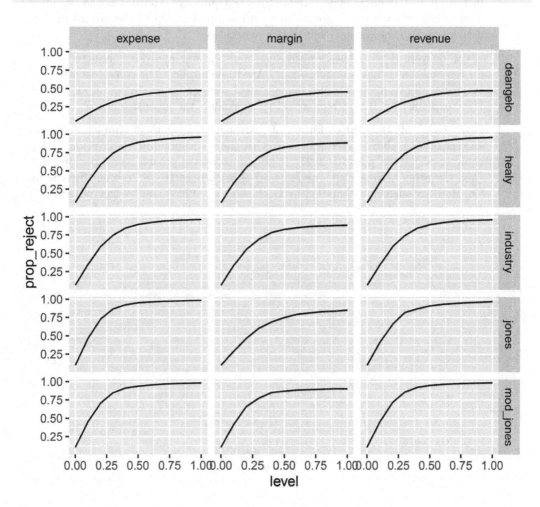

FIGURE 16.1
Power functions for tests of earnings management

16.3.1 Discussion questions

1. How do the results in Figure 16.1 compare with those in Figure 4 of Dechow et al. (1995)?

2. According to the SEC's filing[9] referenced above related to B&L, "B&L recognized, in contravention of GAAP and the Company's own revenue recognition policies, $42.1 million of revenue, resulting in at least a $17.6 million, or 11%, overstatement of the net income originally reported for its 1993 fiscal year." According to a subsequent SEC filing,[10] B&L's total assets for 1994 were $2,457,731,000

---

[9]https://www.sec.gov/litigation/admin/3439329.txt
[10]https://www.sec.gov/Archives/edgar/data/10427/0000010427-95-000006.txt

(it seems reasonable to assume that the 1993 value was not radically different from this). Based on this information (plus any information in the SEC's filing), which of Dechow et al. (1995)'s three categories did B&L's earnings management fall into? What is the approximate magnitude relative to the x-axes of the plots in Figure 4 of Dechow et al. (1995) (or the equivalent above)? Based on these data points, what is the approximate estimated probability of the various models detecting earnings management of this magnitude?

3. What do you view as the implications of the power analysis conducted above for research on earnings management? Are these implications consistent with the extensive literature on earnings management subsequent to Dechow et al. (1995)? If so, explain why. If not, how would you reconcile the inconsistencies?

4. Does each of the three forms of earnings management implemented in `manipulate()` above agree precisely with the corresponding description in Dechow et al. (1995, pp. 201–202)? If not, does one approach seem more correct than the other? (Note that *one* issue arises with negative or zero net income ratio. How are such cases handled by Dechow et al. (1995) and by `manipulate()`?)

Part III

Causal Inference

17

Natural experiments

In this part of the book, we explore a number of issues related to drawing causal inferences from data. Many treatments of this topic dive directly into methods, such as instrumental variables or regression discontinuity design. Here we have consciously chosen to start with the benchmark setting of randomized experiments and gradually build up to settings where statistical methods may (subject to features of the settings) support credible causal inferences. We believe that carefully thinking about a hypothetical randomized experiment and conjectured causal mechanisms sharpens the thinking needed to successfully bring statistical analyses to bear on problems of causal inference.

This chapter introduces the concept of the **randomized controlled trial** and the related idea of the **natural experiment**. While much discussion focuses on the value of random assignment for causal inference, we explain that other features of an experiment are also of critical importance. We explore the question of recognition versus disclosure to illustrate that the framing of the research question itself has important implications for any experiment.

> 💡 Tip
>
> The code in this chapter uses the packages listed below. For instructions on how to set up your computer to use the code found in this book, see Section 1.2. Quarto templates for the exercises below are available on GitHub.[a]
>
> ---
> [a]https://github.com/iangow/far_templates/blob/main/README.md

```
library(dplyr)
library(farr)
library(ggplot2)
```

17.1 Randomized experiments

The randomized experiment is widely regarded to be the "gold standard" of research designs. The idea of a randomized experiment is quite simple: observations are *randomly assigned* to treatment conditions. In the case of a binary treatment variable (i.e., a unit is either *treated* or *not treated*), whether a unit is treated is determined by some randomization mechanism, such as a coin toss. In terms of causal diagrams with X being the treatment variable, such randomization implies that there are no arrows into X and thus no confounders that need to be controlled for in estimating causal effects.

As a concrete example, let's consider Jackson et al. (2009). From the title ("economic consequences of firms' depreciation method choice"), we can infer that the treatment of

DOI: 10.1201/9781003456230-17

interest is firms' depreciation method choice. The specific method choice that Jackson et al. (2009) focus on is accelerated depreciation versus straight-line depreciation, and the outcome ("economic consequence") of interest is capital investment.

In practice, firms are likely to make the choice between accelerated depreciation and straight-line depreciation based on several factors, among which might be the economic depreciation of the relevant assets, the useful life of assets (for assets with shorter lives, the choice is likely to be less important), and various financial reporting incentives (e.g., a growing firm will report higher income in the near term if it uses straight-line depreciation). If any of these conjectured factors affecting depreciation method choice also affects capital investment, then it would confound causal inferences of the kind sought in Jackson et al. (2009).

However, if we somehow had the ability to randomly assign firms to either the accelerated depreciation condition or the straight-line depreciation condition, we could draw causal inferences by comparing the capital investment of firms based on the treatment condition to which they were assigned.

There would obviously be challenges in conducting such an experiment. First, we'd likely need to be a regulator with sufficient authority to compel firms to accept the assigned depreciation method. Even if it had such authority, without a strong desire to understand the effect of depreciation method choice, it's not clear why a regulator would run such an experiment; and regulators generally have bigger fish to fry. Some experiments might only require the cooperation of a single firm, but this does not seem to apply for this research question (we would need, say, a firm with *many* divisions, each with independent investment authority).

In practice, the closest we can get to an experiment is a so-called **natural experiment**, which relies on forces outside the control of the researcher to perform assignment to treatment conditions in a way that is random or, in terms used by Dunning (2012), **as-if random**.

In this chapter, we first study the randomized controlled trial, a benchmark form of the randomized experiment. We then compare the randomized controlled trial with other forms of experiment, including the natural experiment. To make our discussion of randomized experiments more concrete, we focus on a setting of sustained interest to practitioners and researchers alike: recognition versus disclosure. With this setting in mind, we then turn to a plausible natural experiment used in Michels (2017) and consider what credible inferences this setting supports.

17.1.1   Randomized controlled trials as a benchmark

A randomized experiment is often implemented in medical sciences as a **randomized controlled trial** or RCT. Akobeng (2005, p. 837) describes an RCT as "a type of study in which participants are randomly assigned to one of two or more clinical interventions. The RCT is the most scientifically rigorous method of hypothesis testing available, and is regarded as the gold standard trial for evaluating the effectiveness of interventions."

Note that in its idealized form, an RCT involves more than simply random assignment to treatment. Drawing on Akobeng (2005), we can enumerate the following as features of an ideal RCT:

1. The treatments to be compared are specified in advance.
2. The outcomes of interest are specified in advance.
3. Proposed analyses are specified in advance.

4. The required number of participants is identified (using **power calculations**) in advance.

5. Participants are recruited after required ethical approvals have been obtained.

6. Participants are randomly assigned to treatment groups (we can consider the control as one of the treatments).

7. Participants do not know which treatment group they have been assigned to (**concealment**).

8. Those administering treatment (e.g., doctors, nurses, researchers) do not know which treatment group participants have been assigned to (**blinding**).

So random assignment treatment is only one of several components of an RCT. Even if we can argue that a candidate natural experiment offers as-if random assignment to treatment, a natural experiment will lack other features of an RCT.

First, the treatment is specified not by the researcher, but instead by firms, regulators, or "nature", which we interpret broadly to include not only nature, but also complex economic forces perhaps not well understood by the researcher. (We will often put the word "nature" in scare quotes to reinforce the broad interpretation the term is given in the context of natural experiments.) This means that a researcher who "discovers" a natural experiment either needs to hope that the treatments that "nature" assigns are of interest to researchers or that other researchers can be persuaded that these treatments are of interest.

Second, as a consequence of this, it is not meaningful in general to specify outcomes and analysis procedures in advance of the "experiment" being run. In many cases, the natural experiment is only identified after the experiment has been conducted. As a result, it is difficult to prevent **p-hacking**, which refers to processes in which researchers, whether consciously or not, consider a variety of outcomes and analytical procedures until "statistically significant" results are obtained. We discuss this limitation of natural experiments in more detail in Chapter 19.

Third, in general, natural experiments will not provide concealment of treatment assignment.[1]

However, we should note that concealment is not always meaningful in business research settings. RCTs are often used to evaluate the drugs or other medical interventions that are expected to operate without conscious action by the participants. In fact, it is because researchers are often most interested in the direct effect of the treatment unconfounded by other actions that the control group involves a **placebo** treatment as a control. But if the treatment of interest is an accounting standard, an incentive-compensation contract component, or financing constraints, then it's not meaningful to speak of a direct, unconscious effect: if managers are unaware of features of their compensation contracts, then they are not going to take actions in response to them. In general, it is likely that researchers are interested in the total effects of a treatment.

Concealment of the fact of random assignment might be particularly important in settings involving **signalling**, as the response of the receiver of a signal is predicated on the signal being an endogenous response to the sender's circumstances. If a signal is known by the receiver to have been randomly assigned, then there is no signalling value. As discussed by Armstrong et al. (2022), signalling has been used to explain patterns in voluntary disclosure, accounting choice, dividend policy, insider stock purchases, corporate social responsibility, and the decision to get an audit.[2]

---

[1] Blinding is difficult to achieve if nature is administering the treatment.

[2] For more on signalling theory, see Chapter 17 of Kreps (1990).

17.1.2   Identifying the treatment of interest

Even when using an RCT, it can be difficult to ensure that the treatment applied is the one we are most interested in. For example, suppose we're interested in studying the effectiveness of KN95 masks in reducing the spread of Covid-19. While a simple perspective might be that all we need to do for an experiment is to divide participants into treatment and control groups, in reality complexities exist.

One approach for the treatment group would be to issue all participants in this group with a supply of masks to be used during the study period and have participants go about their daily lives in other respects. This implies that the treatment is "receive KN95 masks" rather than "wear KN95 masks", even though the latter might be the treatment we are more interested in evaluating.

If we are interested in "wear KN95 masks" as the treatment, we might implement mechanisms for measuring whether masks are actually worn in practice. While this might be accomplished by sending out monitors to note whether masks are being worn properly, this would entail significant additional costs and it is unlikely that such data recording would be complete or even feasible (e.g., observing a participant at home or at work). Another approach would be to ask participants to maintain a log of their mask-wearing behaviour and rely on participants to be honest and diligent in doing so.

As the recording of behaviour may itself influence behaviour (e.g., making a log may provide a reminder to wear a mask), this effectively alters the treatment to "receive KN95 masks and keep a log of mask-wearing", which may not be of interest unless widespread logging of behaviour is considered part of the policy repertoire.

If we can only randomly assign participants to "receive KN95 masks" or not, and we are interested in "wear KN95 masks" as the treatment, an issue we need to consider is the reality that the conscientiousness with which participants wear masks is not random. Fortunately, with the right data, we can conduct an **intention-to-treat** analysis to evaluate the effect of "wear KN95 masks" even if we only have the ability to assign participants to "receive KN95 masks" as the treatment. This analysis uses **instrumental variable** techniques, which are the focus of Chapter 20, and further discussion of this approach is found in Chapters 4 and 5 of Dunning (2012).

While it may be tempting to think of the control as simply the "no treatment" group, there are actually significant choices to be made in specifying the treatment depending on the specific treatment effect we wish to estimate. For example, wearing a mask may have an effect on Covid-19 transmission not because the mask physically prevents the transmission of the virus, but because wearing a mask may affect behaviour. Wearing a mask might cause the wearer to be more cautious in interacting with others or, by making conversations more awkward, masks might reduce social interaction, or masks might cause others to give the wearer a wide berth (though this effect might have been more plausible in the pre-pandemic era). Alternatively, masks might give wearers (and others) a sense of comfort and lead to increased social interaction. And, as discussed above, if "receive KN95 masks and keep a log of mask-wearing" is the treatment, the control might be "receive *no* KN95 masks and keep a log of mask-wearing" on the assumption that some people will wear their own masks if none are provided.[3]

If we are interested in estimating the direct effect of *KN95* masks on Covid-19 transmission, we might want to use a **double-blind** protocol (i.e., one with concealment and blinding)

---

[3]If such people exist, then some way of tracking mask-wearing in the control group is perhaps necessary to use an intention-to-treat analysis to estimate the effect of wearing masks.

and specify the control treatment as "wear less effective masks" (in which case, the easy-to-estimate effect would be that of KN95 masks over less effective masks) or "wear completely ineffective marks" (in which case, the easy-to-estimate effect would be that of KN95 masks over ineffective masks, which may proxy for "no masks" in some ways).[4]

17.1.3 Identifying the outcome of interest

Continuing our discussion of the hypothetical RCT with KN95 masks, the outcome of interest might seem obvious. Perhaps we just need to track participants and measure the incidence of Covid-19 cases that are recorded. We are implicitly counting on certain things here, such as a reasonable baseline risk of getting Covid-19 with or without KN95 masks. Conducting a study with participants in an environment where Covid-19 has been eliminated during the study period is not likely to be helpful, as no one will get Covid-19 in either treatment condition. It is also important that the time over which symptoms appear is within the time frame of the study. While Covid-19 infections may show up as infections within days, things are much more difficult with diseases whose symptoms show up years after exposure to risk factors.

But if masks affect not only the incidence of cases, but also their seriousness of cases that occur, then we likely need to track additional indicators, such as hospitalizations with Covid-19, serious health issues, and deaths.

More complicated is the issue that the outcome of interest might be the effect of mask-wearing on the incidence or severity of disease in others. Adopting this as the outcome of interest implies significant changes to the treatment-assignment approach and measures tracked, as it would likely be impractical to assign treatment to participants and track the health outcomes of the people with whom they interact. Instead, it is likely that outcomes would be narrowed in scope (e.g., the transmission of disease in the office or at school rather than in general) and treatment assignment done at a level higher than individuals (e.g., schools or offices).

17.1.4 Laboratory experiments

The hypothetical experiments described above are analogous to **field experiments**, which involve real decision-makers making real decisions in the relevant context. More common in accounting research is the **laboratory experiment**. A typical laboratory experiment involves participants of convenience (e.g., undergraduate students or online survey participants) making "decisions" in a highly stylized setting.

The analogy with RCTs can be seen by considering how challenging it would be to evaluate KN95 masks or Covid vaccines in a pure laboratory setting. While one could certainly issue masks or administer vaccines in a laboratory setting, a highly controlled laboratory environment is not a setting in which participants would be expected to encounter the SARS-CoV-2 virus, making it useless as a setting for evaluating the effectiveness of prophylactic measures such as masks or vaccines. Furthermore, the period over which one needs to be potentially exposed to the SARS-CoV-2 virus is measured in weeks and months, likely further ruling out a laboratory setting.[5] As a result, most RCTs would involve field experiment elements with participants entering their normal environment after treatment assignment and administration.

---

[4]Of course, each of these alternatives presents ethical issues that we gloss over, as we are simply trying to illustrate the complexity of specifying treatments.

[5]The closest equivalent to laboratory studies involving human subjects and medical interventions might be **human challenge trials**, which present ethical issues and generally involve small sample sizes.

While laboratory experiments might be useful for understanding generic features of human decision-making, it is a huge leap to go from the reactions of undergraduates to hypothetical accounting policy variables and hypothetical investment decisions to conclusions about real-world business decisions. This is arguably true for almost all research questions examined by accounting researchers. Consistent with this, the impact of research using laboratory experiments on empirical accounting research—the overwhelming majority of research in the top accounting journals (see Gow et al., 2016)—seems fairly limited.

17.2   Natural experiments

Natural experiments occur when observations are assigned by nature (or some other force outside the control of the researcher) to treatment and control groups in a way that is random or "as if" random (Dunning, 2012). If such assignment is (as-if) random, then natural experiments can function much like field experiments for the purposes of causal inference.

Dunning (2012, p. 3, emphasis added) argues that the appeal of natural experiments "may provoke **conceptual stretching**, in which an attractive label is applied to research designs that only implausibly meet the definitional features of the method." As we discuss below, such conceptual stretching seems common in accounting research.

One dimension along which stretching concerns claims of as-if random assignment. In some cases, ignorance of the assignment mechanism appears to substitute for a careful evaluation of how random it is, which requires a deep understanding of that mechanism. While the process by which a coin lands on heads or tails is mysterious, the randomness of the coin toss is predicated on a deep understanding that the coin is fair (i.e., equally likely to come up heads or tails).

The discussion of RCTs above was not intended to be a primer on conducting randomized controlled trials or field experiments. Instead, the goal was to flag how merely randomizing treatment assignment is only one element—albeit a critical one—of a well-designed field experiment. In evaluating a natural experiment, a researcher needs to consider the issues raised in field experiments, even if these elements cannot be adjusted through choices of the researcher.

For example, what "nature" (seemingly) randomizes often is not exactly the variable researchers might be interested in studying. As we saw above, even in field experiments, careful consideration of the precise treatments to consider is required and it seems unlikely that "nature" is going to make choices that align with what a researcher would make. One response to this is for researchers to argue that what nature has randomized *is* the thing of interest. Another response to suggest that the thing randomized is equivalent to something else that is of interest. Finally, in some cases, the variable randomized by "nature" might satisfy the requirements of an **instrumental variable**, something we examine in Chapter 20.

17.2.1   Natural experiments in accounting research

In a survey of accounting research in 2014, Gow et al. (2016) identified five papers that exploited either a "natural experiment" or an "exogenous shock" to identify causal effects.

But Gow et al. (2016) suggest that closer examination of these papers "reveals how none offers a plausible natural experiment."

The main difficulty is that most "exogenous shocks" (e.g., SEC regulatory changes or court rulings) do not randomly assign units to treatment and control groups and thus do not qualify as natural experiments. For example, an early version of the Dodd-Frank Act contained a provision that would force US companies to remove a staggered board structure.[6] It might be tempting to use this event to assess the valuation consequences of having a staggered board by looking at excess returns for firms with and without a staggered board around the announcement of this Dodd-Frank provision. Although potentially interesting, this purported "natural experiment" does not randomly assign firms to treatment and control groups.

In addition, it is important to carefully consider the choice of explanatory variables in studies that rely on natural experiments. In particular, researchers sometimes inadvertently use covariates that are affected by the treatment in their analysis. As noted by Imbens and Rubin (2015, p. 116), including such post-treatment variables as covariates can undermine the validity of causal inferences.[7]

One plausible natural experiment flagged by Gow et al. (2016) is Li and Zhang (2015). Li and Zhang (2015, p. 80) study a regulatory experiment (Reg SHO) in which the SEC "mandated temporary suspension of short-sale price tests for a set of randomly selected pilot stocks." Li and Zhang (2015, p. 79) conjecture "that managers respond to a positive exogenous shock to short selling pressure ... by reducing the precision of bad news forecasts." But if the treatment affects the properties of these forecasts, and Li and Zhang (2015) sought to condition on such properties, they would risk undermining the "natural experiment" aspect of their setting. We examine Reg SHO in more detail in Chapter 19.

Michels (2017) is another plausible natural experiment identified by Gow et al. (2016) and we investigate it further below.

17.3 Recognition versus disclosure

One of the longest-standing questions in accounting research is whether it matters whether certain items are recognized in financial statements or merely disclosed in, say, notes to those financial statements. As the following vignettes illustrate, debates about recognition versus disclosure have been some of the most heated in financial reporting.

17.3.1 Stock-based compensation

Accounting for stock-based compensation is one of the most controversial topics ever addressed by accounting standard-setters. In 1973, the US Financial Accounting Standards Board (FASB) issued APB 25, which required firms to measure the expense as the difference between the stock price and the option exercise price on the grant date (or "intrinsic value"), which equals zero for most employee options. In 1993, the FASB issued an exposure draft proposing that firms measure the expense based on the options' grant date fair value. While APB 25 predated Black and Scholes (1973), approaches to measuring the fair value

---

[6]See Larcker et al. (2011).

[7]See the discussion of mediators in Chapter 4 above.

of options were widely accepted and understood by 1993. This exposure draft met with fierce resistance by firms, and the US Congress held hearings on whether the FASB should be permitted to finalize the standard as proposed.

In 1995, the FASB issued SFAS 123, which preferred measurement of the expense using the grant-date fair value, but allowed firms to *recognize* the expense using the APB 25 measurement approach and only *disclose* what net income would have been had the expense been measured using the grant-date fair value approach. In its basis for conclusions, the FASB admitted that it did not require expense recognition because the severity of the controversy was such that doing so might have threatened accounting standard-setting in the private sector.

Almost all firms applied the measurement approach of APB 25 until the summer of 2002 when a small number of firms adopted the grant date value approach (Aboody et al., 2004; Brown and Lee, 2011). In light of the financial reporting failures of 2001, the FASB revisited accounting for stock-based compensation and, in 2004, the FASB issued SFAS 123R, which took effect for fiscal years beginning on or after June 15, 2005. The primary effect of SFAS 123R was to require recognition of stock-based compensation expense using the grant-date fair value.

But, even after SFAS 123R was adopted, controversy continued. Several prominent persons, including former US cabinet secretaries and three Nobel laureates, reiterated arguments that recognizing the expense is improper because the value of employee stock options does not represent an expense of the firm (Hagopian, 2006) and filed a petition with the SEC in 2008 alleging that the SEC failed in its duties by permitting the FASB to issue SFAS 123R.

17.3.2 Accounting for leases

A lease is an agreement between a **lessor** who owns an asset and a **lessee** that grants the lessee the right to use the asset for a period of time in exchange for (typically periodic) payments. Leasing is a huge business around the globe, and leases are a common way for firms to acquire the right to use assets such as real estate, aircraft, and machinery.

Leasing an asset is often an alternative to buying that asset. If a firm only needs an asset for a short period of time, then leasing the asset may reduce transaction costs and eliminate risks associated with the value of the asset at the end of the lease. Lessors are often firms with specialist expertise in managing the acquisition, leasing, and disposal of assets and understanding how their value changes over their useful lives.

In some circumstances, the economic risk and benefits of asset ownership are borne by the lessee. For example, if a lease includes a *bargain purchase option* that a rational lessee will almost certainly exercise or if title to the asset simply transfers to the lessee at the end of the lease, then the lessee owns the asset in any relevant economic sense.

However, like any contract, a lease has two sides. Generally lessee acquires economic ownership of an asset in exchange for a stream of payments over time. In many leases, the payments are periodic and fixed. Anyone who has mortgaged a house or financed a vehicle purchase will likely recognize an obligation for fixed and periodic payments as a loan.

Thus, in some cases, a lease is *economically equivalent* to the combination of two transactions: first, borrowing money on a loan that requires fixed, periodic repayments and, second, use of the loan proceeds to acquire the leased asset.

SFAS 13 (later relabelled as ASC 840), which was issued by the US Financial Accounting Standards Board in 1976, required leases with features that made them more akin to purchases of assets financed by loans to be accounted for in a manner essentially equivalent to loan-backed asset purchases. SFAS 13 specified criteria, such as the transfer of asset ownership at the end of the lease or the presence of a bargain purchase option, that would trigger this **capital lease accounting** treatment.

Under capital lease accounting, a leased asset and a lease liability would be *recognized* on the balance sheet. The leased asset would be depreciated (or "amortized" in the language of SFAS 13) "in a manner consistent with the lessee's normal depreciation policy" over the lease term. At the same time, "each minimum lease payment shall be allocated between a reduction of the obligation and interest expense so as to produce a constant periodic rate of interest on the remaining balance of the obligation."

SFAS 13 provided that leases not meeting the criteria for capital lease accounting should be accounted for as **operating leases**. For such leases, no asset or liability would be recognized and, for most leases, the standard provided that "rental on an operating lease shall be charged to expense over the lease term as it becomes payable."

In 2004, Jonathan Weil wrote[8] in the *Wall Street Journal* that "companies are still allowed to keep off their balance sheets billions of dollars of lease obligations that are just as real as financial commitments originating from bank loans and other borrowings." While SFAS 13 classified leases that were *obviously* economically equivalent to debt-financed asset purchases, its criteria allowed firms to structure leases so that they were economically very close to such purchases, but which were classified as operating leases for financial reporting purposes. Weil flagged US retail pharmacy chain Walgreen, which "shows no debt on its balance sheet, but it is responsible for $19.3 billion of operating-lease payments mainly on stores over the next 25 years" and stated that the off-balance-sheet operating-lease commitments, as revealed in the footnotes to their financial statements for companies in the Standard & Poor's 500-stock index totalled US$482 billion.

In 2006, the FASB began work on a new lease accounting standard intended to close "loopholes" like those described by Weil. Given the extensive information about operating leases provided under SFAS 13, we can view SFAS 13 as providing *disclosure* of operating leases, whereas many called for their *recognition* on the balance sheet.

In 2016, FASB published a new lease accounting standard, ASC 842, and the IASB issued a similar standard, IFRS 16. These accounting standards require all leases to be brought onto the balance sheet (except short-term leases).

17.3.3 Academic research

The two examples above serve two purposes. First, they highlight the significance of recognition-versus-disclosure questions in financial reporting. Second, they provide concrete settings for thinking about some subtle issues faced by researchers in studying the topic of recognition versus disclosure. As we shall see, some of these subtleties would exist even if we could run any field experiment we could conceive.

What do "recognition" and "disclosure" mean? Simplistically, the choice between recognition and disclosure is a choice between including a given amount in the financial statements or merely disclosing the amount, typically in the footnotes to the financial statements. However, as pointed out by Bernard and Schipper (1994), the choice is significantly more complicated than this simple binary "in or out" election. FASB's Statement of Financial Accounting Concepts No 5 (SFAC 5) defines recognition as "the process of formally incorporating an

---

[8]https://www.wsj.com/articles/SB1095808870299124246

item in the financial statements of an entity as an asset, liability, revenue, expense, or the like. A recognized item is depicted in both words and numbers, with the amount included in the statement totals."

Paragraph 5.1 of the IASB Conceptual Framework contains a similar definition, but clarifies certain elements of the FASB definition. First, recognition involves the statements of financial position and financial performance; putting an item in the statement of cash flows alone would not seem to constitute recognition. Second, "or the like" is clarified to mean "equity", as this rounds out the list of elements of financial statements. Third, "numbers" mean monetary amounts, as this is how elements are put in financial statements. Implicit in both definitions is that the monetary amounts can be added up in some meaningful fashion, as financial statement items are included as components of sums.

Bernard and Schipper (1994) note that "recognition does not appear to have a formal definition in the FASB's official pronouncements" [p. 4], and the same appears to be true of the IASB Conceptual Framework. Instead, Bernard and Schipper (1994) suggest that "one could view recognition as a form of disclosure with special characteristics" [p. 5].

Can recognition and disclosure be viewed as alternatives? Bernard and Schipper (1994) suggest that if we take the conceptual frameworks as binding constraints on standard-setters, then recognition versus disclosure cannot be viewed as a choice. Either an item meets the recognition criteria and thus must be recognized, or it does not and thus must not be recognized. But it is unclear whether the conceptual frameworks determine standards to this degree. Nothing in SFAC 5 changed between SFAS 123 and SFAS 123R, even though the two standards came out on opposite sides with regard to recognition of stock-based compensation expense.[9]

The case of SFAS 123 above is arguably one of the sharpest instances of recognition versus disclosure. Under SFAS 123, a firm that used the APB 25 approach to measuring the cost of employee stock options was required to provide a pro forma disclosure of the expense in the footnotes to its financial statements.[10] In effect, a firm must provide the information that it would have provided had it recognized an expense using the fair-value approach. While the pro forma financial statements only include the statement of financial performance, there is no aggregate impact of applying the SFAS 123 approach on the statement of financial position, as the debit to expense, which would flow through net income to retained earnings, is offset by an addition to additional paid-in capital.

The case of lease accounting is perhaps more typical of recognition-versus-disclosure settings. First, firms cannot elect to apply operating lease accounting instead of capital lease accounting. If a lease is a capital lease, then capital lease accounting must be applied. If a lease is not a capital lease, then operating lease accounting must be applied. (That said, under SFAS 13, firms would often structure their leases such that a minor tweak to the lease would cause it to switch from one accounting treatment to the other.)

Second, there was no equivalent requirement under SFAS 13 to the pro forma disclosures provided under SFAS 123. Such pro forma disclosures would need to be extensive, as lease accounting under SFAS 13 affected the statement of financial position, the statement of financial performance, and the statement of cash flows. Instead, firms were required to

---

[9]Indeed, SFAS 123 gave firms the choice to recognize amounts that most firms merely disclosed without reference to any distinctions related to recognition criteria. One possible counterargument to this example is that SFAS 123 was not really about recognition as understood in the conceptual frameworks, but *measurement*. By measuring the cost of stock-based compensation as zero, there is nothing to recognize. This would seem to be too artful a dodge of the issue and we don't pursue it further here.

[10]An illustrative disclosure is found in Microsoft's 2002 annual report: https://go.unimelb.edu.au/s7d8.

disclose rental expense related to operating leases, minimum rental payments in aggregate and for each of the five succeeding fiscal years and a general description of leasing arrangements, such as details about contingent rental payments, renewal or purchase options, and escalation clauses.

Why might recognition versus disclosure matter? Focusing for the moment on the implications of recognition versus disclosure for capital markets, there are two theories at opposites ends of a spectrum.

According to Watts (1992), "the **mechanistic hypothesis** posits that stock prices react mechanically to the reported earnings number, regardless of how the number is calculated." (A natural generalization of the mechanistic hypothesis would be to apply it to reported balance sheet numbers and financial ratios based on reported numbers in financial statements.) At the other end of the spectrum is the **efficient markets hypothesis** (EMH), which in its **semi-strong form** posits that capital markets efficiently process all publicly available information and impound that information into security prices.

Under the mechanistic hypothesis, if numbers are disclosed in footnotes but not recognized in financial statements, the market will not respond to them. In contrast, under the EMH, markets will react to information whether it is recognized or disclosed.

Implicit in the last sentence is the assumption that the information content is unaffected by the recognition-versus-disclosure decision. But this may not be true for a number of reasons.

First, the information content may differ due to the aggregation and conversion of disclosed amounts into monetary amounts that occur as part of the recognition process. In the case of operating leases, future lease payments are discounted and added up to form the liability recognized on the balance sheet. Additionally, assets are depreciated and reported at net carrying value. These calculations are difficult for an outsider to replicate precisely using disclosed information (e.g., due to imprecise information about the timing of rental payments and discount rates), and their results may convey useful information to market participants.

Second, even if the notional information content does differ between recognition and disclosure, the recognition-versus-disclosure decision may affect the information properties of the information due to differences in behaviour across the two conditions. For example, managers may view recognized information as more consequential and thus may exert more effort to produce more accurate numbers (or more effort on earnings management). Auditors may also view their obligations to provide assurance for recognized numbers to be greater than that for their merely disclosed equivalents.

Even if the EMH holds, if managers or others do not believe that it holds and instead assume that the mechanistic hypothesis better describes reality, they act accordingly. In the lead-up to SFAS 123R, many firms accelerated the vesting of stock options to avoid reporting expenses. The economic consequences of this action were fairly clear (essentially a wealth transfer from firms to employees), and Choudhary et al. (2009) find that the market reacted accordingly when acceleration decisions were announced. Yet Choudhary et al. (2009) quote the perspective of one firm (Central Valley Community Bancorp) that is firmly grounded in the mechanistic view:

> The Board believes it was in the best interest of the shareholders to accelerate these options, as it will have a positive impact on the earnings of the Company over the previously remaining vested period of approximately 3 years.

Designing experiments to study recognition versus disclosure. Given the various ways in which recognition versus disclosure can matter, a regulator seeking to run field experiments to understand the merits of various policies is confronted with a variety of possible experiments to run. Knowing that recognition matters relative to disclosure is not very useful if it is not coupled with some understanding of *why* it matters, which is tantamount to understanding the causal mechanism. This is important because the appropriate experimental design will vary according to the causal mechanism that is sought to be tested.

The best way to see how experimental design should vary by the conjectured mechanism is to consider some specific cases. If we conjecture that recognition matters because of difficulties that investors have in processing disclosed-but-not-recognized footnote information into pro forma financial statement information, we might divide firms into three groups: firms that recognize in financial statements; and firms that do not recognize but disclose pro forma information in the footnotes (*à la* SFAS 123); and firms that disclose information in the footnotes that requires additional processing or detail to calculate pro forma financial information (similar to disclosure of operating leases under SFAS 13).

17.4   Michels (2017)

In this chapter so far, we have covered experiments, including natural experiments, and the broader question of recognition versus disclosure. Now we turn to Michels (2017), who studies one aspect of recognition versus disclosure using a plausible natural experiment.

Michels (2017) exploits the difference in disclosure requirements for significant events that occur before financial statements are issued. Because the timing of these events (e.g., fires and natural disasters) relative to balance sheet dates is plausibly random, the assignment to the disclosure and recognition conditions is plausibly random.

The data set `michels_2017` from the `farr` package provides information about the 423 observations in Michels (2017). For 343 of these observations, the natural disaster occurred *after* the relevant filing for the *previous* financial period was made, and thus the financial effects are recognized in the current period. For the remaining 80 cases, the natural disaster occurred *before* the relevant filing for the *previous* financial period was made, and thus the financial effects are disclosed in that filing (and recognized in a subsequent filing containing financials for the current period).

Figure 17.1 depicts the distribution of the number of days between the applicable natural disaster and the next filing date.

```
michels_2017 |>
  mutate(days_to_filing = as.integer(date_filed - eventdate)) |>
  ggplot(aes(x = days_to_filing, fill = recognize)) +
  geom_histogram(binwidth = 1)
```

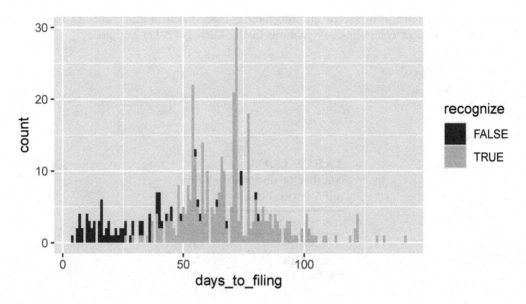

FIGURE 17.1
Distribution of days to filing in Michels (2017) sample

Figure 17.2 depicts the distribution of the number of days between the applicable natural disaster and days to the end of the fiscal period in which it occurs.

```
michels_2017 |>
  mutate(days_to_period_end = as.integer(next_period_end - eventdate)) |>
  ggplot(aes(x = days_to_period_end, fill = recognize)) +
  geom_histogram(binwidth = 1)
```

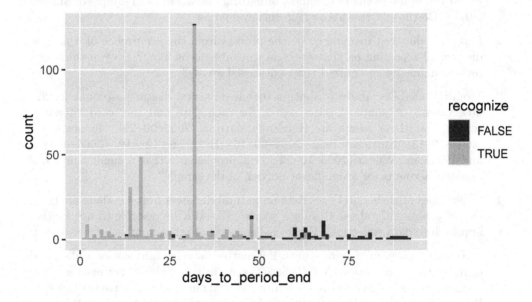

FIGURE 17.2
Distribution of days to period-end in Michels (2017) sample

Obviously major natural disasters can affect more than one firm and Table 17.1 provides data on the five most common disaster dates in the Michels (2017) sample.

```
michels_2017 |>
  count(eventdate) |>
  arrange(desc(n)) |>
  top_n(5)
```

Selecting by n

TABLE 17.1
Number of observations for
top 5 natural disasters

eventdate	n
2005-08-29	131
2008-09-13	54
2004-09-16	23
2004-08-13	11
2005-10-24	9

17.5   Discussion questions

1. One assumption in Michels (2017) is that whether a natural disaster occurs before or after the balance sheet date of the next filing is random. Do the inherent properties of natural disasters ensure that they are random? Why? If not, how would you evaluate the randomness of natural disasters in the sample of Michels (2017)? Do the analyses above help this evaluation?

2. Describe what you imagine to be the process from the occurrence of a natural disaster to reporting on that event in the subsequent filing? Do you think this process differs for recognized versus disclosed events?

3. From the analysis, above it appears that five natural disasters account for 228 observations. A simple Google search for each date and the word "disaster" reveals that these events are Hurricane Katrina (2005-08-29), Hurricane Ike (2008-09-13), Hurricane Ivan (2004-09-16), Hurricane Charley (2004-08-13), and Hurricane Wilma (2005-10-24). Is it problematic that a small number of disasters accounts for a significant portion of the sample?

4. Where does Michels (2017) get data on natural disasters from? Is there anything that is problematic about this data source? Would it be possible to use another approach to data collection? What challenges would that approach face?

5. A recurring question in accounting is if it matters whether information is disclosed or recognized. One view is that, if markets are efficient, it should not matter where the information is disclosed, so recognition should not matter relative to disclosure. What assumptions underlie this view? Are there any reasons to believe that they do or do not hold in the setting of Michels (2017)? What are the implications for the ability of Michels (2017) to deliver clean causal inferences? How might

materiality criteria differ for disclosed and recognized events? Would differences in these criteria affect the empirical analysis of Michels (2017)?

6. What causal inferences does Michels (2017) draw? What (if any) issues do you see with regard to these?

7. Choose a paper that you have seen recently that uses empirical analysis of non-experimental data. (If you cannot choose such a paper, Hopkins et al. (2022) provide one option.) Looking at the abstract of the paper, can you determine whether this paper seeks to draw causal inferences?

8. Choose what you think the authors regard to be the most important causal inference they draw (or would like to draw) in your chosen paper. Which table or tables provide the relevant empirical analyses? Sketch a rough causal diagram for this causal inference using either discussion in the paper or your own background knowledge to identify important variables. How credible do you find the reported causal inferences to be in light of your causal diagram?

18

Causal mechanisms

Gow et al. (2016) argue that, while causal inference is the goal of most accounting research, it is extremely difficult to find settings where straightforward application of statistical methods can produce credible estimates of causal effects (and the remaining chapters of this part arguably support this claim). Does this mean accounting researchers must give up making causal statements? Gow et al. (2016) argue that the answer is "no" and suggest an increased focus on causal mechanisms as one path forward.

In this chapter, we aim to help the reader understand the idea of causal mechanisms, which can suggest empirical analyses that do not rely on, say, random assignment to treatment, which is very rarely found in accounting research. Unlike the other chapters in this part, this chapter is largely qualitative and contains no empirical analysis.

Accounting research is not alone in its reliance on observational data with the goal of drawing causal inferences. For this reason, it is natural to look to other fields using observational data to identify causal mechanisms and ultimately to draw causal inferences. Epidemiology and medicine are two fields that are often singled out in this regard. In the next two sections, we briefly provide examples and highlight the features of the examples that enhanced the credibility of the inferences drawn.[1] A key implication of this discussion is that accounting researchers need to identify clearly and rigorously the causal mechanism that is producing their results.

> Tip
>
> While there is no R code in this chapter, a Quarto template for the discussion questions below is available on GitHub.[a]
>
> ---
> [a]https://github.com/iangow/far_templates/blob/main/README.md

18.1 John Snow and cholera

A widely cited case of successful causal inference is John Snow's work on cholera. As there are many excellent accounts of Snow's work, we will focus on the barest details. As discussed in Freedman (2009, p. 339), "John Snow was a physician in Victorian London. In 1854, he demonstrated that cholera was an infectious disease, which could be prevented by cleaning up the water supply. The demonstration took advantage of a natural experiment. A large area of London was served by two water companies. The Southwark and Vauxhall company distributed contaminated water, and households served by it had a death rate 'between

---
[1]This material is largely based on material in Gow et al. (2016).

eight and nine times as great as in the houses supplied by the Lambeth company,' which supplied relatively pure water." But there was much more to Snow's work than the use of a convenient natural experiment. First, Snow's reasoning (much of which was surely done before "the arduous task of data collection" began) was about the mechanism through which cholera spread. Existing theory suggested "odors generated by decaying organic material." Snow reasoned qualitatively that such a mechanism was implausible. Instead, drawing on his medical knowledge and the facts at hand, Snow conjectured that "a living organism enters the body, as a contaminant of water or food, multiplies in the body, and creates the symptoms of the disease. Many copies of the organism are expelled with the dejecta, contaminate water or food, then infect other victims" (Freedman, 2009, p. 342).

With a hypothesis at hand, Snow then needed to collect data to prove it. His data collection involved a house-to-house survey in the area surrounding the Broad Street pump operated by Southwark and Vauxhall. As part of his data collection, Snow needed to account for anomalous cases (such as the brewery workers who drank beer, not water). It is important to note that such qualitative reasoning and diligent data collection were critical elements in establishing (to a modern reader) the "as if" random nature of the treatment assignment mechanism provided by the Broad Street pump. Snow's deliberate methods contrast with a shortcut approach, which would have been to argue that in his data he had a natural experiment.

Another important feature of this example is that widespread acceptance of Snow's hypothesis did not occur until compelling evidence of the precise causal mechanism was provided. "However, widespread acceptance was achieved only when Robert Koch isolated the causal agent (*Vibrio cholerae*, a comma-shaped bacillus) during the Indian epidemic of 1883" (Freedman, 2009, p. 342). Only once persuasive evidence of a plausible mechanism was provided (i.e., direct observation of micro-organisms now known to cause the disease) did Snow's ideas become widely accepted.

We expect the same might be true in the accounting discipline if researchers carefully articulate the assumed causal mechanism for their observations. It is, of course, necessary for researchers to show that a proposed mechanism is consistent with observed behaviour in the institutional setting being examined. As discussed below, detailed descriptive studies of institutional phenomena provide an important part of the information used to evaluate a proposed mechanism.

18.2   Smoking and heart disease

A more recent illustration of plausible causal inference is discussed by Gillies (2011). Gillies (2011) focuses on the paper by Doll and Peto (1976), which studied the mortality rates of male doctors between 1951 and 1971. The data of Doll and Peto (1976) showed "a striking correlation between smoking and lung cancer" (Gillies, 2011, p. 111). Gillies (2011) argues that "this correlation was accepted at the time by most researchers (if not quite all!) as establishing a causal link between smoking and lung cancer." Indeed Doll and Peto themselves say explicitly (p. 1535) that "the excess mortality from cancer of the lung in cigarette smokers is caused by cigarette smoking." In contrast, while Doll and Peto (1976) had highly statistically significant evidence of an association between smoking and heart disease, they were cautious about drawing inferences of a direct causal explanation for the association. Doll and Peto (1976, p. 1528) point out that "to say that these conditions were related to smoking does not necessarily imply that smoking caused ... them. The relation may have been secondary in that smoking was associated with some other factor, such as alcohol consumption or a feature of the personality, that caused the disease."

Gillies (2011) then discusses extensive research into atherosclerosis between 1979 and 1989 and concludes that "by the end of the 1980s, it was established that the oxidation of LDL was an important step in the process which led to atherosclerotic plaques." Later research (Morrow et al., 1995, p. 1201) provided "compelling evidence that smoking causes oxidative modification of biologic components in humans."[2] Gillies (2011, p. 120) points out that this evidence alone did not establish a confirmed mechanism linking smoking with heart disease because the required oxidation needs to occur in the artery wall, not in the bloodstream, and it fell to later research to establish this missing piece.[3] Thus, through a process involving multiple studies over two decades, a plausible set of causal mechanisms between smoking and atherosclerosis was established.

Gillies (2011) avers that the process by which a causal link between smoking and atherosclerosis was established illustrates the **Russo-Williamson thesis**. Russo and Williamson (2007, p. 159) suggest that "mechanisms allow us to generalize a causal relation: while an appropriate dependence in the sample data can warrant a causal claim 'C causes E in the sample population,' a plausible mechanism or theoretical connection is required to warrant the more general claim 'C causes E.' Conversely, mechanisms also impose negative constraints: if there is no plausible mechanism from C to E, then any correlation is likely to be spurious. Thus mechanisms can be used to differentiate between causal models that are underdetermined by probabilistic evidence alone." The Russo-Williamson thesis was arguably also at work in the case of Snow and cholera, where the establishment of a mechanism (i.e., *Vibrio cholerae*) was essential before the causal explanation offered by Snow was widely accepted. It also appears in the case of smoking and lung cancer, which was initially conjectured based on correlations, prior to a direct biological explanation being offered.[4]

18.3   Causal mechanisms in accounting research

Gow et al. (2016) suggest that accounting researchers can learn from fields such as epidemiology, medicine, and political science, which grapple with observational data yet eventually draw inferences that are causal. While randomized controlled trials are a gold standard of sorts in epidemiology, it is often unfeasible or unethical to use such trials. For example, in political science, it is impossible to randomly assign countries to treatment conditions such as democracy or socialism. Nevertheless, these fields have often been able to draw plausible causal inferences by establishing clear mechanisms, or causal pathways, from putative causes to putative effects.

One paper that has a fairly compelling identification strategy is Brown et al. (2015), which examines "the influence of mobile communication on local information flow and local

---

[2]This evidence is much higher levels of a new measure (levels of F_2-isoprostanes in blood samples) of the relevant oxidation in the body due to smoking. This conclusion was greatly strengthened by the finding that levels of F_2-isoprostanes in the smokers fell significantly after two weeks of abstinence from smoking" (Morrow et al., 1995, pp. 1201–1202).

[3]"Smoking produced oxidative stress. This increased the adhesion of leukocytes to the ... artery, which in turn accelerated the formation of atherosclerotic plaques" (Gillies, 2011, p. 123).

[4]The persuasive force of Snow's natural experiment, coming decades before the work of Neyman (1923) and Fisher (1935), might be considered greater today.

investor activity using the enforcement of state-wide distracted driving restrictions." The authors find that "these restrictions ... inhibit local information flow and ... the market activity of stocks headquartered in enforcement states." Miller and Skinner (2015) (p. 229) suggest that "given the authors' setting and research design, it is difficult to imagine a story under which the types of reverse causality or correlated omitted variables explanations that we normally worry about in disclosure research are at play." However, notwithstanding the apparent robustness of the research design, the results would be much more compelling if there was more detailed evidence regarding the precise causal mechanism through which the estimated effect occurs, and the authors appear to go to lengths to provide such an account.[5] For example, evidence of trading activity by local investors while driving prior to, but not after, the implementation of distracted driving restrictions would add considerable support to conclusions in Brown et al. (2015).[6]

As another example, many published papers have suggested that managers adopt conditional conservatism as a reporting strategy to obtain benefits such as reduced debt costs. However, as Beyer et al. (2010, p. 317) point out, an ex-ante commitment to such a reporting strategy "requires a mechanism that allows managers to credibly commit to withholding good news or to commit to an accounting information system that implements a higher degree of verification for gains than for losses," yet research has only recently begun to focus on the mechanisms through which such commitments are made (e.g., Erkens et al., 2014).[7] It is very clear that we need a much better understanding of the precise causal mechanisms for important accounting research questions. A clear discussion of these mechanisms will enable reviewers and readers to see what is being assumed and assess the reasonableness of the posited causal mechanisms.

18.4 Discussion questions

18.4.1 Li et al. (2018)

1. Let's think about the causal mechanism in Li et al. (2018).

   Suppose you are a manager involved in the process of preparing information about significant customers for your firm's 10-K. What do you think your main area of expertise would be? If a court upheld or rejected the IDD while you are in this role, would you expect to learn about that in your role? If so, how would you learn this information? If not, how might the court ruling nonetheless have an effect on your decisions regarding what to include in the filing?

---

[5]Brown et al. (2015, pp. 277–278) "argue that constraints on mobile communication while driving could impede or delay the collection and diffusion of local stock information across local individuals. Anecdotal evidence suggests that some individuals use car commutes as opportune times to gather and disseminate stock information via mobile devices. For instance, some commuters use mobile devices to collect and pass on stock information either electronically or by word-of-mouth to other individuals within their social network. Drivers also use mobile devices to wirelessly check stock positions and prices in real-time, stream the latest financial news, or listen to earnings calls."

[6]Note that the authors disclaim reliance on trading while driving: "our conjectures do not depend on the presumption that local investors are driving when they execute stock trades ... [as] we expect such behavior to be uncommon." However, even if not *necessary*, given the small effect size documented in the paper (approximately 1% decrease in volume), a small amount of such activity could be *sufficient* to provide a convincing account in support of their results.

[7]We discuss conditional conservatism in Section 4.1.2.

Assume that the relevant managers are aware of relevant court rulings on IDD, how would you expect this to affect disclosure decisions? When would this effect occur? How long would it last?

2. Note that PepsiCo is headquartered in the state of New York. Reading the discussion of the *PepsiCo Inc. v. Redmond* case (Li et al., 2018, pp. 272–3), does this fact have implications for their research design. How might Li et al. (2018) need to do things differently?

3. Look at the plots in Li et al. (2018, p. 284). What do we expect to see here? Try to imagine the plots without the fitted curves and red vertical dashed line. How compelling is this evidence? Why do you think the authors included this? How might one depict the results reported in Table 2 on the figures in Li et al. (2018, p. 284) (pay attention to magnitudes)?

4. In Section 3.2, we ran regressions with **two-way fixed effects** (i.e., fixed effects for `grade` and for `id`, the student identifier). What variables are analogous to `grade` and for `id` in regressions (3) and (4) in Table 2 of Li et al. (2018)? Why is there no $POST \times TREAT$ coefficient reported in Table 2? How would you code $POST$ for a firm in a state that never has IDD?

5. Compare Tables 2 and 3 of Li et al. (2018). Which regressions are most comparable across the two tables? What insight does Table 3 provide over Table 2?

18.4.2   Burks et al. (2018)

1. Suppose you are a senior executive at a bank. What opportunities might implementation of IBBEA by a state other than your own provide to you? How would you evaluate such opportunities? How long would it take to evaluate them?

2. Suppose you are a senior executive at a bank in state that is about to implement IBBEA. What actions might you take if you anticipate increased interest in your state by interstate banks? Would the actions you take depend on how out-of-state banks enter?

3. Burks et al. (2018) do not explain what a "deposit cap" is. Rice and Strahan (2010) describe it thus: "IBBEA specifies a statewide deposit concentration limitation of 30% with respect to interstate mergers that constitute an initial entry of a bank into a state." What might drive differences in the deposit cap for different states? (See Table 1.) Would this restriction affect behaviour of all banks in an IBBEA-implementing state in the same way?

4. "The results are consistent with disclosure significantly increasing upon introduction of the IBBEA (the main effect) ... we find that the restriction imposing a deposit cap of less than 30% has a significant mitigating effect on disclosure." How many states implemented a cap of less than 30%? (See Table 1.) We checked one of these states (Colorado) and it had a cap of 25%. In 1997, the bank with the largest market share had 17.87%. With these facts, what is the practical difference between 25% and 30% in this case?

18.4.3   Christensen et al. (2017); Glaeser and Guay (2017)

1. Christensen et al. (2017) claim to find a causal effect of requiring disclosure in SEC filings on mine safety. Why is it significant that the information was already

available on MSHA website? How persuasive do you find the evidence of a causal effect to be? What elements do you find particularly persuasive (or problematic)?

2. What is the causal mechanism through which Christensen et al. (2017) argue (or conjecture) this effect occurs?

3. Glaeser and Guay (2017) suggest that use of securities laws to enforce safety regulations is another mechanism through which the causal effect occurs? Does this undermine the validity of the estimate of the causal effect?

19

Natural experiments revisited

In this chapter, we return to the topic of natural experiments. We first discuss the notion of **registered reports**, their purpose, and their limitations. We then focus on an experiment ("Reg SHO") run by the US Securities and Exchange Commission (SEC) and studies that examined the effects of Reg SHO, with a particular focus on one study (Fang et al., 2016) that exploited this regulation to study effects on earnings management.

This chapter provides opportunities to sharpen our skills and knowledge in a number of areas. First, we will revisit the topic of earnings management and learn about some developments in its measurement since Dechow et al. (1995), which we covered in Chapter 16. Second, we further develop our skills in evaluating claimed natural experiments, using Reg SHO and the much-studied setting of broker-closure shocks. Third, we explore the popular **difference-in-differences** approach, both when predicated on random assignment and when based on the so-called **parallel trends assumption** in the absence of random assignment. Fourth, we will have an additional opportunity to apply ideas related to causal diagrams and causal mechanisms (covered in Chapters 4 and 18, respectively). Fifth, we will revisit the topic of statistical inference, using this chapter as an opportunity to consider **randomization inference**. Sixth, we build on the Frisch-Waugh-Lovell theorem to consider issues associated with the use of **two-step regressions**, which are common in many areas of accounting research.

> **Tip**
>
> The code in this chapter uses the packages listed below. We load `tidyverse` because we use several packages from the Tidyverse. For instructions on how to set up your computer to use the code found in this book, see Section 1.2. Quarto templates for the exercises below are available on GitHub.[a]
>
> ---
> [a]https://github.com/iangow/far_templates/blob/main/README.md

```
library(tidyverse)
library(DBI)
library(farr)
library(fixest)
library(modelsummary)
library(furrr)          # future_map(), future_pmap()
library(broom)          # tidy()
```

> **! Important**
>
> This chapter is longer than others in the book, so we have made it easier to run code from one section without having to run all the code preceding it. Beyond that, the code in each of Sections 19.1–19.3 and 19.5 is independent of code in other parts of this chapter and can be run independently of those other sections.[1] Code and exercises in Sections 19.7 and 19.8 depend on code in Section 19.6, so you will need to run the code in Section 19.6 before running the code in those sections.

19.1 A replication crisis?

A Financial Times article by Robert Wigglesworth[2] discusses an alleged "replication crisis" in finance research. Wigglesworth quotes Campbell Harvey, professor of finance at Duke University, who suggests that "at least half of the 400 supposedly market-beating strategies identified in top financial journals over the years are bogus."

Wigglesworth identified "the heart of the issue" as what researchers call **p-hacking**, which is the practice whereby researchers search for "significant" and "positive" results. Here "significant" refers to statistical significance and "positive" refers to results that reject so-called "null hypotheses" and thereby (purportedly) push human knowledge forward. Harvey (2017) cites research suggesting that 90% of published studies report such "significant" and "positive" results. Reporting "positive" results is important not only for getting published, but also for attracting citations, which can drive behaviour of both researchers and journals.

Simmons et al. (2011, p. 1359) describe what they term **researcher degrees of freedom**. "In the course of collecting and analyzing data, researchers have many decisions to make: Should more data be collected? Should some observations be excluded? Which conditions should be combined and which ones compared? Which control variables should be considered? Should specific measures be combined or transformed or both?" Simmons et al. (2011, p. 1364) identify another well-known researcher degree of freedom, namely that of "reporting only experiments that 'work'", which is known as the **file-drawer problem** (because experiments that don't "work" are put in a file-drawer).

To illustrate the power of researcher degrees of freedom, Simmons et al. (2011) conducted two hypothetical studies based on experiments with live subjects. They argue that these studies "demonstrate how unacceptably easy it is to accumulate (and report) statistically significant evidence for a false hypothesis" [p. 1359]. Simmons et al. (2011, p. 1359) conclude that "flexibility in data collection, analysis, and reporting dramatically increases actual false-positive rates."

Perhaps in response to concerns similar to those raised by Simmons et al. (2011), the *Journal of Accounting Research* (JAR) conducted a trial for its annual conference held in May 2017.[3] According to the JAR website, at this conference "authors presented papers developed through a Registration-based Editorial Process (REP). The goal of the conference was to

---

[1]While there are two data sets calculated in Section 19.2 that are used later in the chapter, it is not necessary to run the code in this section to use the code later in the chapter, as we have provided those data sets—sho_r3000_sample and sho_r3000_gvkeys—as part of the farr package.

[2]https://www.ft.com/content/9025393f-76da-4b8f-9436-4341485c75d0

[3]https://go.unimelb.edu.au/pzw8

see whether REP could be implemented for accounting research, and to explore how such a process could be best implemented. Papers presented at the conference were subsequently published in May 2018.

According to Bloomfield et al. (2018, p. 317), "under REP, authors propose a plan to gather and analyze data to test their predictions. Journals send promising proposals to one or more reviewers and recommend revisions. Authors are given the opportunity to review their proposal in response, often multiple times, before the proposal is either rejected or granted in-principle acceptance ... regardless of whether [subsequent] results support their predictions."

Bloomfield et al. (2018, p. 317) contrast REP with the Traditional Editorial Process ("TEP"). Under the TEP, "authors gather their data, analyze it, and write and revise their manuscripts repeatedly before sending them to editors." Bloomfield et al. (2018, p. 317) suggest that "almost all peer-reviewed articles in social science are published under ... the TEP."

The REP is designed to eliminate questionable research practices, including those identified by Simmons et al. (2011). For example, one form of p-hacking is **HARKing** (from "Hypothesizing After Results are Known"). In its extreme form, HARKing involves searching for a "significant" correlation and *then* developing a hypothesis to "predict" it. To illustrate, consider the spurious correlations website[4] provided by Tyler Vigen. This site lists a number of evidently spurious correlations, such as the 99.26% correlation between the divorce rate in Maine and margarine consumption or the 99.79% correlation between US spending on science, space, and technology and suicides by hanging, strangulation, and suffocation. The correlations are deemed spurious because normal human beings have strong prior beliefs that no underlying causal relation explains these correlations. Instead, these are regarded as mere coincidence.

However, a creative academic can probably craft a story to "predict" any correlation. Perhaps increasing spending on science raises its perceived importance to society. But drawing attention to science only serves to highlight how the United States has inevitably declined in relative stature in many fields, including science. While many Americans can carry on notwithstanding this decline, others are less sanguine about it and may go to extreme lengths as a result ... . This is clearly a silly line of reasoning, but if one added some references to published studies and fancy terminology, it would probably read a lot like the hypothesis development sections of some academic papers.

Bloomfield et al. (2018, p. 326) examine "the strength of the papers' results" from the 2017 JAR conference in their section 4.2 and conclude that "of the 30 predictions made in the ... seven proposals, we count 10 as being supported at $p \leq 0.05$ by at least one of the 134 statistical tests the authors reported. The remaining 20 predictions are not supported at $p \leq 0.05$ by any of the 84 reported tests. Overall, our analysis suggests that the papers support the authors' predictions far less strongly than is typical among papers published in JAR and its peers."[5]

19.1.1   Discussion questions

1. Simmons et al. (2011) provide a more in-depth examination of issues with the TEP discussed in Bloomfield et al. (2018, pp. 318–319). Do you think the two

---

[4]http://tylervigen.com/spurious-correlations

[5]Note that while Bloomfield et al. (2018) discuss "$p \leq 0.05$", it is more conventional to look for $p < 0.05$, with ties effectively going to the null hypothesis; in practice, this makes no difference.

experiments studied by Simmons et al. (2011) are representative of how accounting research works in practice? What differences are likely to exist in empirical accounting research using archival data?

2. Bloomfield et al. (2018, p. 326) say "we exclude Hail et al. (2018) from our tabulation [of results] because it does not state formal hypotheses." Given the lack of formal hypotheses, do you think it made sense to include the proposal[6] from Hail et al. (2018) in the 2017 JAR conference? Does the REP have relevance to papers without formal hypotheses? Does the absence of formal hypotheses imply that Hail et al. (2018) were not testing hypotheses? Is your answer to the last question consistent with how Hail et al. (2018, p. 650) discuss results reported in Table 5 of that paper?

3. According the analysis of Bloomfield et al. (2018), there were 218 tests of 30 hypotheses and different hypotheses had different numbers of tests. In the following analysis, we assume 30 hypotheses with each having 7 tests (for a total of 210 tests). Does this analysis suggest an alternative possible interpretation of the results than the "far less strongly than is typical" conclusion offered by Bloomfield et al. (2018). Does choosing a different value for `set.seed()` alter the tenor of the results from this analysis? How might you make the analysis below more definitive?[7]

```
set.seed(2021)
results <-
  expand_grid(hypothesis = 1:30, test = 1:7) |>
  mutate(p = runif(nrow(pick(everything()))),
         reject = p < 0.05)

results |>
  group_by(hypothesis) |>
  summarize(reject_one = any(reject), .groups = "drop") |>
  count(reject_one)
```

```
# A tibble: 2 x 2
  reject_one      n
  <lgl>       <int>
1 FALSE          19
2 TRUE           11
```

4. Bloomfield et al. (2018, p. 326) argue "it is easy to imagine revisions of several conference papers would allow them to report results of strength comparable to those found in most papers published under TEP." For example, "Li and Sandino (2018) yielded no statistically significant support for their main hypotheses. However, they found significant results in their planned additional analyses that are consistent with informal predictions included in the accepted proposal. ... [In light of this evidence] we are not ready to conclude that the studies in the issue actually provide weaker support for their predictions than most studies published under TEP." (2018, p. 326). Can these results instead be interpreted as saying something about the strength of results of studies published under TEP?

---

[6]https://go.unimelb.edu.au/8zw8
[7]We explained the use of `pick(everything())` in Chapter 13.

5. Do you believe that it would be feasible for REP to become the dominant research paradigm in accounting research? What challenges would such a development face?

6. A respondent to the survey conducted by Bloomfield et al. (2018, p. 337) provided the remark quoted below. Comment on this remark. What do you think the respondent has in mind with regard to the "learning channel"? Do you agree that the REP shuts down this channel?

"I do not find the abundance of 'null results' surprising. It could have been discovered from one's own experience. Research is an iterative process and it involves learning. I am not sure if there is anything useful that we discover in the research process by shutting down the learning channel; especially with the research questions that are very novel and we do not know much about."

19.2   The Reg SHO experiment

To better understand the issues raised by the discussion above in a real research setting, we focus on the Reg SHO experiment, which has been the subject of many studies. In July 2004, the SEC adopted Reg SHO, a regulation governing short-selling activities in equity markets. Reg SHO contained a pilot program in which stocks in the Russell 3000 index were ranked by trading volume within each exchange and every third one was designated as a pilot stock. From 2 May 2005 to 6 August 2007, short sales on pilot stocks were exempted from price tests, including the tick test for exchange-listed stocks and the bid test for NASDAQ National Market stocks.

In its initial order, the SEC stated that "the Pilot will allow [the SEC] to study trading behavior in the absence of a short sale price test." The SEC's plan was to "examine, among other things, the impact of price tests on market quality (including volatility and liquidity), whether any price changes are caused by short selling, costs imposed by a price test, and the use of alternative means to establish short positions."

19.2.1   The SHO pilot sample

The assignment mechanism in the Reg SHO experiment is unusually transparent, even by the standards of natural experiments. Nonetheless care is needed to identify the treatment and control firms and we believe it is instructive to walk through the steps needed to do so, as we do in this section. (Readers who find the code details tedious could easily skip ahead to Section 19.2.3 on a first reading. We say "first reading" because there are subtle issues with natural experiments that this section helps to highlight, so it may be worth revisiting this section once you have read the later material.)

The SEC's website[8] provides data on the names and tickers of the Reg SHO pilot firms. These data have been parsed and included as the `sho_tickers` data set in the `farr` package.

```
sho_tickers
```

```
# A tibble: 986 x 2
   ticker co_name
```

---

[8]https://www.sec.gov/rules/other/34-50104.htm

```
   <chr>    <chr>
 1 A        AGILENT TECHNOLOGIES INC
 2 AAI      AIRTRAN HOLDINGS INC
 3 AAON     AAON INC
 4 ABC      AMERISOURCEBERGEN CORP
 5 ABCO     ADVISORY BOARD CO
 6 ABCW     ANCHOR BANCORP INC
 7 ABGX     ABGENIX INC
 8 ABK      AMBAC FINANCIAL GRP INC
 9 ABMD     ABIOMED INC
10 ABR      ARBOR REALTY TRUST INC
# i 976 more rows
```

However, these are just the pilot firms and we need to use other sources to obtain the identities of the control firms. It might seem perverse for the SEC to have published lists of treatment stocks, but no information on control stocks.[9] One explanation for this choice might be that because special action (i.e., elimination of price tests) was only required for the treatment stocks (for the control stocks, it was business as usual), no lists of controls were needed for the markets to implement the pilot. Additionally, because the SEC had a list of the control stocks that it would use in its own statistical analysis, it had no reason to publish lists for this purpose. Fortunately, while the SEC did not identify the control stocks, it provides enough information for us to do so, as we do below.

First, we know that the pilot stocks were selected from the Russell 3000, the component stocks of which are found in the `sho_r3000` data set from the `farr` package.

`sho_r3000`

```
# A tibble: 3,000 x 2
   russell_ticker russell_name
   <chr>          <chr>
 1 A              AGILENT TECHNOLOGIES INC
 2 AA             ALCOA INC
 3 AACC           ASSET ACCEPTANCE CAPITAL
 4 AACE           ACE CASH EXPRESS INC
 5 AAI            AIRTRAN HOLDINGS INC
 6 AAON           AAON INC
 7 AAP            ADVANCE AUTO PARTS INC
 8 AAPL           APPLE COMPUTER INC
 9 ABAX           ABAXIS INC
10 ABC            AMERISOURCEBERGEN CORP
# i 2,990 more rows
```

While the Russell 3000 contains 3,000 securities, the SEC[10] and Black et al. (2019) tell us that, in constructing the pilot sample, the SEC excluded 32 stocks in the Russell 3000 index that, as of 25 June 2004, were not listed on the Nasdaq National Market, NYSE, or AMEX "because short sales in these securities are currently not subject to a price test." The SEC also excluded 12 stocks that started trading after 30 April 2004 due to IPOs or spin-offs. And, from Black et al. (2019), we know there were two additional stocks that stopped trading after 25 June 2004 but before the SEC constructed its sample on 28 June

---

[9]There were multiple lists of treatment stocks, as there was a list for each exchange. These lists were updated over time as stocks changed status, a detail we gloss over for the purposes of this chapter.

[10]https://www.sec.gov/rules/other/34-50104.htm

2004. We can get the data for each of these criteria from CRSP, but we need to first merge the Russell 3000 data with CRSP to identify the right PERMNO for each security. For this purpose, we will use data from the five CRSP tables below:

```
db <- dbConnect(RPostgres::Postgres(), bigint = "integer")

mse <- tbl(db, Id(schema = "crsp", table = "mse"))
msf <- tbl(db, Id(schema = "crsp", table = "msf"))
stocknames <- tbl(db, Id(schema = "crsp", table = "stocknames"))
dseexchdates <- tbl(db, Id(schema = "crsp", table = "dseexchdates"))
ccmxpf_lnkhist <- tbl(db, Id(schema = "crsp", table = "ccmxpf_lnkhist"))
```

One thing we note is that some of the tickers from the Russell 3000 sample append the class of stock to the ticker. We can detect these cases by looking for a dot (.) using regular expressions. Because a dot has special meaning in regular expressions (regex), we need to escape it using a backslash (\). (For more on regular expressions, see Chapter 9 and references cited there.) Because a backslash has a special meaning in strings in R, we need to escape the backslash itself to tell R that we mean a literal backslash. In short, we use the regex \\. to detect dots in strings.

```
sho_r3000 |>
  filter(str_detect(russell_ticker, "\\."))
```

```
# A tibble: 12 x 2
   russell_ticker russell_name
   <chr>          <chr>
 1 AGR.B          AGERE SYSTEMS INC
 2 BF.B           BROWN FORMAN CORP
 3 CRD.B          CRAWFORD & CO
 4 FCE.A          FOREST CITY ENTRPRS
 5 HUB.B          HUBBELL INC
 6 JW.A           WILEY JOHN & SONS INC
 7 KV.A           K V PHARMACEUTICAL CO
 8 MOG.A          MOOG INC
 9 NMG.A          NEIMAN MARCUS GROUP INC
10 SQA.A          SEQUA CORPORATION
11 TRY.B          TRIARC COS INC
12 VIA.B          VIACOM INC
```

In these cases, CRSP takes a different approach. For example, where the Russell 3000 sample has `AGR.B`, CRSP has `ticker` equal to `AGR` and `shrcls` equal to B.

The other issue is that some tickers from the Russell 3000 data have the letter E appended to what CRSP shows as just a four-letter ticker.

```
sho_r3000 |>
  filter(str_length(russell_ticker) == 5,
         str_sub(russell_ticker, 5, 5) == "E")
```

```
# A tibble: 4 x 2
  russell_ticker russell_name
  <chr>          <chr>
1 CVNSE          COVANSYS CORP
2 SONSE          SONUS NETWORKS INC
```

```
3 SPSSE          SPSS INC
4 VXGNE          VAXGEN INC
```

A curious reader might wonder how we identified these two issues with tickers, and how we know that they are exhaustive of the issues in the data. We explore these questions in the exercises at the end of this section.

To address these issues, we create two functions: one (`clean_ticker()`) to "clean" each ticker so that it can be matched with CRSP, and one (`get_shrcls()`) to extract the share class (if any) specified in the Russell 3000 data.

Both functions use a regular expression to match cases where the text ends with either "A" or "B" (`[AB]$` in regex) preceded by the a dot (`\\.` in regex, as discussed above). The expression uses capturing parentheses (i.e., `(` and `)`) to **capture** the text before the dot from the beginning of the string (`^(.*)`) to the dot and to capture the letter "A" or "B" at the end (`([AB])$`).

```
regex <- "^(.*)\\.([AB])$"
```

The `clean_ticker()` function uses `case_when()`, which first handles cases with an `E` at the end of five-letter tickers, then applies the regex to extract the "clean" ticker (the first captured text) from cases matching `regex`. Finally, the original ticker is returned for all other cases.

```
clean_ticker <- function(x) {
  case_when(str_length(x) == 5 & str_sub(x, 5, 5) == "E" ~ str_sub(x, 1, 4),
            str_detect(x, regex) ~ str_replace(x, regex, "\\1"),
            .default = x)
}
```

The `get_shrcls()` function extracts the second capture group from the regex (the first value returned by `str_match()` is the complete match, the second value is the first capture group, so we use `[, 3]` to get the second capture group).

```
get_shrcls <- function(x) {
  str_match(x, regex)[, 3]
}
```

We can use `clean_ticker()` and `get_shrcls()` to construct `sho_r3000_tickers`.

```
sho_r3000_tickers <-
  sho_r3000 |>
  select(russell_ticker, russell_name) |>
  mutate(ticker = clean_ticker(russell_ticker),
         shrcls = get_shrcls(russell_ticker))

sho_r3000_tickers |>
  filter(russell_ticker != ticker)
```

```
# A tibble: 16 x 4
  russell_ticker russell_name          ticker shrcls
  <chr>          <chr>                 <chr>  <chr>
1 AGR.B          AGERE SYSTEMS INC     AGR    B
2 BF.B           BROWN FORMAN CORP     BF     B
3 CRD.B          CRAWFORD & CO         CRD    B
```

```
 4 CVNSE        COVANSYS CORP              CVNS   <NA>
 5 FCE.A        FOREST CITY ENTRPRS        FCE    A
 6 HUB.B        HUBBELL INC                HUB    B
 7 JW.A         WILEY JOHN & SONS INC      JW     A
 8 KV.A         K V PHARMACEUTICAL CO      KV     A
 9 MOG.A        MOOG INC                   MOG    A
10 NMG.A        NEIMAN MARCUS GROUP INC    NMG    A
11 SONSE        SONUS NETWORKS INC         SONS   <NA>
12 SPSSE        SPSS INC                   SPSS   <NA>
13 SQA.A        SEQUA CORPORATION          SQA    A
14 TRY.B        TRIARC COS INC             TRY    B
15 VXGNE        VAXGEN INC                 VXGN   <NA>
16 VIA.B        VIACOM INC                 VIA    B
```

Now that we have "clean" tickers, we can merge with CRSP. The following code proceeds in two steps. First, we create `crsp_sample`, which contains the `permno`, `ticker`, and `shrcls` values applicable on 2004-06-25, the date on which the Russell 3000 that the SEC used was created.

```
crsp_sample <-
  stocknames |>
  mutate(test_date = as.Date("2004-06-25")) |>
  filter(test_date >= namedt, test_date <= nameenddt) |>
  select(permno, permco, ticker, shrcls) |>
  distinct() |>
  collect()
```

Second, we merge `sho_r3000_tickers` with `crsp_sample` using `ticker` and then use `filter()` to retain cases where, if a share class is specified in the SEC-provided ticker, it matches the one row in CRSP with that share class, while retaining all rows where no share class is specified in the SEC-provided ticker.

```
sho_r3000_merged <-
  sho_r3000_tickers |>
  inner_join(crsp_sample, by = "ticker", suffix = c("", "_crsp")) |>
  filter(shrcls == shrcls_crsp | is.na(shrcls)) |>
  select(russell_ticker, permco, permno)
```

Unfortunately, this approach results in some tickers being matched to multiple PERMNO values.

```
sho_r3000_merged |>
  group_by(russell_ticker) |>
  filter(n() > 1) |>
  ungroup()
```

```
# A tibble: 40 x 3
   russell_ticker permco permno
   <chr>           <int>  <int>
 1 AGM             28392  80169
 2 AGM             28392  80168
 3 BDG             20262  55781
 4 BDG             20262  77881
 5 BIO               655  61516
```

```
 6 BIO                655  61508
 7 CW               20546  18091
 8 CW               20546  89223
 9 EXP              30381  80415
10 EXP              30381  89983
# i 30 more rows
```

In each case, these appear to be cases where there are multiple securities (`permno` values) for the same company (`permco` value). To choose the security that is the one most likely included in the Russell 3000 index used by the SEC, we will keep the one with the greatest dollar trading volume for the month of June 2004. We collect the data on dollar trading volumes in the data frame `trading_vol`.

```
trading_vol <-
  msf |>
  filter(date == "2004-06-30") |>
  mutate(dollar_vol = coalesce(abs(prc) * vol, 0)) |>
  select(permno, dollar_vol) |>
  collect()
```

We can now make a new version of the table `sho_r3000_merged` that includes just the `permno` value with the greatest trading volume for each ticker.

```
sho_r3000_merged <-
  sho_r3000_tickers |>
  inner_join(crsp_sample, by = "ticker", suffix = c("", "_crsp")) |>
  filter(is.na(shrcls) | shrcls == shrcls_crsp) |>
  inner_join(trading_vol, by = "permno") |>
  group_by(russell_ticker) |>
  filter(dollar_vol == max(dollar_vol, na.rm = TRUE)) |>
  ungroup() |>
  select(russell_ticker, permno)
```

Black et al. (2019) identify 32 stocks that are not listed on the Nasdaq National Market, NYSE, or AMEX firms "using historical exchange code (`exchcd`) and Nasdaq National Market Indicator (`nmsind`) from the CRSP monthly stock file" (in practice, these 32 stocks are smaller Nasdaq-listed stocks). However, `exchcd` and `nmsind` are not included in the `crsp.msf` file we use. Black et al. (2019) likely used the CRSP monthly stock file obtained from the web interface provided by WRDS, which often merges in data from other tables.

Fortunately, we can obtain `nmsind` from the CRSP monthly events file (`crsp.mse`). This file includes information about delisting events, distributions (such as dividends), changes in NASDAQ information (such as `nmsind`), and name changes. We get data on `nmsind` by pulling the latest observation on `crsp.mse` on or before 2004-06-28 where the event related to NASDAQ status (`event == "NASDIN"`).

```
nmsind_data <-
  mse |>
  filter(date <= "2004-06-28", event == "NASDIN") |>
  group_by(permno) |>
  filter(date == max(date, na.rm = TRUE)) |>
  ungroup() |>
  select(permno, date, nmsind) |>
  collect()
```

We can obtain `exchcd` from the CRSP stock names file (`crsp.stocknames`), again pulling the value applicable on 2004-06-28.[11]

```
exchcd_data <-
  stocknames |>
  filter(exchcd > 0) |>
  mutate(test_date = as.Date("2004-06-28")) |>
  filter(between(test_date, namedt, nameenddt)) |>
  select(permno, exchcd) |>
  distinct() |>
  collect()
```

According to its website,[12] the SEC "also excluded issuers whose initial public offerings commenced after April 30, 2004." Following Black et al. (2019), we use CRSP data to identify these firms. Specifically, the table `crsp.dseexchdates` includes the variable `begexchdate`.

```
ipo_dates <-
  dseexchdates |>
  distinct(permno, begexchdate) |>
  collect()
```

Finally, it appears that there were stocks listed in the Russell 3000 file likely used by the SEC (created on 2004-06-25) that were delisted prior to 2004-06-28, the date on which the SEC appears to have finalized the sample for its pilot program. We again use `crsp.mse` to identify these firms.

```
recent_delistings <-
  mse |>
  filter(event == "DELIST",
         between(date, "2004-06-25", "2004-06-28")) |>
  rename(delist_date = date) |>
  select(permno, delist_date) |>
  collect()
```

Now, we put all these pieces together and create variables `nasdaq_small`, `recent_listing`, and `delisted` corresponding to the three exclusion criteria discussed above.

```
sho_r3000_permno <-
  sho_r3000_merged |>
  left_join(nmsind_data, by = "permno") |>
  left_join(exchcd_data, by = "permno") |>
  left_join(ipo_dates, by = "permno") |>
  left_join(recent_delistings, by = "permno") |>
  mutate(nasdaq_small = coalesce(nmsind == 3 & exchcd == 3, FALSE),
         recent_listing = begexchdate > "2004-04-30",
         delisted = !is.na(delist_date),
         keep = !nasdaq_small & !recent_listing & !delisted)
```

As can be seen in Table 19.1, we have a final sample of 2954 stocks that we can merge with `sho_tickers` to create the `pilot` indicator.

---

[11]We exclude non-positive values of `exchcd` as these appear to be rows flagging special temporary trading statuses that create duplicate rows.

[12]https://www.sec.gov/rules/other/34-50104.htm

```
sho_r3000_permno |>
  count(nasdaq_small, recent_listing, delisted, keep)
```

TABLE 19.1
Reconciliation of sample with Russell 3000

nasdaq_small	recent_listing	delisted	keep	n
FALSE	FALSE	FALSE	TRUE	2954
FALSE	FALSE	TRUE	FALSE	2
FALSE	TRUE	FALSE	FALSE	12
TRUE	FALSE	FALSE	FALSE	32

```
sho_r3000_sample <-
  sho_r3000_permno |>
  filter(keep) |>
  rename(ticker = russell_ticker) |>
  left_join(sho_tickers, by = "ticker") |>
  mutate(pilot = !is.na(co_name)) |>
  select(ticker, permno, pilot)
```

As can be seen the number of treatment and control firms in **sho_r3000_sample** corresponds exactly with the numbers provided in Black et al. (2019, p. 42).

```
sho_r3000_sample |>
  count(pilot)
```

```
# A tibble: 2 x 2
  pilot       n
  <lgl>   <int>
1 FALSE    1969
2 TRUE      985
```

Finally, we will want to link these data with data from Compustat, which means linking these observations with GVKEYs. For this, we use **ccm_link** (as used and discussed in Chapter 7) to produce **sho_r3000_gvkeys**, the sample we can use in later analysis.

```
ccm_link <-
  ccmxpf_lnkhist |>
  filter(linktype %in% c("LC", "LU", "LS"),
         linkprim %in% c("C", "P")) |>
  rename(permno = lpermno) |>
  mutate(linkenddt = coalesce(linkenddt, max(linkenddt, na.rm = TRUE))) |>
  select(gvkey, permno, linkdt, linkenddt)
```

Because we focus on a single "test date", our final link table includes just two variables: gvkey and **permno**.[13]

```
gvkeys <-
  ccm_link |>
  mutate(test_date = as.Date("2004-06-28")) |>
```

---

[13]In other words, we don't need date ranges in the **gvkeys** table.

```
filter(between(test_date, linkdt, linkenddt)) |>
select(gvkey, permno) |>
collect()
```

Finally, we can add `gvkey` to `sho_r3000_sample` to create `sho_r3000_gvkeys`.

```
sho_r3000_gvkeys <-
  sho_r3000_sample |>
  inner_join(gvkeys, by = "permno")
```

```
sho_r3000_gvkeys
```

```
# A tibble: 2,951 x 4
   ticker permno pilot gvkey
   <chr>   <dbl> <lgl> <chr>
 1 A       87432 TRUE  126554
 2 AA      24643 FALSE 001356
 3 AACC    90020 FALSE 157058
 4 AACE    78112 FALSE 025961
 5 AAI     80670 TRUE  030399
 6 AAON    76868 TRUE  021542
 7 AAP     89217 FALSE 145977
 8 AAPL    14593 FALSE 001690
 9 ABAX    77279 FALSE 024888
10 ABC     81540 TRUE  031673
# i 2,941 more rows
```

To better understand the potential issues with constructing the `pilot` indicator variable, it is useful to compare the approach above with that taken in Fang et al. (2016). To construct `sho_data` as Fang et al. (2016) do, we use `fhk_pilot` from the `farr` package.[14] We compare `sho_r3000_sample` and `sho_r3000_gvkeys` with `sho_data` in the exercises below.

```
sho_data <-
  fhk_pilot |>
  select(gvkey, pilot) |>
  distinct() |>
  group_by(gvkey) |>
  filter(n() == 1) |>
  ungroup() |>
  inner_join(fhk_pilot, by = c("gvkey", "pilot"))
```

19.2.2 Exercises

1. Before running the following code, can you tell from output above how many rows this query will return? What is this code doing? At what stage would code like this have been used in process of creating the sample above? Why is code like this not included above?

---

[14]These data frames are derived from SAS files provided by the authors of Fang et al. (2016), as can be seen in the source code of the `farr` package: https://go.unimelb.edu.au/ezw8.

```
sho_r3000 |>
  anti_join(crsp_sample, join_by(russell_ticker == ticker)) |>
  collect()
```

2. Focusing on the values of `ticker` and `pilot` in `fhk_pilot`, what differences do you observe between `fhk_pilot` and `sho_r3000_sample`? What do you believe is the underlying cause for these discrepancies?

3. What do the following observations represent? Choose a few observations from this output and examine whether these reveal issues in the `sho_r3000_sample` or in `fhk_pilot`.

```
sho_r3000_sample |>
  inner_join(fhk_pilot, by = "ticker", suffix = c("_ours", "_fhk")) |>
  filter(permno_ours != permno_fhk)
```

```
# A tibble: 37 x 6
   ticker permno_ours pilot_ours gvkey  permno_fhk pilot_fhk
   <chr>        <int> <lgl>      <chr>       <int> <lgl>
 1 AGM          80169 FALSE      015153      80168 FALSE
 2 AGR.B        89400 TRUE       141845      88917 TRUE
 3 BDG          55781 TRUE       002008      77881 TRUE
 4 BF.B         29946 TRUE       002435      29938 TRUE
 5 BIO          61516 TRUE       002220      61508 TRUE
 6 CRD.B        27618 TRUE       003581      76274 TRUE
 7 CW           18091 FALSE      003662      89223 FALSE
 8 EXP          80415 FALSE      030032      89983 FALSE
 9 FCE.A        31974 TRUE       004842      65584 TRUE
10 GEF          83233 TRUE       005338      83264 TRUE
# i 27 more rows
```

4. In constructing the `pilot` indicator, FHK omit cases (`gvkey` values) where there is more than one distinct value for the indicator. A question is: Who are these firms? Why is there more than one value for `pilot` for these firms? And does omission of these make sense? (*Hint*: Identify duplicates in `fhk_pilot` and compare `sho_r3000_gvkeys` for these firms.)

5. What issue is implicit in the output from the code below? How could you fix this issue? Would you expect a fix for this issue to significantly affect the regression results? Why or why not?

```
sho_data |>
  count(gvkey, ticker) |>
  arrange(desc(n))
```

```
# A tibble: 2,993 x 3
  gvkey  ticker     n
  <chr>  <chr>  <int>
1 001076 RNT        2
2 002008 BDG        2
3 002220 BIO        2
```

```
 4 002435 BF.B        2
 5 002710 STZ         2
 6 003581 CRD.B       2
 7 003662 CW          2
 8 003708 TRY.B       2
 9 004842 FCE.A       2
10 005284 GTN         2
# i 2,983 more rows
```

19.2.3   Early studies of Reg SHO

The first study of the effects of Reg SHO was conducted by the SEC's own Office of Economic Analysis. The SEC study[15] examines the "effect of pilot on short selling, liquidity, volatility, market efficiency, and extreme price changes" [p. 86].

The authors of the 2007 SEC study "find that price restrictions reduce the volume of executed short sales relative to total volume, indicating that price restrictions indeed act as a constraint to short selling. However, in neither market do we find significant differences in short interest across pilot and control stocks. ... We find no evidence that short sale price restrictions in equities have an impact on option trading or open interest. ... We find that quoted depths are augmented by price restrictions but realized liquidity is unaffected. Further, we find some evidence that price restrictions dampen short term within-day return volatility, but when measured on average, they seem to have no effect on daily return volatility."

The SEC researchers conclude "based on the price reaction to the initiation of the pilot, we find limited evidence that the tick test distorts stock prices—on the day the pilot went into effect, Listed Stocks in the pilot sample underperformed Listed Stocks in the control sample by approximately 24 basis points. However, the pilot and control stocks had similar returns over the first six months of the pilot."

In summary, it seems fair to say that the SEC found that exemption from price tests had relatively limited effect on the market outcomes of interest, with no apparent impact on several outcomes.

Alexander and Peterson (2008, p. 84) "examine how price tests affect trader behavior and market quality, which are areas of interest given by the [SEC] in evaluating these tests." Alexander and Peterson (2008, p. 86) find that NYSE pilot stocks have similar spreads, but smaller trade sizes, more short trades, more short volume, and smaller ask depths. With regard to Nasdaq, Alexander and Peterson (2008, p. 86) find that the removed "bid test is relatively inconsequential."

Diether et al. (2009, p. 37) find that "while short-selling activity increases both for NYSE- and Nasdaq-listed Pilot stocks, returns and volatility at the daily level are unaffected."

19.2.4   Discussion questions and exercises

1. Earlier we identified one feature of a randomized controlled trial (RCT) as that "proposed analyses are specified in advance", as in a registered reports process. Why do you think the SEC did not use a registered report for its 2007 paper? Do

---

[15]https://www.sec.gov/news/studies/2007/regshopilot020607.pdf

you think the analyses of the SEC would be more credible if conducted as part of a registered reports process? Why or why not?

2. Do you have concerns that the results Alexander and Peterson (2008) have been p-hacked? What factors increase or reduce your concerns in this regard?

3. Evaluate the hypotheses found in the section of Diether et al. (2009, pp. 41–45) entitled *Testable Hypotheses* with particular sensitivity to concerns about HARKing. What kind of expertise is necessary in evaluating hypotheses in this way?

4. How might the SEC have conducted Reg SHO as part of a registered reports process open to outside research teams, such as Alexander and Peterson (2008) and Diether et al. (2009)? How might such a process have been run? What challenges would such a process face?

19.3 Analysing natural experiments

Both Alexander and Peterson (2008) and Diether et al. (2009) use the difference-in-differences estimator ("DiD") of the causal effect that we saw in Chapter 3. The typical approach to DiD involves estimating a regression of the following form:

$$Y_{it} = \beta_0 + \beta_1 \times POST_t + \beta_2 \times TREAT_i + \beta_3 \times POST_t \times TREAT_i + \epsilon_{it} \qquad (19.1)$$

In this specification, the estimated treatment effect is given by the fitted coefficient $\hat{\beta}_3$.

While DiD is clearly popular among researchers in economics and adjacent fields, it is important to note that it is not obvious that it is the best choice in every experimental setting and that credible alternatives exist.

Another approach would be to limit the sample to the post-treatment period and estimate the following regression:

$$Y_{it} = \beta_0 + \beta_1 \times TREAT_i + \epsilon_{it} \qquad (19.2)$$

In this specification, the estimated treatment effect is given by the fitted coefficient $\hat{\beta}_1$. This approach is common in drug trials, which are typically conducted as RCTs. For example, in the paxlovid trial "participants were randomised 1:1, with half receiving paxlovid and the other half receiving a placebo orally every 12 hours for five days. Of those who were treated within three days of symptom onset, 0.8% (3/389) of patients who received paxlovid were admitted to hospital up to day 28 after randomization, with no deaths. In comparison, 7% (27/385) of patients who received placebo were admitted, with seven deaths." (Mahase, 2021, p. 1). For the hospital admission outcome, it would have been possible to incorporate prior hospitalization rates in a difference-in-differences analysis, but this would only make sense if hospitalization rates in one period had a high predictive power for subsequent hospitalization rates.[16]

---

[16] In contrast, death is one outcome where difference-in-differences analysis makes no sense for almost all individuals.

Yet another approach would include pre-treatment values of the outcome variable as a control:

$$Y_{it} = \beta_0 + \beta_1 \times TREAT_i + \beta_2 \times Y_{i,t-1} + \epsilon_{it} \tag{19.3}$$

To evaluate each of these approaches, we can use simulation analysis. The following analysis is somewhat inspired by Frison and Pocock (1992), who use different assumptions about their data more appropriate to their (medical) setting and who focus on mathematical analysis instead of simulations.

Frison and Pocock (1992) assume a degree of correlation in measurements of outcome variables for a given unit (e.g., patient) that is independent of the time between observations. A more plausible model in many business settings would be correlation in outcome measures for a given unit (e.g., firm) that fades as observations become further apart in time. Along these lines, we create `get_outcomes()` below to generate data for outcomes in the absence of treatment. Specifically, we assume that, if there are no treatment or period effects, the outcome in question follows the autoregressive process embedded in `get_outcomes()`, which has the key parameter ρ (`rho`).[17]

```r
get_outcomes <- function(rho = 0, periods = 7) {
  e <- rnorm(periods)
  y <- rep(NA, periods)
  y[1] <- e[1]
  for (i in 2:periods) {
    y[i] <- rho * y[i - 1] + e[i]
  }
  tibble(t = 1:periods, y = y)
}
```

The `get_sample()` function below uses `get_outcomes()` for n firms for given values of `rho`, `periods` (the number of periods observed for each firm), and `effect` (the underlying size of the effect of treatment on y). Here treatment is randomly assigned to half the firms in the sample and the effect is added to y when both `treat` and `post` are true. We also add a time-specific effect (`t_effect`) for each period, which is common to all observations (a common justification for the use of DiD is the existence of such period effects).

```r
get_sample <- function(n = 100, rho = 0, periods = 7, effect = 0) {
  treat <- sample(1:n, size = floor(n / 2), replace = FALSE)

  t_effects <- tibble(t = 1:periods, t_effect = rnorm(periods))

  f <- function(x) tibble(id = x, get_outcomes(rho = rho,
                                               periods = periods))
  df <-
    map(1:n, f) |>
    list_rbind() |>
    inner_join(t_effects, by = "t") |>
    mutate(treat = id %in% treat,
```

---

[17]Note that the fact that the variance of y_t equals $\frac{1}{1-\rho^2}$—so $\rho = 1$ is technically problematic—and a variant of `get_outcomes()` might specify `e <- rnorm(periods, sd = sqrt(1 - rho^2))` to hold the variance of y constant as ρ changes. Doing this does not change the tenor of the analysis, so we do not implement this here.

```
            post = t > periods / 2,
            y = y + if_else(treat & post, effect, 0) + t_effect) |>
    select(-t_effect)
}
```

The `est_effect()` function below applies a number of estimators to a given data set and returns the estimated treatment effect for each estimator. The estimators we consider are the following (the labels *POST*, *CHANGE*, and *ANCOVA* come from Frison and Pocock, 1992):

- *DiD*, the difference-in-differences estimator estimated by regressing y on the treatment indicator, `treat` interacted with the post-treatment indicator, `post` (with the `lm()` function automatically including the main effects of `treat` and `post`), as in Equation 19.1.

- *POST*, which is based on OLS regression of y on `treat`, but with the sample restricted to the post-treatment observations, as in Equation 19.2.
- *CHANGE*, which is based on OLS regression of the change in the outcome on `treat`. The change in outcome (`y_change`) is calculated as the mean of post-treatment outcome value (`y_post`) minus the mean of the pre-treatment outcome value (`y_pre`) for each unit.
- *ANCOVA*, which is a regression of `y_post` on `treat` and `y_pre`, as in Equation 19.3.

```
est_effect <- function(df) {

  fm_DiD <- lm(y ~ treat * post, data = df)

  df_POST <-
    df |>
    filter(post) |>
    group_by(id, treat) |>
    summarize(y = mean(y), .groups = "drop")

  fm_POST <- lm(y ~ treat, data = df_POST)

  df_CHANGE <-
    df |>
    group_by(id, treat, post) |>
    summarize(y = mean(y), .groups = "drop") |>
    pivot_wider(names_from = "post", values_from = "y") |>
    rename(y_pre = `FALSE`,
           y_post = `TRUE`) |>
    mutate(y_change = y_post - y_pre)

  fm_CHANGE <- lm(I(y_post - y_pre) ~ treat, data = df_CHANGE)
  fm_ANCOVA <- lm(y_post ~ y_pre + treat, data = df_CHANGE)

  tibble(est_DiD = fm_DiD$coefficients[["treatTRUE:postTRUE"]],
         est_POST = fm_POST$coefficients[["treatTRUE"]],
         est_CHANGE = fm_CHANGE$coefficients[["treatTRUE"]],
         est_ANCOVA = fm_ANCOVA$coefficients[["treatTRUE"]])
}
```

The `run_sim()` function below calls `get_sample()` for supplied parameter values to create a data set, and returns a data frame containing the results of applying `est_effect()` to that data set.

```
run_sim <- function(i, n = 100, rho = 0, periods = 7, effect = 0) {
  df <- get_sample(n = n, rho = rho, periods = periods, effect = effect)
  tibble(i = i, est_effect(df))
}
```

To facilitate running of the simulation for various values of **effect** and **rho**, we create a data frame (**params**) with effect sizes running from 0 to 1 and $\rho \in \{0, 0.18, 0.36, 0.54, 0.72, 0.9\}$.

```
rhos <- seq(from = 0, to = 0.9, length.out = 6)
effects <- seq(from = 0, to = 1, length.out = 5)
params <- expand_grid(effect = effects, rho = rhos)
```

The `run_sim_n()` function below runs 1000 simulations for the supplied values of **effect** and **rho** and returns a data frame with the results.

```
run_sim_n <- function(effect, rho, ...) {
  n_sims <- 1000
  set.seed(2021)

  res <-
    1:n_sims |>
    map(\(x) run_sim(x, rho = rho, effect = effect)) |>
    list_rbind()

  tibble(effect, rho, res)

}
```

❗ Important

The following code takes several minutes to run. Using `future_pmap()` from the `furrr` package in place of `pmap()` reduces the time needed to run the simulation significantly. Fortunately, nothing in the subsequent exercises requires that you run either variant of this code, so only do so if you have time and want to examine **results** directly.

```
plan(multisession)

results <-
  params |>
  future_pmap(run_sim_n,
              .options = furrr_options(seed = 2021)) |>
  list_rbind() |>
  system_time()
```

```
  user  system elapsed
13.688   1.065 322.511
```

With `results` in hand, we can do some analysis. The first thing to note is that `est_CHANGE` is equivalent to `est_DiD`, as all estimates are within rounding error of each other for these two methods.

```
results |> filter(abs(est_DiD - est_CHANGE) > 0.00001) |> nrow()
```

```
[1] 0
```

Thus we just use the label *DiD* in subsequent analysis.

The second thing we check is that the methods provide unbiased estimates of the causal effect. Figure 19.1 suggests that the estimates are very close to the true values of causal effects for all three methods.

```
results |>
  pivot_longer(starts_with("est"),
               names_to = "method", values_to = "est") |>
  mutate(method = str_replace(method, "^est.(.*)$", "\\1")) |>
  group_by(rho, method) |>
  summarize(bias = mean(est - effect), .groups = "drop") |>
  filter(method != "CHANGE") |>
  ggplot(aes(x = rho, y = bias,
             colour = method, linetype = method)) +
  geom_line() +
  ylim(-0.1, 0.1)
```

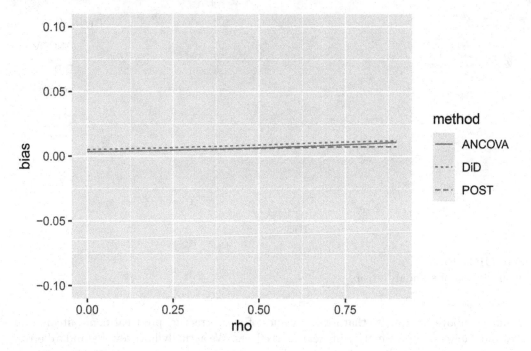

FIGURE 19.1
Comparison of bias

Having confirmed that there is no apparent bias in any of the estimators in this setting, we next consider the empirical standard errors for each method. Because we get essentially identical plots with each value of the true effect, we focus on `effect == 0.5` in the following analysis. Here we rearrange the data so that we have a `method` column and an `est` column for the estimated causal effect. We then calculate, for each `method` and value of `rho`, the standard deviation of `est`, which is the empirical standard error we seek. Finally, we plot the values for each value of `rho` in Figure 19.2.

```
results |>
  filter(effect == 0.5) |>
  pivot_longer(starts_with("est"),
               names_to = "method", values_to = "est") |>
  mutate(method = str_replace(method, "^est.(.*)$", "\\1")) |>
  filter(method != "CHANGE") |>
  group_by(method, rho) |>
  summarize(se = sd(est), .groups = "drop") |>
  ggplot(aes(x = rho, y = se,
             colour = method, linetype = method)) +
  geom_line()
```

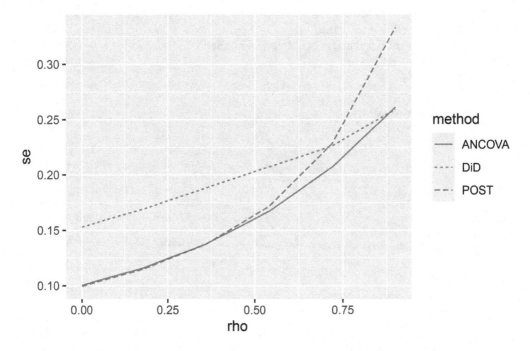

FIGURE 19.2
Comparison of standard errors

From the above, we can see that for low values of ρ, subtracting pre-treatment outcome values adds noise to our estimation of treatment effects. We actually have lower standard errors when we throw away the pre-treatment data and just compare post-treatment outcomes. But for higher levels of ρ, we see that _DiD_ outperforms _POST_; by subtracting pre-treatment outcome values, we get a more precise estimate of the treatment effect. However, we see that both _DiD_ and _POST_ are generally outperformed by _ANCOVA_, which in effect allows for a flexible, data-driven relation between pre- and post-treatment outcome values.

In short, notwithstanding its popularity, it is far from clear that DiD is the best approach to use for all analyses of causal effects based on experimental data. Even in the context of the Reg SHO experiment, the appropriate method may depend on the outcome of interest. For a persistent outcome, *DiD* may be better than *POST*, but for a less persistent outcome, *POST* may be better than *DiD*. And *ANCOVA* may be a better choice than either *POST* or *DiD* unless there are strong *a priori* reasons to believe that *DiD* or *POST* is more appropriate (and such reasons seem more likely to hold for *POST* than for *DiD*).

19.4 Evaluating natural experiments

Because of the plausibly random assignment mechanism used by the SEC, Reg SHO provides a very credible natural experiment. However, in many claimed natural experiments, it will be "nature" who is assigning treatment. In Michels (2017), it was literally nature doing the assignment through the timing of natural disasters. While natural disasters are clearly not completely random, as hurricanes are more likely to strike certain locations at certain times of year, this is not essential for the natural experiment to provide a setting from which causal inferences can be credibly drawn. What is necessary in Michels (2017) is that treatment assignment is **as-if random** with regard to the timing of the natural disaster before or after the end of the fiscal period and discussion questions in Chapter 17 explored these issues.

More often in claimed natural experiments, it will be economic actors or forces, rather than nature, assigning treatment. While such economic actors and forces are unlikely to act at random, again the critical question is whether treatment assignment is *as-if* random. To better understand the issues, we consider a well-studied setting, that of brokerage closures as studied in Kelly and Ljungqvist (2012).

Kelly and Ljungqvist (2012, p. 1368) argue that "brokerage closures are a plausibly exogenous source of variation in the extent of analyst coverage, as long as two conditions are satisfied. First, the resulting coverage terminations must correlate with an increase in information asymmetry. ... Second, the terminations must only affect price and demand through their effect on information asymmetry." Interestingly, these are essentially questions 2 and 3 that we ask in evaluating instrumental variables in Section 20.3. The analogue with instrumental variables applies because Kelly and Ljungqvist (2012) are primarily interested in the effects of changes in information asymmetry, not the effects of brokerage closures *per se*. In principle, brokerage closures could function much like an instrumental variable, except that Kelly and Ljungqvist (2012) estimate **reduced-form** regressions whereby outcomes are related directly to brokerage closures, such as in Table 3 (Kelly and Ljungqvist, 2012, p. 1391). But the first of the three questions from Section 20.3 remains relevant, and this is the critical question for evaluating any natural experiment: **Is treatment assignment (as-if) random?**

Like many researchers, Kelly and Ljungqvist (2012) do not address the (as-if) randomness of treatment assignment directly. Instead, Kelly and Ljungqvist (2012) focus on whether brokerage closure-related terminations of analyst coverage "constitute a suitably exogenous shock to the investment environment". Kelly and Ljungqvist (2012) argue that they do "*unless* brokerage firms quit research because their analysts possessed negative private information about the stocks they covered." But this reasoning is incomplete. For sure, brokerage firms not quitting research for the reason suggested is a *necessary* condition for a natural experiment (otherwise the issues with using brokerage closures as a natural experiment are quite apparent). But it is not a *sufficient* condition. If the firms encountering brokerage closure-related terminations of analyst coverage had different trends in information asymmetry for other reasons, the lack of private information is inadequate to give us a natural experiment.

In general, we should be able to evaluate the randomness of treatment assignment much as we would do so with an explicitly randomized experiment. Burt (2000) suggest that "statisticians will compare the homogeneity of the treatment group populations to assess the distribution of the pretreatment demographic characteristics and confounding factors." With explicit randomization, statistically significant differences in pretreatment variables might prompt checks to ensure that, say, there was not "deliberate alteration of or noncompliance with the random assignment code" or any other anomalies. Otherwise, we might have greater confidence that randomization was implemented effectively, and hence that causal inferences might reliably be drawn from the study.

So, a sensible check with a natural experiment would seem to be to compare various pretreatment variables across treatment groups to gain confidence that "nature" has indeed randomized treatment assignment. In this regard, Table 1 of Kelly and Ljungqvist (2012) is less than assuring. Arguably, one can only encounter brokerage closure-related terminations of analyst coverage if one has analyst coverage in the first place; so the relevant comparison is arguably between the first and third columns of data. There we see that the typical firm in the terminations sample (column 1) is larger, has higher monthly stock turnover, higher daily return volatility, and more brokers covering the stock than does the typical firm in the universe of covered stocks in 2004 (column 3). So clearly "nature" has not randomly selected firms from the universe of covered stocks in 2004.

However, we might come to a similar conclusion if we compared the Reg SHO pilot stocks with the universe of traded stocks or some other broad group. Just as it was essential to correctly identify the population that the SEC considered in randomizing treatment assignment, it is important to identify the population that "nature" considered in assigning treatment in evaluating any natural experiment. While the SEC provided a statement detailing how it constructed the sample, "nature" is not going to do the same and researchers need to consider carefully which units were considered for (as if) random assignment to treatment.

In this regard, even assuming that the controls used in Table 2 of Kelly and Ljungqvist (2012, p. 1388) were the ones "nature" herself considered, it seems that the natural experiment did not assign treatment in a sufficiently random way. Table 2 studies four outcomes: bid-ask spreads, the Amihud illiquidity measure, missing- and zero-return days, and measures related to earnings announcements. In each case, there are pre-treatment differences that sometimes exceed the DiD estimates. For example, pre-treatment bid-ask spreads for treatment and control firms are 1.126 and 1.089, a 0.037 difference that is presumably statistically significant given that the smaller DiD estimate of 0.020 has a p-value of 0.011.[18] In light of this evidence, it seems that Kelly and Ljungqvist (2012) need to rely on the parallel trends assumption to draw causal inferences and we evaluate the plausibility of this assumption in the next section.

---

[18] It is not clear that the matching procedure correctly identified the group of firms considered for treatment. Kelly and Ljungqvist (2012, pp. 1386–1387) "match [control] firms [for each treatment firm] on the Fama and French (1993) pricing factors by using the Daniel et al. (1997) algorithm. Specifically, we choose as controls for terminated stock i five firms in the same size and book-to-market quintile in the preceding month of June, subject to the condition that control firms did not themselves experience a termination in the quarter before or after." But this is the appropriate matching procedure for causal inference only if "nature" herself selected firms for consideration as treatment or control firms based on such criteria (or ones highly correlated with these).

It is important to recognize that the shortcomings of broker closures as a natural experiment do not completely undermine the ability of Kelly and Ljungqvist (2012) to draw causal inferences. There appears to be an unfortunate tendency to believe, on the one hand, that without some kind of natural experiment, one cannot draw causal inferences. On the other hand, there is an apparent tendency to view natural experiments as giving *carte blanche* to researchers to draw all kinds of causal inferences, even when the alleged identification strategies do not, properly understood, support such inferences.

In the case of Kelly and Ljungqvist (2012), it seems the authors would like to believe that they have a natural experiment that allows them to draw inferences about the effects of broker closures on information asymmetry (Table 2) and, because broker closures only affect stock prices through their effects on information asymmetry, to conclude from the evidence in Table 3 that increases in information asymmetry reduce stock prices. But Table 2 could have been based on a bullet-proof identification strategy without implying that broker closures only affect stock prices through their effects on information asymmetry. There is really no evidence offered for this claim, one that is arguably very difficult to support.

At the same time, it is conceptually possible that Kelly and Ljungqvist (2012) could provide compelling evidence that the only plausible explanation for the abnormal returns in Table 3 is reductions in information asymmetry, even if the results in Table 2 were irredeemably corrupted (e.g., because of failure of parallel trends). Evidence that firms did not "quit research because their analysts possessed negative private information about the stocks they covered" might support drawing certain inferences from Table 3 even without a credible equivalent of Table 2.

19.5 The parallel trends assumption

Examining the studies of the direct effects of Reg SHO discussed in Section 19.2.3, we see that randomization generally provided balance in pre-treatment outcome values. In such settings, we see that DiD can provide unbiased estimates of causal effects, but that it has little appeal when treatment assignment is random. Indeed, if there is little to distinguish treatment and control observations in terms of pre-treatment outcome values, DiD will differ little from simply comparing differences in post-treatment means (the *POST* estimator discussed above).

But in many settings, such as that in Kelly and Ljungqvist (2012), differences in pre-treatment outcome values exist, suggesting that random assignment is not an appropriate assumption. In such settings, it will therefore be necessary to rely on a different assumption to justify the use of DiD for causal inferences. This **parallel trends assumption** posits that, but for treatment, the expected change in the outcome variable for the treated observations would equal that for control observations. Using this assumption, we can attribute any difference in the change in the outcome variable between the treated and control observations to a treatment effect and random variation and use standard tools of statistical inference to evaluate the null hypothesis that any difference is due to random variation.

That DiD can be predicated on an assumption other than (as-if) random assignment may explain its popularity. Cunningham (2021, p. 406) suggests that DiD "has become the single most popular research design in the quantitative social sciences" and Armstrong et al. (2022, p. 4) find rapid "increase in [the number of] papers using quasi-experimental methods to draw causal inferences, that more than 75% of such papers use variations of the classic difference-in-differences design."

Arguably DiD is more often used in settings that *without* (as-if) random assignment of treatment. For example, one of the most highly cited papers using DiD is Card and Krueger (1994), which compares the change in employment in the fast-food industry in New Jersey and Philadelphia before and after an increase in the minimum wage in New Jersey. In this case, treatment assignment was very clearly non-random—it was a function of being located in New Jersey.

Unlike the assumption of random assignment, the parallel trends assumption is not implied by a reasonable description of a physical or economic process and thus is of a fundamentally different nature. Random assignment is widely regarded as a reasonable description of, say, a coin toss or the generation of pseudo-random numbers using a computer. In contrast, it is difficult to think of a mechanism for imposing parallel trends on the data. Because DiD is—unlike instrumental variables or regression discontinuity designs—generally *not* predicated on "as-if random variation in the explanatory variable of interest", it is not correct to consider DiD as a **quasi-experimental method** (Armstrong et al., 2022, p. 3).[19]

Instead the basis for the parallel trends assumption appears to be that it is *the* assumption that is necessary (and sufficient) for the DiD estimator to provide an unbiased estimator of the causal effect of treatment. But "assuming a can-opener"[20] seems to be a weak foundation for an approach as widespread as DiD.

On the one hand, as discussed above, there is no obvious economic basis for the parallel trends assumption with general applicability. On the other hand, there are often reasons to believe that the parallel trends assumption will *not* hold for various outcomes. The parallel trends assumption will be dubious when the outcome variable tends to be mean-reverting. For example, it is well known that accounting-based measures of operating performance tend to revert towards the mean. So if treatment and control observations have different levels of pre-treatment operating performance, the parallel trends assumption will be a highly dubious basis for causal inference.

Another reason to doubt the parallel trends assumption is the fact that the measurement of outcomes is often arbitrary. For example, Li et al. (2018) examine the effect of legal changes on disclosure of customer identities using a variant of DiD.[21] One primary outcome measure considered by Li et al. (2018) is `ratio`, the proportion of significant customers whose identities are not disclosed. But if the parallel trends assumption holds in `ratio` then, so long as there are pre-treatment differences between treatment and control observations, it is not mathematically possible for parallel trends to hold in $\log(1 + \texttt{ratio})$, which is the measure used in the regression analysis in Li et al. (2018).

The apparent flimsiness of the parallel trends assumption underlying DiD analysis in non-randomized settings is perhaps reinforced by the treatment of DiD in textbooks. Imbens and Rubin (2015), a significant recent tome on causal inference, buries DiD in endnotes, merely noting that DiD is "widely used" (2015, p. 44) before directing the reader to Angrist and Pischke (2008). While Angrist and Pischke (2008) discuss DiD and its assumptions, and relate it to fixed-effects regressions and panel data methods, they do little to justify the parallel trends assumption. Cunningham (2021) is much more cautious in discussing the parallel trends assumption, which he notes "is by definition untestable since we cannot observe this counterfactual conditional expectation [of post-treatment outcomes absent treatment]".

---

[19]Armstrong et al. (2022) define "quasi-experimental methods" as "methods that seek to exploit as-if random variation in observational settings", but at other times extend this definition to include DiD.

[20]https://en.wikipedia.org/wiki/Assume_a_can_opener

[21]We study Li et al. (2018) closely in Chapter 21.

Two popular approaches to address the parallel trends assumption are discussed by Cunningham (2021) and Huntington-Klein (2021). The first approach compares the trends in pre-treatment outcome values for treatment and control observations. If these trends are similar before treatment, it is perhaps reasonable to assume they are similar after treatment. But Cunningham (2021, p. 426) notes that "pre-treatment similarities are neither necessary nor sufficient to guarantee parallel counterfactual trends" and this seems an especially dubious assumption if treatment is endogenously selected.

The second approach is the placebo test, variants of which are discussed by Cunningham (2021) and Huntington-Klein (2021). One placebo test involves evaluating the treatment effect in a setting where prior beliefs hold that there should be no effect. Another approach involves a kind of random assignment of a pseudo-treatment. In either case, not finding an effect is considered as providing support for the parallel trends assumption in the analysis of greater interest to the researcher. Of course, one might be sceptical of such placebo tests in light of the concerns raised at the start of this chapter. If applying DiD to state-level data on spending on science, space, and technology provides evidence of an effect on suicides by hanging, strangulation, and suffocation, not finding an effect on deaths by drowning after falling out of a canoe or kayak may provide limited assurance.

To illustrate, we now apply a kind of placebo test to evaluate the parallel trends assumption for bid-ask spreads, one of the variables considered in the DiD analysis of Table 2 of Kelly and Ljungqvist (2012) (we choose spreads in part because it is easy to calculate).

We first create the data set **spreads**, which contains data on the average spread for stocks over three-month periods—aligning with one measure used Table 2 of Kelly and Ljungqvist (2012)—for a sample period running from Q1, 2001 (the first quarter of 2001) to Q1, 2008, which is the sample period in Kelly and Ljungqvist (2012). We will conduct a study of a pseudo-treatment that we will assume applies for periods beginning in Q1, 2004, which is roughly halfway through the sample period and we code **post** accordingly.

```
db <- dbConnect(RPostgres::Postgres(), bigint = "integer")

dsf <- tbl(db, Id(schema = "crsp", table = "dsf"))

spreads <-
  dsf |>
  mutate(spread = 100 * (ask - bid) / ((ask + bid) / 2),
         quarter = as.Date(floor_date(date, 'quarter'))) |>
  group_by(permno, quarter) |>
  summarize(spread = mean(spread, na.rm = TRUE), .groups = "drop") |>
  mutate(post = quarter >= "2004-01-01") |>
  filter(!is.na(spread),
         between(quarter, "2000-01-01", "2008-01-01")) |>
  collect()

dbDisconnect(db)
```

We now randomize treatment assignment. Because we want to evaluate the parallel trends assumption and completely randomized treatment assignment *implies* a trivial version of the parallel trends assumption, we specify a small difference in the probability of receiving treatment for observations whose pre-treatment spread exceeds the median ($p = 0.55$) from those whose pre-treatment spread is below the median ($p = 0.45$). This ensures that we have pre-treatment differences and thus need to rely on the parallel trends assumption in a meaningful way.[22]

```
set.seed(2021)

treatment <-
  spreads |>
  filter(!post) |>
  group_by(permno) |>
  summarize(spread = mean(spread, na.rm = TRUE), .groups = "drop") |>
  mutate(treat_prob = if_else(spread > median(spread), 0.55, 0.45),
         rand = runif(n = nrow(pick(everything()))),
         treat = rand < treat_prob) |>
  select(permno, treat)
```

Obviously the null hypothesis of zero treatment effect holds with this "treatment", but the question is whether the parallel trends assumption holds for `spread`. If we find evidence of a "treatment effect", the only sensible interpretation is a failure of the parallel trends assumption for `spread`.

Merging in the `treatment` data set, we estimate a DiD regression (and cluster standard errors by `permno` for reasons that will be apparent after reading the discussion below). Results are reported in Table 19.2.

```
reg_data <-
  spreads |>
  inner_join(treatment, by = "permno")

fm <- feols(spread ~ post * treat, vcov = ~ permno, data = reg_data)
```

```
modelsummary(fm,
             estimate = "{estimate}{stars}",
             gof_map = c("nobs"),
             stars = c('*' = .1, '**' = 0.05, '***' = .01))
```

TABLE 19.2
Test of parallel trends in bid-ask spreads

	(1)
(Intercept)	2.830***
	(0.049)
postTRUE	−2.130***
	(0.044)
treatTRUE	0.489***
	(0.077)
postTRUE × treatTRUE	−0.343***
	(0.069)
Num.Obs.	220,822

---

[22]The particular approach taken also implies a 50-50 split between treatment and control in expectation. We explained the use of `pick(everything())` in Chapter 13.

Because we find a statistically significant effect of -0.343 with a t-statistic of -4.96 with this meaningless "treatment", we can conclude with some confidence that the parallel trends assumption simply does not hold for `spread` in this sample. Given that we might have passed this placebo test even if the parallel trends assumption did not hold for a particular treatment, say due to endogenous selection, it seems reasonable to view this test as being better suited to detecting a failure of parallel trends (as it does here) than it is to validation of that assumption.

Cunningham (2021) and Huntington-Klein (2021) provide excellent pathways to a recent literature examining DiD. However, it is important to recognize that some variant of the scientifically flimsy parallel trends assumption imbues all of these treatments. It would seem to be productive for researchers to discard the "quasi-experimental" pretence attached to DiD and to apply techniques appropriate to what some call **interrupted time-series designs** (e.g., Shadish et al., 2002).[23]

While a randomized experiment provides a sound basis for attributing observed differences in outcomes to either treatment effects or sampling variation, without such randomization, it is perhaps more appropriate to take a more **abductive** approach of identifying causal mechanisms, deeper predictions about the timing and nature of causal effects, explicit consideration of alternative explanations, and the like (Armstrong et al., 2022; Heckman and Singer, 2017). Some evidence of this is seen in the discussion of specific papers in Cunningham (2021), perhaps reflecting reluctance to lean too heavily on the parallel trends assumption.

19.6 Indirect effects of Reg SHO

The early studies of Reg SHO discussed above can be viewed as studying the more *direct* effects of Reg SHO. As a policy change directly affecting the ability of short-sellers to trade in securities, the outcomes studied by these earlier studies are more closely linked to the Reg SHO pilot than are the outcomes considered in later studies. Black et al. (2019, pp. 2–3) point out that "despite little evidence of direct impact of the Reg SHO experiment on pilot firms, over 60 papers in accounting, finance, and economics report that suspension of the price tests had wide-ranging *indirect* effects on pilot firms, including on earnings management, investments, leverage, acquisitions, management compensation, workplace safety, and more."

One indirect effect of short-selling that has been studied in subsequent research is that on earnings management. To explore this topic, we focus on Fang et al. (2016, p. 1251), who find "that short-selling, or its prospect, curbs earnings management."

19.6.1 Earnings management after Dechow et al. (1995)

In Chapter 16, we saw that early earnings management research used firm-specific regressions in estimating standard models such as the Jones (1991) model. Fang et al. (2016) apply subsequent innovations in measurement of earnings management, such the

---

[23]The elision of DiD with some notion of "natural experiments" seems to be a good example of the conceptual stretching discussed by Dunning (2012, p. 3). Huntington-Klein (2021) uses the term **event studies** to describe settings where treatment is applied during a sample period, but there is no random assignment. We avoid that term given its more specialist meaning in accounting and finance (see Chapter 13).

performance-matched discretionary accruals measure developed in Kothari et al. (2005).

Kothari et al. (2005) replace the firm-specific regressions seen in Dechow et al. (1995) with regressions by industry-year, where industries are defined as firms grouped by two-digit SIC codes. Kothari et al. (2005) also add an intercept term to the Jones Model and estimate discretionary accruals under the Modified Jones Model by applying the coefficients from the Jones Model to the analogous terms of the Modified Jones Model.[24]

To calculate performance-matched discretionary accruals, Kothari et al. (2005, p. 1263) "match each sample firm with the firm from the same fiscal year-industry that has the closest return on assets as the given firm. The performance-matched discretionary accruals ... are then calculated as the firm-specific discretionary accruals minus the discretionary accruals of the matched firm." This is the primary measure of earnings management used by Fang et al. (2016), but note that Fang et al. (2016) use the 48-industry Fama-French groupings, rather than the two-digit SIC codes used in Kothari et al. (2005).

19.6.2   FHK: Data steps

To construct measures of discretionary accruals, Fang et al. (2016) obtain data primarily from Compustat, along with data on Fama-French industries from Ken French's website and data on the SHO pilot indicator from the SEC's website. The following code is adapted from code posted by the authors[25] of Fang et al. (2016), which was used to produce the results found in Tables 15 and 16 of Fang et al. (2019). The bulk of the Compustat data used in Fang et al. (2016) come from `comp.funda`. Following Fang et al. (2019), the code below collects data from that table for fiscal years between 1999 and 2012, excluding firms with SIC codes between 6000 and 6999 or between 4900 and 4949.

```
db <- dbConnect(RPostgres::Postgres(), bigint = "integer")

funda <- tbl(db, Id(schema = "comp", table = "funda"))

compustat_annual <-
  funda |>
  filter(indfmt == 'INDL', datafmt == 'STD', popsrc == 'D', consol == 'C',
         between(fyear, 1999, 2012),
         !(between(sich, 6000, 6999) | between(sich, 4900, 4949))) |>
  select(gvkey, fyear, datadate, fyr, sich, dltt, dlc, seq,
         oibdp, ib, ibc, oancf, xidoc, at, ppegt, sale,
         rect, ceq, csho, prcc_f) |>
  mutate(fyear = as.integer(fyear)) |>
  collect()

dbDisconnect(db)
```

Some regressions in Fang et al. (2019) consider controls for market-to-book, leverage, and return on assets, which are calculated as `mtob`, `leverage`, and `roa`, respectively, in the following code:

---

[24]That is, the coefficient on $\frac{\Delta REV_{i,t}}{ASSET_{i,t-1}}$ (β_2) from (3) in Kothari et al. (2005, p. 1263) is applied to $\frac{(\Delta REV_{i,t} - \Delta AR_{i,t})}{ASSET_{i,t-1}}$ from (4) on the same page.

[25]https://go.unimelb.edu.au/f7d8

```
controls_raw <-
  compustat_annual |>
  group_by(gvkey) |>
  arrange(fyear) |>
  mutate(lag_fyear = lag(fyear),
         mtob = if_else(lag(ceq) != 0,
                          lag(csho) * lag(prcc_f) / lag(ceq), NA),
         leverage = if_else(dltt + dlc + seq != 0,
                             (dltt + dlc) / (dltt + dlc + seq), NA),
         roa = if_else(lag(at) > 0, oibdp / lag(at), NA)) |>
  filter(fyear == lag(fyear) + 1) |>
  ungroup() |>
  select(gvkey, datadate, fyear, at, mtob, leverage, roa)
```

Following Fang et al. (2019), we create `controls_filled`, which uses `fill()` to remove many missing values for the controls.

```
controls_filled <-
  controls_raw |>
  group_by(gvkey) |>
  arrange(fyear) |>
  fill(at, mtob, leverage, roa) |>
  ungroup()
```

Following Fang et al. (2019), we create `controls_fyear_avg`, which calculates averages of controls by fiscal year.

```
controls_fyear_avg <-
  controls_filled |>
  group_by(fyear) |>
  summarize(across(c(at, mtob, leverage, roa),
                   \(x) mean(x, na.rm = TRUE)))
```

Like Fang et al. (2019), we use values from `controls_fyear_avg` to replace missing values for controls.

```
df_controls <-
  controls_filled |>
  inner_join(controls_fyear_avg, by = "fyear", suffix = c("", "_avg")) |>
  mutate(at = coalesce(at, at_avg),
         mtob = coalesce(mtob, mtob_avg),
         leverage = coalesce(leverage, leverage_avg),
         roa = coalesce(roa, roa_avg)) |>
  select(gvkey, fyear, at, mtob, leverage, roa)
```

There are multiple steps in the code above and the reasons for the steps involved in calculating `controls_filled` from `controls_raw` and `df_controls` from `controls_filled` are explored in the exercises below.

As discussed above, FHK estimate discretionary-accrual models by industry and year, where industries are based on the Fama-French 48-industry grouping. To create these industries here, we use `get_ff_ind()`, introduced in Chapter 9 and provided by the `farr` package.

```
ff_data <- get_ff_ind(48)
```

We now create functions to compile the data we need to estimate performance-matched discretionary accruals. We use a function to compile the data because (i) doing so is easy in R and (ii) it allows us to re-use code much easily, which will be important for completing the exercises in this chapter.

The first function we create is `get_das()`, which takes as its first argument (`compustat`) a data set derived from Compustat with the requisite data. The second argument (`drop_extreme`) allows to easily tweak the handling of outliers in a way examined in the exercises.

Within `get_das()`, the first data set we compile is `for_disc_accruals`, which contains the raw data for estimating discretionary-accruals models. Following FHK, we require each industry-year to have at least 10 observations for inclusion in our analysis and impose additional data requirements, some of which we explore in the exercises below.

Following FHK, we estimate discretionary-accrual models by industry and year and store the results in the data frame `fm_da`. We then merge the underlying data (`for_disc_accruals`) with `fm_da` to use the estimated models to calculate non-discretionary accruals (`nda`). Because the coefficient on `sale_c_at` is applied to `salerect_c_at`, we cannot use `predict()` or `residuals()` in a straightforward fashion and need calculate `nda` "by hand". We then calculate discretionary accruals (`da = acc_at - nda`) and return the data.

```
get_das <- function(compustat, drop_extreme = TRUE) {

  for_disc_accruals <-
    compustat |>
    inner_join(ff_data,
               join_by(between(sich, sic_min, sic_max))) |>
    group_by(gvkey, fyr) |>
    arrange(fyear) |>
    filter(lag(at) > 0) |>
    mutate(lag_fyear = lag(fyear),
           acc_at = (ibc - (oancf - xidoc)) / lag(at),
           one_at = 1 / lag(at),
           ppe_at = ppegt / lag(at),
           sale_c_at = (sale - lag(sale)) / lag(at),
           salerect_c_at = ((sale - lag(sale)) -
                                (rect - lag(rect))) / lag(at)) |>
    ungroup() |>
    mutate(keep = case_when(drop_extreme ~ abs(acc_at) <= 1,
                            .default = TRUE)) |>
    filter(lag_fyear == fyear - 1,
           keep,
           !is.na(salerect_c_at), !is.na(acc_at), !is.na(ppe_at)) |>
    group_by(ff_ind, fyear) |>
    mutate(num_obs = n(), .groups = "drop") |>
    filter(num_obs >= 10) |>
    ungroup()

  fm_da <-
```

```
      for_disc_accruals |>
      group_by(ff_ind, fyear) |>
      do(model = tidy(lm(acc_at ~ one_at + sale_c_at + ppe_at, data = .))) |>
      unnest(model) |>
      select(ff_ind, fyear, term, estimate) |>
      pivot_wider(names_from = "term", values_from = "estimate",
                  names_prefix = "b_")

  for_disc_accruals |>
    left_join(fm_da, by = c("ff_ind", "fyear")) |>
    mutate(nda = `b_(Intercept)` + one_at * b_one_at + ppe_at * b_ppe_at +
                 salerect_c_at * b_sale_c_at,
           da = acc_at - nda) |>
    select(gvkey, fyear, ff_ind, acc_at, da)
}
```

The next step in the data preparation process is to match each firm with another based on performance. Following FHK, we calculate performance as lagged "Income Before Extraordinary Items" (`ib`) divided by lagged "Total Assets" (`at`) and `perf_diff`, the absolute difference between performance for each firm-year and each other firm-year in the same industry. We then select the firm (`gvkey_other`) with the smallest value of `perf_diff`. We rename the variable containing the discretionary accruals of the matching firm as `da_other` and calculate performance-matched discretionary accruals (`da_adj`) as the difference between discretionary accruals for the target firm (`da`) and discretionary accruals for the matched firm (`da_other`), and return the results. Note that `get_pm()` includes the argument `pm_lag` with default value `TRUE`. If `pm_lag` is set to `FALSE`, then performance for matching is calculated using contemporary values of `ib` and `at` (this option is examined in the exercises below).

```
get_pm <- function(compustat, das, pm_lag = TRUE, drop_extreme = TRUE) {

  das <- get_das(compustat, drop_extreme = drop_extreme)

  perf <-
    compustat |>
    group_by(gvkey) |>
    arrange(fyear) |>
    mutate(ib_at =
      case_when(pm_lag ~ if_else(lag(at) > 0, lag(ib) / lag(at), NA),
                .default = if_else(at > 0, ib / at, NA))) |>
    ungroup() |>
    inner_join(das, by = c("gvkey", "fyear")) |>
    select(gvkey, fyear, ff_ind, ib_at)

  perf_match <-
    perf |>
    inner_join(perf, by = c("fyear", "ff_ind"),
               suffix = c("", "_other")) |>
    filter(gvkey != gvkey_other) |>
    mutate(perf_diff = abs(ib_at - ib_at_other)) |>
    group_by(gvkey, fyear) |>
```

```
    filter(perf_diff == min(perf_diff)) |>
    select(gvkey, fyear, gvkey_other)

  perf_matched_accruals <-
    das |>
    rename(gvkey_other = gvkey,
           da_other = da) |>
    select(fyear, gvkey_other, da_other) |>
    inner_join(perf_match, by = c("fyear", "gvkey_other")) |>
    select(gvkey, fyear, gvkey_other, da_other)

  das |>
    inner_join(perf_matched_accruals, by = c("gvkey", "fyear")) |>
    mutate(da_adj = da - da_other) |>
    select(gvkey, fyear, acc_at, da, da_adj, da_other, gvkey_other)
}
```

The final step is performed in `get_pmdas()`. This function gets the needed data using `get_pm()`, then filters duplicate observations based on (`gvkey`, `fyear`) (the rationale for this step is explored in the discussion questions).

```
get_pmdas <- function(compustat, pm_lag = TRUE, drop_extreme = TRUE) {

  get_pm(compustat,
         pm_lag = pm_lag,
         drop_extreme = drop_extreme) |>
    group_by(gvkey, fyear) |>
    filter(row_number() == 1) |>
    ungroup()
}
```

Finally, we simply pass the data set `compustat_annual` to `get_pmdas()` and store the result in `pmdas`.

```
pmdas <- get_pmdas(compustat_annual)
```

The remaining data set used by FHK is `sho_data`, which we discussed in Section 19.2.1.

```
sho_data <-
  fhk_pilot |>
  select(gvkey, pilot) |>
  distinct() |>
  group_by(gvkey) |>
  filter(n() == 1) |>
  ungroup() |>
  inner_join(fhk_pilot, by = c("gvkey", "pilot"))
```

The final sample `sho_accruals` is created in the following code and involves a number of steps. We first merge data from FHK's `sho_data` with `fhk_firm_years` to produce the sample firm-years and treatment indicator for FHK. As `fhk_firm_years` can contain multiple years for each firm, so we expect each row in `sho_data` to match multiple rows in `fhk_firm_years`. At the same time, some `gvkey` values link with multiple PERMNOs, so some rows in `fhk_firm_years` will match multiple rows in `sho_data`. As such, we set relationship = "many-to-many" in this join below. We then merge the resulting data set with df_controls, which contains data on controls. The final data merge brings in data on performance-matched discretionary accruals from pm_disc_accruals_sorted. Finally, following FHK, we winsorize certain variables using the winsorize() function from the farr package to do this here.[26]

```
win_vars <- c("at", "mtob", "leverage", "roa", "da_adj", "acc_at")

sho_accruals <-
  sho_data |>
  inner_join(fhk_firm_years,
             by = "gvkey",
             relationship = "many-to-many") |>
  select(gvkey, datadate, pilot) |>
  mutate(fyear = year(datadate) - (month(datadate) <= 5)) |>
  left_join(df_controls, by = c("gvkey", "fyear")) |>
  left_join(pmdas, by = c("gvkey", "fyear")) |>
  group_by(fyear) |>
  mutate(across(all_of(win_vars),
                \(x) winsorize(x, prob = 0.01))) |>
  ungroup()
```

19.6.3 Discussion questions and exercises

1. What would be the effect of replacing the code that creates ff_data above with the following code? What changes would we need to make to the code creating for_disc_accruals in get_das() to use this modified version of ff_data?

```
ff_data <-
  get_ff_ind(48) |>
  rowwise() |>
  mutate(sich = list(seq(from = sic_min, to = sic_max))) |>
  unnest(sich)
```

2. What issue is filter(row_number() == 1) addressing in the code above? What assumptions are implicit in this approach? Do these assumptions hold in this case? What would be an alternative approach to address the issue?

3. Why is filter(fyear == lag(fyear) + 1) required in the creation of controls_raw?

4. Does the argument for using salerect_c_at * b_sale_c_at in creating non-discretionary accruals make sense to you? How do Kothari et al. (2005) explain this?

5. Does the code above ensure that a performance-matched control firm is used as a control just once? If so, which aspect of the code ensures this is true? If not, how might you ensure this and does this cause problems? (Just describe the approach in general; no need to do this.)

---

[26]We discuss winsorization and the winsorize() function in Chapter 24.

6. What are FHK doing in the creation of `controls_filled`? (Hint: The key "verb" is `fill()`.) Does this seem appropriate? Does doing this make a difference?

7. What are FHK doing in the creation of `df_controls` from `controls_fyear`? Does this seem appropriate? Does doing this make a difference?

19.6.4 FHK: Regression analysis

FHK consider a number of regression specifications including: with and without controls, with and without firm fixed effects, and with standard errors clustered by firm alone and clustered by firm and year. We make a small function (`reg_year_fe()`) that calculates variables used in the regression (like `during` and `post`) and allows us to specify each of these different options, to change the dependent variable from the default (`dv = "da_adj"`) and to supply a different data set. This function returns a fitted model that is estimated using `feols()` from the `fixest` package.

```
ctrls_list <- c("log(at)", "mtob", "roa", "leverage")

reg_year_fe <- function(df, dv = "da_adj",
                        controls = TRUE, firm_fe = FALSE, cl_2 = TRUE,
                        vcov = NULL) {
  df <-
    df |>
    mutate(year = year(datadate),
           during = year %in% c(2005, 2006, 2007),
           post = year %in% c(2008, 2009, 2010))

  model <- str_c(dv, " ~ pilot * (during + post) ",
                 if_else(controls,
                         str_c(" + ", str_c(ctrls_list,
                                            collapse = " + ")), ""),
                 if_else(firm_fe, "| gvkey + year ", "| year "))
  if (is.null(vcov)) {
    vcov = as.formula(if_else(!cl_2, "~ gvkey ", "~ year + gvkey"))
  }

  feols(as.formula(model),
        vcov = vcov,
        notes = FALSE,
        data = df)
}
```

To facilitate the output of variations, we next make a function that runs regressions with and without controls and with and without firm fixed effects and returns a nicely formatted regression table.

```
make_reg_table <- function(df, dv = "da_adj", cl_2 = TRUE) {
  omit <- str_c("^(", str_c(str_replace_all(c("during", "post", ctrls_list),
                                            "[()]", "."),
                            collapse="|"), ")")

  run_reg <- function(controls, firm_fe) {
```

```
    reg_year_fe(df, dv = dv, controls = controls, firm_fe = firm_fe,
            cl_2 = cl_2)
}

params <- tibble(controls = c(FALSE, TRUE, FALSE, TRUE),
                firm_fe = c(FALSE, FALSE, TRUE, TRUE))

fms <- pmap(params, run_reg)

notes <- tribble(~term,   ~`1`,   ~`2`,   ~`3`,   ~`4`,
                "Firm FEs", "No", "No", "Yes", "Yes",
                "Controls", "No", "Yes", "No", "Yes")

modelsummary(fms,
            estimate = "{estimate}{stars}",
            gof_map = "nobs",
            stars = c('*' = .1, '**' = 0.05, '***' = .01),
            coef_omit = str_c(str_replace_all(ctrls_list, "[()]", "."),
                            collapse = "|"),
            add_rows = notes)
}
```

We now use this function with our version of FHK's data set (`sho_accruals`) to create the regression results reported in Table 19.3.

```
make_reg_table(sho_accruals)
```

TABLE 19.3
Baseline replication of FHK

	(1)	(2)	(3)	(4)
pilotTRUE	0.000	0.000		
	(0.004)	(0.004)		
pilotTRUE × duringTRUE	−0.010**	−0.010**	−0.014**	−0.014**
	(0.004)	(0.004)	(0.005)	(0.005)
pilotTRUE × postTRUE	0.009**	0.008**	0.005	0.005
	(0.004)	(0.004)	(0.004)	(0.004)
Num.Obs.	19,616	19,616	19,616	19,616
Firm FEs	No	No	Yes	Yes
Controls	No	Yes	No	Yes

We next create a function that allows us to plot by-year coefficients for the treatment and control firms. (We leave the details of what this function is doing as an exercise for the reader below.)

```
plot_coefficients <- function(model) {
  tibble(name = names(model$coefficients),
        value = as.vector(model$coefficients)) |>
  filter(str_detect(name, "^year.")) |>
  separate(name, into = c("year", "pilot"), sep = ":", fill = "right") |>
```

```
   mutate(year = as.integer(str_replace(year, "^year", "")),
          pilot = coalesce(pilot == "pilotTRUE", FALSE)) |>
   ggplot(aes(x = year, y = value,
              linetype = pilot, color = pilot)) +
   geom_line() +
   scale_x_continuous(breaks = 2000:2012L) +
   geom_rect(xmin = 2005, xmax = 2007, ymin = -Inf, ymax = Inf,
             color = NA, alpha = 0.01) +
   theme_bw()
}
```

To produce Figure 19.3, we estimate one of the models above by year and feed the fitted model to `plot_coefficients()`.

```
sho_accruals |>
  mutate(year = as.factor(year(datadate))) |>
  feols(da_adj ~ year * pilot - pilot - 1 + log(at) + mtob + roa + leverage,
        vcov = ~ year + gvkey, data = _) |>
  plot_coefficients()
```

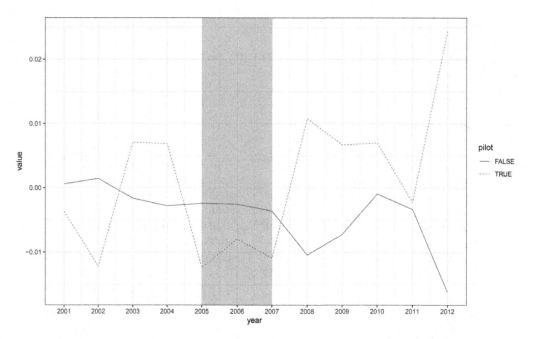

FIGURE 19.3
Baseline by-year replication of FHK

19.6.5 Exercises

1. In words, how does `sho_accruals_alt` (defined below) differ from `sho_accruals`? Does using `sho_accruals_alt` in place of `sho_accruals` affect the regression results?

```
firm_years <-
  controls_raw |>
  select(gvkey, datadate, fyear)

sho_accruals_alt <-
  sho_r3000_gvkeys |>
  inner_join(firm_years, by = "gvkey") |>
  left_join(df_controls, by = c("gvkey", "fyear")) |>
  left_join(pmdas, by = c("gvkey", "fyear")) |>
  group_by(fyear) |>
  mutate(across(all_of(win_vars), winsorize, prob = 0.01)) |>
  ungroup()
```

2. In an online appendix, BDLYY say "FHK winsorize covariates for their covariate balance table at 1/99%. We inferred that they also winsorized accruals at this level. Whether they winsorize across sample years or within each year, they do not specify." The code above winsorized within each year. How would you modify the code to winsorize "across sample years"? Does doing so make a difference?

3. How would you modify the code to winsorize at the 2%/98% level? Does this make a difference to the results? (*Hint*: With the `farr` package loaded, type `?winsorize` in the R console to get help on this function.)

4. How would you modify the code to not winsorize at all? Does this make a difference to the results?

5. Some of the studies discussed by BDLYY exclude 2004 data from the sample. How would you modify the code above to do this here? Does excluding 2004 here make a significant difference?

6. What is the range of values for `year` in `sho_accruals`? Does this suggest any issues with the code `post = year %in% c(2008, 2009, 2010)` above? If so, does fixing any issue have an impact on the results reported above?

7. Would it make sense, in creating `perf` above, if we instead calculated `ib_at` as `if_else(at > 0, ib / at, NA))`? What is the effect on the regression results if we use this modified calculation of `ib_at`? What do Kothari et al. (2005) recommend on this point? (*Hint*: Use `pm_lag = FALSE` where applicable.)

8. Fang et al. (2019, p. 10) follow Fang et al. (2016), who "exclude observations for which the absolute value of total accruals-to-total assets exceeds one. This is a standard practice in the accounting literature because firms with such high total accruals-to-total assets are often viewed as extreme outliers. Nonetheless, the FHK results are robust to winsorizing the accrual measures at the 1% and 99% levels instead of excluding extreme outliers." Does this claim hold up in the reproduction above? What happens if the `filter` on `abs(acc_at) <= 1` is removed from the code above? (*Hint*: Use `drop_extreme = FALSE` where applicable.)

9. Explain what each line of the function `plot_coefficients()` before the line starting with `ggplot()` is doing. (*Hint*: It may be helpful to store the model that is fed to the function above in the variable `model` and then run the function line by line.)

19.7 Statistical inference

One point of difference between FHK and BDLYY concerns clustered standard errors. Fang et al. (2016) generally use "standard errors clustered by year and firm" (2016, p. 1269), while Black et al. (2019) advocate the use of standard errors clustered by firm. Citing Cameron et al. (2008), Black et al. (2019, p. 30) suggest that "clustered standard errors with a small number of clusters can be downward biased." In the context of FHK, there are thousands of firms, but a relatively small number of years, so clustering by year (or firm *and* year) may result in problematic standard error estimates (see Section 5.6.6).

One approach to determining the appropriate clustering is more empirical. In this regard, it is useful to note that cluster-robust standard errors are a generalization of an idea from White (1980). White (1980) provides not only an estimator of standard errors that is robust to heteroskedasticity, but also a test of a null hypothesis of homoskedasticity. Intuitively, if the covariance matrix assuming heteroskedasticity is sufficiently different from that assuming homoskedasticity, then we may reject the null hypothesis of homoskedasticity. With a little algebra, it would be possible to develop a test analogous to that of White (1980) of the null hypothesis of no clustering on variable g. In practice, many researchers will, lacking a formally derived test, compare standard errors with and without clustering on variable g and elect to cluster on variable g when the standard errors when doing so seem significantly higher than when not doing so. This heuristic breaks down in the case of Fang et al. (2016) because standard errors are generally lower when clustering on firm and year than when clustering firm alone. However, if clustering on firm alone is appropriate, standard errors clustering on firm and year will provide noisier estimates than clustering on firm alone, and thus could be lower or higher in any given data set.

A more theoretical approach can be used in the setting of FHK because of our deeper understanding of the assignment mechanism. In this regard, it is important to note that cluster-robust standard errors address correlation in *both X and ϵ* across units within clusters. To explore this (slightly) more formally, recall from Chapter 5 that the cluster-robust covariance matrix is estimated using the following expression:

$$\hat{V}(\hat{\beta}) = (X'X)^{-1}\hat{B}(X'X)^{-1}, \text{ where } \hat{B} = \sum_{g=1}^{G} X_g' u_g u_g' X_g$$

where the observations grouped into G clusters of N_g observations for g in $1, \ldots, G$, X_g is the $N_g \times K$ matrix of regressors, and u_g is the N_g-vector of residuals for cluster g.

If we have a single regressor, demeaned x with no constant term and two firms (i and j) in a cluster, then the contribution of that cluster to $\hat{B}$ will be

$$\begin{bmatrix} x_i & x_j \end{bmatrix} \begin{bmatrix} u_i \\ u_j \end{bmatrix} \begin{bmatrix} u_i & u_j \end{bmatrix} \begin{bmatrix} x_i \\ x_j \end{bmatrix} = \begin{bmatrix} x_i & x_j \end{bmatrix} \begin{bmatrix} u_i^2 & u_i u_j \\ u_i u_j & u_j^2 \end{bmatrix} \begin{bmatrix} x_i \\ x_j \end{bmatrix}$$

$$= \begin{bmatrix} x_i & x_j \end{bmatrix} \begin{bmatrix} x_i u_i^2 + x_j u_i u_j \\ x_i u_i u_j + x_j u_j^2 \end{bmatrix}$$

$$= \begin{bmatrix} x_i^2 u_i^2 + x_i x_j u_i u_j \\ x_i x_j u_i u_j + x_j^2 u_j^2 \end{bmatrix}$$

Now, if x_i and x_j are uncorrelated then, even if ϵ_i and ϵ_j are correlated, this resolves in expectation to

$$\begin{bmatrix} x_i^2 \sigma_i^2 \\ x_j^2 \sigma_j^2 \end{bmatrix}$$

which is the expectation of the analogous component of the heteroskedasticity-robust estimator from White (1980). In the setting of Fang et al. (2016), the "x" of primary interest is the Reg SHO pilot indicator, which is assumed to be randomly assigned, and thus (in expectation) uncorrelated across firms. For this reason, we do not expect cross-sectional dependence to affect standard error estimates on average. On the other hand, the Reg SHO pilot indicator is perfectly correlated over time within firm, so any serial dependence in errors within firm over time will lead to effects of time-series dependence on standard error estimates. This (somewhat loose) theoretical analysis suggests we should cluster by firm (time-series dependence), but not by year (cross-sectional dependence), as suggested by Black et al. (2019, p. 12).

However, the assumed random assignment of treatment allows us to adopt an alternative approach to statistical inference that is agnostic to the form of clustering in the data. This approach is known as **randomization inference** and builds on the **Fisher sharp null** hypothesis of no effect of any kind. This is a "sharp null" because it is more restrictive that a null hypothesis of zero mean effect, which could be true even if half the observations had a treatment effect of $+1$ and half the observations had a treatment effect of -1, in which case the Fisher sharp null would not be true even though null hypothesis of zero mean effect is true.

Under the Fisher sharp null hypothesis and with random assignment to treatment, in principle we can evaluate the distribution of any given test statistic by considering all possible assignments. Focusing on the 2954 firms that the SEC focused on as its initial sample, if assignment to treatment were purely random, then any other assignment of treatment to 985 was as likely as the one chosen. Given that the Fisher sharp null implies that there was no impact of treatment assignment on outcomes, we know what the distribution of the test statistic would have been if the SEC had chosen any one of those alternative assignments because the outcomes would have been exactly the same. With smaller samples, we might proceed to calculate the test statistic for every possible assignment and thereby construct the exact distribution of the test statistic under the Fisher sharp null.[27] But in our case, there will be a huge number of ways to choose 985 treatment firms from 2954 possibilities; so a more feasible approach is to draw a random sample of possible assignments and use the empirical distribution of the test statistic for that random sample as an approximation for the exact distribution.

```
get_coef_rand <- function(i) {
  treatment <-
    sho_accruals |>
    select(gvkey, pilot) |>
    distinct() |>
    mutate(pilot = sample(pilot, size = length(pilot), replace = FALSE))

  reg_data_alt <-
    sho_accruals |>
    select(-pilot) |>
    inner_join(treatment, by = "gvkey")
```

---

[27] See Chapter 5 of Imbens and Rubin (2015) for details on this approach.

```
  reg_data_alt |>
    reg_year_fe(controls = TRUE, firm_fe = TRUE) |>
    broom::tidy() |>
    select(term, estimate) |>
    pivot_wider(names_from = "term", values_from = "estimate") |>
    mutate(iteration = i) |>
    suppressWarnings()
}
```

The test statistic we are interested in here is the coefficient on $PILOT \times DURING$. Below we calculate the p-value of the coefficients on variables involving $PILOT$ using the empirical distribution of coefficients, and the standard errors associated with the coefficients as the standard deviation of those coefficients.

```
set.seed(2021)
rand_results <-
  1:1000 |>
  map(get_coef_rand) |>
  list_rbind() |>
  system_time()
```

```
   user  system elapsed
 36.684   2.544  39.367
```

```
plan(multisession)
```

```
rand_results <-
  1:1000 |>
  future_map(get_coef_rand,
             .options = furrr_options(seed = 2021)) |>
  list_rbind() |>
  system_time()
```

```
   user  system elapsed
  0.659   0.047   9.268
```

In the following, we run regressions with standard errors based on clustering by firm and year, by firm alone, and using randomization inference. We start by running regressions—with controls and firm fixed effects—with standard errors based on clustering by firm (`"CL-i"`) and by firm and year (`"CL-2"`).

```
fms <- list(reg_year_fe(sho_accruals, cl_2 = FALSE),
            reg_year_fe(sho_accruals, cl_2 = TRUE))
```

We extract the variance-covariance matrices for each of these two models and place them in the list `vcovs`.

```
vcovs <- list(vcov(fms[[1]]), vcov(fms[[2]]))
```

Next, we add a third model for which we will calculate standard errors using randomization inference (`"RI"`). The coefficients stored in `fms` for this third model can be taken from either of the two models already stored there.

```
fms[[3]] <- fms[[2]]
```

For the variance-covariance matrix, we use `CL-i` standard errors as the starting point. Then we replace the elements for coefficients on variables involving *PILOT* using the empirical distribution stored in `rand_results`.

```
vcov <- vcovs[[1]]
vcov["pilotTRUE:duringTRUE", "pilotTRUE:duringTRUE"] <-
  var(rand_results[["pilotTRUE:duringTRUE"]])
vcov["pilotTRUE:postTRUE", "pilotTRUE:postTRUE"] <-
  var(rand_results[["pilotTRUE:postTRUE"]])
vcovs[[3]] <- vcov
```

Results of this analysis are provided in Table 19.4.

```
se_notes <- tribble(~term,  ~`1`,   ~`2`,   ~`3`,
                "SEs", "CL-i", "CL-2", "RI")

modelsummary(fms, vcov = vcovs,
        estimate = "{estimate}{stars}",
        gof_map = "nobs",
        stars = c('*' = .1, '**' = 0.05, '***' = .01),
        coef_omit = "^(during|post|pilot)TRUE$",
        add_rows = se_notes)
```

TABLE 19.4
Results with randomization inference

	(1)	(2)	(3)
log(at)	0.001	0.001	0.001
	(0.001)	(0.001)	(0.001)
mtob	−0.001**	−0.001	−0.001*
	(0.000)	(0.000)	(0.000)
roa	−0.009	−0.009	−0.009
	(0.012)	(0.016)	(0.012)
leverage	−0.016***	−0.016**	−0.016***
	(0.004)	(0.006)	(0.004)
pilotTRUE × duringTRUE	−0.010**	−0.010**	−0.010*
	(0.005)	(0.004)	(0.005)
pilotTRUE × postTRUE	0.008	0.008**	0.008
	(0.005)	(0.004)	(0.006)
Num.Obs.	19,616	19,616	19,616
SEs	CL-i	CL-2	RI

19.7.1 Exercises

1. In the function `get_coef_rand()`, we first created the data set `treatment`, then merged this with `reg_data_alt`. Why did we do it this way rather than simply applying the line `mutate(pilot = sample(pilot, size = length(pilot), replace = FALSE))` directly to `reg_data_alt`?

2. Using randomization inference, calculate a p-value for a one-sided alternative hypothesis that $H_1 : \beta < 0$ where β is the coefficient on $PILOT \times DURING$. (*Hint*: You should not need to run the randomization again; modifying the calculation of `p_value` should suffice.)

3. What is the empirical standard error implied by the distribution of coefficients in `rand_results`? Is it closer to the two-way cluster robust standard errors obtained in estimating with `cl_2 = TRUE` or with `cl_2 = FALSE`? Why might it be preferable to calculate p-values under randomization inference using the empirical distribution of the test statistic, instead of calculating these from t-statistics based on the estimated coefficient and the empirical standard error? Would we get different p-values using the former approach?

4. Why did we not use the empirical standard error implied by the distribution of coefficients in `rand_results` to calculate standard errors for the control variables (e.g., `log(at)`)?

19.8   Causal diagrams

It is important to note that we *observe* total accruals, not discretionary accruals. Instead we need to construct *measures* of discretionary accruals. The Jones (1991) model of discretionary accruals "controls for" sales growth and PP&E and the Kothari et al. (2005) model additionally "controls for" performance.

Assuming that the causal diagram below is correct, we get unbiased estimates of causal effects whether we "control for" pre-treatment outcome values (e.g., using DiD) or not (e.g., using *POST*), and it is not clear that we need to control for other factors that drive total accruals. If being a Reg SHO pilot firm leads to a reduction in earnings management, we should observe lower *total* accruals, even if we posit that the effect is through discretionary accruals, which we do not observe directly. If we accept this causal diagram, then the decision as to which factors to control for is—like the choice between *DiD*, *POST*, and *ANCOVA*—a question of statistical efficiency rather than bias.

In this context, it is perhaps useful to consider causal diagrams to sharpen our understanding of the issues, which we explore in the discussion questions below, as matters can be more complicated if the causal diagram in Figure 19.4 is incomplete.

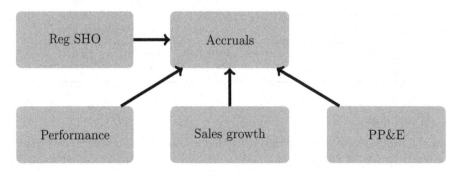

FIGURE 19.4
Effects of Reg SHO on accruals

19.8.1 Discussion questions

1. What features of Figure 19.4 imply that we do not need to control for performance, sales, and PP&E in estimating the causal effect of Reg SHO on accruals? What is the basis for assuming these features in the causal diagram?

2. Black et al. (2024) report that "over 60 papers in accounting, finance, and economics report that suspension of the price tests had wide-ranging indirect effects on pilot firms, including on earnings management, investments, leverage, acquisitions, management compensation, workplace safety, and more (see Internet Appendix, Table IA-1 for a summary)." In light of the Internet Appendix of Black et al. (2024), is there any evidence that Reg SHO might plausibly have an effect on performance, sales growth, or PP&E? If so, how would Figure 19.4 need to be modified to account for these consequences? What would be the implications of these changes on the appropriate tests for estimating the causal effects of Reg SHO on accruals?

3. Produce a regression table like Table 19.3 and a plot like Figure 19.3, but using discretionary accruals *without* performance matching instead of performance-matched discretionary accruals. How do you interpret these results?

4. Produce a regression table and a plot like the ones in the FHK replication above, but using total accruals instead of discretionary accruals and excluding controls (so the coefficients will be simple conditional sample means). How do you interpret these results?

5. Suppose you had been brought in by the SEC to design a study examining the research question examined by FHK in the form of a registered report. What analyses would you conduct to try to understand the best research design? For example, how would you choose between *DiD*, *POST*, *ANCOVA*, and other empirical approaches? What controls would you include? How would you decide how to include controls? (For example, one could control for performance by including performance as a regressor in the model of earnings management, by matching on performance, or by including performance in the main regression specification.) How would you calculate standard errors? Discuss how your proposed empirical test differs from that of FHK. Would you have reported similar results to what FHK reported?

6. Suppose that FHK's empirical analysis had produced a positive effect of Reg SHO on earnings management? Would this imply a lack of support for their hypotheses? Do you believe that publication in the *Journal of Finance* depended on finding a negative effect?

7. What implications would there have been for publication of FHK in the *Journal of Finance* if they had failed to find an effect of Reg SHO on earnings management?

19.9 Causal mechanisms

Black et al. (2024) suggest a number of possible causal channels through which the Reg SHO experiment could have affected the behavior of firms or third parties, including short interest, returns, price efficient, and "manager fear". On the last of these, Black et al. (2024, p. 4) suggest that "even if the Reg SHO experiment did not actually affect short interest or returns, pilot firm managers could have feared being targeted by short sellers and taken pre-emptive actions."

Black et al. (2024, p. 5133) argue that "if firm managers were fearful that relaxing the price tests would affect them, one might expect them to voice concerns in various ways: speaking with business news reporters; writing to the SEC when it sought public comments, or seeking meetings with SEC officials to express opposition. ... We searched the business press during 2003 when the rule was proposed, in 2004 when the experiment was announced, in 2006 when the SEC proposed repeal ... We found no evidence of manager opposition."

Black et al. (2024, p. 5134) suggest that "FHK rely on the manager fear channel. They conjecture that, in response to a greater threat of short selling, pilot firms' managers reduced earnings management to preemptively deter short sellers."

19.9.1   Discussion questions

1. Do you agree with the assertion of Black et al. (2024) that "FHK rely on the manager fear channel"? What causal mechanisms are suggested in Fang et al. (2016)? What evidence do Fang et al. (2016) offer in support of these mechanisms?

2. Evaluate the response of Fang et al. (2019) to Black et al. (2024) as it relates to causal mechanisms?

3. Do you think evidence of causal mechanisms is more or less important when using a natural experiment (i.e., an experiment outside the control of the researcher that is typically analysed after it has been run) than when conducting a randomized experiment? Explain your reasoning given the various issues raised in this chapter.

19.10   Two-step regressions

Chen et al. (2018) examine the question of statistical inference when residuals from one regression are used as a dependent variable in a subsequent regression, which they refer to as "the two-step procedure". For example, discretionary accruals measured using the Jones (1991) model are residuals from a regression of total accruals on changes in sales and PP&E. As we saw in Dechow et al. (1995), which we covered in Chapter 16, many papers examine how Jones (1991) model discretionary accruals relate to various posited incentives for earnings management.

Chen et al. (2018, p. 755) show that "the two-step procedure is likely to generate biased coefficients and t-statistics in many studies" and, drawing on the Frisch-Waugh-Lovell theorem (see Section 3.3), propose using a single regression in place of the two-step procedure. In the case of the Jones (1991) model, this would entail including the regressors from the first step in same regression as the second step and using total accruals in place of discretionary accruals as the dependent variable.

19.10.1 Discussion questions

1. What challenges would exist in implementing the single-regression recommendation of Chen et al. (2018) for a researcher using Kothari et al. (2005) performance-matched discretionary accruals?

2. Do you believe the issues raised by Chen et al. (2018) with regard to two-step procedures also apply if using randomization inference? Why or why not?

20

Instrumental variables

Angrist and Pischke (2008, p. 114) describe **instrumental variables** (IVs) as "the most powerful weapon in the arsenal" of empirical researchers. Accounting researchers have long used instrumental variables in an effort to address concerns about endogeneity (Larcker and Rusticus, 2010; Lennox et al., 2012) and continue to do so.

Unfortunately, faulty reasoning about instruments is common in accounting research. Larcker and Rusticus (2010) lament that "some researchers consider the choice of instrumental variables to be a purely statistical exercise with little real economic foundation" and call for "accounting researchers ... to be much more rigorous in selecting and justifying their instrumental variables." And a review of instrumental variable applications conducted by Gow et al. (2016) suggests that accounting researchers have paid little heed to the suggestions and warnings of Larcker and Rusticus (2010), Lennox et al. (2012), and Roberts and Whited (2013).

As we shall demonstrate in this chapter, the requirements for valid instruments are demanding and the supply of plausibly valid IVs in accounting research to date is arguably zero. As such, the use of IV should be rare, and readers, editors, and reviewers should sceptically apply the requirements discussed below to any proposed IV.

> 💡 Tip
>
> The code in this chapter uses the packages listed below. For instructions on how to set up your computer to use the code found in this book, see Section 1.2. Quarto templates for the exercises below are available on GitHub.[a]
>
> ---
> [a]https://github.com/iangow/far_templates/blob/main/README.md

```
library(dplyr)
library(modelsummary)
library(fixest)
library(purrr)          # For map()
library(MASS)           # For mvrnorm()
```

20.1   The canonical causal diagram

Gow et al. (2016) argue that evaluating instrumental variables requires careful theoretical *causal*, rather than statistical, reasoning. Statistical reasoning is what you will find in most econometrics textbooks, which discuss requirements regarding **probability limits** (e.g., plim $Z^{\mathsf{T}} X \neq 0$ and plim $Z^{\mathsf{T}} \epsilon = 0$). In practice, you are more likely to see researchers

DOI: 10.1201/9781003456230-20

invoking ideas such as the "only through" criterion, which are better understood as causal ideas.

Causal diagrams provide a useful framework for highlighting the requirements for a valid instrument and thus (hopefully) avoiding incorrect reasoning. The causal diagram in Figure 20.1 can be viewed as the "canonical" causal diagram for instrumental variables.

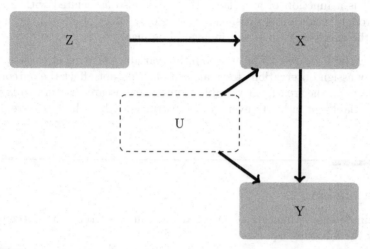

FIGURE 20.1
Canonical causal diagram for IV

Figure 20.1 is similar to the "Z is a confounder" figure we saw in Chapter 4, but for two differences. First, the confounding variable U is not observed. (What was labelled Z in Section 4.2 is now labelled U because of the convention of using Z for instrumental variables, which we follow here.) Our reasoning in Chapter 4 suggested that if we don't condition on U, the coefficients obtained from a regression of Y on X will be biased estimates of the causal effect of Y on X. But with U unobserved, we simply cannot use a strategy of conditioning on U to produce valid causal inferences.

The second difference is that we have a new variable Z, which is our instrumental variable. It turns out that a variable with the attributes that Z has in the causal diagram will (subject to certain additional technical requirements) allow us to get consistent estimates of the causal effect of X on Y.

Figure 20.1 implies three critical features of Z. First, Z is random. This is implied by the absence of any arrows into Z. Second, Z has a causal effect on X, which is indicated by the arrow from Z to X. (Later we will see that the strength of the causal relation between Z and X is important for reliable inferences.) Third, there is no direct causal effect of Z on any variable *other than* X. This is the "only through" criterion we mentioned earlier: the effect (if any) of Z on Y goes "only through" X.

Consistent with the causal perspective we take here, Angrist and Pischke (2008, p. 117) argue that "good instruments come from a combination of institutional knowledge and ideas about the process determining the variable of interest." For example, Angrist (1990) studied a draft lottery used to determinate eligibility for being drafted into military service to estimate effects of such service on long-term labour market outcomes. In the setting of Angrist (1990), it was well understood that the lottery was random and the mapping from the lottery to draft eligibility was well understood. Furthermore, there are good reasons to believe that the draft lottery did not affect anything directly except for draft eligibility.[1]

It is perhaps worth reiterating that arguing that the only effect of an instrument on the outcome variable of interest is via the treatment of interest does not suffice to establish a valid instrument. Even if the claim that Z only affects Y via its effect on X is true, the researcher also needs to argue that variation in the instrument (Z) is as-if random. For example, if Z is a function of a variable W that is also associated with Y, then even if the "only through" criterion is satisfied (i.e., the only effect of Z on Y occurs via X), IV estimates of the effect of X on Y will in general be biased.[2]

Unfortunately, there are few (if any) accounting variables that meet the requirement that they randomly assign observations to treatments *and* do not affect the outcome of interest outside of effects on the treatment variable. Sometimes researchers turn to lagged values of endogenous variables or industry averages as instruments, but these too are problematic.[3]

20.2 Estimation

We now examine the estimation of IV in detail. Suppose that the outcome variable is generated by the following equation:

$$y = X\beta + \epsilon$$

where X is an $n \times k$ matrix and β is a k-element vector of coefficients.

The goal of a researcher is to estimate β, which provides the causal effect of X on y. The OLS estimator is

$$\hat{\beta}_{\text{OLS}} = (X^{\mathsf{T}}X)^{-1}X^{\mathsf{T}}y = (X^{\mathsf{T}}X)^{-1}X^{\mathsf{T}}(X\beta + \epsilon) = \beta + (X^{\mathsf{T}}X)^{-1}X^{\mathsf{T}}\epsilon$$

When X and the other unmeasured, causal variables collapsed into the ϵ term are correlated, the OLS estimator is generally biased and inconsistent for β.

Note that the **predicted values** from an OLS regression are calculated as

$$\hat{y} = X\hat{\beta}_{\text{OLS}} = X(X^{\mathsf{T}}X)^{-1}X^{\mathsf{T}}y = P_X y$$

Where P_X is defined as

$$P_X = X(X^{\mathsf{T}}X)^{-1}X^{\mathsf{T}}$$

Thus we can express the predicted values as

$$\hat{y} = X\hat{\beta} = P_X y$$

Here P_X is an example of a **projection matrix**.[4] For a matrix W and a vector v, the multiple of the projection matrix P_W and v *projects* v into the **linear subspace** defined by

---

[1]Though some have questioned the exclusion restriction even in this case, arguing that the outcome of the draft lottery may have caused some, for example, to move to Canada (see Imbens and Rubin, 2015).

[2]In the case of Angrist (1990), the requirement that Z be random was plausibly satisfied by the use of a lottery.

[3]See Reiss and Wolak (2007) for a discussion regarding the implausibility of general claims that industry averages are valid instruments.

[4]See Appendix A for more on projection matrices.

W. If W is an $n \times k$ matrix, then P_W will be an $n \times n$ matrix that is *symmetric* (meaning that $P_X^\mathsf{T} = P_X$) and *idempotent* (meaning that $P_W \times P_W = P_W$).[5]

Suppose we have a set of r variables Z comprising both exogenous elements of X and instrumental variables where $r \geq k$.

Definition 20.1 (Convergence in probability). A sequence of random variables $\{x_n : n = 1, 2, \dots\}$ *converges in probability* to a constant a if for all $\epsilon > 0$,

$$\mathbb{P}\left(|x_n - a| > \epsilon\right) \to 0 \text{ as } n \to \infty$$

In such a case, we say that a is the *probability limit* (or *plim*) of x_n and write $\operatorname{plim} x_n = a$.

Definition 20.2 (Consistency). Given a model $y = X\beta + \epsilon$, an estimator $\hat{\beta}$ is **consistent** for β if

$$\operatorname{plim} \hat{\beta} = \beta$$

Asymptotic properties of estimators are often sufficiently mathematically tractable that we can often demonstrate that estimators have desirable properties—such as consistency—even when it is difficult to evaluate their finite-sample properties. The conceptual leap we make as applied researchers is that an estimator with desired asymptotic properties will perform well in real-world samples that are sufficiently large, albeit finite.

We can use Z to calculate the **two-stage least squares** (2SLS) estimator as

$$\hat{\beta}_{2\mathrm{SLS}} = (X^\mathsf{T} P_Z X)^{-1} X^\mathsf{T} P_Z y,$$

Denoting $\hat{X}$ as $P_Z X$, we could regress y on $\widehat{X}$ using OLS to get the 2SLS estimator.

$$
\begin{aligned}
\hat{\beta} &= (\hat{X}^\mathsf{T} \hat{X})^{-1} \hat{X}^\mathsf{T} y \\
&= \left((P_Z X)^\mathsf{T} P_Z X\right)^{-1} (P_Z X)^\mathsf{T} y \\
&= \left(X^\mathsf{T} P_Z^\mathsf{T} P_Z X\right)^{-1} (P_Z X)^\mathsf{T} y \\
&= \left(X^\mathsf{T} P_Z X\right)^{-1} X^\mathsf{T} P_Z y
\end{aligned}
$$

(Note that we used the idempotency and symmetry of P_Z in this analysis.)

We can rewrite the underlying model as

$$y = \hat{X}\beta + (X - \hat{X})\beta + \epsilon$$

We can think of $\hat{\beta}_{2\mathrm{SLS}}$ as the estimator from this model (i.e., from regressing y on $\widehat{X}$) with an error term equal to $(X - \hat{X})\beta + \epsilon$.

Here $\hat{X}$ and $(X - \hat{X})$ are, respectively, the fitted values and residuals from an OLS regression of X on Z, so $\hat{X}^\mathsf{T}(X - \hat{X}) = 0$ and $\operatorname{plim} N^{-1}\hat{X}^\mathsf{T}(X - \hat{X})\beta = 0$.

Given that

$$N^{-1}\hat{X}^\mathsf{T}\epsilon = N^{-1}X^\mathsf{T} P_Z \epsilon = N^{-1}X^\mathsf{T} Z(N^{-1}Z^\mathsf{T} Z)^{-1} N^{-1} Z^\mathsf{T}\epsilon$$

we have $\operatorname{plim} N^{-1}\hat{X}^\mathsf{T}\epsilon = 0$ so long as $\operatorname{plim} N^{-1}Z^\mathsf{T}\epsilon = 0$, which is implied by Figure 20.1. This shows that $\hat{\beta}_{2\mathrm{SLS}}$ is a consistent estimator for β. This result depends on the linearity of the model and does not generalize to non-linear models.[6]

---

[5]See Appendix A for more on symmetric and idempotent matrices.

[6]The material in this subsection is based on material on p.189 of Cameron and Trivedi (2005).

A careful reader might note that we have not used the causal diagram from the previous subsection in this analysis. A heuristic analysis of the 2SLS estimator given the canonical causal diagram can be found in Cunningham (2021, pp. 323–329).

20.3   Reasoning about instruments

Larcker and Rusticus (2010, p. 189) find that in a typical accounting research paper, "there is almost no discussion regarding the choice of specific variables for instruments (only slightly more than half of the papers discuss the instruments at all). Researchers do not rigorously discuss why the variables selected as instruments are assumed to be exogenous ... nearly 80% of the papers provide no justification for the choice of instrument whatsoever."

But, even when researchers do attempt to evaluate instruments, confusion and loose reasoning seem common. For example, in the rare cases where instruments are justified, many appear to focus on what Roberts and Whited (2013, p. 514) describe as "the question one should always ask of a potential instrument [namely] does the instrument affect the outcome only via its effect on the endogenous regressor?" But a focus on this question cannot come at the expense of discussion of the requirement that the instrument be (as-if) random.

The discussion of instrument validity of Roberts and Whited (2013, p. 514) actually provides a good case study for accounting researchers, as they are writing to enhance reasoning about endogeneity in finance, a field closely related to accounting research.

The setting explored by Roberts and Whited (2013) is that of Bennedsen et al. (2007), who "find that family successions have a large negative causal impact on firm performance" [p. 647]. While Roberts and Whited (2013) proceed with fairly loose verbal reasoning, we think it is helpful to recapitulate their analysis using a series of questions implied by the requirements for a valid instrument.

One baseline requirement for this analysis is that we need to have a clear idea of what the instrument is, what the treatment variable is, and what the outcome is.

The quoted sentence from Bennedsen et al. (2007) identifies the treatment (family succession) and the outcome of interest (firm performance). The instrument used by Bennedsen et al. (2007) is the gender of the first-born child of a departing CEO.

A second critical element of this analysis is that we use causal reasoning rather than statistical reasoning. While an underlying causal model will imply statistical relations, it is problematic to substitute statistical analysis of observed data for such causal reasoning.

Here are the questions:

1. **Is the instrument (as-if) random?** The question is addressed quite directly by Bennedsen et al. (2007). Our knowledge of human biology tells us that the sex of a child is essentially random at time of conception. And, as "over 80 percent of first-child births [in the sample] occurred prior to 1980, before current techniques to identify the gender of children were widespread", there is little reason to doubt that the sex of live births isn't similarly random.

2. **Does the instrument have a direct causal effect on the treatment variable?** Here Bennedsen et al. (2007) rely on statistical evidence: "We show that the gender of the first-born child of a departing CEO is strongly correlated with

the decision to appoint a family CEO: The frequency of family transitions is 29.4% when the first-born child is female and increases to 39% (a 32.7% increase) when the first-born is male." But it is important that this link be causal. A merely statistical association between Z and X is problematic if we do not have a clear causal explanation. If X and Z are associated for unexplained reasons, it is difficult to rule out the possibility that those unexplained reasons include variables that drive both X and Z or effects of Z on variables other than X that might be associated with Y. Either of these alternatives would undermine the validity of Z as an instrument and thus need to be ruled out.

What Bennedsen et al. (2007) appear to need here is that, for a proportion of the firms in the sample, family succession is *caused* by the first-born child being a son rather than a daughter. The paper mentions "primogeniture" a couple of times and perhaps we can assume that this mechanism exists based on general knowledge of Danish customs, but more discussion beyond the statistics might have been helpful for readers with little knowledge of Danish family dynamics beyond what is depicted in *Hamlet*.

3. **Does the instrument affect the outcome only through its effect on the treatment variable?** Again Bennedsen et al. (2007) rely on statistical evidence: "We find that firms' profitability, age, and size do not differ statistically as a function of the gender of the first child. Moreover, the family characteristics of departing CEOs are comparable." But it is not possible to test the "only through" assumption statistically, instead we need "compelling arguments relying on economic theory and a deep understanding of the relevant institutional details" (Roberts and Whited, 2013, p. 515).

In this regard, the requirements are steep and Bennedsen et al. (2007) arguably fall short. One reason to doubt the "only through" requirement is the sheer amount of time between assignment of the instrument and the assignment of treatment (family succession) and the presumed preference for first-born sons to take over. There might be 40 or 50 or more years between the birth of the first child and CEO succession. The "only through" requirements means that the sex of the first child does not affect anything about the firm other than family succession. This means that nothing about the management of the firm prior to CEO succession is affected by the increased likelihood that the firm will continue to be managed by the family after the CEO leaves. This rules out management assignment decisions that are intended to groom a family member for a CEO role, as these could affect performance through channels other than the mere fact of family succession. It also rules out decisions about the structure or performance of the firm (e.g., which portions of the firm are retained and how much effort is exerted in running it) prior to succession.

In summary, Roberts and Whited (2013) explicitly identify only Question 3 as the critical one, but in a setting where it is Questions 1 and 2 that appear to have the correct answer. But focusing on Question 3 and ignoring Question 1 can be problematic in practice, as we shall see in the next subsection.

An important caveat to the somewhat negative evaluation of the reasoning in Bennedsen et al. (2007) is that just because there are holes in the arguments for an instrument's validity does not mean that one cannot make causal inferences. It just means that the causal reasoning has to be more circumspect and rule out alternative causal pathways from the (plausibly random) instrument to the outcome.

20.4 "Bullet-proof" instruments

While some accounting researchers appear to believe that statistical tests can be used to evaluate whether an instrument is "valid" (see Larcker and Rusticus, 2010), there are no simple statistical tests for the validity of instruments. Indeed, many studies choose to test the validity of their instrumental variables using statistical tests, but such tests of instruments are of little value in practice.

In this section, we examine what we might "achieve" if we relax the requirement for an instrument to be random (i.e., we skip Question 1 above) and rely on statistical tests to address the concerns raised by this hypothetical reviewer.[7]

Suppose that we have $y = X\beta + \epsilon$, with $\rho(X, \epsilon) = 0.2$ and $\beta = 0$. This means that there is no causal relation between X and y, but we have an endogeneity issue and thus might despair our inability to get good inferences about the causal relation between X and y.

```
# Parameters for generating simulated data
n <- 1000
beta <- 0
corr <- 0.2
Sigma <- matrix(c(1, corr, corr, 1), nrow = 2)

# Used to generate instruments
sd_eta <- 0.3

# Used for evaluation of IV
p.cutoff <- 0.05
F.threshold <- 30

# Number of simulations (used later)
k <- 1000
```

Below we create a function—**generate_data()**—to generate data with the properties and parameter values discussed above, then run it to produce `df`. Here we have X and ϵ that are bivariate-normally distributed with mean 0 and variance 1.

```
generate_data <- function() {
  as_tibble(mvrnorm(n, mu = c(0, 0), Sigma = Sigma),
            .name_repair = ~ c("X", "e")) |>
    mutate(y = X * beta + e)
}

set.seed(2019)
df <- generate_data()
```

We know that OLS regression estimates using these data will be biased and therefore will not support credible causal inferences.

But, wait. We have a "solution"! We can simply *create* three instruments: $z_1 = x + \eta_1$, $z_2 = \eta_2$, and $z_3 = \eta_3$, where $\sigma_{\eta_1} = \sigma_{\eta_2} = \sigma_{\eta_3} \sim N(0, 0.09)$ and independent. We can even

---

[7]Much of this section is based on an example that first appeared in Gow et al. (2016).

TABLE 20.1

Regression results

	OLS	IV
(Intercept)	−0.054*	−0.054*
	(0.031)	(0.031)
X	0.184***	0.190***
	(0.031)	(0.032)
Num.Obs.	1,000	1,000
R2	0.035	0.035

make an R function `generate_ivs()` to do this:

```
generate_ivs <- function(df, sd_eta = 0.3) {
  df |>
    mutate(z_1 = X + rnorm(n, sd = sd_eta),
           z_2 = rnorm(n, sd = sd_eta),
           z_3 = rnorm(n, sd = sd_eta))
}
```

We apply `generate_ivs()` to our `data set`df' to create a data set with generated instruments.

```
df_ivs <- generate_ivs(df)
```

Now, do these instruments pass the "only through" requirement? Clearly there is no relation between any of the instruments and *any* other variable *except* for z_1, which is only associated with other variables through its relation with X. So, "yes"!

What happens when we estimate IV using these instruments? We run an OLS regression and an IV regression using our generated instruments.

```
fms <- list("OLS" = lm(y ~ X, data = df),
            "IV" = feols(y ~ 1 | X ~ z_1 + z_2 + z_3, data = df_ivs))
```

The results of these regressions are shown in Table 20.1.

```
modelsummary(fms,
             estimate = "{estimate}{stars}",
             coef_rename = c('fit_X' = 'X'),
             gof_map = c("nobs", "r.squared"),
             stars = c('*' = .1, '**' = 0.05, '***' = .01))
```

There we see that our IV regressions deliver statistically significant positive coefficient on X and, given our use of IV, we might feel comfortable attaching a causal interpretation to this coefficient. (Of course, we *know* in this scenario that the true β is zero; in practice, we would *not* know this.)

Of course, a sceptical reviewer might ask us to justify our generated instruments beyond the "only through" argument we made above. For example, it is common to ask whether our instruments pass a **test of overidentifying restrictions**. Such tests apply when we have more instruments than endogenous variables for which we need instruments and typically posit that at least one instrument is valid and ask whether we can reject the null hypothesis that the other instruments are valid.

The sceptical reviewer might also ask us to provide evidence that we do not have **weak instruments**.[8] Weak instruments lead to bias in small samples (Stock et al., 2002).

We can extract statistics related to standard tests as follows:

```
iv <- fms[["IV"]]
iv_sargan <- iv$iv_sargan
Sargan.stat <- iv_sargan$stat
Sargan.p <- iv_sargan$p
ivf1 <- fitstat(iv, "ivf1", simplify = TRUE)
F.stat <- ivf1$stat
F.p <- ivf1$p
```

Here we have a Sargan test statistic of 1.8 (p-value of 0.41). So we "pass" the test of overidentifying restrictions. Based on a test statistic of 30, which easily exceeds the thresholds suggested by Stock et al. (2002), the null hypothesis of weak instruments is rejected as we have an F-stat of 3884. So we can report to our sceptical reviewer that our instruments are "good". And we have "results". Yay!

But maybe we got lucky. To test this, we create a function `run_simulation()` that we can use to see if this pattern is robust.

```
run_simulation <- function(run) {

  iv <-
    generate_data() |>
    generate_ivs() |>
    feols(y ~ 1 | X ~ z_1 + z_2 + z_3, data = _)

  iv_sargan <- iv$iv_sargan
  ivf1 <- fitstat(iv, "ivf1", simplify = TRUE)

  return(
    tibble(
      run = run,
      coeff = iv$coeftable["fit_X", "Estimate"],
      p.value = iv$coeftable["fit_X", "Pr(>|t|)"],
      Sargan.stat = iv_sargan$stat,
      Sargan.p = iv_sargan$p,
      F.stat = ivf1$stat,
      F.p = ivf1$p))
}
```

The following code runs 1000 simulations and extracts some test statistics.

```
sim_results <-
  1:k |>
  map(run_simulation) |>
  list_rbind()

coeff <- mean(sim_results$coeff)
sig.percent <- mean(sim_results$p.value < p.cutoff) * 100
```

---

[8]See suggestions regarding further reading at the end of the chapter for more about these tests.

```
F.stat <- mean(sim_results$F.stat > F.threshold) * 100
reject.endogeneity <- mean(sim_results$Sargan.p >= p.cutoff) * 100
```

The results of our simulation analysis are as follows:

- The mean estimated coefficient on X is 0.2, which is statistically significant at the 5% level 100% of the time. Note that this coefficient is close to $\rho(X, \epsilon) = 0.2$, which is to be expected given how our data were generated.
- The null hypothesis of weak instruments is rejected 100% of the time using a test statistic of 30, which easily exceeds the thresholds suggested by Stock et al. (2002).
- The test of overidentifying restrictions fails to reject a null hypothesis of valid instruments (at the 5% level) 92.9% of the time.

Again, hooray! Our hypothetical referee should be satisfied almost always. But we *know* that these results are bogus. The problem: the instrument Z is not random; it's a function of X and essentially just as endogenous as X is.

While the generated instruments we used above might seem a bit fanciful, they are probably not very different from instruments such as lagged values or industry averages (z_1) or random items plucked from Compustat (more like z_2 or z_3), which are not unheard of in actual research.

This example shows that completely spurious instruments can easily pass tests for weak instruments and tests of overidentifying restrictions and yet deliver bad inferences.

20.5   Causal diagrams: An application

To illustrate the application of causal diagrams to the evaluation of instrumental variables, we consider Armstrong et al. (2013). Armstrong et al. (2013) study the effect of shareholder voting (*Shareholder support*$_t$) on future executive compensation (*Comp*$_{t+1}$). Because of the plausible existence of unobserved confounding variables that affect both future compensation and shareholder support, a simple regression of *Comp*$_{t+1}$ on *Shareholder support*$_t$ and controls would not allow Armstrong et al. (2013) to obtain an unbiased or consistent estimate of the causal relation.

Among other analyses, Armstrong et al. (2013) use an instrumental variable to estimate the causal relation of interest. Armstrong et al. (2013) claim that their instrument is valid. Their reasoning is represented graphically in Figure 20.2. By conditioning on *Comp*$_{t-1}$ and using Institutional Shareholder Services (ISS) recommendations as an instrument, Armstrong et al. (2013) argue that they can identify a consistent estimate of the causal effect of shareholder voting on *Comp*$_{t+1}$, even though there is an unobserved confounder, namely determinants of future compensation observed by shareholders, but not the researcher.[9]

While the authors note this possibility: "validity of this instrument depends on ISS recommendations not having an influence on future compensation decisions conditional on shareholder support (i.e., firms listen to their shareholders, with ISS having only an indirect impact on corporate policies through its influence on shareholders' voting decisions)",

---

[9]In Figure 20.2, we depict the unobservability of this variable (to the researcher) by putting it in a dashed box. Note that we have omitted the controls included by Armstrong et al. (2013) for simplicity, though a good causal analysis would consider these carefully.

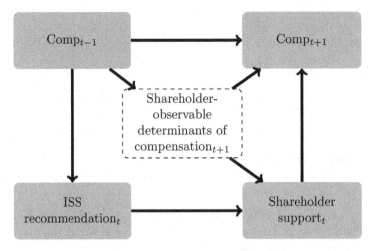

FIGURE 20.2
Identifying effects of shareholder support on compensation

they are unable to test the assumption (Armstrong et al., 2013, p. 912). Unfortunately, this assumption seems inconsistent with the findings of Gow et al. (2013), who provide evidence that firms calibrate compensation plans (i.e., factors that directly affect $Comp_{t+1}$) to comply with ISS's policies so as to get a favourable recommendation from ISS. As depicted in Figure 20.3, this implies a back-door path from $Comp_{t+1}$ into *ISS recommendation$_t$*, suggesting that the instrument of Armstrong et al. (2013) is not valid.[10]

The careful reader will note that we did not need to address the randomness of the instrument (i.e., "Question 1" from above) to conclude that the instrument is problematic. It is implicitly assumed in the first causal diagram that *ISS recommendation$_t$* is random conditional on $Comp_{t-1}$, which would need to be included in the regression even based on the first causal diagram. A fuller analysis of the first causal diagram is beyond the scope of the current version of this course, as it would require results not covered herein.

20.6 Further reading

Given the difficulty of identifying credible instruments in accounting research, we have eschewed the replication of actual papers in this chapter. The challenges with IVs seem more apparent with idealized settings provided by simulation analysis. We direct readers interested in textbook treatments of IVs at a similar or slightly more advanced level to Cunningham (2021) and Adams (2020), each of which provides an up-to-date treatment of issues related to IV estimation. Chapter 4 of Angrist and Pischke (2008) provides a good introduction to IVs with several examples. More on instrumental variables and causal diagrams can be found in Chapter 7 of Pearl (2009b).

---

[10]Armstrong et al. (2013) recognize the possibility that the instrument they use is not valid and conduct sensitivity analysis to examine the robustness of their result to violation of the exclusion restriction assumptions. This analysis suggests that their estimate is highly sensitive to violation of this assumption.

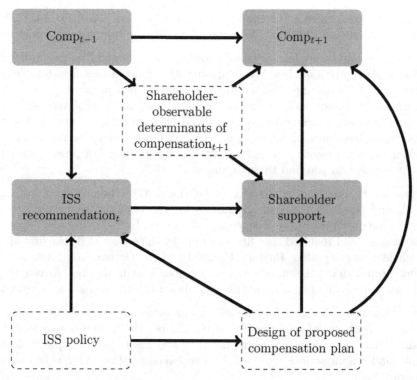

FIGURE 20.3
Alternative causal diagram for Armstrong et al. (2013)

20.7 Discussion questions and exercises

1. In the 1960s and early 1970s, American men were at risk of being drafted for military service. Each year from 1970 to 1972, draft eligibility was prioritized based on the results of a lottery over birthdays. A random number was assigned to the birthdays of 19-year-olds. Those whose birthday was associated with a random number below the cut-off were draft-eligible; the other 19-year-olds were not draft-eligible. Angrist and Pischke (2008) point out that, "in practice, many draft-eligible men were still exempted from service for health or other reasons, while many men who were draft-exempt nevertheless volunteered for service." Using an indicator for draft lottery number being below the cut-off as an instrumental variable, Angrist (1990) finds that those who served in Vietnam earned significantly less than their peers even in 1981. Try to draw the causal diagram for this analysis and apply the three questions outlined above to this setting. What assumptions are needed for the answer to each question to be "yes"? How might these assumptions be violated?

2. Ahern and Dittmar (2012) study a rule requiring firms in Norway to appoint female directors and "use the pre-quota cross-sectional variation in female board representation to instrument for exogenous changes to corporate boards following the quota." What is the significance of the discussion suggesting that "the quota

was implemented without the consent of business leaders" (2012, p. 145)? What about the discussion (2012, p. 155) suggesting that the rule was not anticipated?

3. You and a co-author are interested in studying the effect of independent directors on financial reporting quality. You conjecture that independent directors demand better financial reporting, as they are less likely to benefit from obfuscation of performance than non-independent directors, who are often employees of the firm. However, you are concerned that causation may run in the opposite direction; firms with better financial reporting quality may have more independent directors on their boards. Provide some arguments for the existence of reverse causation. How persuasive do you find these arguments?

4. In response to a request by the SEC in February 2002, the major US exchanges (NYSE and Nasdaq) proposed changes to listing standards that would require firms to have a majority of independent directors. In 2003, the SEC approved these changes and required that firms comply by the earlier of (i) the first annual shareholder meeting after January 15, 2004 and (ii) October 31, 2004. Your co-author argues that this change was "exogenous", as it was not driven by the decisions of individual firms, and this fact allows you to estimate a causal effect.

   Does "exogeneity" as used by your co-author mean the same thing as exogeneity for the purposes of econometric analysis? Does the former notion imply the latter? Think of "examples" to support your arguments (an example could be a simple model, a numerical argument, or a verbal description of a real or imagined scenario).

   More specifically, your co-author argues that the change in the number of independent directors imposed on the firm by the new rules is exogenous and thus could be used as an instrument for your study. Armstrong et al. (2014) examine the changes in listing standards above and use "the minimum required percentage change in independent directors, `Min % change ID`, as an instrument" in studying the effect of independent directors on firm transparency. What are the parallels between the setting of Armstrong et al. (2014) and that of Ahern and Dittmar (2016)? Do the two papers use the same basic approach, or are there important differences between their approaches?

5. The objective of Cohen et al. (2013) "is to investigate how governance regulations in SOX and the exchanges are associated with chief executive officers' incentives and risk-taking behavior." What is the treatment of interest in Cohen et al. (2013)? What are the outcomes of interest? Cohen et al. (2013) state that "our dependent variables, namely, investment and executive incentive compensation, are likely to be determined jointly. As such, the parameter estimates from ordinary least squares (OLS) are likely to be biased. Our empirical analyses address the issue by using simultaneous equations models." What "exclusion restrictions" are implied by Panel A of Table 4? What does the implied causal diagram look like? How persuasive do you find the approach of Cohen et al. (2013) to be?

Listing 20.1 Code to generate the Sneetch world

```
# Set parameters
n <- 10000; max_d <- 10
set.seed(1); z <- rlnorm(n)
a <- 1
gamma <- 3; b <- gamma * z
alpha <- 100; beta <- 10; sd_epsilon <- 5

d_opt_fun <- function(a, b, min_d = 0) {

  v <- function(x) -a * x^2 + b * x

  opts <- min_d:max_d
  vals <- sapply(opts, v)
  d_opt <- opts[vals == max(vals)]
  d_opt
}

d_opt_fun <- Vectorize(d_opt_fun)

# Calculate firm value, etc., at t = 0
df <-
  tibble(a, b) |>
  mutate(epsilon_0 = rnorm(n, sd = sd_epsilon),
         V_0 = alpha + beta * b + epsilon_0,
         d_0 = d_opt_fun(a, b))

# Add firm value, etc., at t = 1
min_d <- 3
df_1 <-
  df |>
  mutate(epsilon_1 = rnorm(n, sd = sd_epsilon),
         V_1 = alpha + beta * b + epsilon_1,
         d_1 = d_opt_fun(a, b, min_d = min_d))
```

6. Suppose the existence of a world of Sneetches.[11] There are two kinds of Sneetches:

> *Now, the Star-Belly Sneetches*
> *Had bellies with stars.*
> *The Plain-Belly Sneetches*
> *Had none upon thars.*
> —Dr. Seuss

In the Sneetch world, there are 10,000 (**n**) firms, each with a board of 10 (**max_d**) directors. The number of Plain-Belly directors d_i is set by firm i as

$$d_i = \underset{x \in [\underline{d}, \overline{d}]}{\arg\max} v_i(x) = -a_i x^2 + b_i x$$

---

[11]The interested reader can learn more about Sneetches in Dr Seuss's *The Sneetches and Other Stories*, available on (for example) Amazon: https://www.amazon.com/dp/0394800893.

where $a_i = 1$ and $b = \gamma z_i$, $\gamma = 3$ and z_i is log-normally distributed with default parameters of $\mu = 0$ and $\sigma = 1$. We start with $\underline{d} = 0$ and $\overline{d} = 10$. The remaining $10 - d_i$ directors would be Star-Belly Sneetches.

The value of firm i at time $t = 0$ is given by the equation:

$$V_{i0} = \alpha + \beta \times b_i + \epsilon_0$$

where $\alpha = 100$, $\beta = 10$ and $\epsilon \sim N(0, \sigma_\epsilon)$, where $\sigma_\epsilon = 5$.

The Sneetches then pass new legislation requiring every board to have at least three Plain-Belly directors at time $t = 1$. The value of firm i at time $t = 1$ is given by the equation:

$$V_{i1} = \alpha + \beta \times b_i + \epsilon_{i1}$$

The code in Listing 20.1 starts by setting the parameters discussed above and generating γ_i (gamma). The code then generates the value of each of the 10,000 firms at time $t = 0$ (V_0), as well as the number of Plain-Belly directors chosen by each firm at $t = 0$ to maximize $v_i(\cdot)$ (d_0). These values are stored in df.

The code in Listing 20.1 then sets min_d to reflect the requirement of the legislation coming into force for $t = 1$. The code then generates the value of each firm at $t = 1$ (V_1) and the number of Plain-Belly directors chosen by each firm at $t = 1$ (d_1), which involves maximization of $v_i(\cdot)$ subject to the new constraint that $d_i \geq 3$. We assume that neither a_i nor b_i is observable to the researcher.

a. Expressed in terms of a_i and b_i, what value of x maximizes $v_i(x)$ when there is no constraint (as is true at $t = 0$)?

b. What issue do we have in using this maximum to select the number of Plain-Belly directors on a board?

c. What does d_opt_fun() in Listing 20.1 do to handle this issue?

d. What does the function Vectorize() do in Listing 20.1? (*Hint*: Use ? Vectorize in R to learn more about this function.)

e. Using OLS and data in df_1 after running the code in Listing 20.1, estimate the relationship between firm value at time $t = 1$ (V_1) and the number of Plain-Belly directors on the board (d_1).

f. Using the data in df_1, estimate an IV regression using the strategy of Ahern and Dittmar (2012).

g. Using the data in df_1, estimate an IV regression using the strategy of Armstrong et al. (2014) (note that rather than Min % change X, you might use something like Min change X, as the denominator is 10 for all observations). *Hint*: The pmin() function in R gives the point-wise minimum of two vectors of values (e.g., pmin(x, y) will return the lower of two for each pair.)

h. What is the true causal effect of Plain-Belly directors on firm value? Do the strategies of either Ahern and Dittmar (2012) or Armstrong et al. (2014) correctly estimate this effect?

21

Panel data

The typical data set used in accounting research comprises **panel data**, namely means repeated observations on multiple units. Typically units are companies and observations are available for several points of time or periods. As we saw in Chapter 3, in certain conditions, researchers can exploit panel data sets to gain insights into causal effects.

In this chapter, we explore two approaches commonly used with panel data sets. The first approach uses **difference in differences** to measure causal effects. The second approach uses **multi-way fixed effects** to account both for what is commonly referred to as "unobserved time-invariant heterogeneity" between units (i.e., unit-specific effects) and period-specific effects. In practice, these approaches are often combined. We discussed difference in differences in Chapter 19 and recommend that you read that chapter before this one.

> Tip
>
> The code in this chapter uses the packages listed below. Rather than invoking several Tidyverse packages separately, we load the **tidyverse** package. For instructions on how to set up your computer to use the code found in this book, see Section 1.2. Quarto templates for the exercises below are available on GitHub.[a]
>
> ---
> [a]https://github.com/iangow/far_templates/blob/main/README.md

```
library(tidyverse)
library(DBI)
library(farr)
library(modelsummary)
library(fixest)
```

21.1 Analysis of simulated data

In Chapter 3, we explored the `test_scores` data set. There we used a number of approaches, including one where we estimated the effect size using a difference-in-differences estimator with grade and individual fixed effects as 15.736.

We can now reveal that the `test_scores` and `camp_attendance` data sets were simulated using the function `get_test_scores()`, with the default values for each of its arguments: `effect_size = 15`, `n_students = 1000`, and `n_grades = 4`.[1]

---
[1]Note that the function `get_test_scores()` returns a version of the data that merges `test_scores` and `camp_attendance` and calculates the `post` and `treat` variables we added in Chapter 3.

DOI: 10.1201/9781003456230-21

TABLE 21.1
Regression analysis: Simulated data

	(1)	(2)
I(treat * post)	0.736**	0.736
	(0.319)	(0.995)
Num.Obs.	4,000	4,000
R2	0.923	0.923

The data-generating process embedded in `get_test_scores()` produces scores using the following equation:

$$y_{ig} = \alpha_i + \beta x_{ig} + \gamma_g + \epsilon_{ig}$$

where i and g denote individuals and grades, respectively. Denoting the grade after application of treatment (i.e., 7 in the `test_scores` data) as G, the treatment indicator x_{ig} is as follows:

$$x_{ig} = \begin{cases} 1 \text{ if } i \text{ is treated and } g \geq G \\ 0 \text{ otherwise} \end{cases}$$

Thus, as we will see, this is precisely the setting where multi-way fixed effects are appropriate. The individual fixed effect picks up α_i and the grade fixed effect picks up γ_g. The estimated effect size of 15.736 seems close to the true value $\beta = 15$, but is it close enough? To better examine this issue, we generate a version of the data where the treatment effect is zero (i.e., $\beta = 0$).

```
set.seed(2021)
test_scores_alt <- get_test_scores(effect_size = 0)
```

We then estimate regressions with grade- and individual-level fixed effects. One regression uses "OLS" standard errors, the other uses standard errors clustered by `grade` and `id` (see Section 5.6 for discussion of alternative approaches to calculating standard errors).

```
fms <- list(feols(score ~ I(treat * post) | grade + id,
              vcov = "iid", data = test_scores_alt),
          feols(score ~ I(treat * post) | grade + id,
              vcov = ~ grade + id, data = test_scores_alt))
```

Because we set the seed to the same value (2021) as used to generate the original data, we get the same random draws and our estimated treatment effect is exactly as before (0.736), but reduced by 15 and the estimated standard error (*without* clustering) is exactly as we saw earlier (0.319). However, in Table 21.1, we see evidence that the coefficient is not "close enough" to the true value, as we would reject the *true* null at the 5% level (p-value of 0.0211).

```
modelsummary(fms,
          estimate = "{estimate}{stars}",
          gof_map = c("nobs", "r.squared"),
          stars = c('*' = .1, '**' = 0.05, '***' = .01))
```

What has "gone wrong" in this case? Some possible explanations are:

1. Our standard errors are too low.

2. We had bad luck (i.e., we don't expect similar results for a different seed value).
3. Our fixed-effects estimator is biased due to its small-sample properties.
4. Our estimator is biased due to endogenous selection.

Let's take these explanations one at a time. For the first, the fact that cluster-robust standard errors are significantly higher than OLS standard errors seems consistent with this explanation. But recall that the standard error is essentially an estimate of the standard deviation of the relevant coefficient over independent random draws of the data. So a more rigorous way to test this explanation is to simulate the data and examine whether the OLS standard errors tend to underestimate the variability of the coefficients.

```
extract_stats <- function(fm, model_name) {
  tibble(model = model_name,
         estimate = coef(fm),
         se = se(fm),
         t = coef(fm) / se(fm),
         `p-value` = pvalue(fm))
}
```

The following function generates a data set using the `get_test_scores()` function, then runs two regressions—one using "OLS" standard errors and one using two-way cluster-robust standard errors ("CL-2")—then returns statistics from the fitted models.

```
sim_run <- function(i, ...) {
  df <- get_test_scores(effect_size = 0, ...)

  fm1 <- feols(score ~ I(treat * post) | grade + id,
               vcov = "iid", data = df)
  fm2 <- feols(score ~ I(treat * post) | grade + id,
               vcov = ~ grade + id, data = df)

  bind_rows(extract_stats(fm1, model_name = "OLS"),
            extract_stats(fm2, model_name = "CL-2"))
}
```

The following code runs `sim_run()` 1,000 times and stores the results in `sim_results`.

```
num_sims <- 1000L

sim_results <- list_rbind(map(1:num_sims, sim_run))
```

Given these data, we want to compare the standard deviation of the estimated coefficient with the mean of the estimated standard error for the two approaches ("OLS" and "CL-2"):

```
sim_results |>
  group_by(model) |>
  summarize(se_obs = sd(estimate),
            se_est = mean(se))
```

```
# A tibble: 2 x 3
  model se_obs se_est
  <chr> <dbl>  <dbl>
1 CL-2  0.335  0.984
2 OLS   0.335  0.315
```

We can see that the two-way cluster-robust standard errors are actually far too high, while the OLS standard errors are pretty close to the true standard deviation of the coefficients. This seems to rule out the first explanation.

To evaluate the second explanation ("bad luck"), we can use the same data. Let's consider critical p-values of 1% and 5% and count how many of the OLS test statistics would lead to rejection of the null hypothesis ($\beta = 0$) at each size. (We focus on the "OLS" t-statistics, as we have established that OLS produces better standard error estimates.)

```
rejection_stats <-
  sim_results |>
  filter(model == "OLS") |>
  summarize(prop_01 = mean(`p-value` <= 0.01),
            prop_05 = mean(`p-value` <= 0.05))

rejection_stats
```

```
# A tibble: 1 x 2
  prop_01 prop_05
    <dbl>   <dbl>
1   0.863   0.951
```

So, we reject a *true* null hypothesis 86% of the time at the 1%-level and 95% of the time at the 5%-level, even with what appear to be good standard errors. This would be an extraordinary degree of bad luck, suggesting that there is bias in our coefficient estimates.

The properties of fixed-effect estimators are derived asymptotically, that is as the number of observations and estimated fixed effects approaches infinity. However, in a typical panel data set, we have a fairly small number of time periods. In our analysis above, we have just four grades of data. But, because these are simulated data and not real people, we can easily expand the number of grades of data that we consider. Let's consider `n_grades = 12`, the maximum handled by `get_test_scores()`.[2]

```
sim_results_wide <- list_rbind(map(1:num_sims, sim_run, n_grades = 12))
```

We collect data on rejection rates for both $\alpha = 0.01$ and $\alpha = 0.05$. Based on our analysis above, we focus on the "OLS" model.

```
rejection_stats <-
  sim_results_wide |>
  filter(model == "OLS") |>
  summarize(prop_01 = mean(`p-value` <= 0.01),
            prop_05 = mean(`p-value` <= 0.05))

rejection_stats
```

```
# A tibble: 1 x 2
  prop_01 prop_05
    <dbl>   <dbl>
1   0.329   0.574
```

Here we see evidence of a reduction of the bias, but the bias is not eliminated. We still reject a *true* null hypothesis 33% of the time at the 1%-level and 57% of the time at the 5%-level.

---

[2]This maximum exists not only because it seems inhumane to require even simulated students to attend school for longer but also because the grade effects are hard-coded and only for 12 years.

So let's consider the final explanation of the bias, which is due to non-random selection. It turns out that assignment to treatment in `get_test_scores()` depends on test scores in the year prior to the camp, but conditional on these scores is completely random.[3] This is termed **selection on observables**, which is often suggested as a basis on which causal inference can be justified when using approaches such as propensity-score matching.

So what happens if we use completely random assignment to treatment? We can request such assignment by setting the `random_assignment = TRUE` in `get_test_scores()`.

```
sim_results_rand <- list_rbind(map(1:num_sims, sim_run,
                                   random_assignment = TRUE))
```

```
rejection_stats <-
  sim_results_rand |>
  filter(model == "OLS") |>
  summarize(prop_01 = mean(`p-value` <= 0.01),
            prop_05 = mean(`p-value` <= 0.05))

rejection_stats
```

```
# A tibble: 1 x 2
  prop_01 prop_05
    <dbl>   <dbl>
1   0.009   0.046
```

Here we see evidence of elimination of the bias. We reject a true null hypothesis 0.9% of the time at the 1%-level and 4.6% of the time at the 5%-level.

Thus, it appears that random assignment is critical in this setting for achieving valid causal inferences. This is somewhat concerning as, beyond non-random treatment assignment, the basic assumptions underlying causal inference appear to be satisfied, as we have a plausible basis for the **parallel trends assumption**. First, we have grade effects that are the same for both treated and untreated observations. Second, we have individual effects that, while different between the treated and untreated observations, remain constant over the sample period and therefore do not undermine the parallel-trends assumption.[4] What this suggests is that subtle biases can enter difference-in-differences analyses even in settings that are unrealistically simple. It seems reasonable to expect that biases exist—and are plausibly worse—in more complex settings of actual research.

21.2 Voluntary disclosure

One paper that uses a multi-way fixed-effect structure like that we analysed above is Li et al. (2018), who "seek to provide causal evidence on the **proprietary cost hypothesis**"

---

[3] If you are curious, you can see this in the source code for the function: https://github.com/iangow/farr/blob/main/R/get_test_scores.R.

[4] The parallel trends assumption is likely subtly violated because of regression-to-the mean effects in test scores, but not in a way likely to be detected using the usual "parallel trends plots" used in the literature. A researcher seeing the underlying data-generating process is likely to find it very easy to imagine the "can-opener" of parallel trends we discussed in Chapter 19. To paraphrase Macbeth: "Is this a [can-opener] which I see before me, The handle toward my hand? Come, let me clutch thee. I have thee not, and yet I see thee still. Art thou not, fatal vision, sensible To feeling as to sight? ... Thou marshal'st me the way that I was going [i.e., to use DiD], And such an instrument I was to use."

(2018, p. 266). While "proprietary costs" are commonly assumed to be those caused by use of information by competitors, Verrecchia (1983, p. 181) uses the term more broadly.

Many researchers focus on settings where firms want to disclose favourable information to investors. For example, Verrecchia (1983, p. 181) writes of the "proprietary cost associated with releasing information which is unfavorable to a firm (e.g., a bank would be tempted to ask for repayment of its loan)." However, Verrecchia (1983, p. 182) points out that in some situations firms may prefer to disclose bad news: "One recent example of this is the response of the UAW (United Auto Workers) for fewer labor concessions in the face of an announcement by Chrysler Corporation's chairman that that firm's fortunes had improved."

While Verrecchia (1983) considers the "reluctance of managers in certain highly competitive industries ... to disclose favorable accounting data", in his model, he assumes that the cost of disclosure is constant and independent of the disclosed value.

Verrecchia (1983) can be viewed as providing a theoretical foundation for the proprietary cost hypothesis. The model in Verrecchia (1983) posits a capital market–driven incentive for disclosure of favourable information; without such an incentive, there would be no disclosure in the Verrecchia (1983) setting. Also in Verrecchia (1983), the firms with favourable news are the ones disclosing, as they have the greater capital-market benefit from doing so.

Li et al. (2018) exploit the **staggered implementation** of the inevitable disclosure doctrine (IDD), which was adopted by state courts as part of the common law of their respective states at different times (some state courts later rejected the doctrine after adopting it). IDD provides an employer with injunctive relief to prevent a current or former employee from working for another company if doing so will lead to the *inevitable disclosure* of trade secret information. Li et al. (2018) argue that IDD increases the marginal benefits of non-disclosure, which they interpret as an increase in the cost of disclosure.

Li et al. (2018) focus their analysis of disclosure choice on the disclosure of customer identities in 10-Ks and assume that the state of a firm's headquarters governs the applicability of IDD to the firm. We will conduct an approximate replication of certain analyses of Li et al. (2018) to understand the empirical approaches of this chapter. This requires combining data on customer disclosures with data on the states of companies' headquarters. This data set is then combined with data on the dates of adoption of IDD by states.

21.2.1   Customer disclosures

We will use three data sets in the replication analysis. The first data set is `compseg.seg_customer`, which contains data derived from companies' disclosures regarding significant customers. We use the second (`compseg.names_seg`) to link companies on Compustat with CIKs, which we use to link with data on headquarters locations. Finally, we use `comp.funda` for data on total sales.

```
db <- dbConnect(RPostgres::Postgres(), bigint = "integer")

seg_customer <- tbl(db, Id(schema = "compseg", table = "seg_customer"))
names_seg <- tbl(db, Id(schema = "compseg", table = "names_seg"))
funda <- tbl(db, Id(schema = "comp", table = "funda"))
```

It turns out that `compseg.seg_customer` contains data on a number of different types of segment, including geographic regions and markets.

```
seg_customer |>
  count(ctype) |>
  arrange(desc(n)) |>
  collect()
```

```
# A tibble: 7 x 2
  ctype          n
  <chr>      <dbl>
1 COMPANY   364802
2 GEOREG    150601
3 MARKET    108389
4 GOVDOM     39524
5 GOVFRN      5101
6 GOVSTATE    1456
7 GOVLOC       574
```

We are interested in data in "COMPANY" segments, which may or may not name individual customers.

```
seg_customer |>
  filter(ctype == "COMPANY") |>
  count(cnms) |>
  arrange(desc(n)) |>
  collect()
```

```
# A tibble: 44,045 x 2
  cnms               n
  <chr>          <dbl>
1 Not Reported   68829
2 NOT REPORTED   31162
3 2 Customers     5981
4 3 Customers     5263
5 5 Customers     4676
6 9 Customers     4199
# i 44,039 more rows
```

To better understand this setting, we examine one example chosen somewhat at random from the last year of the sample period. In its 10-K for the year ended 31 December 2010, Advanced Micro Devices, Inc. (AMD), a global semiconductor company, disclosed the following:

> In 2010, Hewlett-Packard Company accounted for more than 10% of our consolidated net revenues. Sales to Hewlett-Packard consisted primarily of products from our Computing Solutions segment. Five customers, including Hewlett-Packard, accounted for approximately 55% of the net revenue attributable to our Computing Solutions segment. In addition, five customers accounted for approximately 46% of the net revenue attributable to our Graphics segment. A loss of any of these customers could have a material adverse effect on our business.

Elsewhere in its 10-K, AMD discloses the following:

> In 2010, the Company had one customer that accounted for more than 10% of the Company's consolidated net revenues. Net sales to this customer were approximately $1.4 billion, or 22% of consolidated net revenues, and were primarily attributable to the Computing Solutions segment.

Note that the Graphics and Computing Solutions segments had sales of $1,663 million and $4,817 million, respectively, and therefore made up 99.8% of AMD's total sales of $6,494 million.

Thus, 46% of Graphics represents $765 million, while 55% of Computing Solutions is $2,650 million. After subtracting $1,400 million of these sales to Hewlett-Packard (HP), Compustat appears to ascribe the remaining $1,250 to "4 Customers". The basis for assigning $1400 to "Not Reported" is unclear (while the sentence above merely says "one customer", it is quite clear that this is HP and Compustat elsewhere assumes as much). (In fact, it seems almost certain that there is double-counting in this case, making addition of the numbers in `salecs` problematic.)

```
seg_customer |>
  filter(gvkey == "001161", datadate == "2010-12-31") |>
  select(cnms, ctype, salecs, stype) |>
  collect()
```

```
# A tibble: 8 x 4
  cnms                ctype    salecs stype
  <chr>               <chr>     <dbl> <chr>
1 HEWLETT-PACKARD CO  COMPANY    1400 BUSSEG
2 4 Customers         COMPANY    1250. BUSSEG
3 Not Reported        COMPANY    1400 BUSSEG
4 5 Customers         COMPANY     765 BUSSEG
5 Not Reported        COMPANY      NA BUSSEG
6 Not Reported        COMPANY      NA BUSSEG
7 Not Reported        COMPANY      NA BUSSEG
8 International        GEOREG    5715. BUSSEG
```

Focusing on this case perhaps help us to understand the disclosure decision faced by AMD. Even if AMD had not disclosed the identity of its largest customer, it perhaps would have been easy to infer that "Customer #1" was indeed HP, as HP was the largest personal computer manufacturer at the time.[5]

The second data set we use is `undisclosed_names`, which contains the values of `cnms` considered to be non-disclosures and is part of the `farr` package.[6]

```
undisclosed_names
```

```
# A tibble: 460 x 2
  cnms         disclosed
  <chr>        <lgl>
1 Not Reported FALSE
2 NOT REPORTED FALSE
3 2 CUSTOMERS  FALSE
4 6 Customers  FALSE
5 4 Customers  FALSE
6 3 Customers  FALSE
# i 454 more rows
```

---

[5]See Gartner (2011): https://go.unimelb.edu.au/yww8.

[6]See the source code for the `farr` package for details on the somewhat manual construction of this data set: https://github.com/iangow/farr/blob/main/data-raw/create_undisclosed_names.R. This likely mirrors a process used for Li et al. (2018) itself.

We can then perform a `left_join()` of the customer-related data in `seg_customer` to identify disclosures (and non-disclosures) of customer identities. Following Li et al. (2018), we focus on the period from 1994 to 2010.

```
sample_start <- as.Date("1994-01-01")
sample_end <- as.Date("2010-12-31")

disclosure_raw <-
  seg_customer |>
  filter(ctype == "COMPANY") |>
  filter(between(datadate, sample_start, sample_end)) |>
  collect() |>
  left_join(undisclosed_names, by = "cnms") |>
  mutate(disclosed = coalesce(disclosed, TRUE)) |>
  select(gvkey, datadate, cnms, salecs, disclosed)
```

Table 21.2 lists the largest disclosed customers in the data. Many of these observations relate to Walgreens and CVS, two large US pharmacy chains.

```
disclosure_raw |>
  filter(disclosed) |>
  arrange(desc(salecs)) |>
  select(-disclosed) |>
  head(n = 10)
```

TABLE 21.2

Largest disclosed customers

gvkey	datadate	cnms	salecs
002751	2010-06-30	CVS Health Corp	23,641
002751	2009-06-30	WALGREEN CO	22,888
118122	1999-12-31	General Motors Corp	22,302
002751	2010-06-30	WALGREEN CO	21,671
002751	2009-06-30	CVS Caremark Corp	20,898
118122	2000-12-31	General Motors Corp	20,665
002751	2008-06-30	CVS Caremark Corp	20,040
002751	2007-06-30	CVS Caremark Corp	18,239
012136	2005-06-30	GENERAL ELECTRIC CO	17,712
002751	2008-06-30	Walgreen Co	17,307

As a shortcut way of imposing the sample requirements used in Li et al. (2018) (e.g., not financial services), we restrict our analysis to firms (GVKEYs) in the sample of Li et al. (2018). The `llz_2018` data set from the `farr` package contains these GVKEYs.[7]

To get a feel for the data, we calculate the proportion of total sales that are made to the largest disclosed customer.

```
sales <-
  funda |>
  filter(indfmt == "INDL", datafmt == "STD",
```

---

[7]See the source code for the `farr` package for the code used in creating this data set: https://github.com/iangow/farr/blob/main/data-raw/create_llz_2018_gvkeys.R.

```
        consol == "C", popsrc == "D") |>
  select(gvkey, datadate, sale) |>
  collect()

biggest_customers <-
  disclosure_raw |>
  filter(disclosed) |>
  inner_join(sales, by = c("gvkey", "datadate")) |>
  group_by(gvkey, datadate) |>
  mutate(max_customer = max(salecs) / sale)
```

Figure 21.1 provides a look at the proportion of sales attributed to the largest customer. Note that there are some cases where the largest customer appears to account for more than 100% of sales; many of these appear to be data-entry errors. But there are hundreds of firm-years with revenue from just one customer.

There is also an interesting jump at 10% of sales. While this is likely driven by the 10% cut-off for mandatory disclosure, interestingly many firms disclose significant customer information even when the largest customer is below 10% of sales, as seen in Figure 21.1.

```
biggest_customers |>
  filter(between(max_customer, 0, 1)) |>
  ggplot(aes(max_customer)) +
  geom_histogram(binwidth = 0.02) +
  scale_x_continuous(breaks = seq(0, 1, by = 0.1))
```

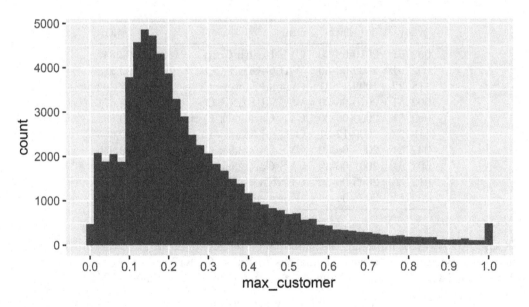

FIGURE 21.1
Histogram of proportion of sales from largest disclosed customer

Interestingly, Li et al. (2018) only include firm-year observations where at least one customer is disclosed (whether identified or not) as having at least 10% of total sales. We construct a data set (`prin_custs`) to identify these firms.

```
prin_custs <-
  disclosure_raw |>
  inner_join(sales, by = c("gvkey", "datadate")) |>
  group_by(gvkey, datadate) |>
  filter(!is.na(salecs), sale > 0) |>
  summarize(prin_cust = max(salecs / sale, na.rm = TRUE),
            .groups = "drop") |>
  mutate(has_prin_cust = prin_cust >= 0.1)
```

We next restrict the data to the `llz_2018` sample and calculate the two measures of disclosure choice used in Li et al. (2018), one measuring the proportion of significant customers
whose identities are not disclosed and another weighting that proportion by sales.

```
disclosure <-
    disclosure_raw |>
    inner_join(sales, by = c("gvkey", "datadate")) |>
    semi_join(llz_2018, by = "gvkey") |>
    group_by(gvkey, datadate) |>
    summarize(ratio = mean(!disclosed),
              ratio_sale = sum((!disclosed) * salecs) / sum(salecs),
              .groups = "drop") |>
    mutate(year = year(datadate))
```

Table 21.3 provides summary statistics for the two measures of disclosure choice. These are
fairly similar to those disclosed in Li et al. (2018) (e.g., Panel C of Table 1 reports a mean
of 0.447 for *Ratio 1*, the measure we call `ratio`).

```
disclosure |>
  pivot_longer(cols = c("ratio", "ratio_sale"),
               names_to = "measure") |>
  group_by(measure) |>
  summarize(mean = mean(value, na.rm = TRUE),
            median = median(value, na.rm = TRUE),
            sd = sd(value, na.rm = TRUE))
```

TABLE 21.3

Means of key disclosure measures

measure	mean	median	sd
ratio	0.459	0.333	0.449
ratio_sale	0.461	0.316	0.465

21.2.2   Data on adoption of IDD

The next data set we use is `idd_periods` which is returned by `get_idd_periods()` from
the `farr` package. This function relies on `idd_dates`, a data set in the `farr` package derived
from data reported in Klasa et al. (2018) and reproduced in Li et al. (2018).

```
idd_dates
```

```
# A tibble: 24 x 3
  state idd_date   idd_type
```

```
   <chr> <date>      <chr>
1 AR     1997-03-18 Adopt
2 CT     1996-02-28 Adopt
3 DE     1964-05-05 Adopt
4 FL     1960-07-11 Adopt
5 FL     2001-05-21 Reject
6 GA     1998-06-29 Adopt
# i 18 more rows
```

While `idd_dates` represents dates of either adoption or rejection of the inevitable disclosure doctrine, the `idd_periods` table takes a sample period (defined using `min_date` and `max_date`) and breaks that sample period into three sub-periods by state: pre- and post-adoption and post-rejection.

```
idd_periods <- get_idd_periods(min_date = sample_start,
                               max_date = sample_end)

idd_periods
```

```
# A tibble: 65 x 4
  state period_type   start_date end_date
  <chr> <chr>         <date>     <date>
1 AK    Pre-adoption  1994-01-01 2010-12-31
2 AL    Pre-adoption  1994-01-01 2010-12-31
3 AR    Pre-adoption  1994-01-01 1997-03-18
4 AR    Post-adoption 1997-03-18 2010-12-31
5 AZ    Pre-adoption  1994-01-01 2010-12-31
6 CA    Pre-adoption  1994-01-01 2010-12-31
# i 59 more rows
```

21.2.3   Data on state headquarters

The final piece of the puzzle in terms of data is `state_hq` from the `farr` package, which contains the headquarters location used by Li et al. (2018).[8] The `state_hq` table reports the range of dates for SEC filings for which each combination of CIK (`cik`) and state of headquarters (`ba_state`) applies.[9]

```
state_hq
```

```
# A tibble: 53,133 x 4
  cik        ba_state min_date   max_date
  <chr>      <chr>    <date>     <date>
1 0000066382 MI       1994-01-04 2018-10-10
2 0000070415 NY       1994-01-04 2007-03-14
3 0000084129 PA       1994-01-05 2018-10-04
4 0000832922 OH       1994-01-05 2001-01-09
5 0000909832 CA       1994-01-05 1996-12-20
6 0000004911 PA       1994-01-06 1996-01-05
# i 53,127 more rows
```

---

[8]This data frame is derived from data provided by Bill McDonald. See Bill McDonald's website for the original data: https://sraf.nd.edu/data/augmented-10-x-header-data/.

[9]Any gaps are filled in by extending the `min_date` back to the date after the preceding `max_date`.

To use these data, we need to link CIKs with GVKEYs, which we do using the table `compseg.names_seg` provided alongside Compustat's segment data.[10]

```
ciks <-
  names_seg |>
  filter(!is.na(cik)) |>
  select(gvkey, cik) |>
  collect()
```

The data frame `state_hq_linked` provides the state of headquarters applicable to each firm-year (i.e., combination of `gvkey` and `datadate`).[11]

```
state_hq_linked <-
  state_hq |>
  inner_join(ciks, by = "cik") |>
  inner_join(disclosure,
             join_by(gvkey, between(y$datadate, x$min_date, x$max_date))) |>
  select(gvkey, datadate, ba_state) |>
  rename(state = ba_state)
```

21.2.4 Regression analysis

Finally, we pull all these pieces together. Like Li et al. (2018), we delete "post-rejection" observations and log-transform the dependent variable.

```
reg_data <-
  disclosure |>
  inner_join(prin_custs, by = c("gvkey", "datadate")) |>
  filter(has_prin_cust) |>
  inner_join(state_hq_linked, by = c("gvkey", "datadate")) |>
  inner_join(idd_periods,
             join_by(state,
                     datadate >= start_date, datadate <= end_date)) |>
  filter(period_type != "Post-rejection") |>
  mutate(post = period_type == "Post-adoption",
         ln_ratio = log(1 + ratio),
         ln_ratio_sale = log(1 + ratio_sale)) |>
  select(-start_date, -end_date)

reg_data |> count(period_type)
```

```
# A tibble: 2 x 2
  period_type       n
  <chr>         <int>
1 Post-adoption 15831
2 Pre-adoption  12914
```

---

[10]Note that `names_seg` provides just one CIK for each GVKEY, though firms can change CIKs over time, which means that we will lose observations related to firms that previously used a different CIK from that found on `names_seg`.

[11]Note that we use `x$` and `y$` to indicate that `datadate` comes from `disclosure`, supplied to `inner_join()` as the y argument, and that `min_date` and `max_date` come from the x argument.

To keep the analysis simple, we do not include controls in the regressions below (apart from firm and year fixed effects). In a typical regression analysis with firm and year fixed effects, omission of controls will often not have a material impact on coefficient estimates.

```
fms <- list(feols(ln_ratio ~ post | gvkey + year,
                  vcov = "iid", data = reg_data),
            feols(ln_ratio_sale ~ post | gvkey + year,
                  vcov = "iid", data = reg_data))
```

We then run the analysis again, but focused on a subset of firms. We explore this analysis more deeply in the discussion questions.

```
switchers <-
  reg_data |>
  distinct(gvkey, post) |>
  group_by(gvkey) |>
  filter(n() > 1) |>
  select(gvkey) |>
  distinct() |>
  ungroup()

reg_data_switchers <-
  reg_data |>
  semi_join(switchers, by = "gvkey")

fms[[3]] <- feols(ln_ratio ~ post | gvkey + year,
                  vcov = "iid", data = reg_data_switchers)
fms[[4]] <- feols(ln_ratio_sale ~ post | gvkey + year,
                  vcov = "iid", data = reg_data_switchers)
```

The results reported in Table 21.4 are similar to those in Table 2 of Li et al. (2018).

```
modelsummary(fms,
             estimate = "{estimate}{stars}",
             gof_map = c("nobs", "r.squared"),
             stars = c('*' = .1, '**' = 0.05, '***' = .01))
```

TABLE 21.4
Effect of IDD adoption

	(1)	(2)	(3)	(4)
postTRUE	0.016**	0.024***	0.014	0.019**
	(0.008)	(0.008)	(0.009)	(0.010)
Num.Obs.	28,745	24,569	3,617	3,164
R2	0.672	0.711	0.624	0.661

We next conduct analysis similar to that reported in Table 3 of Li et al. (2018), which examines the effect of IDD adoption on disclosure by year related to adoption.

We first make `factor_t()`, a small function that takes a vector of numbers and returns them as a factor that has nicely formatted strings ordered by the original numbers.[12] This will be useful for presenting "nice" regression results.[13]

```
factor_t <- function(t) {
  t <- relevel(as.factor(t), ref = "-1")
  levels(t) <- str_replace(levels(t), "^([0-9]+)", "+\\1")
  levels(t) <- str_replace_all(levels(t), "([0-9]+)", " \\1")
  levels(t) <- str_replace(levels(t), "^", " ")
  levels(t) <- str_replace(levels(t), "-", "–")
  t
}
```

We next make `year_diff()`, which turns a vector of year-difference values into a factor—with values outside $(-5, +5)$ collapsed into -5 and $+5$ as appropriate—and labels the result in a way that facilitates regression output. Note that if a firm is never-treated, it's value of t will be missing and we set the value for these years to `-Inf` (R's way of saying $-\infty$).

```
year_diff <- function(t) {
  t <- case_when(t < -5 ~ -5,
                 t > 5 ~ 5,
                 is.na(t) ~ -Inf,
                 .default = t)
  factor_t(t)
}
```

We construct `switch_years` to store the first IDD adoption year for each GVKEY.

```
switch_years <-
  reg_data |>
  group_by(gvkey) |>
  arrange(datadate) |>
  filter(period_type == "Post-adoption",
         lag(period_type) == "Pre-adoption") |>
  group_by(gvkey) |>
  summarize(adoption_year = min(year), .groups = "drop")
```

We next construct our data set for this set of regressions (`reg_data_t`) by merging in data on adoption years and then calculating `t` using the `year_diff()` function we created above.

```
reg_data_t <-
  reg_data |>
  left_join(switch_years, by = "gvkey") |>
  mutate(t = year_diff(year - adoption_year))
```

Again we conduct two regressions for each of the two disclosure measures.

```
fms <- list(feols(ln_ratio ~ t | gvkey + year, vcov = "iid",
                  data = reg_data_t),
            feols(ln_ratio_sale ~ t | gvkey + year, vcov = "iid",
                  data = reg_data_t),
```

---

[12]Examine the output of `as.character(factor_t(c(-2, -1, 0, 1, 2)))` to see what `factor_t()` is doing.

[13]The penultimate line of `factor_t()` replaces hyphens (-) with proper minus signs (–) to improve the appearance of Table 21.5.

```
        feols(ln_ratio ~ t | gvkey + year, vcov = "iid",
                data = filter(reg_data_t, !str_detect(t, "Inf"))),
        feols(ln_ratio_sale ~ t | gvkey + year, vcov = "iid",
                data = filter(reg_data_t, !str_detect(t, "Inf")))))
```

Results from these regressions are provided in Table 21.5.

```
modelsummary(fms,
                estimate = "{estimate}{stars}",
                align = "cdddd",
                gof_map = "none",
                stars = c('*' = .1, '**' = 0.05, '***' = .01))
```

TABLE 21.5
Effect of IDD adoption by year

	(1)	(2)	(3)	(4)
t − 5	−0.005	0.001	−0.017	−0.009
	(0.019)	(0.021)	(0.027)	(0.029)
t − 4	−0.017	−0.030	−0.026	−0.038
	(0.022)	(0.023)	(0.025)	(0.026)
t − 3	−0.030	−0.029	−0.031	−0.031
	(0.020)	(0.020)	(0.021)	(0.022)
t − 2	−0.008	−0.003	−0.014	−0.010
	(0.018)	(0.018)	(0.019)	(0.019)
t + 0	−0.001	0.016	0.007	0.020
	(0.015)	(0.015)	(0.016)	(0.017)
t + 1	0.009	0.025	0.021	0.031
	(0.016)	(0.017)	(0.018)	(0.020)
t + 2	−0.007	0.010	0.009	0.017
	(0.017)	(0.018)	(0.021)	(0.023)
t + 3	0.023	0.021	0.038	0.026
	(0.019)	(0.020)	(0.025)	(0.027)
t + 4	0.028	0.039*	0.041	0.042
	(0.019)	(0.021)	(0.027)	(0.030)
t + 5	0.044***	0.048***	0.046	0.039
	(0.015)	(0.016)	(0.032)	(0.036)

Next we make `plot_coefs()`, to plot the estimated coefficients over time. The `plot_coefs()` function first arranges the coefficients for each period as a data frame and turns the coefficient labels (e.g., $t + 1$) into values (i.e., 1). In making the plot, we add in the reference period $(t - 1)$, which has a coefficient value of 0 by construction, and add 95% confidence intervals for the coefficients.

```
plot_coefs <- function(fm, ref = -1) {

  coefs <-
    tibble(t = names(coef(fm)),
           value = coef(fm),
           se = se(fm)) |>
    mutate(t = str_replace(t, "-", "-")) |>
```

```
    mutate(t = as.integer(str_replace_all(t, "[t ]+", ""))) |>
    filter(t != -Inf)

ci <- 0.95
mult <- qnorm(1 - (1 - ci) / 2)

tibble(t = ref, value = 0, se = 0) |>
  bind_rows(coefs) |>
  mutate(mean = value,
         top = value + mult * se,
         bot = value - mult * se) |>
  ggplot(aes(x = t, y = mean)) +
  geom_errorbar(aes(ymin = bot, ymax = top), width = .1) +
  geom_line() +
  geom_point() +
  scale_x_continuous(breaks = seq(from = min(coefs$t),
                                  to = max(coefs$t)))
}
```

We apply `plot_coefs()` to model (1) of Table 21.5 to create Figure 21.2.[14]

```
plot_coefs(fms[[1]])
```

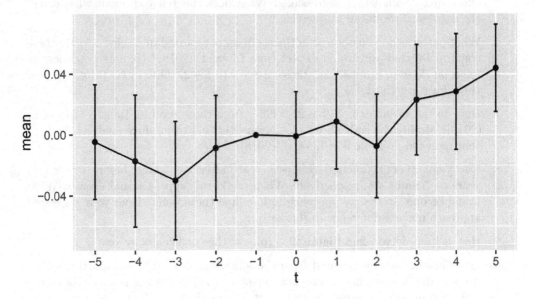

FIGURE 21.2
Plot of coefficients by year

21.2.5 Discussion questions

1. The proprietary cost hypothesis tested by Li et al. (2018) posits that increased cost of disclosure will lead to reduced supply of disclosure? What rival theories

---

[14]Here `align = "cdddd"` centres the first column (c) and aligns the others by the decimal point (d).

exist that would make alternative predictions? Are there other elements of disclosure theory that might be tested? Why do you think Li et al. (2018) focused on this specific element of the proprietary cost hypothesis?

2. In the analysis above, we do not include the control variables considered by Li et al. (2018). For example, Li et al. (2018) "include R&D expenditures to sales, advertisement [sic] expenditure to sales, and intangible assets scaled by total assets to control for a firm's proprietary costs of disclosure." Using the approach outlined in Chapter 4, in what circumstances would it be necessary to control for "a firm's proprietary costs of disclosure" in this way? Do these circumstances seem applicable in this setting, where the treatment is a (presumably exogenous) shock to disclosure costs and the outcome is disclosure choices?

3. What differs between the regressions reported in columns (1) and (3), and (2) and (4) of Table 21.4, respectively? Does this tell you anything about what drives the regression results in this setting? What happens if you omit the year fixed effects from both sets of regressions? What does this tell you about the role of the "non-switchers" (i.e., firms not in the `switchers` data frame) in the regression?

4. Would you expect the inclusion of controls (see the question above) to have a significant impact on the regression results? Why or why not?

5. What differs between the regressions reported in columns (1) and (3), and (2) and (4) of Table 21.5, respectively? What happens if you omit the year fixed effects from both sets of regressions? What does this tell you about what drives the regression results in this setting?

6. What patterns do you observe in the coefficients reported in Table 21.5? Do these conform to what you would expect from Li et al. (2018)? (It may be easiest to focus on these in groups, e.g., those in $t-5$ to $t-2$, those in $t+0$ to $t+3$ and those for $t+4$ and $t+5$.)

7. How do the variables in Table 21.5 differ from those used in Table 3 of Li et al. (2018)? Modify the code above (e.g., `year_diff()`) to produce analysis closer to that reported in Table 3 of Li et al. (2018).

8. The `year_diff()` function collapses years after $t+5$ and before $t-5$ into years $t+5$ and $t-5$, respectively. When would this approach make sense? What would be one alternative approach to handling these years? Does your suggested approach make a difference in this setting?

9. How helpful do you find Figure 21.2 (plot of the coefficients by year)?

10. Describe the data set created from the following code. What proportion of the firms in the data set have `same_state` equal to `TRUE`? For the purposes of empirical analysis, do the firms with `same_state` equal to `FALSE` enhance, or detract from, our ability to draw causal inferences about the effect of adoption of IDD?

```
switch_years <-
  reg_data_switchers |>
  group_by(gvkey) |>
  arrange(datadate) |>
  mutate(same_state = state == lag(state),
         adoption_year = period_type == "Post-adoption" &
           lag(period_type) == "Pre-adoption") |>
```

```
    filter(adoption_year) |>
    ungroup()
```

11. What issues might be implied by the following data? How might you address these?

```
reg_data_t |> count(t)
```

```
# A tibble: 12 x 2
      t          n
   <fct>      <int>
 1 " - 1"       322
 2 " -Inf"    25635
 3 " - 5"       232
 4 " - 4"       117
 5 " - 3"       154
 6 " - 2"       219
 7 " + 0"       385
 8 " + 1"       306
 9 " + 2"       233
10 " + 3"       192
11 " + 4"       168
12 " + 5"       782
```

21.3 Further reading

Chapter 5 of Angrist and Pischke (2008) provides a good introduction to the topic of fixed effects and panel data. Chapters 8 and 9 of Cunningham (2021) go beyond the topics here and provide an excellent pathway to recent developments in the fixed-effects literature. Chapters 16 and 18 of Huntington-Klein (2021) also provide excellent coverage, including discussion of some issues not covered in this book.

22

Regression discontinuity designs

Over the last twenty years or so, the regression discontinuity design (RDD) has seen a dramatic rise in popularity among researchers. Cunningham (2021) documents a rapid increase in the number of papers using RDD after 1999 and attributes its popularity to its ability to deliver causal inferences with "identifying assumptions that are viewed by many as easier to accept and evaluate" than those of other methods. Lee and Lemieux (2010, p. 282) point out that RDD requires "seemingly mild assumptions compared to those needed for other non-experimental approaches ... and that causal inferences from RDD are potentially more credible than those from typical 'natural experiment' strategies."

Another attractive aspect of RDD is the availability of quality pedagogical materials explaining its use. The paper by Lee and Lemieux (2010) is a good reference, as are chapters in Cunningham (2021) and Angrist and Pischke (2008). Given the availability of these materials, we have written this chapter as a complementary resource, focusing more on practical issues and applications in accounting research. The goal of this chapter is to provide a gateway for you to either using RDD in your own research (should you find a setting with a discontinuity in a treatment of interest) or in reviewing papers using (or claiming to use) RDD.

A number of phenomena of interest to accounting researchers involve discontinuities. For example, whether an executive compensation plan is approved is a discontinuous function of shareholder support (e.g., Armstrong et al., 2013) and whether a debt covenant is violated is typically a discontinuous function of the realization of accounting ratios. So it is not surprising that RDD has attracted some interest of accounting researchers, albeit in a relatively small number of papers. Unfortunately, it is also not surprising that RDD has not always yielded credible causal inferences in accounting research, as we discuss below.

> 💡 Tip
>
> The code in this chapter uses the packages listed below. We load **tidyverse** because we use several packages from the Tidyverse. For instructions on how to set up your computer to use the code found in this book, see Section 1.2. Quarto templates for the exercises below are available on GitHub.[a]
>
> ---
> [a]https://github.com/iangow/far_templates/blob/main/README.md

```
library(tidyverse)
library(DBI)
library(farr)
library(dbplyr)          # copy_inline()
library(fixest)
library(rdrobust)
library(modelsummary)
```

22.1 Sharp RDD

Sharp RDD exploits the use of sharp, somewhat arbitrary, cut-offs by various decision-makers or regulations. For example, a university might admit students reaching a certain GPA or a regulator might require compliance for all firms with sales of more than $25 million. In some cases, the decision-maker in question is effectively using the cut-off to determine the assignment of observations to a treatment that may be of interest to researchers. Denoting a treatment indicator for observation i as D_i, we might have all cases where x_i is at or above some threshold x_0 receiving treatment, while all other cases do not.

$$D_i = \begin{cases} 1 & x_i \geq x_0 \\ 0 & x_i < x_0 \end{cases}$$

For example, the first application of RDD (Thistlethwaite and Campbell, 1960) examined the effect of National Merit Scholarships, which were awarded to students who scored 11 or higher on a qualifying exam.[1]

Apart from the sharp threshold for treatment assignment, the basic identifying assumption of RDD is encapsulated in the continuity of the function $f(\cdot)$ in the following expression relating the **running variable** (x_i) and treatment (D_i) to the outcome of interest (y_i).

$$y_i = f(x_i) + \rho D_i + \eta_i$$

Given a small positive number Δ, we can approximate the value of y_{0i} (the outcome when not treated) when $x_i = x_0$ by averaging over values just to the left of x_0

$$\mathbb{E}[y_{0i}|x_i = x_0] \approx \mathbb{E}[y_i|x_0 - \Delta < x_i < x_0]$$

Similarly, we can approximate the value of y_{1i} (the outcome when treated) when $x_i = x_0$ by averaging over values just to the right of x_0.

$$\mathbb{E}[y_{1i}|x_i = x_0] \approx \mathbb{E}[y_i|x_0 \leq x_i < x_0 + \Delta]$$

This approximation gets tighter and tighter if we move Δ closer to zero and we can conclude that

$$\lim_{\Delta \to 0} \left(\mathbb{E}[y_i|x_0 \leq x_i < x_0 + \Delta] - \mathbb{E}[y_i|x_0 - \Delta < x_i < x_0] \right) = \mathbb{E}[y_{1i}|x_i = x_0] - \mathbb{E}[y_{0i}|x_i = x_0]$$

$$= \mathbb{E}[y_{1i} - y_{0i}|x_i = x_0]$$

$$= \rho$$

There are two noteworthy points for this analysis. First, we have relied on the continuity of the expectation to the left and right of x_0, which is given by the continuity of $f(x_i)$ in x_i to the left of x_0 and the continuity of $f(x_i) + \rho$ in x_i to the right of x_0. Second, we recover an estimate of the effect of interest ρ even though we have no random assignment.

RDD exploits the fact that whether an observation is just to the left or right of x_0 is effectively random. In practice, this implicit assumption of our analysis may not hold and

---

[1] While Thistlethwaite and Campbell (1960) seems to be very commonly cited as an example of sharp RDD, it appears that only a minority of students who scored higher than the cut-off were awarded National Merit Scholarships, making it strictly an example where fuzzy RDD should be applied if the treatment of interest is receipt of a National Merit Scholarship.

below we discuss how we might detect deviations from this assumption and what they mean for causal analysis.

Of course, another noteworthy feature of the analysis above is that we assumed we could set Δ arbitrarily close to zero and still have enough data to estimate the two conditional expectations we used to estimate ρ. In practice, we will not have this luxury and thus there is a trade-off between getting precise estimates (using more data) and reducing bias (restricting analysis to data as close to x_0 as possible).

22.1.1   Use of polynomial regressions

Studies exploiting RDD have often used polynomial functions to approximate $f(x_i)$ with regressions of the following form:

$$y_i = \alpha + \rho D_i + \beta_1 x_i + \beta_2 x_i^2 + \dots \beta_p x_i^p$$
$$+ \gamma_2 D_i x_i + \gamma_2 D_i x_i^2 + \dots \gamma_p D_i x_i^p$$

In other words, separate polynomials of degree p are fitted on either side of the threshold and the estimated effect is given by ρ. Discussion of this approach can be found in Angrist and Pischke (2008, p. 255) and Lee and Lemieux (2010, p. 326).

However, more recent work by Gelman and Imbens (2019) suggests that the use of polynomial-based regressions of the form above "is a flawed approach with three major problems: it leads to noisy estimates, sensitivity to the degree of the polynomial, and poor coverage of confidence intervals. We recommend researchers instead use estimators based on local linear or quadratic polynomials or other smooth functions." Here "poor coverage of confidence intervals" means excessive rejection of true null hypotheses, which is of particular concern with "noisy estimates", as the rejection of null hypotheses is precisely what many researchers seek in methods.

Fortunately, canned regression routines for implementing the methods recommended by Gelman and Imbens (2019) are widely available (e.g., the `rdrobust` package that we use below is available for both R and Stata) and easy to use.

22.1.2   Manipulation of the running variable

While RDD makes relatively mild assumptions, in practice these assumptions may be violated. In particular, manipulation of the **running variable** (i.e., the variable that determines whether an observation is assigned to treatment) may occur and lead to biased estimates of treatment effects.

For example, Hoekstra (2009) studies the effect of attending a flagship state university on future income using a discontinuous probability of admission and enrolment around an admission cut-off in terms of SAT points. A concern that might arise in the setting of Hoekstra (2009) is that if students knew the precise cut-off, then the more energetic and determined students who just missed out on the first attempt might retake the SAT to cross the threshold. This would create problems for causal inference, as the students just to the right of the cut-off could be significantly different from those just to the left of the cut-off in terms of variables (i.e., energy and determination) plausibly associated with the outcome variable (future income). These differences would undermine the validity of the estimates of effects from RDD.[2]

---

[2]Hoekstra (2009, p. 719) takes care to explain why this is not a concern in his setting.

Fortunately, methods exist for researchers to examine their data for this possibility (see, e.g., Listokin, 2008; McCrary, 2008), which would manifest as "too few" students just to the left of the cut-off and "too many" just to the right.[3]

22.2 Fuzzy RDD

Fuzzy RDD applies when the cut-off x_0 is associated not with a sharp change from "no treatment for any" to "treatment for all", but with a sharp increase in the probability of treatment:

$$P(D_i = 1|x_i) = \begin{cases} g_1(x_i) & x_i \geq x_0 \\ g_0(x_i) & x_i < x_0 \end{cases}, \text{ where } g_1(x_0) > g_0(x_0)$$

Note that sharp RDD can be viewed as a limiting case of fuzzy RDD in which $g_1(x_i) = 1$ and $g_0(x_i) = 0$.

The best way to understand fuzzy RDD is as instrumental variables meets sharp RDD. Suppose that shifting from just below to x_0 to x_0 or just above increases the probability of treatment by π, then x_0 is effectively random in that small range (thus addressing the first requirement for a valid instrument), it has an effect on treatment so long as π is sufficiently different from 0 (the second requirement) and its only effect on y over that small range occurs only through its effect on treatment (the third requirement).

The value for y_i just to the right of x_0 will be approximately $f(x_0) + D_i + \eta_i$ if treatment occurs and approximately $f(x_0) + \eta_i$ if treatment does not occur. Given that treatment occurs with probability approximately equal to $g_1(x_0)$ at or just to the right of x_0, we have $\mathbb{E}[y_i|x_0 \leq x_i < x_0 + \Delta] \approx f(x_0) + g_1(x_0)D_i + \eta_i$. The value for y_i just to the left of x_0 will be approximately $f(x_0) + D_i + \eta_i$ if treatment occurs and approximately $f(x_0) + \eta_i$ if treatment does not occur. Given that treatment occurs with probability approximately equal to $g_0(x_0)$ just to the left of x_0, we have $\mathbb{E}[y_i|x_0 - \Delta < x_i < x_0] \approx f(x_0) + g_0(x_0)D_i + \eta_i$. If $g_1(x_0) - g_0(x_0) = \pi$, we have

$$\lim_{\Delta \to 0} (\mathbb{E}[y_i|x_0 \leq x_i < x_0 + \Delta] - \mathbb{E}[y_i|x_0 - \Delta < x_i < x_0]) = \mathbb{E}[y_{1i}|x_i = x_0] - \mathbb{E}[y_{0i}|x_i = x_0]$$

$$= \mathbb{E}[y_{1i} - y_{0i}|x_i = x_0]$$

$$= \pi\rho$$

We can estimate the effect of treatment, ρ, by noting that

$$\rho = \lim_{\Delta \to 0} \frac{\mathbb{E}[y_i|x_0 \leq x_i < x_0 + \Delta] - \mathbb{E}[y_i|x_0 - \Delta < x_i < x_0]}{\mathbb{E}[D_i|x_0 \leq x_i < x_0 + \Delta] - \mathbb{E}[D_i|x_0 - \Delta < x_i < x_0]}$$

Hoekstra (2009) is an excellent paper for building intuition with regard to fuzzy RDD and we recommend you read it. Figure 1 of Hoekstra (2009) demonstrates a sharp jump in the probability of enrolment (by 0.388) at the flagship state university as students SAT score crosses the threshold for admission. Figure 2 of Hoekstra (2009) depicts an increase in future annual earnings (by 0.095) as a student's SAT score crosses the threshold for admission. This suggests that the estimated effect of enrolment on future earnings will be *approximately* $0.095 \div 0.388 = 0.245$, which is actually quite close to the estimate of 0.223 in Table 1 of Hoekstra (2009).

---

[3]Here notions about "too few" and "too many" reflect expectations given the natural distribution of data and not some evaluation of the merits of having students on one side of the cut-off or the other.

22.3   Other issues

One issue with RDD is that the effect estimated is a local estimate (i.e., it relates to observations close to the discontinuity). This effect may be very different from the effect at points away from the discontinuity. For example, in designating a public float of $75 million as a threshold for applying the requirements of the Sarbanes-Oxley Act (SOX), the SEC may have reasoned that at that point the benefits of SOX were approximately equal to the costs of complying with it. If true, we would expect to see an estimate of approximately zero effect, even if there were substantial benefits of the law for shareholders of firms having a public float well above the threshold.

Another critical element of RDD is the bandwidth used in estimation (i.e., in effect how much weight is given to observations according to their distance from the cut-off). Gow et al. (2016) encourage researchers using RDD to employ methods that exist to estimate optimal bandwidths and the resulting estimates of effects (e.g., Imbens and Kalyanaraman, 2012). The `rdrobust` package used above provides various approaches to estimating optimal bandwidths and makes it easy to estimate RDD with alternative bandwidths. Credible results from RDD are likely to be robust to alternative bandwidth specifications.

Finally, one strength of RDD is that the estimated relation is often effectively univariate and easily plotted. As suggested by Lee and Lemieux (2010), it is highly desirable for researchers to plot both the underlying data and the fitted regression functions around the discontinuity. This plot will enable readers to evaluate the strength of the results. If there is a substantive impact associated with the treatment, this should be obvious from a plot of the actual data and the associated fitted function. Inspection of the plots in Hoekstra (2009) provides some assurance that the estimated effect is "there". In contrast, the plots below using data from Bloomfield (2021) seem somewhat less compelling.

22.4   Sarbanes-Oxley Act

The Sarbanes-Oxley Act (SOX), passed by the US Congress in 2002 in the wake of scandals such as Enron and WorldCom, was arguably the most significant shift in financial reporting in recent history, with effects not only in the United States, but elsewhere in the world as measures taken in the United States were adopted in other jurisdictions. Iliev (2010, p. 1163) writes "the law's main goal was to improve the quality of financial reporting and increase investor confidence."

A key element of SOX is section 404 (SOX 404). Section 404(a) requires the Securities and Exchange Commission (SEC) to prescribe rules requiring management to include in annual reports "an assessment of ... effectiveness of the internal control ... for financial reporting." Section 404(b) requires the company's external auditor to "attest to, and report on, the assessment made by the management of the issuer."

As discussed in Iliev (2010), SOX 404 generated significant controversy, with particular concerns being expressed about the costs of compliance, especially for smaller firms. In response to such concerns. the SEC required only companies whose **public float** exceeded $75 million in either 2002, 2003, or 2004 to comply with SOX 404 for fiscal years ending on or after 15 November 2004.[4] Smaller companies were not required to submit management reports until fiscal 2007 and these did require auditor attestation until 2010.

Iliev (2010) exploits the discontinuity in treatment around $75 million in public float and RDD to evaluate the effect of SOX 404 on audit fees and earnings management, finding evidence that SOX 404's required management reports increased audit fees, but reduced earnings management.

Iliev (2010) "hand-collected" data on public float, and whether a firm was an accelerated filer, and whether it provided a management report under SOX 404 directly from firms' 10-K filings.[5]

Some of these data are included in the `iliev_2010` data frame included with the `farr` package.

`iliev_2010`

```
# A tibble: 7,214 x 9
   gvkey  fyear fdate      pfdate     pfyear publicfloat mr    af    cik
   <chr>  <int> <date>     <date>      <dbl>       <dbl> <lgl> <lgl> <dbl>
 1 028712  2001 2001-12-31 2002-03-21   2002        21.5 FALSE FALSE 908598
 2 028712  2002 2002-12-31 2002-06-28   2002        11.5 FALSE FALSE 908598
 3 028712  2003 2003-12-31 2003-06-30   2003        13.9 FALSE FALSE 908598
 4 028712  2004 2004-12-31 2004-06-30   2004        68.2 FALSE FALSE 908598
 5 028712  2005 2005-12-31 2005-06-30   2005        26.4 FALSE FALSE 908598
 6 013864  2001 2001-03-31 2001-07-06   2001        35.2 FALSE FALSE 819527
 7 013864  2002 2002-03-31 2002-06-27   2002        31.8 FALSE FALSE 819527
 8 013864  2003 2003-03-31 2002-09-30   2002        20.2 FALSE FALSE 819527
 9 013864  2004 2004-03-31 2003-09-30   2003        18.3 FALSE FALSE 819527
10 013864  2005 2005-03-31 2004-09-30   2004        9.04 FALSE FALSE 819527
# i 7,204 more rows
```

The variables `af` and `mf` are indicators for a firm being an accelerated filer and for a firm filing a management report under SOX 404, respectively.

Iliev (2010, p. 1169) explains the criteria for a SOX 404 management report being required as follows:

> All accelerated filers with a fiscal year ending on or after November 15, 2004 had to file a MR and an auditor's attestation of the MR under Section 404. I denote these firms as MR firms. Companies that were not accelerated filers as of their fiscal year ending on or after November 15, 2004 did not have to file an MR in that year (non-MR firms). Those were companies that had a public float under $75 million in their reports for fiscal years 2002 (November 2002 to October 2003), 2003 (November 2003 to October 2004), and 2004 (November 2004 to October 2005).

The code producing the following data frame `mr_req_df` attempts to encode these requirements.

```
mr_req_df <-
  iliev_2010 |>
```

---

[4]Iliev (2010, p. 1165) describes public float as "the part of equity not held by management or large shareholders, as reported on the first page of the company 10-K." See also https://www.sec.gov/news/press/2004-158.htm.

[5]Here "hand collection" includes using web-scraping the SEC EDGAR repository. Web-scraping might use code applying regular expressions to text, such as we saw in Chapter 9.

```
filter(!is.na(fdate)) |>
filter(fdate >= "2002-11-01", fdate <= "2005-10-31") |>
group_by(gvkey, fdate) |>
mutate(max_float = max(publicfloat), .groups = "drop") |>
filter(fdate >= "2004-11-15") |>
mutate(mr_required = max_float >= 75) |>
ungroup()
```

Interestingly, as seen in Table 22.1, there appear to be a number of firms that indicated that they were not accelerated filers and did not have to file a management report, despite appearing to meet the criteria.

```
mr_req_df |>
  filter(mr_required, !mr, !af) |>
  select(cik, fdate, pfdate, publicfloat) |>
  arrange(publicfloat) |>
  top_n(6)
```

```
Selecting by publicfloat
```

TABLE 22.1

Sample of firm-years appearing to meet criteria, but no SOX 404 report

cik	fdate	pfdate	publicfloat
1003472	2004-12-31	2005-03-09	141.9853
943861	2004-12-31	2005-03-24	168.5346
906780	2004-12-31	2005-03-01	171.6621
1085869	2004-12-31	2005-02-28	172.0916
29834	2004-12-31	2005-03-16	184.9020
1028205	2005-06-30	2005-08-15	279.1016

Use the firm with a CIK 1022701 as an example, we see[6] that the values included in `iliev_2010` are correct. However, digging into the details of the requirements for accelerated filers[7] suggests that the precise requirement was a float of "$75 million or more as of the last business day of its most recently completed *second* fiscal quarter." Unfortunately, this number may differ from the number reported on the first page of the 10-K introducing possible measurement error in classification based solely on the values in `publicfloat`. Additionally, there may have been confusion about the precise criteria applicable at the time.

Note that Iliev (2010) uses either the variable `mr` or (for reasons explained below) public float values from 2002, and so does not rely on a precise mapping from public float values over 2002–2004 to `mr` in his analysis.

---

[6]https://www.sec.gov/Archives/edgar/data/0001022701/000114420405006759/v013795_10ksb.txt
[7]https://go.unimelb.edu.au/t7d8

22.4.1 Bloomfield (2021)

The main setting we will use to understand the application of RDD is that of Bloomfield (2021), who predicts that reducing reporting flexibility will lower managers' ability to obfuscate poor performance, thus lowering *risk asymmetry*, measured as the extent to which a firm's returns co-vary more with negative market returns than with positive market returns. Bloomfield (2021, p. 869) uses SOX 404 as a constraint on managers' ability to obfuscate poor performance and "to justify a causal interpretation" follows Iliev (2010) in using "a regression discontinuity design ('RDD') to implement an event study using plausibly exogenous variation in firms' exposure to the SOX 404 mandate."

Bloomfield (2021) conducts two primary analyses of the association between SOX 404 adoption and risk asymmetry, which are reported in Tables 2 and 4.

Table 2 of Bloomfield (2021) reports results from regressing risk asymmetry on an indicator for whether a firm is a SOX 404 reporter, based on the variable `auopic` found on `comp.funda`.[8] However, Bloomfield (2021, p. 884) cautions that "these analyses do not exploit any type of plausibly exogenous variation in SOX reporting, and should therefore be interpreted cautiously as evidence of an association, but not necessarily a causal relation."

Table 4 reports what Bloomfield (2021) labels the "main analysis". In that table, Bloomfield (2021, p. 884) "follow[s] Iliev's [2010] regression discontinuity methodology and use a difference-in-differences design to identify the causal effect of reporting flexibility on risk asymmetry. ... [and] use[s] firms' 2002 public floats—before the threshold announcement—as an instrument for whether or not the firm will become treated at the end of 2004."

All analyses in Table 4 of Bloomfield (2021) include firm and year fixed effects. Panel B includes firm controls, such as firm age and size, while Panel A omits these controls. Like many papers using such fixed-effect structures, inferences across the two panels are fairly similar and thus here we focus on the analysis of the "full sample" without controls (i.e., column (1) of Panel A of Table 4).

To reproduce the result of interest, we start with `bloomfield_2021`, which is provided in the `farr` package based on data made available by Bloomfield (2021). The data frame `bloomfield_2021` contains the fiscal years and PERMCOs for the observations in Bloomfield (2021)'s sample. We then compile data on public floats in 2002, which is the proxy for treatment used in Table 4 of Bloomfield (2021). For this Bloomfield (2021) uses data based on that supplied in `iliev_2010`. Because we will conduct some analysis inside the database, we copy these data to the database.

Because we do not have write access to the WRDS PostgreSQL server, we use `copy_inline()`, which allows us to create a virtual table even without such access.

```
db <- dbConnect(RPostgres::Postgres())

bloomfield_2021 <- copy_inline(db, df = farr::bloomfield_2021)
iliev_2010 <- copy_inline(db, df = farr::iliev_2010)
```

---

[8]According to the Compustat manual, the variable `auopic`, when not `NA`, takes the value 0 if there is "No Auditor's Report", 1 if "Effective (No Material Weakness)", 2 if "Adverse (Material Weakness Exists)", 3 if "Disclaimer (Unable to Express Opinion)", and 4 if "Delayed Filing": https://wrds-www.wharton.upenn.edu/data-dictionary/comp_na_daily_all/funda/auopic.

We can create the data used by Bloomfield (2021) by translating the Stata code made available on the site[9] maintained by the *Journal of Accounting Research* into the following R code.

```
float_data <-
  iliev_2010 |>
  group_by(gvkey, fyear) |>
  filter(publicfloat == min(publicfloat, na.rm = TRUE),
         pfyear %in% c(2004, 2002)) |>
  group_by(gvkey, pfyear) |>
  summarize(float = mean(publicfloat, na.rm = TRUE), .groups = "drop") |>
  pivot_wider(names_from = pfyear, values_from = float,
              names_prefix = "float")
```

To use the data in `float_data` we need to link PERMCOs to GVKEYs and use `ccm_link` (as defined below) to create `float_data_linked`. To maximize the number of successful matches, we do not condition on link-validity dates.

```
ccmxpf_lnkhist <- tbl(db, Id(schema = "crsp", table = "ccmxpf_lnkhist"))
```

```
ccm_link <-
  ccmxpf_lnkhist |>
  filter(linktype %in% c("LC", "LU", "LS"),
         linkprim %in% c("C", "P")) |>
  rename(permco = lpermco) |>
  select(gvkey, permco) |>
  distinct()
```

```
float_data_linked <-
  float_data |>
  inner_join(ccm_link, by = "gvkey")
```

To help us measure risk asymmetry over the twelve months ending with a firm's fiscal year-end, we get data on the precise year-end for each firm-year from `comp.funda` and store it in `firm_years`. (We also collect data on `sox` as this is also found on `comp.funda`.)

```
funda <- tbl(db, Id(schema = "comp", table = "funda"))
```

```
funda_mod <-
  funda |>
  filter(indfmt == "INDL", datafmt == "STD",
         consol == "C", popsrc == "D")
```

```
firm_years <-
  funda_mod |>
  mutate(sox = coalesce(as.double(auopic) > 0, FALSE)) |>
  select(gvkey, fyear, datadate, sox)
```

We then link this table with `bloomfield_2021` using `ccm_link` (again) to create `risk_asymm_sample`, which is simply `bloomfield_2021` with `datadate` replacing `fyear`. As `bloomfield_2021` can contain multiple years for each firm, we expect each row in `ccm_link` to match multiple rows in `bloomfield_2021`. At the same time, some `permco` values link

---

[9]https://research.chicagobooth.edu/arc/journal-of-accounting-research/online-supplements/volume-59

with multiple GVKEYs, so some rows in `bloomfield_2021` will match multiple rows in `ccm_link`. As such, we set `relationship = "many-to-many"` in the join below.

```
risk_asymm_sample <-
  bloomfield_2021 |>
  inner_join(ccm_link, by = "permco",
             relationship = "many-to-many") |>
  inner_join(firm_years, by = c("fyear", "gvkey")) |>
  select(permco, datadate) |>
  distinct() |>
  mutate(period_start = datadate - months(12) + days(1))
```

To calculate risk asymmetry using the approach described in Bloomfield (2021), we need data on market returns. Bloomfield (2021, pp. 877–878) "exclude[s] the highest 1% and lowest 1% of market return data. ... This trimming procedure improves the reliability of the parameters estimates." To effect this, we draw on the `truncate()` from the `farr` package.[10] To apply the `truncate()` function, we need to bring the data into R using `collect()`.

But we will want to merge these data with data from CRSP (`crsp.dsf`), so we copy the data back to the PostgreSQL database. As we do not have write access to WRDS PostgreSQL, we use the `copy_inline()` function here.

```
factors_daily <- tbl(db, Id(schema = "ff", table = "factors_daily"))

mkt_rets <-
  factors_daily |>
  select(date, rf, mktrf) |>
  collect() |>
  mutate(mktrf = truncate(mktrf)) |>
  filter(!is.na(mktrf)) |>
  copy_inline(db, df = _)
```

```
dsf <- tbl(db, Id(schema = "crsp", table = "dsf"))
```

Finally, we calculate risk asymmetry using the approach described in Bloomfield (2021). Note that we use `regr_slope()` and `regr_count()` functions, which are database functions (available in both PostgreSQL and DuckDB) rather than R functions. Because the data we are using are in the database, we have access to these functions. Performing these calculations inside the database delivers much faster performance than we would get if we brought the data into R. After calculating the regression statistics (`slope` and `nobs`) grouped by `sign_ret` (and `permco` and `datadate`), we use `pivot_wider()` so that we can bring `beta_plus` and `beta_minus` into the same rows.

```
risk_asymmetry <-
  dsf |>
  inner_join(risk_asymm_sample,
             join_by(permco,
                     between(date, period_start, datadate))) |>
  inner_join(mkt_rets, by = "date") |>
  mutate(retrf = ret - rf,
         sign_ret = mktrf >= 0) |>
  group_by(permco, datadate, sign_ret) |>
```

---

[10]We explain `truncate()` in Chapter 24.

```
summarize(slope = regr_slope(retrf, mktrf),
          nobs = regr_count(retrf, mktrf),
          .groups = "drop") |>
pivot_wider(id_cols = c("permco", "datadate"),
          names_from = "sign_ret",
          values_from = c("slope", "nobs")) |>
mutate(beta_minus = slope_FALSE,
       beta_plus = slope_TRUE,
       nobs = nobs_FALSE + nobs_TRUE) |>
select(permco, datadate, beta_minus,
       beta_plus, nobs)
```

We combine all the data from above into a single data frame for regression analysis. Bloomfield (2021, p. 878) says that "because of the kurtosis of my estimates, I winsorize $\hat{\beta}$, $\hat{\beta}^+$, and $\hat{\beta}^-$ at 1% and 99% before constructing my measures of risk asymmetry." To effect this, we use `winsorize()` from the `farr` package.[11]

```
cutoff <- 75

reg_data <-
  risk_asymmetry |>
  left_join(float_data_linked, by = "permco") |>
  left_join(firm_years, by = c("datadate", "gvkey")) |>
  mutate(treat = coalesce(float2002 >= cutoff, TRUE),
         post = datadate >= "2005-11-01",
         year = year(datadate)) |>
  collect() |>
  mutate(across(starts_with("beta"), winsorize)) |>
  mutate(risk_asymm = beta_minus - beta_plus)
```

We then run two regressions: one with firm and year fixed effects and one without. The former aligns with what Bloomfield (2021) reports in Table 4, while the latter better aligns with the RDD analysis we conduct below.

```
fms <- list(feols(risk_asymm ~ I(treat * post), data = reg_data),
            feols(risk_asymm ~ I(treat * post) | permco + year,
                  data = reg_data))

modelsummary(fms,
             estimate = "{estimate}{stars}",
             gof_map = c("nobs", "r.squared"),
             notes = "Column (2) includes firm and year fixed effects",
             stars = c('*' = .1, '**' = 0.05, '***' = .01))
```

The results reported above confirm the result reported in Bloomfield (2021). While our estimated coefficient of (-0.331) is not identical to that in Bloomfield (2021) (-0.302), it is close to it and has the same sign and similar statistical significance.[12] However, the careful reader will note that Table 4 presents "event study difference-in-differences results"

---

[11]We discuss winsorization and the `winsorize()` function in Chapter 24.

[12]Note that Bloomfield (2021) calculates "standard errors clustered by industry and fiscal year" while we just use unclustered standard errors, as we did not collect industry data to keep things simple.

TABLE 22.2
Effect of SOX 404 on risk asymmetry

	(1)	(2)
(Intercept)	0.267***	
	(0.023)	
I(treat * post)	−0.181***	−0.331***
	(0.066)	(0.101)
Num.Obs.	1,849	1,849
R2	0.004	0.194

Column (2) includes firm and year fixed effects

rather than a conventional RDD analysis. Fortunately, we have the data we need to perform conventional RDD analyses ourselves.

First, we ask whether the 2002 float (`float2002`) being above or below $75 million is a good instrument for treatment (the variable `sox` derived from Compustat). This is analogous to producing Figure 1 in Hoekstra (2009), which shows a clear discontinuity in the probability of treatment (enrolment at the flagship state university) as the running variable (SAT points) crosses the threshold. Figure 1 in Hoekstra (2009) is essentially an RDD analysis in its own right, but with the probability of treatment being represented on the y-axis.

In the case of Bloomfield (2021), we have treatment represented by `sox`, which is derived from Compustat's `auopin` variable. One difference between Iliev (2010) and Bloomfield (2021) is that the data in Iliev (2010) extend only to 2006, while data in the sample used in Table 4 of Bloomfield (2021) extend as late as 2009. This difference between the two papers is potentially significant for the validity of `float2002` as an instrument for `sox`, as whether a firm has public float above the threshold in 2002 is likely to become increasingly less predictive of treatment as time goes on.

In the following analysis, the `rdplot()` function from the `rdrobust` package divides the post-SOX data from the sample of Bloomfield (2021) into bins based on values of `float2002`. The `rdplot()` function then calculates and plots the average of the treatment indicator (`sox`) in each bin, along with a fitted relation between `float2002` and `sox` for each.

```
reg_data_fs <-
  reg_data |>
  filter(post)

rdplot(reg_data_fs$sox, reg_data_fs$float2002, c = cutoff,
       masspoints = "off",
       y.label = "SOX 404 report indicator",
       x.label = "Public float in 2002")
```

What we observe in Figure 22.1 is that, in contrast to Figure 1 of Hoekstra (2009), there is no obvious discontinuity in the probability of treatment (`sox`) around the cut-off, undermining the notion that `float2002` is a viable instrument for `sox`.

What happens if we use `float2004` as the partitioning variable?

```
rdplot(reg_data_fs$sox, reg_data_fs$float2004, c = cutoff,
       masspoints = "off")
```

From Figure 22.2, we see that things now look a little better and we confirm this statistically next:

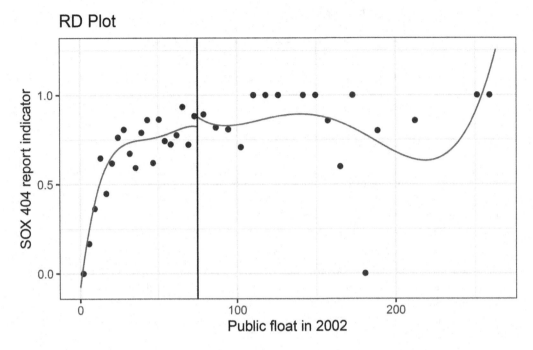

FIGURE 22.1
SOX 404 report indicator and 2002 public float

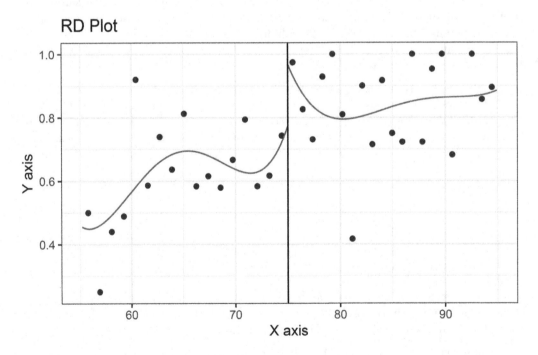

FIGURE 22.2
SOX 404 report and 2004 public float

```
rd_fs <- rdrobust(reg_data_fs$sox, reg_data_fs$float2004,
                  c = cutoff, masspoints = "off")
summary(rd_fs)
```

Sharp RD estimates using local polynomial regression.

```
Number of Obs.                       866
BW type                            mserd
Kernel                        Triangular
VCE method                            NN

Number of Obs.                       479              387
Eff. Number of Obs.                  193              154
Order est.  (p)                        1                1
Order bias  (q)                        2                2
BW est.  (h)                       6.988            6.988
BW bias  (b)                      10.036           10.036
rho (h/b)                          0.696            0.696
```

Method	Coef.	Std. Err.	z	P>|z|	[95% C.I.]
Conventional	0.241	0.065	3.706	0.000	[0.113 , 0.368]
Robust	-	-	3.273	0.001	[0.100 , 0.400]

The probability jump at the cut-off is 0.241 with a standard error of 0.065. This suggests we might run fuzzy RDD with float2004 as the running variable (we defer concerns about the appropriateness of this to the discussion questions). We can do this using rdrobust() as follows:

```
rd_ss <- rdrobust(reg_data_fs$risk_asymm, reg_data_fs$float2004,
                  fuzzy = reg_data_fs$sox,
                  c = cutoff, masspoints = "off")
summary(rd_ss)
```

Fuzzy RD estimates using local polynomial regression.

```
Number of Obs.                       866
BW type                            mserd
Kernel                        Triangular
VCE method                            NN

Number of Obs.                       479              387
Eff. Number of Obs.                  211              169
Order est.  (p)                        1                1
Order bias  (q)                        2                2
BW est.  (h)                       7.837            7.837
BW bias  (b)                      11.938           11.938
rho (h/b)                          0.656            0.656
```

First-stage estimates.

```
================================================================================
     Method     Coef. Std. Err.          z     P>|z|       [ 95% C.I. ]
================================================================================
Conventional    0.240    0.061      3.925     0.000     [0.120 , 0.360]
      Robust        -        -      3.409     0.001     [0.106 , 0.392]
================================================================================
```

Treatment effect estimates.

```
================================================================================
     Method     Coef. Std. Err.          z     P>|z|       [ 95% C.I. ]
================================================================================
Conventional   -0.778    0.677     -1.148     0.251     [-2.105 , 0.550]
      Robust        -        -     -0.850     0.395     [-2.212 , 0.874]
================================================================================
```

The estimated effect at the cut-off is −0.778, which is not significantly different from zero.

We can also run a kind of **intent-to-treat** RDD as follows and can also plot this as in Figure 22.3.

```
rd_rf <- rdrobust(reg_data_fs$risk_asymm, reg_data_fs$float2004,
            c = cutoff, masspoints = "off")
summary(rd_rf)
```

Sharp RD estimates using local polynomial regression.

```
Number of Obs.                866
BW type                     mserd
Kernel                  Triangular
VCE method                     NN

Number of Obs.               479         387
Eff. Number of Obs.          174         142
Order est. (p)                 1           1
Order bias  (q)                2           2
BW est. (h)                5.909       5.909
BW bias (b)                8.490       8.490
rho (h/b)                  0.696       0.696
```

```
================================================================================
     Method     Coef. Std. Err.          z     P>|z|       [ 95% C.I. ]
================================================================================
Conventional   -0.211    0.177     -1.194     0.233     [-0.558 , 0.136]
      Robust        -        -     -1.018     0.309     [-0.615 , 0.195]
================================================================================
```

```
rdplot(reg_data_fs$risk_asymm, reg_data_fs$float2004,
       c = cutoff, masspoints = "off")
```

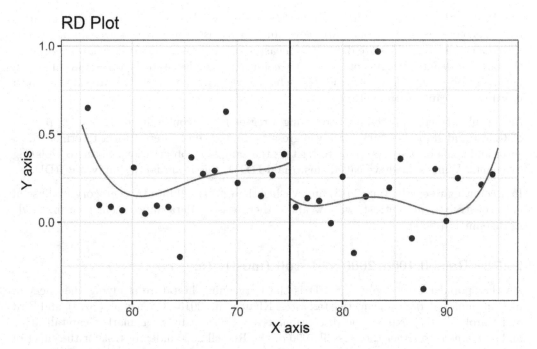

FIGURE 22.3
Risk asymmetry and 2004 public float

However, in neither the fuzzy RDD nor the intent-to-treat RDD do we reject the null hypothesis of no effect of SOX on risk asymmetry.[13]

22.5 RDD in accounting research

While RDD has seen an explosion in popularity since the turn of the century, relatively few papers in accounting research have used it. Armstrong et al. (2022) identify seven papers using RDD (which they flag as 5% of papers using what they term "quasi-experimental methods") and we identify an additional five papers not included in their survey, suggesting that papers using RDD represent less than 1% of papers published in top accounting journals since 1999. In this subsection, we examine each of these papers, most of which fall into one of the categories below, to illustrate and discuss how RDD is used in practice in accounting research.

22.5.1 Quasi-RDD

In some settings, RDD would appear to be applicable because there is sharp assignment to treatment based on the realization of a continuous running variable exceeding a predetermined threshold. For example, debt covenants are often based on accounting-related variables that are continuous, but covenant violation is a discontinuous function of those

---

[13]Note that we see even weaker evidence of an effect if we replace `fuzzy2004` with `fuzzy2002` in the intent-to-treat analysis.

variables. In this setting, manipulation of the running variable would be one issue (e.g., firms may manage earnings to avoid costly covenant violations). But a bigger issue in practice is that the running variables are difficult to measure precisely (e.g., many covenants use ratios that are not defined in terms of standard accounting variables) and the thresholds are often not easily observed (e.g., covenants that change over time in ways that are difficult for a researcher to capture precisely).

As a result, papers that claim to be "using a regression discontinuity design" (e.g., p. 2114 of Chava and Roberts, 2008) simply regress the outcome of interest on a treatment indicator and some controls, perhaps restricting the sample to observations closer to violating covenants. While some have called this **quasi-RDD**, this is arguably simply *not*-RDD.

In accounting research, Tan (2013) and Vashishtha (2014) fall into this category because of data constraints. In contrast, we saw above there was no data impediment to using RDD in Bloomfield (2021).

22.5.2   Russell 1000/2000 and stuff (mostly taxes)

Another popular area applying RDD is the burgeoning literature studying the effect of institutional ownership on effective tax rates. Khan et al. (2016), Chen et al. (2019), and Bird and Karolyi (2017) exploit the discontinuous relationship between market capitalization and assignment to either the Russell 1000 or the Russell 2000 index to test for the effect of institutional ownership on tax planning. These papers build on the setting studied earlier in Boone and White (2015) (see discussion questions below).

Unfortunately, subsequent research (e.g., Appel et al., 2024) shows that the data on index assignments supplied by Russell yielded significant differences in pre-treatment outcome variables on either side of the notional discontinuity going from 1,000 to 1,001 in the rank of market capitalization. Young (2018) even shows that this year's "plausibly exogenous" index assignment actually "causes" lower tax rates in *past* years, violating the laws of causation and suggesting that the results in Bird and Karolyi (2017) are spurious (Bird and Karolyi, 2017 was later retracted by *The Accounting Review*).

Another paper using the Russell 1000/2000 setting is Lin et al. (2017), which relates index assignment to "increases [in] industry *peers'* likelihood and frequency of issuing management forecasts" (emphasis added).

22.5.3   Other papers using RDD

Kajüter et al. (2018) and Manchiraju and Rajgopal (2017) use RDD in the context of event studies of the classic capital-market kind. Manchiraju and Rajgopal (2017) is covered in the discussion questions below, so we will briefly discuss just Kajüter et al. (2018) here.

Kajüter et al. (2018) "exploit a regulatory change in Singapore to analyze the capital market effects of mandatory quarterly reporting. The listing rule implemented in 2003 has required firms with a market capitalization above S\$75 million—but not firms with a market capitalization below this threshold—to publish quarterly financial statements." The main analyses of Kajüter et al. (2018) appear to be those in Table 3 (univariate analysis) and Table 4 (polynomial regressions). Given that analysis for Table 3 has only 54 observations (25 of which were treatment firms), overly sophisticated analyses may have asked too much of the data, but the authors do include a sensitivity of the estimated effect to alternative bandwidths in Figure 2. So, it is not clear why the authors do not provide a plot of the underlying 54 data points (given the size of the data set, it could have been supplied as a table).

Two additional papers in accounting research apply RDD in corporate governance settings. Armstrong et al. (2013) also use RDD to evaluate the effectiveness of stock-based compensation plans being voted down but report no statistically significant effects of the treatment in their setting. Ertimur et al. (2015) examine the market reaction to shareholder votes that provide majority support for switching from a plurality voting standard for directors to the more stringent majority voting standard and we provide some discussion questions related to this paper below.

Finally, Figure 1 of Li et al. (2018), which we covered in Chapter 21, presents RDD plots and we include some discussion questions on this analysis below.

22.6 Further reading

As discussed above, this chapter is intended to complement existing materials, and we focus more on practical issues and applications in accounting research. The quality of pedagogical materials for RDD is high. Lee and Lemieux (2010) remains an excellent reference as is Chapter 6 of Angrist and Pischke (2008). Chapter 6 of Cunningham (2021) and Chapter 20 of Huntington-Klein (2021) cover RDD.

22.7 Discussion questions

There are many discussion questions below and we expect that instructors will only assign a subset of these. You should read enough of the papers to be able to answer questions you have been assigned below.

22.7.1 Hoekstra (2009)

1. What is the treatment in Hoekstra (2009)? What alternative (control) is it being compared with? Is identifying the "treatment" received by the control group always as difficult as it is in Hoekstra (2009)? Provide examples from other papers or settings in your answer.

2. Which approach makes the most sense in Hoekstra (2009)? Sharp RDD? Or fuzzy RDD? Why?

3. RDD inherently estimated a "local" treatment effect. Which group of potential students is the focus of Hoekstra (2009)? Can you think of other groups that we might be interested in learning more about? How might the actual treatment effects for those groups differ and why?

22.7.2 Bloomfield (2021)

1. Compare the data in `iliev_2010` and `float_data` (as used in Bloomfield (2021)) for the two firms shown in the output below. What choices has Bloomfield (2021)

made in processing the data for these two firms? Do these choices seem to be the best ones? If not, what alternative approach could be used?

```
iliev_2010 |>
  filter(gvkey == "001728")
```

gvkey	fyear	fdate	pfdate	pfyear	publicfloat	mr	af	cik
001728	2001	2001-12-31	2002-03-01	2002	96.9818	FALSE	FALSE	225051
001728	2002	2002-12-31	2002-06-28	2002	69.0750	FALSE	FALSE	225051
001728	2003	2003-12-31	2003-06-27	2003	71.2024	FALSE	FALSE	225051
001728	2005	2005-01-01	2004-07-03	2004	105.8267	TRUE	TRUE	225051
001728	2005	2005-12-31	2005-07-01	2005	87.1456	TRUE	TRUE	225051

```
float_data |>
  filter(gvkey == "001728")
```

gvkey	float2004	float2002
001728	NA	83.0284

```
iliev_2010 |>
  filter(gvkey == "028712")
```

gvkey	fyear	fdate	pfdate	pfyear	publicfloat	mr	af	cik
028712	2001	2001-12-31	2002-03-21	2002	21.523	FALSE	FALSE	908598
028712	2002	2002-12-31	2002-06-28	2002	11.460	FALSE	FALSE	908598
028712	2003	2003-12-31	2003-06-30	2003	13.931	FALSE	FALSE	908598
028712	2004	2004-12-31	2004-06-30	2004	68.161	FALSE	FALSE	908598
028712	2005	2005-12-31	2005-06-30	2005	26.437	FALSE	FALSE	908598

```
float_data |>
  filter(gvkey == "028712")
```

gvkey	float2004	float2002
028712	68.161	16.4915

2. The code `treat = coalesce(float2002 >= cutoff, TRUE)` above is intended to replicate Stata code used by Bloomfield (2021): `generate treat = float2002 >= 75`.[14] Why does the R code appear to be more complex? What does Stata do that R does not do? (*Hint*: If you don't have access to Stata, you may find Stata's documentation[15] helpful.)

---

[14]Note that we have adapted Bloomfield (2021)'s code to reflect the variable names we use above.
[15]https://www.stata.com/support/faqs/data-management/logical-expressions-and-missing-values/

3. Bloomfield (2021)'s Stata code for the `post` indicator reads `generate post = fyear - (fyr > 5 & fyr < 11) >= 2005`, where `fyear` is fiscal year from Compustat (see Section 14.1) and `fyr` represents the month of the fiscal-year end (e.g., May would be 5).[16] The code above sets `post = datadate >= "2005-11-01"`. Are the two approaches equivalent? Do we seem to get the right values from `post` using either approach?

4. In the text of the paper, Bloomfield (2021) claims to "use firms' 2002 public floats ... as an instrument for whether or not the firm will become treated" and to "follow Iliev's [2010] regression discontinuity methodology". Evaluate each of these claims, providing evidence to support your position.

5. Bloomfield (2021), inspired by Iliev (2010), uses `float2002` rather than `float2004` as the running variable for treatment. What issues would you be concerned about with `float2004` that might be addressed using `float2002`? Provide some evidence to test your concerns. (*Hint:* Iliev (2010) uses plots, McCrary (2008) suggests some tests.) What implications do you see for the fuzzy RDD analysis we ran above using `float2004` as the running variable?

6. Why do you think Bloomfield (2021) did not include RDD analyses along the lines of the ones we have done above in his paper?

7. In Table 4, Bloomfield (2021, p. 884) uses "a difference-in-differences design to identify the causal effect of reporting flexibility on risk asymmetry." As we say in Chapter 21, a difference-in-differences estimator adjusts differences in post-treatment outcome values by subtracting differences in pre-treatment outcome values. Why might differences in pre-treatment outcome values between observations on either side of the threshold be particularly problematic in RDD? Does the use of firm and year fixed effects (as Bloomfield (2021) does in Table 4) address this problem? Or does it just suppress it?

22.7.3 Boone and White (2015)

1. What is the treatment in Boone and White (2015)? (*Hint*: Read the title.) Most of the analyses in the paper use a "sharp RD methodology" (see Section 4). Does this make sense given the treatment? Why or why not?

2. In Section 5, Boone and White (2015) suggest that while "pre-index assignment firm characteristics are similar around the threshold, one concern is that there could be differences in other unobservable firm factors, leading to a violation of the necessary assumptions for the sharp RD methodology." Is the absence of "differences in other unobservable firm factors" the requirement for sharp (rather than fuzzy) RD?

3. What implications, if any, does the discussion on pp. 94–95 of Bebchuk et al. (2017) have for the arguments of Boone and White (2015)?

4. What is the treatment variable implied by the specification in Equation (1) in Boone and White (2015)?

---

[16]Note that we have adapted Bloomfield (2021)'s code to reflect the variable names we use above.

412 *Regression discontinuity designs*

22.7.4   Manchiraju and Rajgopal (2017)

1. Identify some issues in applying RDD in Manchiraju and Rajgopal (2017)? What steps do the authors take to address these issues?

2. What is the treatment in Manchiraju and Rajgopal (2017)? Is this the same treatment variable as analysed in prior research?

22.7.5   Ertimur et al. (2015)

1. Consider Figure 3. How persuasive do you find this plot as evidence of a significant market reaction to majority support in shareholder proposals on majority voting? What aspects of the plot do you find persuasive or unpersuasive?

2. If shareholders react to successful shareholder proposals on majority voting so positively, why do so many shareholders vote against such proposals (e.g., a proposal that gets 51% support has 49% of shareholders voting against it)?

3. Ertimur et al. (2015, p. 38) say "our analyses suggest that high votes withheld do not increase the likelihood of a director losing their seat but often cause boards to respond to the governance problems underlying the vote, suggesting that perhaps director elections are viewed by shareholders as a means to obtain specific governance changes rather than a channel to remove a director." How do you interpret this statement? Do you find it convincing?

22.7.6   Li et al. (2018)

1. Figure 1 of Li et al. (2018) presents RDD plots. How does the running variable in Figure 1 differ from that in other RDD analyses you have seen? What would you expect to be the relation between the running variable and the outcome variable? Would this vary from the left to the right of the cut-off? Do you agree with the decision of Li et al. (2018, p. 283) to "include high-order polynomials to allow for the possibility of nonlinearity around the cut off time"?

2. What is the range of values of "distance to IDD adoption" reported in Figure 1? What is the range of possible values given the sample period of Li et al. (2018) and data reported in Appendix B of Li et al. (2018, p. 304)?

3. Li et al. (2018, p. 283) say that "the figures in both panels show a clear discontinuity at the date of IDD adoption." Do you agree with this claim?

Part IV

Additional Topics

23

Beyond OLS

This part of the book explores some additional topics that are too advanced to be covered in Part I and that do not quite fit in Part III. These include generalized linear models (this chapter), extreme values and sensitivity analysis (Chapter 24), matching (Chapter 25), and a chapter on prediction that also introduces machine learning approaches (Chapter 26). Apart from Chapter 24 building on this chapter, the material in each of these chapters is largely independent of that in the others in this section. Material in this chapter and in Chapters 24 and 25 does build on ideas in Part III to some extent, but likely could be covered before Part III. The material in Chapter 26 is fairly discrete and could be tackled by students with the level of understanding provided by the first few chapters of Part I.

In Chapter 3, we studied the fundamentals of regression analysis with a focus on OLS regression. While we have used OLS as the foundation for almost all regression analyses so far, OLS will at times appear not to be appropriate. For example, if the dependent variable is binary, some aspects of OLS may be difficult to interpret and OLS may be econometrically inefficient. Examples of binary dependent variables include an indicator variable for accounting fraud (see Chapter 26) or for using a Big Four auditor (see Chapter 25).

Similar issues apply to inherently non-negative dependent variables, such as total assets or **count variables**, such as the *number* of management forecasts, the *number* of 8-K filings, or the *number* of press releases, as seen in Guay et al. (2016), the focus of this chapter.

In this chapter, we cover **generalized linear models (GLMs)**, a class of models that includes **probit** and **logit** models—which assume a binary dependent variable—and **Poisson** regression—which is most naturally motivated by count variables.

A second theme of this chapter is the introduction of additional data skills relevant for applied researchers. In previous chapters we have eschewed larger data sets beyond those available through WRDS. However, most researchers need to know how to manage larger data sets, so we introduce some skills for doing so here.

In particular, we collect metadata for every filing made on the SEC EDGAR system since 1993, producing a data frame that occupies 3 GB if loaded into R. While 3 GB certainly is not "big data", it is large enough that more efficient ways of working have a significant impact on the facility of manipulating and analysing the data. In the analysis below, we process the data into **parquet** files that we can access using the DuckDB database package.

> Tip
>
> The code in this chapter uses the packages listed below. For instructions on how to set up your computer to use the code found in this book, see Section 1.2. Quarto

templates for the exercises below are available at https://github.com/iangow/far_te
mplates/blob/main/README.md.

```
library(tidyverse)
library(DBI)
library(farr)
library(modelsummary)
library(arrow)          # write_parquet(), read_parquet()
library(googledrive)    # drive_download(), drive_deauth()
library(fixest)         # feols()
library(lmtest)         # coeftest()
library(sandwich)       # vcovHC()
library(httr2)          # request(), req_*(), resp_body_html()
library(rvest)          # html_elements(), html_table()
```

23.1 Complexity and voluntary disclosure

Like Li et al. (2018) (covered in Chapter 21), Guay et al. (2016) examine the phenomenon of voluntary disclosure. A core idea of Guay et al. (2016) is that users of a more complex firm's disclosures require more information to understand the business sufficiently well to invest in it. Guay et al. (2016, p. 235) posit that "economic theory suggests that managers will use other disclosure channels to improve the information environment" when their reporting is more complex. A number of "other disclosure channels" are explored in Guay et al. (2016), including the issuance of management forecasts (the primary measure), filing of 8-Ks, and firm-initiated press releases.

Guay et al. (2016) find evidence of greater voluntary disclosure by firms with more complex financial statements. For example, Table 3 of Guay et al. (2016) presents results of OLS regressions with two different measures of financial statement complexity, and with and without firm fixed effects.

23.1.1 Discussion questions

1. Guay et al. (2016, p. 252) argue that the "collective results from our two quasi-natural experiments ... validate that our text-based measures of financial statement complexity reflect, at least in part, the complexity of the underlying accounting rules." Which specific table or figure of Guay et al. (2016) provides the most direct evidence in support of this claim? Can you suggest any alternative tests or presentations to support this claim?

2. What is the causal diagram implied by the regression specification in Table 3 of Guay et al. (2016)? Provide arguments for and against the inclusion of *ROA*, *SpecialItems* and *Loss* as controls.

3. Assuming that the causal effect of *FS_complexity* implied by the results in Table 3 exists, provide an explanation for the changes in the coefficients when firm fixed effects are added to the model. When might you expect these changes to have the opposite sign?

4. Guay et al. (2016, p. 234) conclude that "collectively, these findings suggest managers use voluntary disclosure to mitigate the negative effects of complex financial statements on the information environment." Suppose you wanted to design a field experiment to test the propositions embedded in this sentence and were given the support of a regulator to do so (e.g., you can randomize firms to different regimes, as was done with Reg SHO discussed in Section 19.2). How would you implement this experiment? What empirical tests would you use? Are some elements of the hypothesis more difficult than others to test?

5. Clearly Guay et al. (2016) did not have the luxury of running a field experiment. Do you see any differences between the analyses you propose for your (hypothetical) field experiment and those in Guay et al. (2016)? What do you think best explains any differences?

23.2 Generalized linear models

In this section, we introduce the idea of **generalized linear models** (**GLMs**), which provides a powerful framework for analysing a wide class of settings beyond OLS. We begin by discussing maximum likelihood as an organizing principle for estimation before applying that to understand regression analysis with binary dependent variable.

After discussing the concept of marginal effects, we finish up this section with a discussion of GLM regressions with count variables.

23.2.1 Maximum likelihood

The concept of **maximum likelihood estimation** provides a useful framework for thinking about regression, including GLMs. As the name "ordinary least squares" suggests, OLS regression is often motivated as the linear estimator that minimizes the squared residuals in a setting in which $y = X\beta$.

However, we can take another perspective. Suppose that $y_i = X_i\beta + \epsilon_i$ where $\epsilon_i \sim N(0, \sigma^2)$ and ϵ_i is independent across individuals (i). As such, the density for y_i is given by

$$f(y_i | X_i\beta) = \frac{1}{\sigma\sqrt{2\pi}} \exp\left(-\frac{(y_i - X_i\beta)^2}{2\sigma^2}\right)$$

But mathematically, the y_i and X_i can be seen as given and this function can be viewed instead as a function of β and relabelled as a **likelihood function**.

$$\mathcal{L}(\beta; X_i, y_i) = \frac{1}{\sigma\sqrt{2\pi}} \exp\left(-\frac{(y_i - X_i\beta)^2}{2\sigma^2}\right)$$

Given independence, the density for the full sample $y = (y_1, \dots, y_N)$ is simply the product of the densities of each observation and we can consider the likelihood function for the full sample in the same way.

$$\mathcal{L}(\beta; X, y) = \prod_{i=1}^{N} \frac{1}{\sigma\sqrt{2\pi}} \exp\left(-\frac{(y_i - X_i\beta)^2}{2\sigma^2}\right)$$

We can also write the **log-likelihood** as follows

$$\log \mathcal{L}(\beta; X, y) = \sum_{i=1}^{N} \left(\log \left(\frac{1}{\sigma\sqrt{2\pi}} \right) + \left(-\frac{(y_i - X_i\beta)^2}{2\sigma^2} \right) \right)$$

$$= \sum_{i=1}^{N} \log \left(\frac{1}{\sigma\sqrt{2\pi}} \right) + \sum_{i=1}^{N} \left(-\frac{(y_i - X_i\beta)^2}{2\sigma^2} \right)$$

The idea of **maximum likelihood estimation** (MLE) is to choose the values $\hat{\beta}$ and $\hat{\sigma}$ that maximize the likelihood function. But note that maximizing the log-likelihood will yield the same solution as maximizing the likelihood (because $\log(x) > \log(y) \Leftrightarrow x > y$) and that only the second term in the sum above is affected by the value of β.

This means that maximizing the log-likelihood with respect to β is achieved by maximizing the second term. But the second term is nothing but the negative of the sum of square residuals. So the least-squares estimator *is* the maximum likelihood estimator in this setting.

So why introduce MLE in the first place? First, this equivalence will not hold in general. Merely assuming heteroskedasticity (i.e., that $\sigma_i^2 \neq \sigma_j^2$ for $i \neq j$) is enough to create a wedge between least-squares and MLE.

Second, MLE can be shown to have desirable econometric properties in a broad class of settings. According to Wooldridge (2010, p. 385), "MLE has some desirable efficiency properties: it is generally the most efficient estimation procedure in the class of estimators that use information on the distribution of the endogenous variables given the exogenous variables." Additionally, MLE has "a central role" (Wooldridge, 2010, p. 386) in estimation of models with limited dependent variables, which we cover in the next section.

23.2.2   Binary dependent variables

If we have a setting in which $y_i \in \{0, 1\}$ (i.e., the dependent variable is binary), then a generic framework has the conditional density of y_i as

$$g(y_i|X_i; \beta) = \begin{cases} h(X_i \cdot \beta), & y_i = 1 \\ 1 - h(X_i \cdot \beta), & y_i = 0 \\ 0, & \text{otherwise} \end{cases}$$

where $X = (1, x_{i1}, x_{i2}, \dots, x_{ik})$, $\beta = (\beta_0, \beta_1, \dots, \beta_k)$, and $h(\cdot)$ is a known function $\mathbb{R} \to [0, 1]$.

With this setup, we define the **logit model** in terms of $h(\cdot)$:

$$h(X\beta) = \frac{e^{X\beta}}{1 + e^{X\beta}}$$

The **probit model** is defined using the standard normal distribution function $\Phi(\cdot)$:

$$h(X\beta) = \Phi(X\beta)$$

If we assume that one of the above functions describes the underlying data-generating process, then we can construct the corresponding maximum-likelihood estimator of β. For example, the probit model implies that

$$f_{Y|X} = g(Y, X; \beta) = \Phi(X\beta)^Y (1 - \Phi(X\beta))^{1-Y}$$

$$\mathcal{L}(\beta|Y,X) = \prod_{i=1}^{n} \Phi(X_i\beta)^{Y_i}(1 - \Phi(X_i\beta))^{1-Y_i}$$

$$\log \mathcal{L}(\beta|Y,X) = \log \left(\prod_{i=1}^{n} \Phi(X_i\beta)^{Y_i}(1 - \Phi(X_i\beta))^{1-Y_i} \right)$$

$$= \sum_{i=1}^{n} \log \left(\Phi(X_i\beta)^{Y_i}(1 - \Phi(X_i\beta))^{1-Y_i} \right)$$

$$= \sum_{i=1}^{n} Y_i \log \Phi(X_i\beta) + (1 - Y_i) \log(1 - \Phi(X_i\beta))$$

Thus, given Y and X, we can obtain the ML estimate of β using a computer to find

$$\hat{\beta}_{ML} = \underset{\beta \in \mathbb{R}^{k+1}}{\arg\max} \sum_{i=1}^{n} Y_i \log \Phi(X_i\beta) + (1 - Y_i) \log(1 - \Phi(X_i\beta))$$

A similar analysis can be conducted for logit regression.

But rarely would a researcher have good reasons to believe strongly that the correct model is the probit model, rather than the logit model, or vice versa. This raises questions about the consequences of estimation using MLE, but with the wrong model. In other words, what happens if we use probit regression when the correct model is logit? Or logit regression when the correct model is probit? And what happens if we simply use OLS to fit a model? (The term **linear probability model** is used to describe the fitting of OLS with a binary dependent variable, as the fitted values can be interpreted as probabilities that $y = 1$, even though these fitted values can fall outside the $[0, 1]$ interval.)

A natural way to examine these questions is with simulated data. We make a small function (`get_probit_data()`) that generates data using the assumptions underlying the probit model. We generate random x and then calculate $h(X\beta) = \Phi(X\beta)$, which gives us a vector h representing the probability that $y_i = 1$ for each i. We then compare $h(X_i\beta)$ with u_i, a draw from the uniform distribution over $[0, 1]$.[1] If $h(X_i\beta) > u_i$, then $y = 1$; otherwise $y = 0$. If $h(X_i\beta) = 0.9$ then we have a 90% chance that $h(X_i\beta) > u_i$. Finally, the `get_probit_data()` function returns the generated data.

```
get_probit_data <- function(beta_0 = -2, beta_1, n = 1000) {
  x <- rnorm(n)
  h <- pnorm(beta_0 + x * beta_1)
  y <- runif(n = n) < h
  tibble(x, y)
}
```

23.2.3 Marginal effects

An important concept with non-linear models such as those examined in this chapter is the **marginal effect**. Given a causal model $\mathbb{E}[y|X] = f(X)$, the partial derivative of x_j at $X = x$—that is,

$$\left. \frac{\partial \mathbb{E}[y|X]}{\partial x_i} \right|_{X=x}$$

---

[1]Some sources will formulate the data-generating process in a different, but equivalent, fashion. For example, Angrist and Pischke (2008) posit a latent variable $y_i^* = X_i\beta - \nu_i$, where $\nu_i \sim N(0, \sigma_\nu^2)$.

for $j \in (1, \ldots, k)$—can be viewed as the **marginal effect** of a change in x_j on the value of $\mathbb{E}[y|X]$ evaluated at $X = x$.

If y is a binary variable, then $\mathbb{E}[y_i|X_i] = \mathbb{P}[y_i = 1|X_i]$. If we have a probit model, then $\mathbb{P}[y_i = 1|X_i] = \Phi(X_i\beta)$. Assuming that X_i is a continuous variable, we then have the following:[2]

$$\frac{\partial \mathbb{E}[y_i|X_i]}{\partial x_{ij}} = \frac{\partial \mathbb{P}[y_i = 1|X_i]}{\partial x_{ij}}$$

$$= \frac{\partial \Phi(X_i\beta)}{\partial x_{ij}}$$

$$= \frac{\partial \Phi(X_i\beta)}{\partial(X_i\beta)} \frac{\partial(X_i\beta)}{\partial x_{ij}}$$

$$= \phi(X_i\beta)\beta_j$$

That is, the marginal effect is the product of the probability density function and the coefficient β_j. Note that we used the facts that the first derivative of the cumulative normal distribution $\Phi(\cdot)$ is the normal density function $\phi(\cdot)$ and that the partial derivative of $X_i\beta = \beta_0 + x_{i1}\beta_1 + \cdots + x_{ij}\beta_j + \cdots + x_{ik}\beta_k$ with respect to x_{ij} is β_j.

With linear models such as those fitted using OLS where x_j only enters the model in a linear fashion, the marginal effect of x_j is simply β_j and does not depend on the value x_j at which it is evaluated. But in models in which x_j interacts with other variables or enters the model in a non-linear fashion, the marginal effect of x_j can vary. For example, in the following model, the marginal effect of x_1 will depend on the values of x_1 and x_2 at which it is evaluated.[3]

$$y_i = \beta_0 + x_{i1}\beta_1 + x_{i2}\beta_2 + x_{i1}x_{i2}\beta_3 + x_{i2}^2\beta_4 + \epsilon_i$$

With non-linear models the marginal effect of x_j will generally depend on the value x at which it is evaluated. One common approach evaluates the marginal effect at the sample mean of $\overline{X} = (\overline{x}_1, \ldots, \overline{x}_k)$. Another approach would be to report the mean of the marginal effects evaluated for each X in the sample, but this approach may be difficult to interpret if the signs of the marginal effects vary across the sample distribution.

$$\mathbb{E}\left[\frac{\partial \mathbb{E}[y_i|X_i]}{\partial x_{ij}}\right] = \mathbb{E}\left[\phi(X_i\beta)\right]\beta_j$$

The following function implements the latter computation for probit, logit, and OLS models. While a more sophisticated version of this function would infer the model's type from its attributes, this function is adequate for our needs here. The `mfx` package[4] offers estimates of marginal effects for a variety of models.

```
get_mfx <- function(fm, type) {
  pdf <-
    case_when(type == "probit" ~ mean(dnorm(predict(fm, type = "link"))),
              type == "logit" ~ mean(dlogis(predict(fm, type = "link"))),
              type == "Poisson" ~ mean(predict(fm, type = "response")),
              type == "OLS" ~ 1)
```

---

[2]Note that we have suppressed the evaluation of these quantities at $X = x$ for reasons of space and clarity.

[3]Note that this is still a linear model because it is linear in its *coefficients*.

[4]https://cran.r-project.org/web/packages/mfx/index.html

```
  mfx <- pdf * coef(fm)
  mfx
}
```

The following `fit_model()` convenience function fits a model, extracts statistics from that model, then returns a data frame containing those statistics.[5]

```
fit_model <- function(data = df, type = "OLS") {
  family <- ifelse(type %in% c("probit", "logit"), binomial,
                   ifelse(type == "Poisson", poisson, gaussian))

  link <- case_when(type %in% c("probit", "logit") ~ type,
                    type == "Poisson" ~ "log",
                    .default = "identity")

  fm <- glm(y ~ x, family(link = link), data = data)
  coefs <- summary(fm)$coefficients

  tibble(type,
         coef = coefs[2, 1],
         se = coefs[2, 2],
         t_stat = coefs[2, 3],
         p = coefs[2, 4],
         mfx = (get_mfx(fm, type))[["x"]])
}
```

We then create a function to run an iteration of the simulation: it gets the data using `get_probit_data()`, extracts statistics using `fit_model()` for three different model types, then returns a data frame with the results.

```
run_probit_sim <- function(i, beta_1) {
  df <- get_probit_data(beta_1 = beta_1)

  fms <- list(fit_model(df, type = "probit"),
              fit_model(df, type = "logit"),
              fit_model(df, type = "OLS"))

  bind_cols(tibble(i),
            beta_1_true = beta_1,
            list_rbind(fms))
}
```

Finally, we run the simulation `n_iters` times for each of two values of β_1. We consider $\beta_1 = 0$, so that the null hypothesis of no relation between x_1 and y is true. We also consider $\beta_1 = 0.3$, so that the null hypothesis of no relation between x_1 and y is false. The former allows us to check that the rejection rate matches the size of the test ($\alpha = 0.05$ or $\alpha = 0.01$). The latter provides insights into the power of the tests.

```
set.seed(2024)
n_iters <- 1000L
```

---

[5]We use `ifelse()` from base R here because `if_else()` expects vectors in positions where we have functions. See Chapter 12 of *R for Data Science* for more details: https://r4ds.hadley.nz/logicals.

```
results <-
  list(
    map(1:n_iters, \(x) run_probit_sim(x, beta_1 = 0.0)),
    map(1:n_iters, \(x) run_probit_sim(x, beta_1 = 0.3))) |>
  bind_rows() |>
  system_time()
```

```
   user  system elapsed
 16.646   0.684  17.408
```

To understand the relationships between the various approaches, we focus on the case where $\beta_1 \neq 0$ (so there is an underlying relationship between x and y to estimate).

```
rel_df <-
  results |>
  filter(beta_1_true != 0) |>
  pivot_wider(names_from = type, values_from = coef:mfx)
```

We examine the relation between the estimated coefficients for the various models using the probit model as the benchmark (after all, we know that the underlying data-generating process *is* a probit model).

```
fms <- list(
  "Logit" = lm(coef_logit ~ coef_probit, data = rel_df),
  "OLS" = lm(coef_OLS ~ coef_probit, data = rel_df),
  "mfx_logit" = lm(mfx_logit ~ mfx_probit, data = rel_df),
  "mfx_OLS" = lm(coef_OLS ~ mfx_probit, data = rel_df))
```

The first thing to note from our simulation results is that there are strong relationships between the coefficients and marginal effects across the various models. In Table 23.1, we can see a high correlation (measured using the R^2) between probit, logit, and OLS coefficients and that this correlation is even higher when we move to estimates of the marginal effects. This strong relationship is also evident in Figure 23.1.

```
modelsummary(fms,
             estimate = "{estimate}{stars}",
             statistic = "statistic",
             gof_map = c("nobs", "r.squared"),
             stars = c('*' = .1, '**' = 0.05, '***' = .01))
```

TABLE 23.1

Relationships between model coefficients

	Logit	OLS	mfx_logit	mfx_OLS
(Intercept)	0.034***	0.001***	0.000	0.000
	(9.937)	(2.868)	(−0.191)	(−0.653)
coef_probit	2.172***	0.057***		
	(202.626)	(54.862)		
mfx_probit			0.997***	0.998***
			(309.056)	(235.526)
Num.Obs.	1,000	1,000	1,000	1,000
R2	0.976	0.751	0.990	0.982

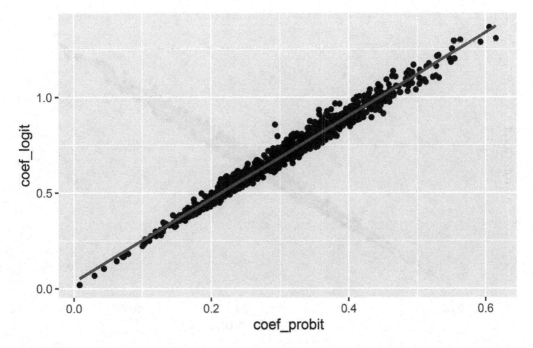

FIGURE 23.1
Relationship between probit and logit model coefficients

```
rel_df |>
  ggplot(aes(x = coef_probit, y = coef_logit)) +
  geom_point() +
  geom_smooth(formula = "y ~ x", method = "lm", se = FALSE)
```

Of course, very little attention is paid to the magnitude of the coefficients from estimated probit or logit models in accounting research. While this partly reflects the focus on the statistical significance and sign of the coefficients in the NHST paradigm, it also reflects the difficulty of attaching economic meaning to the values of the coefficients. In this regard, more meaning can be placed on the marginal effect estimates.

Table 23.2 reveals that in our simulation analysis there is little difference in the estimated marginal effect sizes from the three approaches, even though only one (probit) is the correct model. The strong relationship between probit and logit fixed effects is also clear in Figure 23.2.

```
rel_df |>
  ggplot(aes(x = mfx_probit, y = mfx_logit)) +
  geom_point() +
  geom_smooth(formula = "y ~ x", method = "lm", se = FALSE)
```

But even marginal effect sizes receive little attention in most accounting research papers, as the focus is generally on the coefficients' signs and statistical significance. Table 23.2 provides insights on these issues too. When $\beta_1 = 0$ (i.e., the null hypothesis is true), we see that the rejection rates are quite close to the size of the tests (either 5% or 1%) for OLS, logit, and probit and very close to each other. When $\beta_1 = 0.3$ (i.e., the null hypothesis is

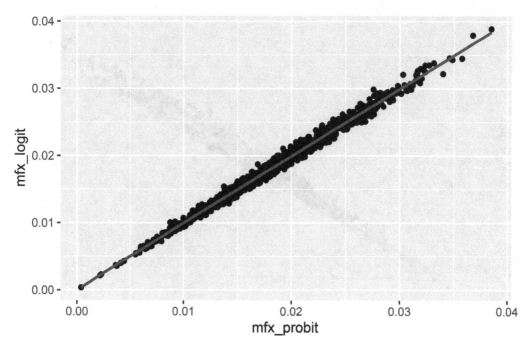

FIGURE 23.2
Relationship between probit and logit marginal effect estimates

false), we see that the rejection rates for OLS, logit, and probit are very similar. We have power over 90% at the 5% level and around 80% at the 1% level. As such, there is arguably little value in this setting in using probit or logit rather than OLS regression, which yields coefficients that are much easier to interpret.

```
results |>
  group_by(type, beta_1_true) |>
  summarize(mean_coef = mean(coef),
            mean_mfx = mean(mfx),
            rej_rate_5 = mean(p < 0.05),
            rej_rate_1 = mean(p < 0.01),
            .groups = "drop") |>
  arrange(beta_1_true, type)
```

TABLE 23.2
Probit simulation results

type	beta_1_true	mean_coef	mean_mfx	rej_rate_5	rej_rate_1
OLS	0.0	0.000	0.000	0.035	0.008
logit	0.0	0.006	0.000	0.033	0.008
probit	0.0	0.002	0.000	0.035	0.008
OLS	0.3	0.018	0.018	0.938	0.805
logit	0.3	0.696	0.018	0.938	0.800
probit	0.3	0.305	0.018	0.934	0.801

Of course, the analysis above should not be assumed to apply in all settings. But it seems reasonable for deviations from straightforward OLS analysis in specific settings to be justified. Fortunately, simulation analysis like that above is relatively straightforward to conduct.

23.2.4 Count variables

Another case in which OLS may be regarded as "inadequate" is when the dependent variable is a count variable. One member of the class of GLMs that models such a dependent variable is **Poisson** regression. In this model, we have $y \sim \text{Pois}(\mu)$ where $\mu = \exp(\beta_0 + \beta_1 x)$, where $\text{Pois}(\mu)$ is the Poisson distribution with parameters μ.

As above, we start by making a small function (`get_pois_data()`) that generates data using the assumptions underlying the Poisson model.

```
get_pois_data <- function(beta_0 = -2, beta_1, n = 1000) {
  x <- rnorm(n)
  h <- exp(beta_0 + x * beta_1)
  y <- rpois(n = n, lambda = h)
  tibble(x, y)
}
```

We then create a function to run an iteration of the simulation: get the data using `get_pois_data()`, extract statistics using `fit_model()` for both Poisson and OLS, then return a data frame with the results.

```
run_pois_sim <- function(i, beta_1) {
  df <- get_pois_data(beta_1 = beta_1)

  fm_pois <- fit_model(df, type = "Poisson")
  fm_ols  <- fit_model(df, type = "OLS")

  bind_cols(tibble(i),
            beta_1_true = beta_1,
            bind_rows(fm_pois, fm_ols))
}
```

Finally, we run the simulation `n_iters` times for each of two values of β_1. As above, we consider $\beta_1 = 0$ and $\beta_1 = 0.3$.

```
set.seed(2024)
n_iters <- 1000L

results_pois <-
  list(
    map(1:n_iters, \(x) run_pois_sim(x, beta_1 = 0.0)),
    map(1:n_iters, \(x) run_pois_sim(x, beta_1 = 0.3))) |>
  bind_rows() |>
  system_time()
```

```
   user  system elapsed
 10.397   0.244  10.677
```

Table 23.3 reveals that in our simulation analysis there is little difference in the estimated marginal effect sizes between Poisson and OLS regression. When $\beta_1 = 0$ (i.e., the null hypothesis is true), we see that the rejection rates are quite close to the size of the tests (either 5% or 1%) and to each other. When $\beta_1 = 0.3$ (i.e., the null hypothesis is false), we see that the rejection rates for Poisson and OLS are very similar. Again, we have power over 90% at the 5% level and over 80% at the 1% level. As such, similar to what we saw in the analysis of probit analysis above, there is arguably little value in this setting in using Poisson regression rather than OLS regression. This seems especially true given that OLS yields coefficients that are much easier to interpret.

```
results_pois |>
  group_by(type, beta_1_true) |>
  summarize(mean_coef = mean(coef),
            mean_mfx = mean(mfx),
            rej_rate_5 = mean(p < 0.05),
            rej_rate_1 = mean(p < 0.01),
            .groups = "drop") |>
  arrange(beta_1_true, type)
```

TABLE 23.3
Poisson simulation results

type	beta_1_true	mean_coef	mean_mfx	rej_rate_5	rej_rate_1
OLS	0.0	0.000	0.000	0.059	0.011
Poisson	0.0	0.001	0.000	0.063	0.010
OLS	0.3	0.043	0.043	0.942	0.849
Poisson	0.3	0.302	0.043	0.940	0.850

23.3 Application: Complexity and voluntary disclosure

To enrich our understanding of GLMs and to examine some additional data skills, we return to a setting similar to that considered in Guay et al. (2016). Like Guay et al. (2016), we will examine the association between complexity of financial reporting and subsequent voluntary disclosures. Our purpose here is not to conduct a close replication of Guay et al. (2016). In place of the measures of financial reporting complexity and of voluntary disclosure used in Guay et al. (2016), we use measures that are readily available. Additionally, we do not restrict ourselves to the sample period used in Guay et al. (2016).

This is the first chapter in which we need to collect a substantial amount of data from sources other than WRDS. To facilitate this, we will perform a number of data steps as quite discrete steps. This allows the re-running of code for one data set without requiring us to rerun code for other data sets.

To make this approach easier, we create `save_parquet()`, which takes a (possibly remote) data frame and uses the `write_parquet()` function from the `arrow` library to save it as a parquet file. The parquet format is described in *R for Data Science* (Wickham et al., 2023, p. 393) as "an open standards-based format widely used by big data systems." Parquet files provide a format optimized for data analysis, with a rich type system. More details on the parquet format can be found in Chapter 22[6] of *R for Data Science*. We use

---

[6]https://r4ds.hadley.nz/arrow

`write_parquet()` from the `arrow` package to store the data in `sec_index_2022q2` in a file named `sec_index_2022q2.parquet`.

```
save_parquet <- function(df, name, schema = "", path = data_dir) {
  file_path <- file.path(path, schema, str_c(name, ".parquet"))
  write_parquet(collect(df), sink = file_path)
}
```

23.3.1 Measures of reporting complexity

Guay et al. (2016) consider a number of measures of reporting complexity, including measures based on the "readability", measured using the Fog index (Guay et al., 2016, p. 239), and length of a firm's 10-K, measured as the natural logarithm of the number of words. Loughran and McDonald (2014) find "that 10-K document file size provides a simple readability proxy that outperforms the Fog Index, does not require document parsing, facilitates replication, and is correlated with alternative readability constructs." Thus, we use 10-K document file size here.

Data on 10-K document file size can be collected from SEC EDGAR, but gathering these directly from SEC EDGAR would involve costly web scraping. Thus, we simply get these data from the website[7] provided by Loughran and McDonald (2014).

In this book so far, we have avoided substantial data collection efforts of the kind that are often central in conducting research. From a pedagogical perspective, such data collection efforts typically involve messy details and complications that could distract more than they enlighten. However, we feel that it is useful to readers to provide some examples of more substantial data-collection tasks.

We first specify a directory in which we will store the data we will collect and create this directory if it doesn't already exist. While the use of the `DATA_DIR` environment variable does not add anything in the analysis here, it may have a role if we are using local file storage across projects (see Appendix E for more on this point).[8]

> Tip
>
> The following line of code below is only indicative. If you have already set `DATA_DIR` (e.g., if you have set up a data repository as discussed in Appendix E), then you many not need (or want) to run this line. If you are within an RStudio project when running the following line, then it should set `DATA_DIR` to a `data` folder inside that project, which is a reasonable choice for current purposes.

```
Sys.setenv(DATA_DIR = "data")
```

```
data_dir <- Sys.getenv("DATA_DIR")
if (!dir.exists(data_dir)) dir.create(data_dir)
```

Using an approach discussed in Appendix E, we will organize the data in this directory into subdirectories, which we can think of as schemas, much as data on WRDS are organized into schemas.

---

[7]https://sraf.nd.edu

[8]If you are using the parquet-based approach described in Appendix E, you may have already set `DATA_DIR` and you should not need to run this line of code.

Specifically, we imagine that this code is being used for a hypothetical project named "glms" and we put all data specific to this project into a directory with that name. We store the full path to this directory, a subdirectory of `data_dir`, in `project_dir`.

```
project_dir <- file.path(data_dir, "glms")
if (!dir.exists(project_dir)) dir.create(project_dir)
```

It turns out that the Loughran and McDonald (2014) website provides a link to a CSV file stored in Google Drive and we can use the `googledrive` package to download this. This requires supplying the Google Drive ID (stored in `id` below) and a local file name to use (stored in `lm_data` below). The code below only downloads the file if it is not already present in the location stored in `lm_data`.[9]

```
drive_deauth()
lm_data <- file.path(project_dir, "lm_10x_summary.csv")
id <- "1puReWu4AMuVOjfWTrrf8IbzNNEU6kfpo"
if (!file.exists(lm_data)) drive_download(as_id(id), lm_data)
```

Having downloaded this CSV file, we can read it into R. While one way to read this file would use `read_csv()` from the `readr` package, we continue the approach of using database-backed `dplyr` used elsewhere in the book. However, rather than using a PostgreSQL database, we will use a local DuckDB database. DuckDB works much like PostgreSQL, but does not require a server to be set up and comes with excellent facilities for reading common file formats, such as CSV and parquet files (discussed below).

In fact, we can set up a temporary database with a single line of code:

```
db <- dbConnect(duckdb::duckdb())
```

Having created this database connection, we can use DuckDB's native ability to read CSV files provided by its `read_csv()` function to create a database table as follows. We convert all variable names to lower case, convert `cik` to integer type and `filing_date` and `cpr` to date types, and handle the `-99` value used to represent missing values in `cpr`. (According to the source website[10], `cpr` refers to the "conformed period of report" of the filing, which is the last date of the reporting period.)

Note that we force the table to be materialized as a remote temporary table by appending the command `compute()`. This takes less than a second, but means that the conversion from CSV and processing of dates and missing values and the like happens only once rather than each time we call on `lm_10x_summary` in later analyses. Judicious use of `compute()` can also enhance query performance even when a table is not used multiple times by making it easier for the query optimizer in our database to understand the query. Unfortunately, WRDS does not grant its users sufficient privileges to create temporary tables and thus we have not been able to use `compute()` in previous chapters.[11]

```
lm_10x_summary <-
  tbl(db, str_c("read_csv('", lm_data, "')")) |>
  rename_with(tolower) |>
  mutate(cik = as.integer(cik),
```

---

[9]It is possible that by the time you read this, the `id` for the CSV file will have changed because the data have been updated. If so, you should be able to identify the current `id` by visiting Loughran and McDonald's website for "10X Summaries" and extracting the `id` from the URL for the CSV file listed there.

[10]https://sraf.nd.edu/sec-edgar-data/lm_10x_summaries/

[11]Note that, because we end up saving the data to a parquet file before further analysis, the use of `compute()` has little impact in this case.

```
              filing_date = as.character(filing_date),
         cpr = if_else(cpr == -99, NA, as.character(cpr))) |>
  mutate(filing_date = as.Date(strptime(filing_date, '%Y%m%d')),
         cpr = as.Date(strptime(cpr, '%Y%m%d'))) |>
  compute()
```

This remote table `lm_10x_summary` is just like the remote tables we first saw in Section 6.1.

```
lm_10x_summary
```

```
# Source:    table<dbplyr_if0zhyOp7u> [?? x 25]
# Database: DuckDB v1.1.0 [iangow@Darwin 24.0.0:R 4.4.1/:memory:]
     cik filing_date acc_num  cpr        form_type coname    sic ffind n_words
   <int> <date>      <chr>    <date>     <chr>     <chr>   <dbl> <dbl>   <dbl>
1  60512 1993-08-13  000006~  1993-06-30 10-Q      LOUIS~   1311    30    4068
2  66740 1993-08-13  000006~  1993-06-30 10-Q      MINNE~   2670    38    4389
3  60512 1993-10-07  000006~  1992-12-31 10-K-A    LOUIS~   1311    30    8719
4  60512 1993-11-10  000006~  1993-09-30 10-Q      LOUIS~   1311    30    4938
5  11860 1993-11-12  000001~  1993-09-30 10-Q      BETHL~   3312    19    3823
6  20762 1993-11-12  000095~  1993-09-30 10-Q      CLARK~   2911    30    4136
# i more rows
# i 16 more variables: n_unique_words <dbl>, n_negative <dbl>,
#   n_positive <dbl>, n_uncertainty <dbl>, n_litigious <dbl>,
#   n_strongmodal <dbl>, n_weakmodal <dbl>, n_constraining <dbl>,
#   n_negation <dbl>, grossfilesize <dbl>, netfilesize <dbl>,
#   nontextdoctypechars <dbl>, htmlchars <dbl>, xbrlchars <dbl>,
#   xmlchars <dbl>, n_exhibits <dbl>
```

From a quick inspection, we can see that the data begin in 1993, but only appear to be comprehensive from some time in 1995 or 1996.

```
lm_10x_summary |>
  mutate(year = year(filing_date)) |>
  count(year) |>
  arrange(year) |>
  collect()
```

```
# A tibble: 29 x 2
   year      n
  <dbl>  <dbl>
1  1993     13
2  1994   9752
3  1995  21880
4  1996  44531
5  1997  55484
6  1998  55735
# i 23 more rows
```

Finally, we save the data to a parquet file in our `glms` project directory and disconnect from the in-memory DuckDB database.

```
lm_10x_summary |>
  save_parquet(name = "lm_10x_summary", schema = "glms")
```

```
dbDisconnect(db)
```

23.3.2   Data on EDGAR filings

As discussed above, Guay et al. (2016) consider a number of measures of voluntary disclosure.
While measures based on management forecasts require data from IBES and measures of
firm-initiated press releases require data from RavenPack, the measure based on the number
of 8-Ks filed in a period can be collected from SEC EDGAR fairly easily and we focus on
that measure for this reason.

Here we collect filings data from SEC EDGAR. Specifically, we collect index files for each
quarter. These index files include details on each filing in the applicable quarter.

For example, if you visit https://www.sec.gov/Archives/edgar/full-index/2022/QTR2/,
you can see a number of files available to download. As these files are mostly different ways
of presenting the same information in a variety of formats, we will focus on the `company.gz`
files.

The first practical challenge we face is that if we try to use `download.file()` to get
`company.gz`, we will get a "403 Forbidden" error message because the SEC does not allow anonymous downloads. To avoid this, we set the `HTTPUserAgent` option using our email
address (obviously, use your actual email address here).

```
options(HTTPUserAgent = "your_name@some_email.com")
```

The file `company.gz` is a compressed fixed-width format text file of the kind examined in
Chapter 9.

```
url <- str_c("https://www.sec.gov/Archives/edgar/full-index/",
             "2022/QTR2/company.gz")
t <- tempfile(fileext = ".gz")
download.file(url, t)

sec_index_2022q2 <-
  read_fwf(t,
           fwf_cols(company_name = c(1, 62),
                    form_type = c(63, 74),
                    cik = c(75, 86),
                    date_filed = c(87, 98),
                    file_name = c(99, 150)),
           col_types = "cciDc",
           skip = 10,
           locale = locale(encoding = "macintosh"))
```

```
sec_index_2022q2 |> select(company_name, cik)
```

```
# A tibble: 291,566 x 2
  company_name                                        cik
  <chr>                                             <int>
1 006 - Series of IPOSharks Venture Master Fund, LLC 1894348
2 008 - Series of IPOSharks Venture Master Fund, LLC 1907507
3 01 Advisors 03, L.P.                              1921173
4 01Fintech LP                                      1928541
5 07 Carbon Capture, LP                             1921663
6 09 Carbon Capture, LP                             1936048
# i 291,560 more rows
```

```
sec_index_2022q2 |> select(form_type, date_filed, file_name)
```

```
# A tibble: 291,566 x 3
  form_type date_filed file_name
  <chr>     <date>     <chr>
1 D         2022-04-08 edgar/data/1894348/0001892986-22-000003.txt
2 D         2022-06-07 edgar/data/1907507/0001897878-22-000003.txt
3 D         2022-05-20 edgar/data/1921173/0001921173-22-000001.txt
4 D         2022-05-13 edgar/data/1928541/0001928541-22-000001.txt
5 D         2022-04-05 edgar/data/1921663/0001921663-22-000001.txt
6 D         2022-06-30 edgar/data/1936048/0001936048-22-000001.txt
# i 291,560 more rows
```

As the file we downloaded above is over 4 MB in size and we will need to download dozens of such files, we likely want to store these data in some permanent fashion. While one approach would be to use a PostgreSQL database to which we have write access, we do not assume here that you have such a database.[12]

Another approach would be to simply download all the `company.gz` files and process them each time we use them. The problem with this approach is that this processing is time-consuming and thus best done once.

Here we use the parquet file format to store the data. As it is easy to imagine using these on projects other than our hypothetical `"glms"` project, we will store the SEC EDGAR filings data discussed in `edgar`, a subdirectory of `DATA_DIR` defined above named.

```
edgar_dir <- file.path(data_dir, "edgar")
if (!dir.exists(edgar_dir)) dir.create(edgar_dir)
sec_index_2022q2 |>
  save_parquet("sec_index_2022q2", schema = "edgar")
```

This parquet file can be read into R in many ways. For example, we could use the `open_dataset()` function from the `arrow` library. However, we will use a DuckDB database connection `db`—created using a single line of code below—and the DuckDB-specific `load_parquet()` function provided by the `farr` package to create a database table.[13]

```
db <- dbConnect(duckdb::duckdb())
```

```
sec_index <- load_parquet(db, "sec_index_2022q2", schema = "edgar")
```

If we were using just a single index file, the benefits of using DuckDB rather than the `open_dataset()` function might be less clear. However, we will soon see how easy it is to work with parquet files using DuckDB and also look to combine these index files with the data in the `lm_10x_summary` data frame we created above and data from Compustat and CRSP.

Our next step is to embed the code we used above to download and process `company.gz` for a single quarter in a function that can download an SEC index file and save it as a parquet file. Armed with this function, we can then process multiple index files in one step. Naturally, the function we create below (`get_sec_index()`) addresses possibilities ignored above, such as download failures.

---

[12]See Appendix D for guidance on how to set up a PostgreSQL server of your own.

[13]The `load_parquet()` uses DuckDB's `read_parquet()` function behind the scenes.

```r
get_sec_index <- function(year, quarter, overwrite = FALSE) {

  pq_path <- str_c(edgar_dir, "/sec_index_",
                   year, "q", quarter, ".parquet")
  if (file.exists(pq_path) & !overwrite) return(TRUE)

  # Download the zipped index file from the SEC website
  url <- str_c("https://www.sec.gov/Archives/edgar/full-index/",
               year,"/QTR", quarter, "/company.gz")

  t <- tempfile(fileext = ".gz")
  result <- try(download.file(url, t))

  # If we didn't encounter an error downloading the file, parse it
  # and save as a parquet file
  if (!inherits(result, "try-error")) {
    temp <-
      read_fwf(t, fwf_cols(company_name = c(1, 62),
                           form_type = c(63, 74),
                           cik = c(75, 86),
                           date_filed = c(87, 98),
                           file_name = c(99, 150)),
               col_types = "ccicc", skip = 10,
               locale = locale(encoding = "macintosh")) |>
      mutate(date_filed = as.Date(date_filed))

    write_parquet(temp, sink = pq_path)
    return(TRUE)
  } else {
    return(FALSE)
  }
}
```

We next create a data frame containing the years and quarters for which we want to collect SEC index files. SEC EDGAR extends back to 1993 and index files for the current quarter will appear after the first day of the quarter. We can use the `crossing()` function from the `tidyr` package to create the **Cartesian product** of the quarters and years.

```r
now <- now(tz = 'America/New_York') - days(1)
current_year <- as.integer(year(now))
current_qtr <- quarter(now)
year <- 1993L:current_year
quarter <- 1:4L

index_files_to_get <-
  crossing(year, quarter) |>
  filter(year < current_year |
           (year == current_year & quarter <= current_qtr))
```

Finally, we can download the SEC index files using the following few lines of code.

> 💡 Tip
>
> The following code takes several minutes to run. This code took us around 4 minutes
> to run, but the time required will depend on things such as the speed of your internet
> connection.

```
index_files_downloaded <-
  index_files_to_get |>
  mutate(available = map2(year, quarter, get_sec_index))
```

After running the code above, we have 127 index files saved as parquet files in `edgar_dir`
occupying 462 MB of *drive space*. This represents much more than 462 MB of *data* because
parquet files are highly compressed.

Reading these data into our database is simple and quick. We can use a wildcard to combine
several parquet files (`"sec_index*.parquet"` here) into a single table.

```
sec_index <- load_parquet(db, "sec_index*", "edgar")
```

We can check that we have data for 127 quarters. As we saw above with the data on 10-Ks,
there are few observations in 1993 and increase from 1994 to 1995.

```
sec_index |>
  mutate(year = year(date_filed),
         quarter = quarter(date_filed)) |>
  count(year, quarter) |>
  arrange(year, quarter) |>
  collect()
```

```
# A tibble: 127 x 3
   year quarter       n
  <dbl>   <dbl>   <dbl>
1  1993       1       4
2  1993       2       4
3  1993       3       7
4  1993       4      20
5  1994       1   20879
6  1994       2   16500
# i 121 more rows
```

Each filing on SEC EDGAR has a `form_type`. As can be seen in Table 23.4, the most
common form type is Form 4, which is filed whenever there is a material change in the
holdings of company insiders. The second most common form type is Form 8-K, which the
SEC describes[14] as "the 'current report' companies must file with the SEC to announce
major events that shareholders should know about."

```
sec_index |>
  count(form_type, sort = TRUE) |>
  collect(n = 10)
```

---

[14]https://www.sec.gov/answers/form8k.htm

TABLE 23.4

Most common form types on SEC EDGAR

form_type	n
4	8,962,439
8-K	1,956,527
SC 13G/A	854,515
3	807,001
424B2	757,350
10-Q	683,426
6-K	523,774
497	511,773
SC 13G	450,673
D	410,707

As mentioned above, we will construct a measure of disclosure based on the number of 8-K filings. While 8-K filings represent the second-most common form of filing, they are still a small share of all filings. Figure 23.3 shows some clear trends in the number of filings over time. Form 8-K filings appear to have peaked in number in 2005 and to have been in slow decline in number since then.

```
sec_index |>
  mutate(year = year(date_filed),
         form_type = if_else(form_type %in% c("4", "8-K"),
                             form_type, "Other")) |>
  filter(year < max(year, na.rm = TRUE)) |>
  count(year, form_type) |>
  mutate(last_year =  year == max(year, na.rm = TRUE),
         label = if_else(last_year, form_type, NA)) |>
  ggplot(aes(x = year, y = n / 1000,
             colour = form_type, linetype = form_type)) +
  geom_line() +
  geom_label(aes(label = label), na.rm = TRUE,
             position = position_nudge(y = -0.4)) +
  scale_y_continuous(name = "Number of filings (thousands)") +
  scale_x_continuous(breaks = seq(1993, current_year, 2)) +
  theme(axis.text.x = element_text(angle = 90),
        legend.position = "none")
```

One thing to note about SEC form types is that there are often variants on basic form types. For example, `8-K/A` indicates an amended 8-K filing. We can identify the variants using regular expressions. In the case of 8-K filings, it appears from Table 23.5 that the vast majority are on the base `8-K` form type.

```
sec_index |>
  filter(str_detect(form_type, "^8-K")) |>
  count(form_type) |>
  mutate(percent = n / sum(n, na.rm = TRUE) * 100) |>
  arrange(desc(n)) |>
  collect()
```

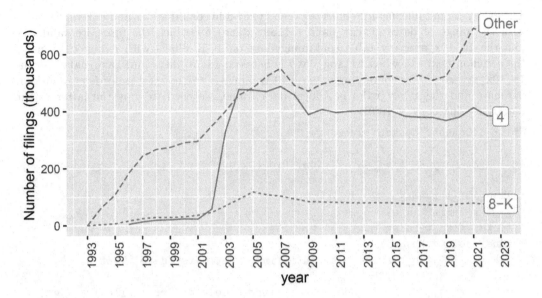

FIGURE 23.3
Number of filings over time by type

TABLE 23.5
Number of 8-K filings by variant

form_type	n	percent
8-K	1,956,527	95.852
8-K/A	83,473	4.089
8-K12G3	637	0.031
8-K12B	335	0.016
8-K12G3/A	115	0.006
8-K15D5	77	0.004
8-K12B/A	37	0.002
8-K15D5/A	2	0.000

```
dbDisconnect(db)
```

23.3.3 Measure of financial statement complexity

Our analysis below will take as the basic unit of observation the filing of a 10-K by a firm and examine the level of voluntary disclosure around that 10-K filing as a function of its complexity.

While `lm_10x_summary` contains many metrics related to filings, we retain only `acc_num`, `cik`, the two date columns (`filing_date` and `cpr`), and the `grossfilesize` column in our table `filing_10k_merged`.

As we will following Guay et al. (2016) in including a number of controls calculated using data from Compustat (and CRSP), we need to link each 10-K filing to a GVKEY. To do so, we will use the `gvkey_ciks` table included in the `farr` package.

The `gvkey_ciks` table includes links between `gvkey-idd` combinations and `cik` values along with the range of dates (`first_date` to `last_date`) for which the links are valid. We join the `lm_10x_summary` table containing data on 10-K filings with `gvkey_ciks` using the common field `cik` when `filing_date` is between `first_date` and `last_date`. As the `lm_10x_summary` table containing data on 10-K filings is a remote table inside our DuckDB database and the `gvkey_ciks` is a local data frame, we need to copy the latter to the database using `copy = TRUE`.

```
db <- dbConnect(duckdb::duckdb())

lm_10x_summary <- load_parquet(db, "lm_10x_summary", schema = "glms")

filing_10k_merged <-
  lm_10x_summary |>
  filter(form_type == "10-K") |>
  inner_join(gvkey_ciks,
             join_by(cik, between(filing_date, first_date, last_date)),
             copy = TRUE) |>
  select(gvkey, iid, acc_num, cik, filing_date, cpr, grossfilesize) |>
  mutate(eomonth = floor_date(cpr, "month") + months(1) - days(1)) |>
  save_parquet(name = "filing_10k_merged", schema = "glms")

dbDisconnect(db)
```

23.3.4 Measure of voluntary disclosure

We can now move on to constructing our measure of voluntary disclosure. We start by creating `filing_8k`, which only includes 8-K filings. We rename `date_filed` to `date_8k` to avoid confusion that might arise when we combine with data on 10-Ks that have their own filing dates.

```
db <- dbConnect(duckdb::duckdb())
sec_index <- load_parquet(db, "sec_index*", "edgar")

filing_8k <-
  sec_index |>
  filter(form_type == '8-K') |>
  rename(date_8k = date_filed)
```

We then combine `filing_10k_merged` with `filing_8k` using CIK and keep all 8-K filings between one year before and one year after the 10-K filing date. We rename `filing_date` as `fdate` in part to shorter lines of code. Having constructed this merged table, we then count the number of 8-K filings within various windows (e.g., `between(datediff, 0, 30)` represents 8-K filings in the 30-day window starting on the date of the 10-K filing.[15]

It is noteworthy that this code runs in less than a second on our computer, illustrating the performance benefits of DuckDB. If the source tables (`filing_10k_merged` and `filing_8k`) are instead converted to local data frames using `collect()`, this query takes nearly a minute.

Because the arguments to `between()` are "out of order" in that `date_8k` comes from `filing_8k`, the second table referred to in `inner_join()`, we indicate this by qualifying

---

[15]Strictly speaking, this is a 31-day window, as it includes day zero.

it with y$ and that **fdate_m1** and **fdate_p1** come from **filing_10k_merged** by qualifying these with x$.

```
filing_10k_merged <- load_parquet(db, "filing_10k_merged", schema = "glms")

vdis_df <-
  filing_10k_merged |>
  rename(fdate = filing_date) |>
  mutate(fdate_p1 = fdate + years(1L),
         fdate_m1 = fdate - years(1L)) |>
  inner_join(filing_8k,
             join_by(cik,
                     between(y$date_8k, x$fdate_m1, x$fdate_p1))) |>
  mutate(datediff = as.double(date_8k - fdate)) |>
  group_by(acc_num, gvkey, iid, cik) |>
  summarize(vdis_p1y = sum(as.integer(datediff > 0)),
            vdis_p30 = sum(as.integer(between(datediff, 0, 30))),
            vdis_p60 = sum(as.integer(between(datediff, 0, 60))),
            vdis_p90 = sum(as.integer(between(datediff, 0, 90))),
            vdis_m1y = sum(as.integer(datediff < 0)),
            vdis_m30 = sum(as.integer(between(datediff, -30, -1))),
            vdis_m60 = sum(as.integer(between(datediff, -60, -1))),
            vdis_m90 = sum(as.integer(between(datediff, -90, -1))),
            .groups = "drop") |>
  group_by(acc_num, gvkey, iid) |>
  filter(vdis_p1y == max(vdis_p1y, na.rm = TRUE)) |>
  ungroup() |>
  save_parquet(name = "vdis_df", schema = "glms")

dbDisconnect(db)
```

23.3.5 Controls

In addition to our measure of voluntary disclosure (in **vdis_df**) and our measure of financial statement complexity (in **filing_10k_merged**), we need data on various controls for our empirical analysis. These controls are based on Guay et al. (2016) and the required data come from Compustat and CRSP.

Thus, as in previous chapters, we will collect these data from the WRDS PostgreSQL database. However, we take a different approach to data collection in this chapter from the approaches we have used in earlier chapters.

For example, in Chapter 12, we had a relatively small amount of data on earnings announcement dates in a local data frame. When we needed to combine those event dates with data from CRSP, we simply copied the data to the WRDS server using the **copy_inline()** function, then summarized the data into a smaller data set using the WRDS PostgreSQL server, then used **collect()** to retrieve the data for further empirical analysis. This approach will not work well when our local data frame is too large to work well with **copy_inline()**.[16]

---

[16]The **copy_inline()** function creates a literal SQL string to represent a table, and it is not efficient when the table is large.

The approach we use in this chapter does not require sending large amounts of data to the CRSP PostgreSQL server. Additionally, we use persistent storage (specifically parquet files) to make it easier to break our analysis into discrete steps that do not require the rerunning of costly earlier data steps each time we start R.

The data sets we need to collect from WRDS are:

- Fundamental data from `comp.funda`.
- Data on event dates for GVKEYs derived from `crsp.dsi` and `crsp.ccmxpf_lnkhist`.
- Data on "trading dates" derived from `crsp.dsi`. (See Chapter 13 for more details.)
- Data on liquidity measures derived from `crsp.dsf`.

For each of these four data sets, we will perform the following steps.

- Connect to the database server.
- Create a remote data frame for the underlying table.
- Process the data as needed using `dplyr` verbs.
- Save the resulting data as a parquet file.
- Close the connection to the database server.

23.3.5.1 Fundamental data from `comp.funda`

We start with fundamental data and indicators for relevance of SFAS 133 and SFAS 157, as discussed in Guay et al. (2016) and papers cited therein, and store these as `compustat`.

```
db <- dbConnect(RPostgres::Postgres())

funda <- tbl(db, Id(schema = "comp", table = "funda"))

compustat <-
  funda |>
  filter(indfmt == 'INDL', datafmt == 'STD',
         popsrc == 'D', consol == 'C') |>
  mutate(mkt_cap = prcc_f * csho,
         size = if_else(mkt_cap > 0, log(mkt_cap), NA),
         roa = if_else(at > 0, ib / at, NA),
         mtb = if_else(ceq > 0, mkt_cap / ceq, NA),
         special_items = if_else(at > 0, coalesce(spi, 0) / at, NA),
         fas133 = !is.na(aocidergl) & aocidergl != 0,
         fas157 = !is.na(tfva) & tfva != 0) |>
  select(gvkey, iid, datadate, mkt_cap, size, roa, mtb,
         special_items, fas133, fas157) |>
  filter(mkt_cap > 0) |>
  mutate(eomonth = floor_date(datadate, "month") + months(1) - days(1)) |>
  save_parquet(name = "compustat", schema = "glms")

dbDisconnect(db)
```

23.3.5.2 Data on event dates for GVKEYs

Next, we create the data frame `event_dates` that will contain `permno`, `start_date` and `end_date` for each of our events (10-K filing dates). The logic here is similar to that we saw in constructing event dates for Apple media events in Chapter 13 and we use the now-familiar `ccm_link` table that we first saw in Section 7.3.

```
pg <- dbConnect(RPostgres::Postgres())
db <- dbConnect(duckdb::duckdb())

ccmxpf_lnkhist <- tbl(pg, Id(schema = "crsp", table = "ccmxpf_lnkhist"))
filing_10k_merged <- load_parquet(db, table = "filing_10k_merged",
                                  schema = "glms")
ccm_link <-
  ccmxpf_lnkhist |>
  filter(linktype %in% c("LC", "LU", "LS"),
         linkprim %in% c("C", "P")) |>
  mutate(linkenddt = coalesce(linkenddt, max(linkenddt, na.rm = TRUE))) |>
  rename(permno = lpermno,
         iid = liid) |>
  copy_to(db, df = _, name = "ccm_link", overwrite = TRUE)

filing_permnos <-
  filing_10k_merged |>
  inner_join(ccm_link,
             join_by(gvkey, iid,
                     between(filing_date, linkdt, linkenddt))) |>
  select(gvkey, iid, filing_date, permno)

event_dates <-
  filing_permnos |>
  distinct(permno, filing_date) |>
  collect() |>
  get_event_dates(pg, permno = "permno",
                  event_date = "filing_date",
                  win_start = -20, win_end = 20) |>
  copy_to(db, df = _, name = "event_dates", overwrite = TRUE) |>
  inner_join(filing_permnos, by = join_by(permno, filing_date)) |>
  save_parquet(name = "event_dates", schema = "glms")

dbDisconnect(pg)
dbDisconnect(db)
```

23.3.5.3 Data on "trading dates"

We first saw the `trading_dates` table in Chapter 12 and we use the `get_trading_dates()` function from the `farr` package to create that using data from the WRDS PostgreSQL database and then save it as a parquet file.

```
db <- dbConnect(RPostgres::Postgres())

trading_dates <-
  get_trading_dates(db) |>
  save_parquet(name = "trading_dates", schema = "glms")

dbDisconnect(db)
```

23.3.5.4 Data on liquidity measures

A core idea of Guay et al. (2016) is that firms use voluntary disclosure to mitigate the deleterious effects of financial statement complexity. One of the costs of greater financial statement complexity is greater *illiquidity* (i.e., less liquidity) in the market for a firm's equity. Guay et al. (2016) examine two primary measures of liquidity: the "Amihud (2002) measure" and the bid-ask spread. The following code calculates these measures using the formulas provided by Guay et al. (2016, p. 240).

> 💡 Tip
>
> The following code takes several minutes to run because a large about of data needs to be downloaded from `crsp.dsf`. This code took us nearly 4 minutes to run, but the time required will depend on things such as the speed of your internet connection.

```
pg <- dbConnect(RPostgres::Postgres())
db <- dbConnect(duckdb::duckdb())

dsf <- tbl(pg, Id(schema = "crsp", table = "dsf"))
filing_10k_merged <- load_parquet(db, "filing_10k_merged", schema = "glms")

first_year <-
  filing_10k_merged |>
  summarize(min(year(filing_date)) - 1L) |>
  pull()

last_year <-
  filing_10k_merged |>
  summarize(max(year(filing_date)) + 1L) |>
  pull()

liquidity <-
  dsf |>
  filter(between(year(date), first_year, last_year)) |>
  mutate(prc = abs(prc),
         spread = if_else(prc > 0, (ask - bid) / prc, NA),
         illiq = if_else(vol * prc > 0, abs(ret) / (vol * prc), NA)) |>
  mutate(spread = spread * 100,
         illiq = illiq * 1e6) |>
  filter(!is.na(spread), !is.na(illiq)) |>
  select(permno, date, spread, illiq) |>
  save_parquet(name = "liquidity", schema = "glms")

dbDisconnect(pg)
dbDisconnect(db)
```

23.3.6 Empirical analysis

Now that we have collected all the data we need, we can run the final analysis. We can load the parquet data created in the previous section using the `load_parquet()` function made available by the `farr` package.

```
db <- dbConnect(duckdb::duckdb())

vdis_df <- load_parquet(db, table = "vdis_df", schema = "glms")
filing_10k_merged <- load_parquet(db, table = "filing_10k_merged",
                                  schema = "glms")
liquidity <- load_parquet(db, table = "liquidity", schema = "glms")
trading_dates <- load_parquet(db, table = "trading_dates", schema = "glms")
event_dates <- load_parquet(db, table = "event_dates", schema = "glms")
compustat <- load_parquet(db, table = "compustat", schema = "glms")
```

We combine event dates with liquidity data.

Because the arguments to `between()` are again "out of order", in joining `event_dates` and `liquidity`, we indicate the source tables with `x$` and `y$`. We then use `trading_dates` to match each `filing_date` and each `date` with a trading date (`td`) value.

```
liquidity_merged <-
  event_dates |>
  inner_join(liquidity,
             join_by(permno,
                     between(y$date, x$start_date, x$end_date))) |>
  inner_join(trading_dates, join_by(filing_date == date)) |>
  rename(filing_td = td) |>
  inner_join(trading_dates, join_by(date)) |>
  mutate(rel_td = td - filing_td) |>
  select(gvkey, iid, permno, filing_date, rel_td, spread, illiq) |>
  compute()
```

Some analyses in Guay et al. (2016) reduce complexity measures to either quintiles or deciles and we use quintiles for the complexity measure here. To allow for secular changes in complexity over time, we form quintiles by the year of `filing_date`.

```
complexity <-
  filing_10k_merged |>
  mutate(year = year(filing_date)) |>
  group_by(year) |>
  mutate(complex_q5 = ntile(grossfilesize, 5))
```

Like Guay et al. (2016), we impose a sample requirement of complete data (i.e., 20 trading days) prior to the filing of the 10-K.

```
complete_cases <-
  liquidity_merged |>
  group_by(gvkey, iid, filing_date) |>
  filter(rel_td < 0) |>
  summarize(num_obs = n(), .groups = "drop") |>
  filter(num_obs == 20) |>
  select(-num_obs) |>
  compute()
```

To better understand the underlying relations in the data, we examine the behaviour of measures of illiquidity around event dates by complexity quintile. We form deciles by year for the two measures of illiquidity. To facilitate plotting, we use `pivot_longer()` and convert `complex_q5` from a numeric to a character variable.[17]

```
plot_data <-
  complexity |>
  inner_join(liquidity_merged,
             by = join_by(gvkey, iid, filing_date)) |>
  semi_join(complete_cases,
            by = join_by(gvkey, iid, filing_date)) |>
  group_by(year) |>
  mutate(spread = ntile(spread, 10),
         illiq = ntile(illiq, 10)) |>
  group_by(rel_td, complex_q5) |>
  summarize(spread = mean(spread, na.rm = TRUE),
            illiq = mean(illiq, na.rm = TRUE),
            num_obs = n(),
            .groups = "drop") |>
  pivot_longer(spread:illiq, names_to = "measure") |>
  mutate(complex_q5 = as.character(complex_q5)) |>
  compute()
```

Once we have structured the data appropriately, creation of the plot seen in Figure 23.4 requires just a few lines of code.

```
plot_data |>
  mutate(last_day =  rel_td == max(rel_td),
         label = if_else(last_day, as.character(complex_q5), NA)) |>
  ggplot(aes(x = rel_td,
             y = value,
             color = complex_q5,
             group = complex_q5)) +
  geom_line() +
  geom_label(aes(label = label), na.rm = TRUE) +
  facet_wrap( ~ measure) +
  theme(legend.position = "none")
```

To conduct regression analysis analogous to that in Table 3 of Guay et al. (2016), we combine our data on voluntary disclosure (`vdis_df`) with our data on financial statement complexity (`filing_10k_merged`) and controls (`compustat`).

```
reg_data_glms <-
  vdis_df |>
  inner_join(filing_10k_merged,
             by = join_by(acc_num, gvkey, iid, cik)) |>
  inner_join(compustat, by = join_by(gvkey, iid, eomonth)) |>
  mutate(ln_grossfilesize = log(grossfilesize)) |>
  collect()
```

We run three regressions. The first two are, following Guay et al. (2016), OLS with and without firm fixed effects. The third regression is Poisson regression.

---

[17]In earlier chapters, we used factors to a similar end, but these data are coming from DuckDB, which does not have the same native support for factors that R does.

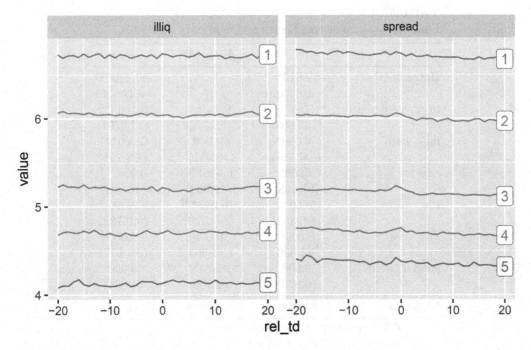

FIGURE 23.4
Behaviour of illiquidity around 10-K filing dates by complexity quintile

```
controls <- c("mkt_cap", "size", "roa", "mtb", "special_items")

model <- str_c("vdis_p1y ~ ",
               str_c(c("ln_grossfilesize", controls),
                     collapse = " + "))

fms <- list(
  "OLS" = feols(as.formula(model), data = reg_data_glms),
  "Firm FE" = feols(as.formula(str_c(model, " | gvkey + iid")),
             data = reg_data_glms),
  "Pois" = glm(as.formula(model), family = "poisson", data = reg_data_glms))
```

We make a small function that returns "robust" standard errors if the model is estimated using `glm()` (as in Poisson regression). Based on papers we examine in Chapter 24, use of robust standard errors seems to be a standard practice with Poisson regression.

```
get_coefs <- function(fm, type = "HC1") {
  if (inherits(fm, "glm")) {
    coeftest(fm, vcov = vcovHC(fm, type = type))
  } else {
    coeftest(fm)
  }
}
```

TABLE 23.6
Financial statement complexity and voluntary disclosure

	OLS	Firm FE	Pois
(Intercept)	−19.613***		−0.826***
	(−107.209)		(−42.136)
ln_grossfilesize	1.786***	1.396***	0.186***
	(135.856)	(52.776)	(132.757)
mkt_cap	0.000	0.000	0.000***
	(−0.518)	(0.051)	(−3.731)
size	0.526***	0.415***	0.053***
	(51.698)	(12.339)	(46.069)
roa	−0.003	−0.027***	0.000
	(−0.852)	(−2.579)	(−0.722)
mtb	0.000*	0.000	0.000
	(−1.826)	(−1.503)	(−1.517)
special_items	0.004	0.023**	0.000
	(0.973)	(2.566)	(0.934)
Num.Obs.	106,988	106,988	106,988

$*p < 0.1$, $**p < 0.05$, $***p < 0.01$

Regression results are shown in Table 23.6.

```
modelsummary(map(fms, get_coefs),
             statistic = "statistic",
             gof_map = "nobs",
             stars = c('*' = .1, '**' = 0.05, '***' = .01))
```

Once we are done with our in-memory DuckDB database, we can disconnect from it.

```
dbDisconnect(db)
```

23.3.7 Exercises

1. Instead of `tbl(db, sql(lm_10x_summary_sql))`, we could use have used `read_csv(lm_data, show_col_types = FALSE)`. What benefits do you see from using the former option? (*Hint*: Try both versions of the code and perhaps use `system.time()` to evaluate processing times.)

2. What primary key could we use for `lm_10x_summary`? Check that this is a valid primary key.

3. What primary key could we use for `filing_10k_merged`? Check that this is a valid primary key.

4. From Figure 23.4, do you observe a change in measures of liquidity around 10-K filings dates? Why (or why not) would we expect to see a change around these events?

5. From Figure 23.4, do you observe a relation between the complexity of financial statements and illiquidity? Is the sign of any relationship consistent with what we would expect from reading Guay et al. (2016)? Does the nature of the relationship affect how you interpret the results of Guay et al. (2016)?

6. Calculate the marginal effects associated with the regression coefficients shown in Table 23.6. Do you observe significant differences across specifications? (*Hint*: The `mfx()` function should be able to do most of the work here.)

23.4 Further reading

Chapter 13 of Wooldridge (2010) provides a more thorough discussion of MLE. Chapter 15 of Wooldridge (2010) covers probit and logit, including MLE and marginal effects. Chapter 19 of Wooldridge (2010) covers count models, including Poisson regression. We demonstrated that OLS works well for practical purposes with simulated data, Angrist and Pischke (2008) also highlight the benefits of OLS and show that it delivers very similar inferences in a particular setting.

For more on the GLMs, including details about estimation of coefficients and standard errors, see Dunn and Smyth (2018). Aronow and Miller (2019) provide a modern take on MLE estimation.

More details on the parquet format can be found in Chapter 22[18] of R4DS. Chapter 19[19] of R4DS discusses Cartesian products.

23.5 Appendix: Maintaining a repository of SEC index files

In Section 23.3.4 we described how one can download the SEC index files and create a repository of parquet files that can be used to identify and locate filings on SEC EDGAR. Because these files are continually being updated, we might want to update our files too. However, ideally we would avoid downloading files that we already have.

If you visit any of pages from which we obtained the filings, such as the page for the second quarter of 2022 that we looked at earlier in the chapter, you can see that the "last modified" timestamp for each file.[20] We can use collect this information and store it and then later on only download files with a more recent "last modified" timestamp.

The `get_last_update()` function scrapes the "last modified" information from the page for a given `year` and `quarter`.

```
get_last_update <- function(year, quarter) {

  url <- str_c("https://www.sec.gov/Archives/edgar/full-index/",
               year, "/QTR", quarter, "/")

  resp <-
    request(url) |>
    req_user_agent(getOption("HTTPUserAgent")) |>
    req_perform() |>
```

---

[18] https://r4ds.hadley.nz/arrow
[19] https://r4ds.hadley.nz/joins.html
[20] See https://www.sec.gov/Archives/edgar/full-index/2022/QTR2/.

```
    resp_body_html() |>
    html_elements("body")

  resp[[1]] |>
    html_table() |>
    filter(Name == "company.gz") |>
    select(`Last Modified`) |>
    mdy_hms(tz = "America/New_York")
}
```

We can apply this function to the data frame of `index_files_to_get` created above to get the `last_modified` data for each quarter. Because we are getting data from 127 web pages, the following code takes about 30 seconds to run.

```
last_modified_scraped <-
  index_files_to_get |>
  rowwise() |>
  mutate(last_modified = get_last_update(year, quarter)) |>
  ungroup() |>
  system_time()
```

```
  user   system elapsed
 2.765   0.107  45.191
```

We next check for the presence of a parquet data file from a previous examination of the SEC EDGAR index files. If this is present, we read the data into R. If it is not present, we create an empty data frame with one row of `NA` values.

```
pq_path <- file.path(edgar_dir, "last_modified.parquet")

if (file.exists(pq_path)) {
  last_modified <- read_parquet(pq_path)
} else {
  last_modified <- tibble(year = NA, quarter = NA,
                          last_modified = NA)
}
```

We then compare `last_modified_scraped` with the `last_modified` value we just read to identify the index files that we either do not have or that need to be updated.

```
to_update <-
  last_modified_scraped |>
  left_join(last_modified,
            by = c("year", "quarter"),
            suffix = c("_new", "_old")) |>
  filter(is.na(last_modified_old) |
           last_modified_new > last_modified_old)
```

Now, we can run code much like that in the text above to download index files, but only for those that we don't already have.

```
index_files_downloaded <-
  to_update |>
  mutate(available = map2(year, quarter, get_sec_index, overwrite = TRUE))
```

Having downloaded the index files we needed, we can save the scraped data on `last_modified` for use the next time we want to update our repository of SEC EDGAR index files.

```
last_modified_scraped |>
  save_parquet(name = "last_modified", schema = "edgar")
```

24

Extreme values and sensitivity analysis

In this chapter, we look at approaches used in handling the kinds of **extreme values** encountered in real-world data sets. We start the chapter with Leone et al. (2019), which provides guidance on how to address extreme values. We then examine Call et al. (2018), which explores a setting in which extreme values are present. We also use this chapter to explore the idea of **sensitivity analysis** using the approach of Frank (2000) seen in Call et al. (2018).

> 💡 Tip
>
> The code in this chapter uses the following packages. We load `tidyverse` because we use several packages from the Tidyverse. For instructions on how to set up your computer to use the code found in this book, see Section 1.2. Quarto templates for the exercises below are available on GitHub.[a]
>
> ---
> [a]https://github.com/iangow/far_templates/blob/main/README.md

```
library(tidyverse)
library(farr)
library(DBI)
library(modelsummary)
library(fixest)
library(dbplyr)          # window_order()
library(furrr)           # future_map()
library(lmtest)          # coeftest()
library(sandwich)        # vcovHC()
library(robustbase)      # lmrob()
```

24.1 Leone et al. (2019)

Leone et al. (2019) show that many papers in accounting research modify or eliminate observations with extreme values, as these can have an outsized influence on regression results. **Winsorization** defines extreme values as those above or below certain quantiles and replaces these with values *at* those quantiles (we saw winsorization in Chapter 25). Another approach is **truncation**, which is like winsorization, except that extreme observations are eliminated from subsequent analysis.

Leone et al. (2019) also cover two alternative approaches to handling extreme values. The first alternative approach is the use of diagnostic statistics to identify influential observations where "such observations are excluded from second-pass (or later) regressions estimated to draw a researcher's 'final' inferences" (p. 343).

The second alternative approach is **robust regression**. While Leone et al. (2019) allude to an extensive literature on robust regression, we follow Leone et al. (2019) in focusing on so-called **MM-estimators**, which are made available via the `robustbase`[1] package.

Leone et al. (2019) provide three kinds of evidence regarding approaches to extreme values and influential observations. First, they survey the literature to gather data describing what researchers do to address extreme observations. The results of this effort are found in Table 1 of Leone et al. (2019).

Second, Leone et al. (2019) replicate three studies and examine how inferences vary by approach. Below we study one of these replications.

Third, Leone et al. (2019) provide simulation evidence on the effectiveness of the various approaches. We replicate this simulation evidence in Section 24.4.

24.1.1   Replication analysis

To better understand the approaches discussed in Leone et al. (2019), we will follow their replication of Chen et al. (2018), which is also the focus of material later in this chapter. The required data come from `comp.company` and `comp.funda`.

```
db <- dbConnect(RPostgres::Postgres(), bigint = "integer")


company <- tbl(db, Id(schema = "comp", table = "company"))
funda <- tbl(db, Id(schema = "comp", table = "funda"))
```

While Chen et al. (2018) focus on fiscal years from 1996 through 2015, we get data for fiscal year 1995, so that lagged values are available for 1996. Note that after collecting the data, we convert `fyear` into a factor, which facilitates estimation of fixed effects and interaction terms.[2]

```
comp <-
  funda |>
  filter(indfmt == "INDL", datafmt == "STD",
         consol == "C", popsrc == "D") |>
  filter(!between(sich, 6000, 6999) | is.na(sich)) |>
  filter(between(fyear, 1995, 2015)) |>
  group_by(gvkey) |>
  window_order(fyear) |>
  mutate(big_n = as.integer(au) %in% 1:8L,
         lag_at = lag(at),
         inv_at = if_else(lag_at > 0, 1 / lag_at, NA),
         ta = if_else(lag_at > 0, (ib - oancf) / lag_at, NA),
         roa = if_else(lag_at > 0, ib / lag_at, NA),
         cfo = if_else(lag_at > 0, oancf / lag_at, NA),
         mkt_cap = prcc_f * csho,
         lag_mkt_cap = lag(mkt_cap),
         size = if_else(lag_mkt_cap > 0, log(lag_mkt_cap), NA),
```

---

[1] https://cran.r-project.org/web/packages/robustbase/index.html

[2] While much of the following code resembles that used in replicating Lawrence et al. (2011) in Chapter 25, there are minor differences.

```
        debt = coalesce(dltt, 0) + coalesce(dlc, 0),
        lev = if_else(lag_at > 0, debt / lag_at, NA),
        mtb = if_else(lag(ceq) > 0, lag_mkt_cap / lag(ceq), NA),
        assets = lag_at,
        d_sale = if_else(lag_at > 0, (revt - lag(revt)) / lag_at, NA),
        d_ar =  if_else(lag_at > 0, (rect - lag(rect)) / lag_at, NA),
        ppe = if_else(lag_at > 0, ppent / lag_at, NA)) |>
  ungroup() |>
  select(gvkey, datadate, fyear, big_n, ta, big_n, roa, cfo, size, lev,
        mtb, inv_at, d_sale, d_ar, ppe) |>
  collect() |>
  mutate(fyear = as.factor(fyear))
```

We create two functions that are quite similar in structure. The `winsorize()` function offers more than we use here, as we stick to the default argument of `prob = 0.01`, which is then mapped into `p_low = 0.01` and `p_high = 0.99`, which implements "the 1 percent and 99 percent winsorization rule common to accounting studies" (and also used in Chen et al., 2018). The first line of the function uses `quantile()` to identify the 1st and 99th percentiles of the distribution of `x` and stores these two values in the vector `cuts`. The `type` argument specifies how ties are resolved and how averages at breakpoints are calculated. The `type` argument should be "an integer between 1 and 9 selecting one of the nine quantile algorithms ... to be used."[3] By selecting `type = 2`, we get the SAS default, which means we get the same values as we would using the SAS macro commonly used by other researchers. The third line of the `winsorize()` function sets all values below `cuts[1]`, which is the `p_low` quantile, equal to `cuts[1]`. The fourth line sets all values above `cuts[2]` (the `p_high` quantile) equal to `cuts[2]`. The last line returns the modified value of `x`.

```
winsorize <- function(x, prob = 0.01, p_low = prob, p_high = 1 - prob) {
  cuts <- quantile(x, probs = c(p_low, p_high), type = 2, na.rm = TRUE)
  x[x < cuts[1]] <- cuts[1]
  x[x > cuts[2]] <- cuts[2]
  x
}
```

The `truncate()` function is almost identical, except that values outside the bounds in `cuts` are set to `NA`, which effectively removes them from later analysis.

```
truncate <- function(x, prob = 0.01, p_low = prob, p_high = 1 - prob) {
  cuts <- quantile(x, probs = c(p_low, p_high), type = 2, na.rm = TRUE)
  x[x < cuts[1]] <- NA
  x[x > cuts[2]] <- NA
  x
}
```

We use `mutate()` with `across()` to apply these functions to all variables between `ta` and `ppe` inclusive (i.e., we do not apply them to `gvkey`, `datadate`, `fyear`, or `big_n`) and store the results in `comp_win` and `comp_trunc`.

```
comp_win <-
  comp |>
  mutate(across(ta:ppe, winsorize))
```

---

[3]See help by typing `?` `quantile` at the R console; R Core Team (2021)

```
comp_trunc <-
  comp |>
  mutate(across(ta:ppe, truncate))
```

We next run regressions. The first model we consider is essentially that found in Chen et al. (2018). As Leone et al. (2019) point out, `roa` and `cfo` can only be included in a model with `ta` as the dependent variable when winsorization breaks the identity `ta = roa - cfo`.[4] In the second model, we use raw data, but drop `roa` to make the model meaningful.

```
fms <- list(
  "Wins, w/ROA" = feols(ta ~ big_n + roa + cfo + size + lev + mtb +
                          fyear  * (inv_at + I(d_sale - d_ar) + ppe),
                        ~ gvkey + fyear,
                        data = comp_win, na.action = na.exclude),
  "Raw" = feols(ta ~ big_n + cfo + size + lev + mtb +
                  fyear  * (inv_at + I(d_sale - d_ar) + ppe),
                ~ gvkey + fyear,
                data = comp, na.action = na.exclude))
```

The third and fourth models tweak the second model (`"Raw"`) by using winsorized and truncated data respectively. We use `update()` to make the relationships between the models clearer and to avoid errors caused by unintended differences between models.

```
fms[["Wins"]] <- update(fms[["Raw"]], data = comp_win)
fms[["Trunc"]] <- update(fms[["Raw"]], data = comp_trunc)
```

Given that we don't use `comp_trunc` other than to estimate `fms[["Trunc"]]`, we could have used the following alternative to avoid creating `comp_trunc` as a separate variable.

```
fms[["Trunc"]] <-
  comp |>
  mutate(across(ta:ppe, truncate)) |>
  update(fms[["Raw"]], data = _)
```

The fifth model we estimate uses Cook's D to identify extreme observations for exclusion. As the `fixest` objects returned by `feols()` do not support the `cooks.distance()` function, we estimate an OLS model and use that to identify extreme observations for exclusion.

```
fm <- lm(ta ~ big_n + cfo + size + lev + mtb +
               fyear  * (inv_at + I(d_sale - d_ar) + ppe),
         data = comp, na.action = na.exclude)

fms[["Cook's D"]] <-
  comp |>
  mutate(cooksd = cooks.distance(fm),
         extreme = cooksd > 4 / nobs(fm),
         ta = if_else(extreme, NA, ta)) |>
  update(fms[["Raw"]], data = _)
```

---

[4]Actually this is the model estimated in column (1) of Table 2 of Leone et al. (2019), as a model estimated with "raw data" would not work. Additionally, Andy Leone confirms that the coefficients on `roa` and `cfo` have been switched in Table 2 of Leone et al. (2019).

The sixth and final model uses robust regression. Leone et al. (2019) use the `robreg` module made available for Stata, which defaults to an asymptotic efficiency of 85% for the bisquare estimator used in the second step of the MM estimation procedure and we mimic this choice here. This efficiency level is obtained by specifying the `tuning.psi` parameter as `3.4437`.[5]

An alternative would be to follow Koller and Stahel (2017) by selecting `control = lmrob.control("KS2014")`. Koller and Stahel (2017) suggest that this approach "allows for ... estimation in designs with factors and interactions between factors and continuous regressors", which describes our setting here due to the interaction of `fyear` (a factor) with `inv_at + I(d_sale - d_ar) + ppe` (continuous variables).

Following the recommendation in Leone et al. (2019), we recover the weights from the fitted `lmrob` object and feed those to an update of the `fixest` object in `fms[["Raw"]]`, as doing so allows us to use cluster-robust standard errors. Note that many values returned by `weights()` will be `NA`, so we use `coalesce(0)` to convert those to zeros.

```
fmrob <- lmrob(formula(fm),
               data = comp, na.action = na.exclude,
               method = "MM",
               control = lmrob.control(tuning.psi = 3.4437))

fms[["Robust"]] <-
  fmrob |>
  weights(type = "robustness") |>
  coalesce(0) |>
  update(fms[["Raw"]], data = comp, weights = _)
```

Results of estimating each of these models are shown in Table 24.1.

```
modelsummary(fms,
             estimate = "{estimate}{stars}",
             statistic = "({statistic})",
             coef_omit = "(fyear|ppe|inv_at|d_sale)",
             gof_map = c("nobs", "r.squared"),
             stars = c('*' = .1, '**' = 0.05, '***' = .01))
```

24.1.2 Discussion questions

1. Table 5 of Leone et al. (2019) presents results from a simulation analysis. Which panel of that table likely best reflects the kind of data researchers encounter in practice?

2. Consider the left half of Table 5 of Leone et al. (2019) (i.e., $\beta_1 = 0$). Using the panel you identified in question 1 above, interpret the implications of the simulation results for researchers in practice. Note that the mean estimates of β_1 using Cook's D or robust regression are very similar. Does Table 5 of Leone et al. (2019) provide strong support for rejecting Cook's D in favour of robust regression?

---

[5]See p.31 of Maronna et al. (2019). The Stata routine `robreg` uses the function `mm_biweight_k` from the `moremata` routine to convert efficiency to the equivalent of the `tuning.psi` parameter: https://go.unimelb .edu.au/wyw8.

TABLE 24.1

Results from various approaches to extreme values

	Wins, w/ROA	Raw	Wins	Trunc	Cook's D	Robust
(Intercept)	−0.031***	0.262	−0.085***	−0.119***	−0.107***	−0.023***
	(−8.934)	(1.204)	(−7.352)	(−14.799)	(−10.300)	(−16.929)
big_nTRUE	0.000	−0.073	0.006	0.009**	0.020***	−0.008***
	(0.264)	(−1.078)	(1.316)	(2.614)	(3.930)	(−12.677)
roa	0.815***					
	(63.785)					
cfo	−0.786***	0.799*	0.257***	0.124***	0.194***	−0.386***
	(−37.425)	(1.762)	(8.932)	(5.573)	(5.237)	(−144.226)
size	0.003***	−0.066*	0.004**	0.007***	0.001	0.003***
	(5.116)	(−1.862)	(2.626)	(5.680)	(0.685)	(13.730)
lev	−0.035**	−0.072	−0.111***	−0.031***	−0.074***	−0.019***
	(−2.641)	(−0.344)	(−5.089)	(−4.478)	(−4.161)	(−22.438)
mtb	−0.001***	−0.003	−0.007***	−0.006***	−0.001***	0.000***
	(−4.393)	(−0.938)	(−8.695)	(−11.031)	(−3.855)	(−4.471)
Num.Obs.	113,994	113,994	113,994	106,584	113,566	82,874
R2	0.912	0.574	0.278	0.095	0.129	0.981

3. In the search for new research questions, accounting researchers increasingly need to study variables whose most plausible effect sizes are consistent with the right half of Table 5 of Leone et al. (2019) (i.e., $\beta_1 = 0$). Using the panel you identified in question 1 above, interpret the implication of the simulation results for researchers in practice.

4. Table 2 of Leone et al. (2019) reports results similar to those given in columns (4), (5), and (6) of Table 24.1. They say, "[when we] perform estimation using RR, the coefficient on *BIG_N* is −0.011 (significant at the 1 percent level), opposite of what Chen et al. (2018) find, and instead consistent with the negative coefficient documented in prior studies using the two-step procedure." In light of the simulation evidence, how persuasive do you find the evidence above in support of a negative coefficient on `big_n`? How probative is the evidence of prior studies if it is based on approaches (e.g., winsorization and two-step procedures) that Leone et al. (2019) and Chen et al. (2018) demonstrate lead to unreliable results?

24.2 Call et al. (2018)

Call et al. (2018, p. 123) examine "whether there is a link between whistleblower involvement and the outcomes of enforcement action" by US regulators. Specifically, they show that "whistleblower involvement is associated with larger monetary penalties for the targeted firms and longer prison sentences for targeted employees" (2018, p. 126).

Kuvvet (2019) points out that a relatively small number of observations account for the bulk of the monetary penalties imposed on firms, employees, and others. The `cmsw_2018` data frame provided with the `farr` package includes data used in Call et al. (2018), which allows us to reproduce Table 2 of Kuvvet (2019) as Table 24.2.[6]

```
cmsw_2018 |>
  filter(tousesox == 1) |>
  pivot_longer(cols = matches("(penalty|mos)"),
               values_to = "amount", names_to = "target") |>
  group_by(target) |>
  arrange(desc(amount)) |>
  mutate(rank = row_number()) |>
  summarize(total = sum(amount),
            total_top = sum(if_else(rank <= 6, amount, 0)),
            .groups = "drop") |>
  mutate(top_perc = round(total_top / total * 100, 1),
         across(total:total_top, \(x) round(x, 0)))
```

TABLE 24.2
Top 1% observations as percentage of total penalties

target	total	total_top	top_perc
emppenalty	21,176	15,877	75.0
empprisonmos	17,181	3,658	21.3
firmpenalty	13,578	7,396	54.5
otherpenalty	11,107	9,684	87.2

24.2.1   Econometric claims in accounting research

Call et al. (2018, p. 134) concede the existence of "two challenges" for empirical analysis in their setting: "the large number of zero-valued observations (i.e., enforcement actions without any resultant penalties or criminal prison sentences) and the severe positive skewness in the dependent variable (i.e., some extremely large penalties)."

Call et al. (2018, p. 134) assert that "whereas other regression techniques ... suffer from potentially severe bias ... prior research shows that the Poisson pseudo-maximum likelihood (PPML) estimator ... is a particularly effective modeling technique for data distributions characterized by a disproportionate number of zeros and severe skewness" and cite seven sources for this claim.

This approach to econometrics is fairly standard in accounting research. While it is rare to see econometric analysis of estimators in accounting research papers, it is common for claims to be made and for readers to be directed to research in economics, finance, or econometrics for support of those claims. Given the prevalence of this approach, we believe it is useful to demonstrate how a reader can evaluate such claims.

We consider each of the seven cited sources as a basis for the claims made in Call et al. (2018). The first of these (Irarrazabal et al., 2013) can be eliminated, as it does not use this estimator, let alone provide econometric support for these claims.[7] Similarly, Cameron and Trivedi (2009) and Wooldridge (2010) are standard econometrics references that say

---

[6]There is a slight discrepancy apparent here for employee penalties, for which Kuvvet (2019) appears to have summed the top 7 values instead of the top 6.

[7]Footnote 13 of Irarrazabal et al. (2013, p. 80) says "the Poisson estimator ... delivers qualitatively similar results (available on request)".

precisely nothing about the ability of Poisson regression to handle "a disproportionate number of zeros and severe skewness" in any setting, let alone in settings using dependent variables other than the count variables that motivate the use of Poisson regression.

The two papers by Santos Silva and Tenreyro seem more relevant.[8] Santos Silva and Tenreyro (2006, p. 642) note that "the pioneering work of Tinbergen (1962) initiated a vast theoretical and empirical literature on the gravity equation for trade. ... In its simplest form, the gravity equation for trade states that the trade flow from country i to country j, denoted by T_{ij}, is proportional to the product of the two countries' GDPs, denoted by Y_i and Y_j, and inversely proportional to their distance, D_{ij}".

This implies the following equation (in stochastic form):

$$T_{ij} = \alpha_0 Y_i^{\alpha_1} Y_j^{\alpha_2} D_{ij}^{\alpha_3} \eta_{ij}$$

where α_0, α_1, α_2, and α_3 are unknown parameters and η_{ij} is an error factor assumed the independent of regressors. Santos Silva and Tenreyro (2006) note that "there is a long tradition in the trade literature of log-linearizing" the equation above and estimating using OLS:

$$\ln T_{ij} = \alpha_0 + \alpha_1 \ln Y_i + \alpha_2 \ln Y_j + \alpha_3 \ln D_{ij} + \ln \eta_{ij}$$

Notwithstanding the popularity of this approach, Santos Silva and Tenreyro (2006) find that "the presence of heteroskedasticity can generate strikingly difference estimates when the gravity equation is log-linearized, rather than estimated in levels [and] ... that inferences drawn on log-linearized regressions can produce misleading conclusions."

Fortunately, Santos Silva and Tenreyro (2006) provide "simulation evidence that the PPML is well behaved in a wide range of situations and is resilient to the presence of a specific type of measurement error of the dependent variable" (Santos Silva and Tenreyro, 2011, p. 220). However, Santos Silva and Tenreyro (2006) provide little evidence on the performance of PPML when the dependent variable is frequently zero and Santos Silva and Tenreyro (2011) seek to address this by simulating a simple structural model in which the number and size of exporters is stochastic in a model that incorporates ideas from the gravity model but yields zero trade levels in most of their simulated samples. Santos Silva and Tenreyro (2011) find that PPML performs very well in this simulated setting.

Before discussing the implications of these results, it is important to note that PPML as used by Santos Silva and Tenreyro (2006) and Santos Silva and Tenreyro (2011) is neither more nor less than Poisson regression. An important property of the Poisson model is that the mean and variance of a Poisson-distributed random variable are equal. While statisticians have developed generalizations of the basic Poisson model to allow for (say) variance greater than the mean (this is called **overdispersion** and is common in real-world data), Santos Silva and Tenreyro (2006, p. 645) argue that such models "might give excessive weight to the observations that are more prone to measurement errors" in the context of trade data. As such, "the Poisson regression emerges as a reasonable compromise, giving less weight to the observations with larger variance than the standard [non-linear least squares] estimator, without giving too much weight to observations more prone to contamination by measurement error" (Santos Silva and Tenreyro, 2006, p. 646).

---

[8]Tenreyro (2007) involves an extension of Santos Silva and Tenreyro (2006) involving instrumental variables and directs readers to Santos Silva and Tenreyro (2006) regarding its efficiency and robustness. The seventh paper (Karolyi and Taboada, 2015) has a more careful linking of prior research to its setting, but does not provide independent evidence of the effectiveness of PPML.

Now that we better understand the results of prior research (Santos Silva and Tenreyro, 2011, 2006), we can return to the claims of Call et al. (2018) provided above. Unfortunately, Call et al. (2018) provide no reason to believe that the results of Santos Silva and Tenreyro (2011) carry over to whatever model seems applicable to the setting of Call et al. (2018). There is no "vast theoretical and empirical literature" to suggest any model of the factors driving firm penalties, let alone that either the gravity model or the Santos Silva and Tenreyro (2011) model is applicable. As such, the results of prior research are arguably of little value in the Call et al. (2018) setting.

The reasoning of Call et al. (2018) seems analogous to saying that "prior research shows that when y is a function of X the best estimator is OLS with OLS standard errors" (in the sense used in Chapter 5) without noting that such results are applicable to the classical linear model with spherical errors and without justifying that model in a setting with different features.

One concern highlighted by the above discussion is the presence of fundamental asymmetries in the making and checking of claims like those in Call et al. (2018). *Making* such a claim is close to costless; a simple internet search is likely to yield papers supporting the use of a particular estimator in a context that shares features with the authors' setting.

Additionally, the econometric knowledge needed to make such claims is close to zero. This is illustrated by footnote 13 of Call et al. (2018, p. 134), where it is claimed that "the primary difference between PPML and conventional Poisson regression is that PPML does not impose the assumption of equality in the first and second moments of the distribution." Yet if the authors looked at their own Stata code, they would see that, as in Santos Silva and Tenreyro (2006), PPML *is* Poisson regression.[9]

In contrast, *checking* such claims is far from costless. One needs to carefully understand what is going on in those papers and how they relate to the current setting and this requires some knowledge of econometrics. Adding to the asymmetry is the matter of incentives: authors have strong incentives to claim the appropriateness and robustness of those methods that yield results, whereas reviewers have very weak incentives of any kind.

Loose econometric analysis in accounting research is not new. Many researchers claimed that FM-NW and Z2 estimators of standard errors had properties that they did not have. These claims were either unsubstantiated or justified by citing papers that did not address them in any way. Once these claims were evaluated rigorously (Gow et al., 2010), they were found to have no merit.

In light of the discussion above, the response of Call et al. (2019) to the concern of Kuvvet (2019) about influential observations that Call et al. (2018) "addresses the role of extreme observations in enforcement actions with an estimator designed specifically to handle skewed data (Poisson pseudo-maximum likelihood)" seems unpersuasive.

24.2.2 Replicating Call et al. (2018)

To better understand the estimator used by Call et al. (2018), we replicate their Table 4. First, we prepare the data. Following Call et al. (2018), we winsorize a subset of the control variables at the 1st and 99th percentiles. Second, we follow Call et al. (2018) in calculating a log-transformation of the dependent variables considered (we will use these transformed variables below). We use `set_names()` so that the result of applying `map()` to `yvars` retains

---

[9]Call et al. (2018) use the `poisson` function from Stata. Even if the authors were using the `ppml` routine provided by Santos Silva and Tenreyro (2006), they would see that it's just Poisson regression by default.

the values of `yvars` in the names of the elements, as we will use these names in the tables. We also convert `ff12` to a factor variable and any logical variables to integers.[10]

```
yvars <- set_names(c("firmpenalty", "emppenalty", "empprisonmos"))

cmsw <-
  cmsw_2018 |>
  mutate(across(c(blckownpct, initabret, pctinddir, mkt2bk, lev),
                winsorize),
         across(any_of(yvars), \(x) log(1 + x), .names = "ln_{.col}"),
         ff12 = as.factor(ff12),
         across(where(is.logical), as.integer)) |>
  filter(tousesox == 1)
```

Next, we create a function `get_poisson_fit()` to estimate Poisson regressions using `glm()`. Because we refer to the same set of controls in other contexts below, we specify `controls` outside `get_poisson_fit()`.

```
x <- "wbflag"
controls <- c("selfdealflag", "blckownpct", "initabret", "lnvioperiod",
              "bribeflag", "mobflag", "deter", "lnempcleveln", "lnuscodecnt",
              "viofraudflag", "misledflag", "audit8flag", "execterrmflag",
              "coopflag", "impedeflag", "pctinddir", "recidivist",
              "lnmktcap", "mkt2bk", "lev", "lndistance", "ff12")

get_poisson_fit <- function(y, df = cmsw) {
  form <- as.formula(str_c(y, " ~ ",
                           str_c(c(x, controls), collapse = " + ")))
  fm <- glm(form, family = "poisson", data = df,
            control = glm.control(maxit = 100))
}
```

We then use `get_poisson_fit()` to get regression results.

```
fms <- map(yvars, get_poisson_fit)
```

We use `coeftest()` to return "robust" standard errors that match those used in Call et al. (2018). In addition to `type = "HC1"`, `vcovHC()` offers `"HC2"` and `"HC3"`. The differences between these options relate to degrees-of-freedom corrections that have the greatest impact with small samples. Chapter 8 of Angrist and Pischke (2008) provides more discussion and explanation of HC1, HC2, and HC3.

```
get_coefs <- function(fm, type = "HC1") {
  coeftest(fm, vcov = vcovHC(fm, type = type))
}
```

Finally, we tabulate the results in Table 24.3. To save space, we omit estimated coefficients for the control variables.[11]

```
modelsummary(map(fms, get_coefs),
             estimate = "{estimate}{stars}",
             statistic = "{statistic}",
```

---

[10]The latter transformation does not affect the substance of the regressions, but does simplify their presentation.

[11]The coefficients on the control variables are reported in Table 4 of Call et al. (2018).

TABLE 24.3

Enforcement outcomes (Table 4 of CMSW)

	firmpenalty	emppenalty	empprisonmos
wbflag	1.250*	1.111***	0.621**
	1.949	2.637	2.372
Num.Obs.	658	658	658

```
coef_map = "wbflag",
gof_map = c("nobs", "r.squared"),
stars = c('*' = .1, '**' = 0.05, '***' = .01))
```

Call et al. (2019) flag concerns raised by Kuvvet (2019) about the "tipster" variable as an example of "claims offered by Kuvvet [that] reflect a misunderstanding of both the enforcement action setting and the whistleblower designations in CMSW". Call et al. (2019) argue that "CMSW's 'tipster' and 'non-tipster' whistleblower designations are defined specifically in relation to the enforcement process and are not an attempt to identify the individual who first uncovered the misconduct." However, Call et al. (2018, p. 147) say "some whistleblowers sound the first alarm about potential violations, bringing to the attention of regulators possible violations that could be investigated. We refer to these whistleblowers as *tipsters*."

24.2.3   Discussion questions

1. Suppose you were a regulator interested in understanding the effects of whistleblowers on enforcement outcomes. How might you design an experiment to examine these effects? What challenges would you expect to face in implementing your experiment? How would the experiment differ from the setting of Call et al. (2018)?

2. As an additional example of "misunderstanding" by Kuvvet (2019), Call et al. (2019) claim that "Kuvvet argues that CMSW's findings speak to correlation rather than causation. The published version of CMSW makes this point clearly throughout the paper." What claim in Kuvvet (2019) are Call et al. (2019) addressing here? Do you agree that "CMSW makes this point clearly throughout the paper"?

3. "The published version of CMSW empirically addresses the role of extreme observations in enforcement actions with an estimator designed specifically to handle skewed data (Poisson pseudo-maximum likelihood) and with additional robustness tests, including one focused on the incidence rather than the magnitude of penalties." As we see in Table 24.3, the estimator used in Call et al. (2018) is standard Poisson regression, which is called in R using the `glm()` function with `family = "poisson"`. How might we use data sets covered in Section 24.1 or Section 24.4 to evaluate the claim that Poisson regression estimator is "designed specifically to handle skewed data"?

4. Could we use approaches covered in Section 24.1 or Section 24.4 to address extreme observations in the setting of Call et al. (2018)?

5. Call et al. (2019) argue that "unlike many other settings in accounting, finance, and economics where the focus is often on the average firm, the enforcement action setting is inherently extreme" and claim that this is another example of misunderstanding" in Kuvvet (2019). Do you agree that "the enforcement action setting is inherently extreme"? Does this inherent extremeness undermine the arguments of Kuvvet (2019)?

6. What claim by Kuvvet (2019) are Call et al. (2019) (quoted above) trying to refute regarding the "tipster" variable? Do you find their response convincing?

24.3 Sensitivity analysis

In general, **sensitivity analysis** is used to assess how outputs differ as inputs or assumptions are varied. If we could assume that assignment of whistleblower status to an enforcement action is (conditional on control variables) unconfounded in the setting of Call et al. (2018), then we can interpret regression coefficients as estimates of causal effects.[12] For this reason, we focus in this section on the analysis of the sensitivity of inferences to posited violations of the unconfoundedness of treatment assignment.

In practice, whistleblowers are not just randomly assigned to enforcement actions. Rather there is the possibility that the factors driving assignment of whistleblowers also affect the enforcement outcomes, making them confounders in the sense studied in Section 4.2. If these factors are not observed by the researcher, then they cannot be controlled for in empirical analysis.

Even in this narrow sense of the term we use here, there are various approaches to sensitivity analysis that can be employed. In this section, we apply the approach proposed by Frank (2000), which involves the calculation of the **impact threshold for a confounding variable** or **ITCV**.

To proceed with the calculation of ITCV, we begin by estimating OLS regressions, as the Frank (2000) approach is predicated on OLS regressions.

Following Call et al. (2018), we use the log-transformed versions of the confounding variable and robust standard errors.

```
get_ols_fit <- function(y, df = cmsw) {
  form <- as.formula(paste0("ln_", y, " ~ ",
                      str_c(c(x, controls), collapse = " + ")))
  lm(form, data = df)
}

fms <- map(yvars, get_ols_fit)
```

Regression results analogous to those in columns (7) through (9) of Panel A of Table 8 of Call et al. (2018) are reported in Table 24.4.

```
modelsummary(map(fms, get_coefs),
            estimate = "{estimate}{stars}",
```

---

[12]Of course, even in such a case we need to be mindful of functional-form issues of the kind explored in Chapter 25.

TABLE 24.4

Enforcement outcomes with OLS (Table 8 of CMSW)

	firmpenalty	emppenalty	empprisonmos
wbflag	0.484***	0.072	0.181
	3.316	0.627	1.118
Num.Obs.	658	658	658

```
statistic = "{statistic}",
coef_map = "wbflag",
gof_map = c("nobs", "r.squared"),
stars = c('*' = .1, '**' = 0.05, '***' = .01))
```

The impact of a confounding variable in this context is the effect it would have on the estimated coefficient of the variable of interest if it were included in the regression analysis. In the context of OLS regression in which we are regressing y on x and controls Z, we know from Section 3.3 that the regression coefficient β on x can be calculated from a regression of the residuals of a regression of y on Z against the residuals of a regression of x on Z. If we consider a confounding variable cv, then its impact comes from (i) its effect on the residuals of a regression of y including both Z and cv and (ii) its effect on the residuals of a regression of x including both Z and cv.

It turns out that these two quantities relate to **partial correlations**. A partial correlation between x and y measures the degree of association between x and y, with the effect of a set of control variables (say Z) removed. The partial correlations of interest will be that between y and the hypothetical confounding variable cv ($r_{y \cdot cv}$) and that between x and the hypothetical confounding variable cv ($r_{x \cdot cv}$).

The ITCV can be expressed in terms of the product $r_{y \cdot cv} \times r_{x \cdot cv}$. In effect, the ITCV represents the minimum value of this product that would turn a positive and statistically significant coefficient into a positive and *just* statistically significant coefficient, with impact values greater than the ITCV leading to statistically insignificant coefficients.

In Table 24.4, we have a statistically significant coefficient in just one regression (the one with firm penalties as the dependent variable), so like Call et al. (2018) we focus on this regression in calculating ITCV.[13]

The `get_itcvs()` function below calculates the ITCV for $\alpha \in \{0.01, 0.05, 0.10\}$. While we refer the reader to Frank (2000) for details on the mathematics, it seems noteworthy that we extract just two quantities from the fitted model: the t-statistic for the coefficient on the variable of interest (`var`) and the degrees of freedom for the model. Lower α values imply higher critical values for the t-statistic, hence lower threshold values.

```
get_itcvs <- function(fm, var) {
  coefs <- get_coefs(fm)
  tstat <- coefs[var, "t value"]
  df <- df.residual(fm)
  numer <- tstat^2 / df
  r_yx <- sqrt(numer / (1 + numer))
```

---

[13]The ITCV is not meaningfully defined when the coefficient of interest is not statistically significant even without the inclusion of a hypothetical confounding variable.

```
  tibble(alpha = c(0.01, 0.05, 0.1)) |>
    mutate(crit_val = qt(1 - alpha / 2, df),
           rhash = crit_val / sqrt(df + crit_val^2),
           itcv = (r_yx - rhash) / (1 - abs(rhash))) |>
    select(-rhash)
}
```

We can now apply the `get_itcvs()` function to the first model (`fms[[1]]`) and the variable of interest (`wbflag`). Results are reported in Table 24.5.[14]

```
get_itcvs(fms[[1]], "wbflag")
```

TABLE 24.5
Impact threshold for a confounding variable (ITCV)

alpha	crit_val	itcv
0.01	2.584	0.032
0.05	1.964	0.058
0.10	1.647	0.070

Evaluating ITCVs requires some sense of plausible magnitudes for $r_{y \cdot cv}$ and $r_{x \cdot cv}$. Lacking prior beliefs about confounding variables, we might consider the control variables included in the regression (Z) to be a set of natural benchmarks. Inclusion of these control variables is (or should be) based on these variables being confounding variables themselves.

As such we can calculate $r_{y \cdot z}$ and $r_{x \cdot z}$ for each $z \in Z$. To do this, we first create a function `pcor()` to calculate partial correlations. This function is adapted from code for the `pcor()` function in the `ppcor` package;[15] it is trimmed down to focus on our limited needs here, which focus on the vector of partial correlations between the first variable in x and the remaining variables.

```
pcor <- function (x) {
  cvx <- cov(as.matrix(x))
  if (det(cvx) < .Machine$double.eps) {
    icvx <- MASS::ginv(cvx)
  }
  else icvx <- solve(cvx)
  pcor <- -cov2cor(icvx)
  diag(pcor) <- 1
  pcor[-1, 1]
}
```

The `pcor()` function requires data in matrix form, which in turn requires that each factor variable be encoded as a set of indicator variables. The `get_factors()` flags the factor variables in a data frame.

---

[14]Here we see a slight difference from the 0.059 reported in Call et al. (2018), perhaps due to their using 1.96 as the critical value in place of the more exact value returned by `qt()` in our code here.
[15]https://cran.r-project.org/package=ppcor

```
get_factors <- function(df) {
  names(which(unlist(map(df, is.factor))))
}
```

The `make_dummies()` function converts `var` in the data frame `df` to a set of dummies.

```
make_dummies <- function(df, var) {
  vals <- as.vector(unique(df[[var]]))
  for (val in vals) {
    df[[str_c(var, "_", val)]] <- as.integer(df[[var]] == val)
  }
  df |> select(-any_of(var)) |> as_tibble()
}
```

Finally `convert_factors()` calls `get_factors()` to identify factor variables then `make_dummies()` to convert each of these.

```
convert_factors <- function(df) {
  vars <- get_factors(df)
  for (var in vars) {
    df <- make_dummies(df, var)
  }
  df
}
```

Finally, we create the function `get_impacts()` to calculate the elements of impact for the control variables.

```
get_impacts <- function(df, y, x, controls) {
  corr_data <-
    df |>
    select(any_of(c(y, x, controls))) |>
    convert_factors() |>
    na.omit()

  r_yz <-
    corr_data |>
    select(-all_of(x)) |>
    pcor()

  r_xz <-
    corr_data |>
    select(-all_of(y)) |>
    pcor()

  var <- names(corr_data |> select(-any_of(c(x, y))))

  tibble(var, r_yz, r_xz) |>
    mutate(impact = r_yz * r_xz) |>
    arrange(desc(impact))
}
```

TABLE 24.6
Impacts for control variables

var	r_yz	r_xz	impact
lnmktcap	0.332	0.274	0.091
bribeflag	0.289	0.035	0.010
lnvioperiod	0.140	0.053	0.007
lev	0.070	0.100	0.007
blckownpct	−0.092	−0.075	0.007
lnuscodecnt	0.158	0.036	0.006
mkt2bk	−0.060	−0.079	0.005
misledflag	−0.051	−0.079	0.004
initabret	0.098	0.040	0.004
recidivist	0.034	0.080	0.003

In Table 24.6, the impact of the 10 most impactful control variables are shown. Only one of these control variables (`lnmktcap`) has an impact exceeding the ITCV for $\alpha = 0.05$, suggesting that the conjectured control variable would have to rank second on this list of controls to have enough impact to drive the p-value above 0.05. As such, the OLS results reported in Table 24.4 for firm penalties might be interpreted as somewhat robust.

```
get_impacts(cmsw,
            y = "ln_firmpenalty",
            x = "wbflag",
            controls = controls) |>
  top_n(n = 10, wt = impact)
```

However, a concern with this analysis is that it may be unrealistic to imagine that a single confounding variable has been omitted from the analysis. This concern seems accentuated in a setting where our understanding of the factors driving assignment to treatment is limited. It seems plausible that the factors driving the presence of whistleblowers are complex and poorly captured by the set of controls used in Call et al. (2018).

Another concern is that there is some arbitrariness to the measurement of controls. For example, z_1, z_2, and z_3 might all be noisy proxies for an underlying confounder z. Including all three variables might make sense to control for z, but the impact of each variable is likely to be low because of multicollinearity, thus understating the impact of z.

One approach to addressing this is to consider the controls as a single variable. The fitted value $\hat{z}$ from a regression of x on $Z = \{z_i\}$ can be used as a single control in a regression where y is the dependent variable without affecting the coefficient on x.[16] As such, we can examine the impact of $\hat{z}$ as a composite control.

We embed this logic in the function `get_combined_impact()`.

```
get_combined_impact <- function(df, y, x, controls) {

  form <- as.formula(str_c(x, " ~ ",
                     str_flatten(controls, collapse = " + ")))
  fm <- lm(form, data = df)
```

---

[16]Consistency might suggest $\hat{x}$ as the better notation, but $\hat{z}$ seems to better capture what we are aiming for here.

```
  df |>
    mutate(z = fitted(fm)) |>
    get_impacts(y = y,
                x = x,
                controls = "z")
}
```

Results from `get_combined_impact()` are provided in Table 24.7. There it can be seen that the impact of the included controls as a set (0.273) is much higher than the ITCV for $\alpha = 0.05$ (0.0578), suggesting that the impact of omitted confounders only has to be a fraction of the impact of included variables for the results to be statistically insignificant.

```
get_combined_impact(cmsw,
                    y = "ln_firmpenalty",
                    x = "wbflag",
                    controls = controls)
```

TABLE 24.7
Combined impacts of control variables

var	r_yz	r_xz	impact
z	0.553	0.494	0.273

24.4 Appendix: Simulation study from Leone et al. (2019)

Here we present code to run simulations like those underlying Table 5 of Leone et al. (2019). One reason to include this simulation is to help the reader better understand the results in Table 5.

An additional reason for presenting this particular simulation is to illustrate a modular approach to simulations that is fostered by use of R functions. This approach potentially offers a template for readers looking to develop simulations.

A big-picture view of the simulation is that the `run_sim()` function uses `get_data()` to generate the data, then feeds these data to `fit_models()`, then sends the returned fitted models to `compile_stats()`. The `get_results()` function feeds parameter values from `params` to the `run_sim()` function.

The first step is to produce the data. If you wanted to change the data-generating process, this could be achieved simply by editing this function. The purpose of this function is to replicate the data-generating process described in Leone et al. (2019).[17] Rather than create separate code for each of the three panels of Table 5, we create one function and use argument values to select the data-generating process used to create the data. For example, the default values (`b_true = 0.8`, `clean = FALSE`, and `random = TRUE`) produce data matching the left half of Panel C of Table 5.

---

[17]We thank Miguel Minutti-Meza for sharing his Stata code with us, but note that the description of the simulation in the Leone et al. (2019) is clear and complete and sufficient to allow a reader to reproduce it without this code.

```
get_data <- function(b_true = 0.8, clean = FALSE,
                     random = TRUE, n = 10000) {
  prop_cont <- if_else(random, 0.02, 0.25)
  tibble(x1 = rnorm(n = n), x2 = rnorm(n = n),
         x3 = rnorm(n = n), v = rnorm(n = n),
         y = b_true * x1 + 0.4 * x2 + 0.2 * x3 + v,
         z = rnorm(n = n, mean = 3, sd = 1)) |>
    mutate(id = row_number(),
           contaminated =
             case_when(!clean & !random ~ id < prop_cont * n & x1 < -1.5,
                       !clean & random  ~ id < prop_cont * n,
                       .default = FALSE),
           cy = if_else(contaminated, y + z, y),
           cyw = winsorize(cy), x1w = winsorize(x1), x2w = winsorize(x2),
           x3w = winsorize(x3), cyt = truncate(cy), x1t = truncate(x1),
           x2t = truncate(x2), x3t = truncate(x3))
}
```

Another benefit of putting `get_data()` in a separate function that returns a data frame is that we can examine the pieces one at a time.

```
get_data()
```

```
# A tibble: 10,000 x 17
       x1     x2     x3      v      y      z   id contaminated     cy    cyw
    <dbl>  <dbl>  <dbl>  <dbl>  <dbl>  <dbl> <int> <lgl>         <dbl>  <dbl>
1  0.0635 -0.952  0.170 -0.647 -0.943  4.21     1 TRUE           3.27   3.27
2  2.08    1.92  -0.963  0.327  2.56   2.46     2 TRUE           5.02   3.76
3  0.491  -1.09  -0.957 -1.43  -1.67   3.09     3 TRUE           1.42   1.42
4  0.757  -0.871  0.481 -1.29  -0.940  1.77     4 TRUE           0.826  0.826
5  1.65   -1.07  -1.73   1.11   1.65   3.13     5 TRUE           4.79   3.76
6 -0.358   0.903  1.70  -0.295  0.119  2.10     6 TRUE           2.22   2.22
# i 9,994 more rows
# i 7 more variables: x1w <dbl>, x2w <dbl>, x3w <dbl>, cyt <dbl>,
#   x1t <dbl>, x2t <dbl>, x3t <dbl>
```

The next function fits the models of interest. Following Leone et al. (2019), we fit seven models and store each model as an element of a list, as this makes manipulation of the models (e.g., extracting statistics) easier.

```
fit_models <- function(df) {
  fms <- list(lm(cy ~ x1, data = df))
  fms[[2]] <- update(fms[[1]], cyw ~ x1w + x2w + x3w)
  fms[[3]] <- update(fms[[1]], cy ~ x1w + x2w + x3w)
  fms[[4]] <- update(fms[[1]], cyt ~ x1t + x2t + x3t)
  fms[[5]] <- update(fms[[1]], cy ~ x1t + x2t + x3t)
  fms[[6]] <-
    df |>
    mutate(cooksd = cooks.distance(fms[[1]]),
           extreme = cooksd > 4 / nobs(fms[[1]])) |>
    filter(!extreme) |>
    update(fms[[1]], data = _)
```

```
  fms[[7]] <- lmrob(formula(fms[[1]]), data = df, method = "MM",
                    control = lmrob.control(tuning.psi = 3.4437))
  names(fms) <- c("Do nothing", "Winsorize all variables",
                  "Winsorize only X", "Truncate all variables",
                  "Truncate only X", "Cook's distance",
                  "Robust regression")
  fms
}
```

For the simulation output, we want to retain estimated coefficients and results of tests of the null hypothesis at sizes of 5% and 1%. We create the `extract_stats()` function to do this, as this makes for leaner, easier-to-debug code.

```
extract_stats <- function(fm, b_null) {
  rdf <- fm$df.residual
  coefs <- summary(fm)$coefficients
  est <- coefs[2, 1]
  se <- coefs[2, 2]
  tval <- (est - b_null) / se
  pval <- 2 * pt(abs(tval), rdf, lower.tail = FALSE)

  p_5 <- pval < 0.05
  p_1 <- pval < 0.01

  tibble(est, p_5, p_1)
}
```

We next create `compile_stats()`, which applies `extract_stats()` to a list of models (such as those returned by `fit_models()`) and puts the results in a data frame with human-friendly labels for each of the models.

```
compile_stats <- function(fms, b_null) {
  fms |>
    map(extract_stats, b_null = b_null) |>
    list_rbind() |>
    mutate(model_num = row_number(),
           model = names(fms))
}
```

Now we create `run_sim()`, which runs one iteration of the simulation and returns a data frame with parameter values and statistics from the simulation.

```
run_sim <- function(sim_num = 1, b_true = 0.8, b_null = b_true,
                    clean = FALSE, random = FALSE) {
  get_data(b_true = b_true, clean = clean, random = random) |>
  fit_models() |>
  compile_stats(b_null = b_null) |>
  mutate(sim_num = sim_num, b_true = b_true, b_null = b_null,
         clean = clean, random = random)
}
```

We now compile a data frame with the parameters we will run in the simulation. We will consider different two different values of each of `b_true`, `clean`, and `random`. Note that

TABLE 24.8
Partial replication of Panel B of Table 5 of Leone et al. (2019)

model	est	p_5	p_1
Do nothing	0.70	1.00	1.00
Winsorize all variables	0.70	1.00	1.00
Winsorize only X	0.72	1.00	1.00
Truncate all variables	0.68	1.00	1.00
Truncate only X	0.71	1.00	1.00
Cook's distance	0.76	0.93	0.83
Robust regression	0.77	0.52	0.28

when `clean` is `TRUE`, the value of `random` has no significance, so we delete the case where `clean` is `TRUE` and `random` is also `TRUE` to avoid running redundant parameter values.

```
params <- expand_grid(b_true = c(0, 0.8),
                      clean = c(TRUE, FALSE),
                      random = c(FALSE, TRUE)) |>
  filter(!(clean & random))
```

Now we finally have almost all the pieces to run the simulation. We will do this using `get_results()`, which uses `future_map()` from the **furrr** package to use multiple threads for substantially shorter run times.

```
get_results <- function(b_true = 0, clean = FALSE, random = FALSE, ...) {
  future_map(1:1000L, run_sim, b_true = b_true,
             clean = clean, random = random,
             .options = furrr_options(seed = 2021)) |>
    list_rbind()
}
```

The following step, which actually runs `get_results()`, takes more than a minute to run and produces data for all panels of Table 5 of Leone et al. (2019).

```
plan(multisession)

all_results <-
  pmap(params, get_results) |>
  list_rbind() |>
  system_time()
```

```
  user  system elapsed
 6.833   0.271  85.484
```

Table 24.8 provides a version of the left half of Panel B of Table 5 of Leone et al. (2019).

```
all_results |>
  filter(b_true == 0.8, !random, !clean) |>
  group_by(b_true, clean, random, model_num, model) |>
  summarize(across(est:p_1, mean), .groups = "drop") |>
  select(model, est, p_5, p_1)
```

25

Matching

In Chapter 4, we discussed three basic causal diagrams and suggested that the concept of nonparametric conditioning on Z is more demanding than simply including Z as another regressor in a linear regression model. This chapter develops this idea more fully and examines practical approaches to properly conditioning on confounding variables.

> 💡 Tip
>
> The code in this chapter uses the packages listed below. We load **tidyverse** because we use several packages from the Tidyverse. For instructions on how to set up your computer to use the code found in this book, see Section 1.2. Quarto templates for the exercises below are available on GitHub.[a]
>
> ---
> [a]https://github.com/iangow/far_templates/blob/main/README.md

```
library(tidyverse)
library(DBI)
library(farr)
library(dbplyr)          # window_order()
library(fixest)
library(modelsummary)
library(MatchIt)         # match.data(), matchit()
library(optmatch)
```

25.1 Background on auditor choice

For concreteness, we will explore the basic ideas of this chapter using the setting of auditor choice and its effects. An extensive literature has examined the question of whether "Big N" auditors produce higher audit quality. This question is examined in two of the papers we will study in this chapter (DeFond et al., 2017; Lawrence et al., 2011).

For data on auditors, the table `comp.funda` has a column labelled `au`, which lines up with `aucd` in `comp.r_auditors`.

```
db <- dbConnect(RPostgres::Postgres(), bigint = "integer")

company <- tbl(db, Id(schema = "comp", table = "company"))
funda <- tbl(db, Id(schema = "comp", table = "funda"))
r_auditors <- tbl(db, Id(schema = "comp", table = "r_auditors"))
```

DOI: 10.1201/9781003456230-25

When Lawrence et al. (2011) refer to the "Big Four", they are actually referring to the current group that DeFond et al. (2017) refer to as the "Big N". The latter term alludes to the prior incarnations of the set of top audit firms, including the "Big Five" (prior to the demise of Arthur Andersen) and the "Big Eight" (prior to mergers in the late 1980s) (see Gow and Kells, 2018 for more on this history).

```
r_auditors |> arrange(aucd) |> collect(n = 12)
```

TABLE 25.1
Values of `aucd` for top auditors

aucd	audesc
0	Unaudited
1	Arthur Andersen
2	Arthur Young
3	Coopers & Lybrand
4	Ernst & Young
5	Deloitte & Touche
6	KPMG Peat Marwick
7	PricewaterhouseCoopers
8	Touche Ross
9	Other
10	Altschuler, Melvoin, and Glasser
11	BDO Seidman

Looking closer at a sample from `r_auditors` in Table 25.1, we see that Arthur Andersen has an `aucd` of 1. Note that Arthur Young is now part of Ernst & Young, Coopers & Lybrand is now part of PricewaterhouseCoopers, and Touche Ross is now part of Deloitte & Touche. So a reasonable approach to "Big N" would appear to be `big4 = aucd %in% 1:8L`.[1] It's commonly understood that most large firms choose a Big Four auditor. But rather than simply accepting that, we can look at the data.

Here we focus on firms with meaningful financial statements (`sale > 0, at > 0`) and fiscal 2019, the latest fiscal year with complete data at the time of first writing this chapter.

```
size_big4 <-
  funda |>
  filter(indfmt == "INDL", datafmt == "STD",
         consol == "C", popsrc == "D") |>
  filter(sale > 0, at > 0, fyear == 2019, !is.na(au)) |>
  mutate(au = as.integer(au)) |>
  select(gvkey, datadate, fyear, au, prcc_f, csho) |>
  mutate(big4 = au %in% 1:8L,
         mkt_cap = prcc_f * csho * 1e6) |>
  arrange(gvkey, datadate) |>
  collect()
```

We calculate market share by "bins" where each bin represents an order of magnitude of market capitalization and show statistics for each bin in Table 25.2. For example, (9, 10]

---

[1]This appears to be the approach used in DeFond et al. (2017), but whether this lines up with the approach used in Lawrence et al. (2011) is hard to say.

includes all firms with market capitalization over \$1 billion (10^9) and less than or equal to \$10 billion (10^{10}).

```
size_big4 |>
  filter(mkt_cap > 1e6) |>
  mutate(log_mkt_cap = log10(mkt_cap)) |>
  mutate(bin = cut(log_mkt_cap, breaks = seq(6, 13, 1))) |>
  group_by(bin) |>
  summarize(big4_perc = 100 * mean(big4),
            big4_num = sum(big4),
            non_big4_num = sum(!big4))
```

TABLE 25.2

Big 4 market share by market capitalization

bin	big4_perc	big4_num	non_big4_num
(6,7]	16.32	63	323
(7,8]	27.53	337	887
(8,9]	58.03	1051	760
(9,10]	89.67	1580	182
(10,11]	96.76	687	23
(11,12]	97.85	91	2
(12,13]	100.00	1	0

Interestingly, one of the two cases of companies with an apparent market capitalization over \$100 billion (10^{11}) and a non–Big Four auditor is Vanjia, which apparently misstated the number of its shares outstanding as 8,550 million rather than 30 million.[2]

```
size_big4 |>
  filter(!big4, mkt_cap > 1e11) |>
  mutate(fyear = as.character(fyear))
```

TABLE 25.3

Two large non-Big 4 firms

gvkey	datadate	fyear	au	prcc_f	csho	big4	mkt_cap
014447	2019-12-31	2019	9	92.988	2,518.262	FALSE	234,168,146,856
026478	2019-12-31	2019	9	14.380	8,550.000	FALSE	122,949,000,000

We can also present the data as a histogram. From either Table 25.2 or Figure 25.1, it is clear that the Big Four have overwhelming market share among the largest firms and have most of the market even among firms with market capitalization in the \$100 million-to-\$1 billion range.

```
size_big4 |>
  filter(mkt_cap > 1e6) |>
```

---

[2]The higher number can be seen in Vanjia's 10-Q filing for Q1 2020 (https://go.unimelb.edu.au/rzw8), but appears to have been quietly corrected in Vanjia's 10-Q filing for Q2 2020 (https://go.unimelb.edu.au/jzw8). The other company is LVMH, the French luxury goods firm, which appears to have been audited in part by Ernst & Young: https://go.unimelb.edu.au/izw8.

```
ggplot(aes(x = log10(mkt_cap), fill = big4)) +
geom_histogram(breaks = seq(6, 12.25, 0.25))
```

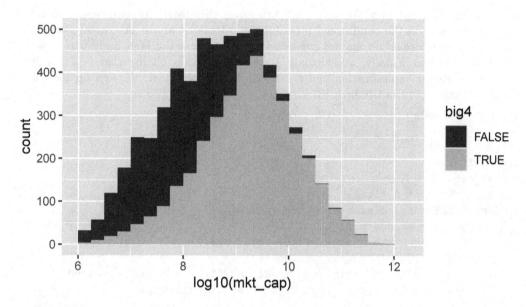

FIGURE 25.1
Histogram of market capitalization with auditor type

25.2   Simulation analysis

One challenge for an empirical researcher studying the effect of having a Big Four auditor is that market capitalization is plausibly a confounding variable, or confounder, of the kind seen in Figure 4.1. That is, having a large market capitalization is likely to cause—or be associated with variables that cause—firms to choose a Big Four auditor and also to affect a variety of economic outcomes that may be of interest, such as accounting quality. We know from our analysis in Chapter 4 that we want to "control for" confounders but we show here that this is not always as simple as including the confounding variable in OLS regression.

In this section, we use simulation analysis to examine more closely how to control for confounding variables. We start with the actual empirical distribution of market capitalization from the data depicted in the histogram above. For each bin in the histogram depicted in Figure 25.1, we calculate the observed conditional probability of a firm in that bin having a Big Four auditor and store the results in `prob_big4`.

```
prob_big4 <-
  size_big4 |>
  filter(mkt_cap > 1e6) |>
  select(big4, mkt_cap) |>
  mutate(log_mkt_cap = log10(mkt_cap)) |>
  mutate(bin = cut(log_mkt_cap, breaks = seq(6, 12.25, 0.25))) |>
  group_by(bin) |>
```

```
  mutate(p_big4 = mean(big4)) |>
  ungroup()
```

We then run our simulation. We draw 5,000 firms from this distribution and assign each firm an auditor based on the conditional probabilities calculated above. A critical assumption here is that, *conditional on market capitalization,* whether a Big Four auditor is chosen is completely random. Finally, we calculate a measure of audit quality that is a non-linear function of market capitalization and a random noise component. We explain the role of `pick(everything())` in Chapter 13.

```
n_firms <- 5000

set.seed(2021)
sim_auditor <-
  prob_big4 |>
  sample_n(size = n_firms) |>
  mutate(rand = runif(nrow(pick(everything()))),
         big4 = rand < p_big4,
         epsilon = rnorm(nrow(pick(everything()))),
         id = 1:nrow(pick(everything()))) |>
  mutate(audit_quality = big4 * 0 + mkt_cap^(1/3) * 0.003 + epsilon) |>
  select(id, big4, mkt_cap, log_mkt_cap, audit_quality)
```

Given this is a simulation, we *know* that the correct causal diagram is that shown in Figure 25.2.

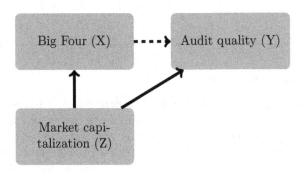

FIGURE 25.2
Market capitalization is a confounder

We draw the arrow connecting X and Y with a dashed line because the true coefficient on `big4` is zero, so *in this simulation* there is no actual causal relation between `big4` and `audit_quality`. Of course, in reality, we would not know this fact and would need to infer it from data and a posited causal diagram.

The good news from this causal diagram is that correctly conditioning on Z should give us an unbiased estimator of the causal effect of X on Y. So let's regress `audit_quality` on `big4` and `mkt_cap`.

```
fm <- lm(audit_quality ~ big4 + I(mkt_cap / 1e9),
         data = sim_auditor)
```

TABLE 25.4

Regression of audit quality using simulated data

	(1)
(Intercept)	1.567***
	(0.052)
big4TRUE	2.328***
	(0.066)
I(mkt_cap/1e+09)	0.053***
	(0.001)
Num.Obs.	5,000
R2	0.542

```
modelsummary(fm,
             estimate = "{estimate}{stars}",
             gof_map = c("nobs", "r.squared"),
             stars = c('*' = .1, '**' = 0.05, '***' = .01))
```

In Table 25.4, we see a clearly positive coefficient on `big4` even though we are "controlling for" `mkt_cap`, and we might be tempted to conclude that we have identified a causal effect. Of course, we know that this conclusion is invalid. The issue is that we are effectively assuming a linear relation between `mkt_cap` and audit quality, whereas the true relation is non-linear in `mkt_cap`. As a result, "controlling for" `mkt_cap` by including `mkt_cap` in a regression does not adequately control for the relation between `mkt_cap` and audit quality. Because there is a non-linear relation between `big4` and `mkt_cap`, the regression specification can use information in `big4` to explain variation in `audit_quality` in addition to that variation explained by a linear function of `mkt_cap`. This additional explanatory power shows up as a statistically significant coefficient on `big4` in the regression above.

Would some kind of matching analysis help?

Given its prominence in accounting research, we will focus on **propensity-score matching** here. In this section, we present propensity-score matching as a recipe of sorts. We then evaluate how well it does in our simulated settings before discussing some aspects of the theory that explains why (and when) we might expect it to work.

The first step in our recipe is the estimation of propensity scores, which are estimates of the probability of receiving treatment as a function of observed variables. The most common approach is to use logistic regression, which we can do using `glm()`, as seen in Chapter 23.

```
sim_fm <- glm(!big4 ~ mkt_cap, data = sim_auditor,
              family = binomial(link = "logit"))
```

We can estimate propensity scores as the fitted values from the estimated logit model (we specify `type = "response"` to get fitted values on the original $[0, 1]$ scale of the dependent variable).

```
sim_match_pscores <-
  sim_auditor |>
  mutate(pscore = predict(sim_fm, type = "response"))
```

The second step in the recipe is to match treatment observations to their nearest counterparts among the control observations, where "nearest" means closest in terms of propensity scores. But even with the propensity score, we need an algorithm. To this end, we can use `matchit()` function the `MatchIt` package. This function returns a `matchit` object to which we can apply `match.data()` to retrieve the resulting matches, which we store in `sim_matches`. Following Lawrence et al. (2011), we use a caliper of 3% and use `!big4` as the dependent variable, as *not* having a Big Four auditor is the less common condition.

```
sim_matches <-
  matchit(!big4 ~ mkt_cap, data = sim_match_pscores, caliper = 0.03) |>
  match.data()
```

Inspecting `sim_matches`, we see that it contains the variable `distance` that exactly matches `pscore` that we estimated above.

```
sim_matches |> count(pscore == distance)
```

```
# A tibble: 1 x 2
  `pscore == distance`      n
  <lgl>                 <int>
1 TRUE                   2348
```

In addition to the original contents of `sim_auditor` for the matched observations, `sim_matches` contains data on the matched pairs in the column `subclass`. Picking observations for three arbitrary values of `subclass`, we can see in Table 25.5 that each value is associated with a pair of observations, one with `big4 == FALSE` and one with `big4 == TRUE` and very similar values of `mkt_cap` and hence very similar `pscore` values.

```
sim_matches |>
  filter(subclass %in% c(220, 481, 333)) |>
  select(id, big4, mkt_cap, pscore, subclass) |>
  arrange(subclass, big4)
```

TABLE 25.5

Sample matched pairs

id	big4	mkt_cap	pscore	subclass
1896	FALSE	3,474,299	0.527	220
2424	TRUE	11,966,370	0.526	220
2387	FALSE	877,821,520	0.444	333
2434	TRUE	877,805,370	0.444	333
2920	FALSE	216,822,120,000	0.000	481
2904	TRUE	384,003,796,090	0.000	481

To examine the success of the match, we can compare two pairs of histograms. The first pair of histograms (Figure 25.3) shows the propensity scores for all observations. Here we see that there are many treatment observations with p-scores above say 0.70, but very few control observations with comparable p-scores to match with these. In this relatively simple setting, this is really just another way of saying there are many large Big Four clients and few similarly sized non–Big Four clients to match with them. This suggests that many observations will be unmatched.

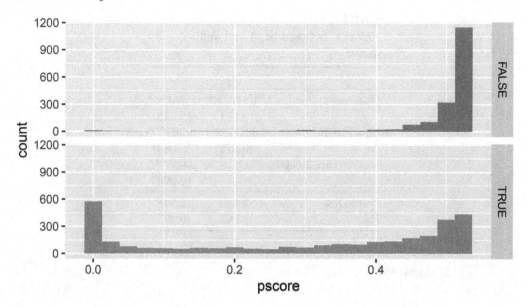

FIGURE 25.3
Propensity scores distribution: All observations

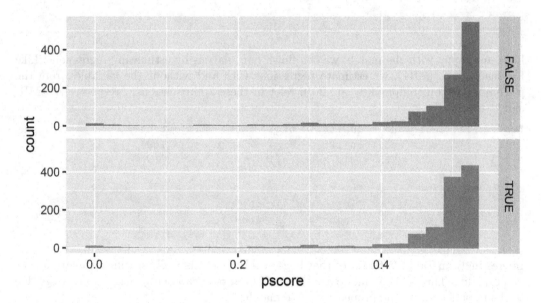

FIGURE 25.4
Propensity scores distribution: Matched observations

The second pair of histograms (Figure 25.4) shows the propensity scores for the *matched* observations. We can see that the distributions of p-scores for treatment and control observations look very similar, suggesting a fairly good match.

```
sim_match_pscores |>
  ggplot(aes(x = pscore, fill = big4)) +
  geom_histogram(binwidth = 0.025) +
```

TABLE 25.6
Relation between `big4` and audit quality
(matched)

	(1)	(2)
(Intercept)	1.870***	1.781***
	(0.051)	(0.045)
big4TRUE	0.235***	0.220***
	(0.072)	(0.064)
I(mkt_cap/1e+09)		0.085***
		(0.003)
Num.Obs.	2,348	2,348
R2	0.004	0.227

```
facet_grid(big4 ~ .) +
theme(legend.position = "none")
```

```
sim_matches |>
  ggplot(aes(x = pscore, fill = big4)) +
  geom_histogram(binwidth = 0.025) +
  facet_grid(big4 ~ .) +
  theme(legend.position = "none")
```

If we are happy with the match, we can finish our analysis by estimating regressions. Like
Lawrence et al. (2011), we estimate regressions with and without the variables from the
p-score regression (Shipman et al., 2016 refer to these approaches as, , respectively, "MR"
and "t-test").

```
fms <- list(feols(audit_quality ~ big4, data = sim_matches),
            feols(audit_quality ~ big4 + I(mkt_cap / 1e9),
                  data = sim_matches))
```

```
modelsummary(fms,
             estimate = "{estimate}{stars}",
             gof_map = c("nobs", "r.squared"),
             stars = c('*' = .1, '**' = 0.05, '***' = .01))
```

Interestingly, in Table 25.6, we see that `big4` still has a statistically significant relation with
audit quality. This is likely due to a residual effect of `mkt_cap` being picked up by `big4` due
to the imperfection of the propensity-score match.

25.2.1 Exercises

1. What happens in the above analyses if the true effect of `big4` on audit quality is
   `10` instead of `0`?

2. What happens in the above analyses if the caliper is reduced to `0.01` from `0.03`
   (but the true effect of `big4` on audit quality is still `0`)?

3. What happens in the above analyses if the caliper is reduced to `0.01` from `0.03`
   and the true effect of `big4` on audit quality is `10`?

4. Given the above, what are your conclusions about the value of propensity-score matching and OLS in causal inference?

5. What regression results do you get if you repeat the analysis shown in Table 25.4, but replace I(mkt_cap / 1e+09) with I(mkt_cap^(1/3) * 0.003)? Does this suggest a solution to the issues we see above? If so, what challenges does such a solution face?

6. Was endogeneity an issue in the simulation above? If not, why not? If so, in what way, and what could you do to address it?

25.3 Replication of Lawrence et al. (2011)

To facilitate discussion, we conduct an approximate replication of Lawrence et al. (2011) and begin by constructing a subset of Compustat with calculations used in that paper.

```
comp <-
  funda |>
  filter(indfmt == "INDL", datafmt == "STD",
         consol == "C", popsrc == "D") |>
  filter(!is.na(sich), !between(sich, 6000, 6999),
         sale > 0, at > 0,
         (dltt >= 0 | is.na(dltt)) & (dlc >= 0 | is.na(dlc)),
         prcc_f * csho > 0, ceq > 0) |>
  select(gvkey, datadate, fyear, au, prcc_f, csho, at, ib,
         dltt, dlc, rect, ppent, sale, act, lct, sich,
         oancf) |>
  mutate(sic2 = floor(sich / 100)) |>
  filter(between(fyear, 1988, 2006)) |>
  group_by(gvkey) |>
  window_order(fyear) |>
  mutate(lag_fyear = lag(fyear),
         avg_at = (lag(at) + at) / 2,
         log_at = log(at),
         aturn = if_else(lag(at) > 0, sale / lag(at), NA),
         inv_avg_at = 1 / avg_at,
         d_sale = sale - lag(sale),
         d_rect = rect - lag(rect),
         d_sale_at = (d_sale - d_rect) / avg_at,
         ppe_at = ppent / avg_at,
         curr = if_else(lct > 0, act / lct, NA),
         lag_curr = lag(curr),
         lev = if_else(avg_at > 0,
                       (dltt + coalesce(dlc, 0)) / avg_at, NA),
         lag_lev = lag(lev),
         roa = if_else(avg_at > 0, ib / avg_at, NA),
         lag_roa = lag(roa),
         accruals = ib - oancf,
         ta = accruals / avg_at,
```

```
        mkt_cap = prcc_f * csho,
        log_mkt = if_else(mkt_cap > 0, log(mkt_cap), NA)) |>
  ungroup() |>
  filter(lag_fyear == fyear - 1) |>
  collect()
```

Lawrence et al. (2011) estimate discretionary accrual models by industry and year, subject to a requirement of at least 10 data points for each model. So we compile a list of industry-years that meet this criterion.

```
industries <-
  comp |>
  group_by(fyear, sic2) |>
  summarize(n_firms = n(), .groups = "drop") |>
  filter(n_firms >= 10)
```

We next construct a function to estimate discretionary accruals along the lines described in footnote 11 of Lawrence et al. (2011). Because we self-join the `results` data frame created in this function in a way that results in multiple matches for each observation, we specify `relationship = "many-to-many"`. As we did in Chapter 14, we use !! to distinguish the values `sic2` and `fyear` supplied to the function from the variables found in `comp`.[3]

```
get_klw_data <- function(sic2, fyear, ...) {

  reg_data <-
    comp |>
    filter(sic2 == !!sic2, fyear == !!fyear)

  fm <- tryCatch(lm(ta ~ inv_avg_at + d_sale_at + ppe_at,
                    data = reg_data, na.action = na.exclude),
                 error = function(e) NULL)

  if (is.null(fm)) return(NULL)

  results <-
    reg_data |>
    select(gvkey, fyear, roa) |>
    bind_cols(da = resid(fm)) |>
    filter(!is.na(da))

  results |>
    inner_join(results, by = "fyear",
               suffix = c("", "_match"),
               relationship = "many-to-many") |>
    filter(gvkey != gvkey_match) |>
    mutate(roa_diff = abs(roa - roa_match)) |>
    group_by(gvkey, fyear) |>
    filter(roa_diff == min(roa_diff, na.rm = TRUE)) |>
    ungroup() |>
    mutate(ada = abs(da - da_match)) |>
```

---

[3]Chapter 19 of Wickham (2019) has more details on this "unquote" operator !!.

```
    select(gvkey, fyear, da, ada)
}
```

We can then **pmap()** this function onto **industries**, our set of industry-years and store the results in **klw_results**.[4]

```
klw_results <-
  industries |>
  pmap(get_klw_data) |>
  bind_rows()
```

We first estimate an OLS regression using the full sample and raw (i.e., unwinsorized) data and store it in a list named **fms**.

```
full_sample <-
  comp |>
  inner_join(klw_results, by = c("gvkey", "fyear")) |>
  filter(avg_at > 0) |>
  mutate(big4 = au %in% 1:8L,
         mkt_cap = prcc_f * csho,
         log_mkt = if_else(mkt_cap > 0, log(mkt_cap), NA))

fms <- list("OLS" = feols(ada ~ big4 + log_mkt + lag_roa +
                            lag_lev + lag_curr | sic2 + fyear,
                          data = full_sample))
```

Lawrence et al. (2011) winsorize several variables in their analysis. We use the **winsorize()** function from the **farr** package to do this here.[5] We then fit a model using the winsorized data and store it in **fms**.

```
win_vars <- c("ada", "log_mkt", "lag_roa", "lag_lev", "lag_curr",
              "roa", "lev", "curr", "mkt_cap")

full_sample_win <-
  full_sample |>
  mutate(across(all_of(win_vars),
                \(x) winsorize(x, prob = 0.01)))

fms[["OLS, win"]] <-
  feols(ada ~ big4 + log_mkt + lag_roa + lag_lev + lag_curr |
          sic2 + fyear,
        ~ gvkey + fyear,
        data = full_sample_win)
```

We next create a matched samples using propensity-score matching. Following Lawrence et al. (2011), we discard matches where the difference in p-scores is greater than 3%. First, we match using the raw data and estimate a model using this matched sample.

```
lmz_match <-
  full_sample |>
  filter(if_all(c(log_at, log_mkt, aturn, curr, lev, roa),
```

---

[4]Here KLW refers to Kothari et al. (2005).
[5]We discuss winsorization and the **winsorize()** function in Chapter 24.

```
              \(x) !is.na(x))) |>
  matchit(!big4 ~ log_at + log_mkt + aturn +
                    curr + lev + roa,
          data = _,
          caliper = 0.0300, std.caliper = FALSE,
          m.order = "largest", discard = "both")

fms[["PSM"]] <-
  lmz_match |>
  match.data() |>
  feols(ada ~ big4 + log_mkt + lag_roa + lag_lev + lag_curr |
             sic2 + fyear, ~ gvkey + fyear,
        data = _)
```

Second, we match using the winsorized data and estimate a model using the matched sample.

```
lmz_match_win <-
  full_sample_win |>
  filter(if_all(c(log_at, log_mkt, aturn, curr, lev, roa),
               \(x) !is.na(x))) |>
  matchit(!big4 ~ log_at + log_mkt + aturn + curr + lev + roa,
          data = _,
          caliper = 0.0300, std.caliper = FALSE,
          m.order = "largest", discard = "both")

lmz_matched_win <- match.data(lmz_match_win)

fms[["PSM, win"]] <-
  feols(ada ~ big4 + log_mkt + lag_roa + lag_lev + lag_curr |
             sic2 + fyear,
        ~ gvkey + fyear,
        data = lmz_matched_win)
```

Regression results from all four models are shown in Table 25.7.

```
modelsummary(fms,
             estimate = "{estimate}{stars}",
             gof_map = c("nobs", "r.squared"),
             stars = c('*' = .1, '**' = 0.05, '***' = .01))
```

25.4   DeFond et al. (2017)

DeFond et al. (2017) re-examine the question of Lawrence et al. (2011), but with some differences in approach.

The first set of differences relates to design choices in the application of propensity-score matching. While Lawrence et al. (2011) focus on a single propensity-score model with five primary covariates, DeFond et al. (2017) augment the propensity-score model with twenty additional variables, namely the squares and cubes of these covariates and the ten

TABLE 25.7

Relation between Big 4 and absolute value of discretionary accruals

	OLS	OLS, win	PSM	PSM, win
big4TRUE	−0.016***	−0.014***	−0.007***	−0.006***
	(0.002)	(0.001)	(0.002)	(0.002)
log_mkt	−0.006***	−0.005***	−0.007***	−0.006***
	(0.001)	(0.000)	(0.001)	(0.001)
lag_roa	−0.108***	−0.124***	−0.111***	−0.135***
	(0.006)	(0.005)	(0.017)	(0.005)
lag_lev	−0.045***	−0.045***	−0.055***	−0.052***
	(0.003)	(0.002)	(0.005)	(0.005)
lag_curr	0.000***	−0.001***	0.000	−0.001*
	(0.000)	(0.000)	(0.000)	(0.000)
Num.Obs.	72,996	72,996	24,217	24,227
R2	0.127	0.150	0.117	0.132

interactions between the five covariates. DeFond et al. (2017) then select 1,500 random subsets from the twenty variables to develop propensity-score models. For each estimated propensity-score model, the caliper is set at a random value (less than 30%) that yields a "balanced matched sample".

A second design-choice variation that DeFond et al. (2017) consider is matching with replacement. While Lawrence et al. (2011) focus on matching one treatment observation with (at most) one control observation without replacement (i.e., once a control observation is matched to a treatment observation, it is not available to match with another treatment observation), DeFond et al. (2017) consider matches *with* replacement and consider matches where each treatment observation is matched to one, two, or three control firms.

The second set of differences between DeFond et al. (2017) and Lawrence et al. (2011) relates to the measures of audit quality considered. Lawrence et al. (2011) consider cost of equity capital and analyst forecast accuracy, while DeFond et al. (2017) omit these two measures on the basis that they have "poor construct validity, and as a result are rarely used in the prior literature" (DeFond et al., 2017, p. 3637). Instead, DeFond et al. (2017) consider three other measures of audit quality: income-increasing discretionary accruals, restatements, and going concern opinions.

25.5 Further reading

The goal of this chapter was to provide a quick introduction to matching and show what matching can and cannot do. Perhaps the most important takeaway is that matching is *not* a panacea for issues arising due to endogenous selection into treatment based on unobservable variables (Shipman et al., 2016).

Cunningham (2021) provides a good introduction to matching with discussion of inverse probability weighting and coarsened exact matching. Guo et al. (2020) provide a book-length treatment of the topic.

We saw in Section 25.2 that, even when selection into treatment is based on observable variables, estimates can be unreliable. Chapter 14 of Huntington-Klein (2021) discusses a number of approaches that can be used to address such cases, including inverse probability weighted regression adjustment and entropy balancing.

25.6   Discussion questions

1. Given the evidence presented in the simulation, how do you interpret the regression results presented above? How do DeFond et al. (2017) appear to interpret the results?

2. Referring back to the basic causal relations described in Section 4.2, what is the causal diagram implied by equation (1) of and Table 2 of Lawrence et al. (2011)? What variables are in *PROXY_CONTROLS* for Table 2? Why is *LOG_ASSETS* not found in Table 2?

3. Do Lawrence et al. (2011) report results from estimating equation (1)? What happens to the "difference in means" between the Full Sample and the Propensity-Score Matched Sample in Table 1? Does this make sense?

4. Why do you think the difference in means is still statistically significant for two variables in the Propensity-Score Matched Sample column of Table 1 in Lawrence et al. (2011)? Do you think this is a concern?

5. Lawrence et al. (2011) evaluate three outcomes as measures of audit quality: discretionary accruals, cost of equity, and analyst forecast accuracy? Evaluate each of these measures. Which do you think is best? What are the strengths and weaknesses of this best measure?

6. Can you think of other measures of audit quality that might make sense?

7. Apply the check list of Shipman et al. (2016, pp. 217–218) and the questions in Panels B and C of Table 1 of Shipman et al. (2016) to Lawrence et al. (2011). How do Lawrence et al. (2011) stack up against these?

26

Prediction

This chapter focuses on prediction using approaches that fall under the broad heading of **statistical learning**, also known as **machine learning**. We begin by discussing what prediction means and how it can be used in research. We then explore methods and concepts related to prediction using the setting of accounting fraud examined in Bao et al. (2020).

> 💡 Tip
>
> The code in this chapter uses the packages listed below. We load `tidyverse` because we use several packages from the Tidyverse. For instructions on how to set up your computer to use the code found in this book, see Section 1.2. Quarto templates for the exercises below are available on GitHub.[a]
>
> ---
> [a]https://github.com/iangow/far_templates/blob/main/README.md

```
library(tidyverse)
library(DBI)
library(farr)
library(furrr)
library(rpart)
library(rpart.plot)
library(glmnet)
```

> ❗ Important
>
> Some of the code in this chapter—especially in Section 26.6—takes a long time to run, even with multiple cores. To make it easier for readers to run just the code they are interested in, here we provide some guidance regarding how code from one section depends on code in other sections.
>
> Running the code in Section 26.2 is a prerequisite for running code in Section 26.3, Section 26.5, and Section 26.6, each of which can be run independently of the others. While some results from Section 26.6 are used in Section 26.7, we have embedded these results in the code for Section 26.7, so that from a reader's perspective the latter section only depends on results of code in Section 26.2. (There is no code in Section 26.1 or Section 26.4.)

DOI: 10.1201/9781003456230-26

26.1   Prediction in research

Prediction can be taken to mean the activity of using known information to estimate or predict unknown events or values. In ordinary usage, prediction suggests that the unknown events or values are in the future, but often prediction is of pre-existing facts which may only become known at a later date (e.g., we may use prediction models to diagnose a disease or to detect misstatements in observed financial statements).

In this chapter we will cover a number of statistical approaches to prediction. Prediction is perhaps the most active area of research in statistics and the methods we cover in this chapter fall under the umbrella of **statistical learning**, also known as **machine learning**.

It seems natural to ask why social science researchers would be interested in prediction. As discussed in Chapter 4, research papers can often be distinguished by their research objectives, which might include prediction, measurement, description, and causal inference.

When prediction is part of the research question, the methods covered in this chapter are clearly relevant. A researcher might be interested in prediction not for its own sake, but to understand how information might be used in practice. For example, Bernard and Thomas (1989) were not interested in predicting post-earnings announcement returns so much as providing evidence of the existence of a predictable element of these returns to understand market efficiency.

Prediction methods are also often used in measurement. For example, papers have developed models for predicting events such as bankruptcy and earnings restatement and these models have often been used in subsequent research as measures of, say, bankruptcy risk.

Finally, an emerging area of research examines the usefulness of prediction for causal inference. While prediction is at the heart of causal inference in that a causal model has predictive value, it has been less clear whether predictive models developed using the methods provided in this chapter offer insights into the existence of causal relations.

Beyond research, prediction is also fundamental to financial accounting in many settings. For example, the provision for loan losses is a prediction of future write-offs. Researchers might seek to understand accounting in such settings as prediction problems and might even simulate such processes using statistical learning approaches like those studied in this chapter.

And beyond financial accounting, prediction models have innumerable uses in practice, from identifying anomalies in accounting records to making movie recommendations to Netflix customers.

26.2   Predicting accounting fraud

To help the reader understand the issues and approaches used in statistical learning, we focus in this chapter on the setting of Bao et al. (2020), which focuses on predicting accounting fraud in publicly traded US firms during the period 2003–2008.

26.2.1 Features

Any prediction problem involves an **outcome** that we want to predict based on a set of **features**. Features (also known as **predictors**) refer to the variables or data that we will use to make predictions.

We start with code to collect data on the features used in this chapter. We follow Bao et al. (2020) in using "28 raw financial data items" drawn from Compustat as our features. While not described in Bao et al. (2020), from the SAS code provided by the authors, four variables (ivao, pstk, ivst, txp) are replaced with zeros if missing. The code from Bao et al. (2020) has the equivalent of `filter(!is.na(at), !is.na(prcc_f))`, but given that the equivalent of `na.omit()` is applied at the end of their code, this filter is unnecessary.

```
db <- dbConnect(RPostgres::Postgres(), bigint = "integer")

funda <- tbl(db, Id(schema = "comp", table = "funda"))

X_vars <- c("act", "ap", "at", "ceq", "che", "cogs", "csho", "dlc",
            "dltis", "dltt", "dp", "ib", "invt", "ivao", "ivst",
            "lct", "lt", "ni", "ppegt", "pstk", "re", "rect", "sale",
            "sstk", "txp", "txt", "xint", "prcc_f")

y_var <- "misstate"

features_all <-
  funda |>
  filter(indfmt == "INDL", datafmt == "STD",
         consol == "C", popsrc == "D") |>
  filter(fyear >= 1991, fyear <= 2008) |>
  mutate(across(c(ivao, pstk, ivst, txp),
                \(x) coalesce(x, 0))) |>
  select(gvkey, datadate, fyear, all_of(X_vars)) |>
  arrange(gvkey, datadate)

features <-
  features_all |>
  collect() |>
  na.omit()
```

While the features that are used in building a prediction model and even how they are encoded can "have a significant impact on model performance" (Kuhn and Johnson, 2013, p. 27), we only consider the features used by Bao et al. (2020) in our analysis here and explore issues related to features in discussion questions.

26.2.2 Outcome

An outcome might be measured as a continuous variable (e.g., life expectancy) or it might be a categorical variable (e.g., "no restatement", "minor restatement", or "major restatement"). A categorical variable will often naturally be a **binary variable** taking one of just two values, but in principle, any variable can be coded as a binary variable (e.g., "hot dog" or "not hot dog"[1]).[2]

---

[1] https://www.youtube.com/watch?v=ACmydtFDTGs

[2] Obviously, some information is lost in recoding a variable in this way and this may not be helpful in many decision contexts.

In this chapter, we follow Bao et al. (2020) and focus on a single binary categorical variable taking values of "no fraud" and "fraud", which might be represented using $\{0,1\}$ (i.e., an indicator variable), but in statistical learning one often sees such variables represented using $\{-1,1\}$.[3] Bao et al. (2020) define a fraud as the existence of an Accounting and Auditing Enforcement Release (AAER) released by the SEC affecting the firm.[4] While the focus on binary variables might seem limiting, it facilitates our presentation of some core ideas of statistical learning within the confines of a single chapter.

Data on the outcome are found in the data set `aaer_firm_year` from the `farr` package.

`aaer_firm_year`

```
# A tibble: 415 x 4
   p_aaer gvkey  min_year max_year
   <chr>  <chr>     <int>    <int>
 1 1033   021110     1992     1992
 2 1037   008496     1992     1993
 3 1044   028140     1993     1993
 4 1047   012455     1994     1994
 5 1053   025927     1993     1993
 6 1067   023126     1993     1993
 7 1071   024712     1992     1993
 8 1104   028450     1994     1995
 9 1106   002711     1992     1993
10 1110   010185     1994     1995
# i 405 more rows
```

As can be seen above, for each AAER (identified by `p_aaer`), `aaer_firm_year` contains information on the GVKEY of the affected firm and the first and last fiscal years affected by accounting fraud (`min_year` and `max_year`, respectively).

In practice, careful thinking about outcome measures is required. The outcome selected and the approach to measuring it are likely to depend on the purpose for which prediction is being made. While Bao et al. (2020) aim to predict "accounting fraud", it is important to note that the variable `p_aaer` indicates an Accounting and Auditing Enforcement Release (AAER) by the SEC. But accounting fraud might not result in AAERs, either because it is never detected, or because it is detected but does not rise to the level that leads to an AAER, or even because the fraud is so profound that an AAER is somewhat irrelevant.

With regard to the last category, consider Enron, the public company at the heart of one of the most notorious cases of accounting fraud this century. While the CEO (Jeffrey Skilling) and CFO (Andrew Fastow) of Enron ended up serving time in prison, there is no AAER related to either Skilling or Fastow (many AAERs relate to individuals). There is not even an AAER directed at Enron, perhaps because it entered bankruptcy shortly after fraud was detected. The one AAER in the Bao et al. (2020) sample related to Enron has `p_aaer` of 1821.[5] However, this AAER[6] related to the order for Citigroup to pay an amount in a settlement arising because "Citigroup assisted [Enron and Dynegy] in enhancing artificially

---

[3] In many respects, the specific representation does not matter. However, a number of statistical learning algorithms will be described presuming one representation or another, in which case care is needed to represent the variable in the manner presumed.

[4] See the SEC website for more information: https://www.sec.gov/divisions/enforce/friactions.htm.

[5] Use Enron's GVKEY 006127 to find this: `aaer_firm_year |> filter(gvkey == "006127")`.

[6] https://www.sec.gov/litigation/admin/34-48230.htm

their financial presentations through a series of complex structured transactions ... to allow those companies to report proceeds of financings as cash from operating activities".

Even if we accept that AAERs are the measure of interest, there are several wrinkles related to timing. To see the issues here, consider the AAER with `p_aaer` equal to 1500, which affected fiscal years 1998, 1999, and 2000.

```
aaer_firm_year |>
  filter(p_aaer == "1500")
```

```
# A tibble: 1 x 4
  p_aaer gvkey  min_year max_year
  <chr>  <chr>     <int>    <int>
1 1500   028958     1998     2000
```

Suppose we were an analyst and had data on features up to the end of fiscal 1999 and were looking to train a model that could be used to predict future AAERs related to fiscal 1999. This would place us somewhere in 2000 (fiscal 1999 includes firms with fiscal years ending as late as 2000-05-31) and the largest meaningful model-building data set could include fiscal years up to 1998. Sitting in 2018 or 2022, we might be tempted to code `p_aaer` as 1 for the firm-year with gvkey of 028958 and `fyear` of 1998, but it turns out that AAERs are only disclosed *after* the end of the last fiscal year that they affect and this would mean that the `p_aaer == '1500'` AAER would not be observed by our hypothetical analyst in 2000, who would instead see `misstate` as 0 for that observation. To simulate the approach taken by our hypothetical analyst, we need to code `misstate` for this case as 0.

At some point it would be reasonable to assume that the analyst can see the AAER just as we can and use it in training a model, but the question is "when?". Bao et al. (2020, p. 209) "require a gap of 24 months between the financial results announcement of the last training year and the results announcement of a test year ... because Dyck et al. (2010) find that it takes approximately 24 months, on average, for the initial disclosure of the fraud." For the `p_aaer == '1500'`AAER, Bao et al. (2020)'s approach would mean that the hypothetical analyst could observe the AAER by the end of fiscal 2002 (two years after the last affected period of 2000).

While Bao et al. (2020) relied on the broad observation of Dyck et al. (2010), data on AAER dates are easily obtained from the SEC's website for AAERs since late 1999 and included in the data set `aaer_dates` made available with the `farr` package.[7]

```
head(aaer_dates |> select(aaer_num, aaer_date))
```

```
# A tibble: 6 x 2
  aaer_num  aaer_date
  <chr>     <date>
1 AAER-1212 1999-12-14
2 AAER-1211 1999-12-03
3 AAER-1210 1999-11-19
4 AAER-1209 1999-11-16
5 AAER-1208 1999-11-15
6 AAER-1206 1999-11-10
```

---

[7]The code used to create this table can be found in source code for the `farr` package: https://github.com/iangow/farr/blob/main/data-raw/get_aaer_dates.R.

Using these data, we can examine the actual distribution of the time between the end of the latest affected fiscal year and the AAER being released by the SEC.

```
aaer_days <-
  aaer_dates |>
  mutate(p_aaer = str_replace(aaer_num, "^AAER-", "")) |>
  inner_join(aaer_firm_year, by = "p_aaer") |>
  mutate(fyear = max_year) |>
  inner_join(features, by = c("gvkey", "fyear")) |>
  select(p_aaer, gvkey, datadate, aaer_date) |>
  mutate(days_to_aaer = as.integer(aaer_date - datadate)) |>
  filter(!is.na(days_to_aaer))
```

Figure 26.1 reveals a lot of variation that is ignored by a blanket "24 months" assumption (represented by the dashed vertical line). In addition, the median time to release of an AAER after the last affected fiscal year is about 3.7 years (1,350 days).

```
aaer_days |>
  ggplot(aes(x = days_to_aaer)) +
  geom_histogram(binwidth = 30) +
  geom_vline(xintercept = 365 * 2, linetype = "dashed", color = "red")
```

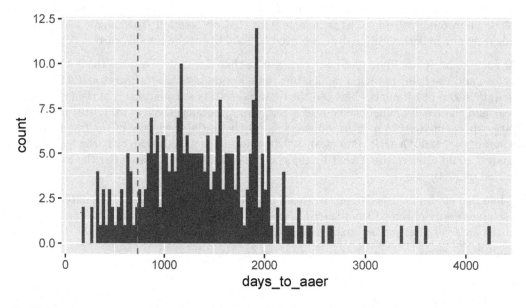

FIGURE 26.1
Days to AAER announcement from end of last affected period

How might we use these data? It would seem appropriate to imagine our analyst looking to apply a model to a fiscal year where that model has been trained on features drawn from prior fiscal years and AAERs revealed to date. For example, if our analyst were now looking to apply a model to data from fiscal 2001, she would be situated somewhere in late 2002 (fiscal 2001 includes firms with fiscal years ending as late as 2002-05-31) and thus would only see AAERs released prior to that time. Assuming the conventional four months for financial data to be released, we might set a "test date" of 2002-09-30 for fiscal 2001, so AAERs with an `aaer_date` prior to 2002-09-30 would be observable to the analyst.[8]

We start by creating `aaer_long`, a data frame with the fiscal years affected by misstatements (i.e., all years between `min_year` and `max_year`).

```
aaer_long <-
  aaer_firm_year |>
  rowwise() |>
  mutate(fyear = list(seq(min_year, max_year, by = 1))) |>
  unnest(fyear) |>
  select(gvkey, p_aaer, fyear, min_year, max_year)
```

In `get_aaers()`, we construct `misstate` so that it can be presumed observable to the analyst on `test_date` and include `misstate_latent`, which includes accounting fraud observable to the researcher, but *not* to the analyst.

```
get_aaers <- function(test_date, include_latent = FALSE) {
  min_aaer_date <- min(aaer_dates$aaer_date, na.rm = TRUE)

  df <-
    aaer_dates |>
    mutate(p_aaer = str_replace(aaer_num, "^AAER-", "")) |>
    select(p_aaer, aaer_date) |>
    distinct() |>
    right_join(aaer_long, by = "p_aaer") |>
    mutate(aaer_date = coalesce(aaer_date, min_aaer_date)) |>
    mutate(misstate = aaer_date <= test_date) |>
    select(gvkey, fyear, misstate) |>
    distinct()

  if (include_latent) {
    df |> mutate(misstate_latent = TRUE)
  } else {
    df
  }
}
```

For concreteness, let's assume a "test date" of 2002-09-30 and call `get_aaers()` to make a data set of AAERs.

```
aaers <- get_aaers(test_date = "2002-09-30", include_latent = TRUE)
```

Our hypothetical analyst could use `misstate` from `aaers` to train a model to predict accounting fraud that, if present, will be revealed at some point in the future. In doing so, our analyst could fit models using data as late as fiscal 2000.[9]

---

[8] Ironically given the attention we are paying to the distribution of AAER dates, we are ignoring the variation in the time between the fiscal period-end and the filing of financial statements. Additionally, we are implicitly assuming that the analyst is constrained to wait until all firms have reported in a fiscal year before applying a prediction model. However, it would be feasible to apply a model to prediction accounting fraud as soon as a firm files financial statements, which might be in March 2002 for a firm with a 2001-12-31 year-end. This might be a better approach for some prediction problems because of the need for more timely action than that implied by the "fiscal year completist" approach assumed in the text.

[9] Actually, it may be that *some* restatements affecting fiscal 2001 have already been revealed and these could be used to fit a model. But we assume there would be few such cases and thus focus on fiscal 2000

Related to this issue, Bao et al. (2020, p. 212) "recode all the fraudulent years in the training period to zero for those cases of serial fraud [accounting fraud spanning multiple consecutive reporting periods] that span both the training and test periods." Because we require the release of an AAER before coding `misstate` to one in the training period, and all AAERs are released after any affected fiscal periods (including any that might be considered part of a test period), this kind of recoding is not necessary with our approach.[10]

However, just because an analyst *could* use data from fiscal 2000 doesn't mean she *should*. If we accept `misstate_latent` as "truth", then `misstate` can be viewed as measuring true accounting fraud with error.[11] Sitting many years in the future (from the perspective of 2001), we can get a handle on the rate of errors assuming a test date at the end of fiscal 2001:

```
aaers |>
  group_by(fyear) |>
  summarize(error_rate = mean(misstate != misstate_latent)) |>
  filter(fyear <= 2000)
```

fyear	error_rate
1990	0.000
1991	0.000
1992	0.074
1993	0.033
1994	0.130
1995	0.182
1996	0.242
1997	0.512
1998	0.702
1999	0.829
2000	0.900

Based on this analysis, it seems plausible that including fiscal 2000 restatements in the data used to train a model for application in fiscal 2001 might *increase* prediction errors because of the extremely high measurement error in `misstate`. In fact, the same *may* even be true of fiscal years 1999 and 1998. Examining this question carefully would require detailed analysis beyond the scope of this chapter.[12] Instead, we will table this issue and broadly follow the

---

as the latest plausible year for training. As we will see, measurement error seems sufficiently high for fiscal 2000 that using data fiscal 2001 for training is unlikely to be helpful.

[10] It might be suggested that we shouldn't wait for an AAER to be released before encoding an AAER as 1. For example, we don't need to wait until 2003-07-28 to code the affected Enron firm-years as misstated. But we argue that this would embed one prediction problem inside another (that is, we'd be predicting AAERs using general knowledge to use them in predicting AAERs using a prediction model). Additionally, it was far from guaranteed that there would be an AAER for Enron and in less extreme cases, it would be even less clear that there would be an AAER. So some predicted AAERs that we'd code would not eventuate. We argue that the only reasonable approach is to assume that the prediction models are trained only with confirmed AAERs at the time they are trained.

[11] We put *truth* in "scare quotes" because even `misstate_latent` has issues as a measure of AAERs, because some AAERs may emerge after the sample is constructed by the researcher, which in turn may be some years after the test period.

[12] Also, this analysis could not be conducted by our hypothetical analyst sitting at the end of fiscal 2001, because she would not have access to `misstate_latent`.

approach of Bao et al. (2020), who exclude the data on the fiscal year prior to the test year from their training data. That is, for a test year of 2001, we only consider AAERs affecting fiscal 1999 and earlier.[13]

Before moving on, we consider one more timing issue related to the purpose for which predictions of future AAERs are being made. For concreteness, we will consider two possible roles that our hypothetical analyst may have.

The first role is that of a regulator, perhaps even the SEC itself. The SEC might posit that everything that should become an AAER eventually does become an AAER, but yet want to predict future AAERs so as to focus its investigation and enforcement efforts, perhaps to bring AAERs to light sooner rather than later.[14] For this purpose, it would seem sensible to want to predict future AAERs whenever they might occur and calibrate models that best serve that end.[15]

The second role might be that of a short-seller. For example, McCrum (2022) provides numerous examples of short-sellers being interested in information about accounting fraud at Wirecard. One short-seller discussed in McCrum (2022) is Matt Earl, who declared a short position in Wirecard in November 2015 (2022, p. 95), when the stock price was €46.50. Given his short position, Earl might have expected to profit if a regulator announced the equivalent of an AAER, as this might be expected to cause the stock price to fall.[16] Unfortunately, there was no enforcement release announced by August 2018, when the stock price was at a high of €227.72.[17] While Earl's prediction of accounting fraud was correct, he likely lost a lot of money betting on a stock price fall. By April 2018, Earl was no longer shorting Wirecard (2022, p. 208). What Earl could have profited from was an accurate prediction that Wirecard would be subject to an enforcement action related to accounting fraud *in the next n months*. Because there was no such action prior to June 2020, his prediction that accounting fraud would be revealed was unprofitable. Of course, a short-seller is not necessarily a passive participant in the flow of information to capital markets and regulators. Earl did produce a report under the name of Zatarra Research and Investigations (2022, p. 100), but was evidently unable to trigger an effective investigation by regulators.[18]

Aside from timing, another set of considerations for our hypothetical users of a prediction model relates to resource constraints and loss functions. A model that predicts "fraud" at hundreds of firm-years, only some of which are truly affected by fraud, is likely to be difficult for a regulator to act on if there are resource constraints. If a regulator only has the resources to conduct investigations into 10 firms, then it might want a model that assesses the probability of fraud so that it can investigate the 10 firms most likely to be affected. A short-seller is likely constrained by human and capital resources, so may want to identify the firms most likely to be affected so that additional investigation and short-selling can be

---

[13]It is important to note that our rationale for doing this differs from that of Bao et al. (2020). Bao et al. (2020) exclude AAERs affecting fiscal years after 1999 on the basis that these would not be known when hypothetically testing their models; for this purpose we use data on AAER dates that Bao et al. (2020) did not collect. Our rationale for excluding AAERs affecting fiscal year 2000 and released prior to the test date is one about measurement error.

[14]There are problems here in terms of having a well-formed prediction problem if the prediction model itself changes the set of future AAERs due to changes in SEC behaviour. But we ignore these here.

[15]Of course, the analysis we do below is limited to publicly available information, so perhaps not a good model of the prediction problem that is actually faced by the SEC, which presumably has access to private information related to future AAERs.

[16]As a German company not listed in the United States, Wirecard would likely not be the subject of an AAER, but presumably there is some rough equivalent in the German context.

[17]See https://go.unimelb.edu.au/xzw8 for details.

[18]The Wirecard fraud—including Earl's role in trying to uncover it—is the subject of a Netflix documentary: https://www.netflix.com/watch/81404807.

focused on those firms. These considerations can be incorporated into the objective function used in calibrating the prediction model, as we discuss below.

26.2.3 Merged data

We can now compile our full data set. (Note that we drop `misstate_latent`, as we would not know that variable when building our model.)

```
df <-
  aaers |>
  select(-misstate_latent) |>
  right_join(features, by = c("gvkey", "fyear")) |>
  mutate(misstate = coalesce(misstate, FALSE))
```

We extract the model-building data set (`data_train`) that we will use in later sections by selecting all observations from `df` with fiscal years in or before 2001.

```
data_train <-
  df |>
  filter(fyear <= 2001)
```

26.2.4 Sample splits

In statistical learning, it is common to divide a data set used for developing a statistical prediction model into three disjoint data sets. The first of these is the **training** data set, which is used to train the model (e.g., estimate coefficients). The second data set is the **validation** data, which is used to calibrate so-called meta-parameters (discussed below) and choose among alternative modelling approaches. The final data set is the **test** data which is used to evaluate the performance of the model chosen using the training and validation data sets.

We believe it is helpful to consider the process of training and validating a model as together forming the **model-building process** and the data set made available to this process as the **model-building** data.

In principle, if each observation in the data set available to the analyst is independent, these three data sets could be constructed by randomly assigning each observation to one of the data sets.[19]

But in the current context, we are simulating a forecasting process, making time an important consideration, as we want to make sure that the process is feasible. For example, it is not realistic to use data from 2008 to train a model that is then tested by "forecasting" a restatement in 1993. Instead, in testing forecasts from a model, it seems more reasonable to require that the model be based on information available at the time the forecast is (hypothetically) being made.

In this regard, we follow Bao et al. (2020) in adapting the model-building data set according to the test set and in defining test sets based on fiscal years. Bao et al. (2020) consider a number of test sets and each of these is defined as observations drawn from a single fiscal

---

[19]Note that the word **sample** is frequently used in the statistical learning literature to refer to a single observation, a term that may seem confusing to researchers used to thinking of a sample as a set of observations.

year (the test year). They then specify the model-building sample for each test set as the firm-year observations occurring in fiscal years at least two years before the test year.

An important point is that the test data set should not be used until the model is finalized. The test data set is meant to be **out-of-sample** data so as to model the arrival of unseen observations needing classification. If modelling choices are made based on out-of-sample performance, then the test data are no longer truly out of sample and there is a high risk that the performance of the model will be worse when it is confronted with truly novel data.

Having defined the model-building data set, it is necessary to work out how this sample will be split between training and validation data sets. For many models used in statistical learning, there are parameters to be chosen that affect the fitted model, but whose values are not derived directly from the data. These **meta-parameters** are typically specific to the kind of model under consideration. In penalized logistic regression, there is the penalty parameter λ. In tree-based models, we might specify a cost parameter or the maximum number of splits.

The validation sample can be used to assess the performance of a candidate modelling approach with different meta-parameter values. The meta-parameter values that maximize performance on the validation sample will be the ones used to create the final model that will be used to predict performance on the test set.

There are different approaches that can be used to create the validation set. Bao et al. (2020) use fiscal 2001 as their validation data set and limit the training data for this purpose to observations with fiscal years between 1991 and 1999.

A more common approach is what is known a n-**fold cross-validation**. With 5-fold cross-validation ($n = 5$), a model-building sample is broken into 5 sub-samples and the model is trained 5 times using 4 sub-samples and evaluated using data on the remaining sub-sample. Values of n commonly used are 5 and 10.

Bao et al. (2020, p. 215) say that "because our fraud data are intertemporal in nature, performing the standard n-fold cross validation is inappropriate." In this regard, Bao et al. (2020) appear to confuse model validation and model testing. It is generally not considered appropriate to test a model using cross-validation, instead cross-validation is used (as the name suggests) for validation.

But Bao et al. (2020) do raise a valid concern with using standard cross-validation due to what they call serial fraud. This issue is probably best understood with an example. Suppose we expanded the feature set to include an indicator for the CFO's initials being "AF" (perhaps a clue as to "*a*ccounting *f*raud", but in reality unlikely to mean anything).

```
aaer_firm_year |>
  filter(p_aaer == "1821")
```

```
# A tibble: 1 x 4
  p_aaer gvkey  min_year max_year
  <chr>  <chr>    <int>    <int>
1 1821   006127    1998     2000
```

With Enron in the sample for 1998–2000 (when Andrew Fastow was CFO), it's quite possible that the model would detect the apparent relation between our new feature and accounting fraud and include this in the model. However, this would not improve out-of-sample model performance because Enron would not be in the test sample. So, to maximize out-of-sample performance, we do not want to include this feature in the model and thus want to calibrate a model in a way that discourages the training process from using this feature. If we are not careful in splitting our model-building sample and have Enron in 1998 and 1999 in the training set, and Enron in 2000 in the validation set, then it is going to appear that this feature is helpful and our model will underperform when taken to the test data.

The phenomenon described in the previous paragraph is an instance of **overfitting**, which refers to the tendency of models excessively calibrated to fit model-building data to perform worse when confronted with truly novel data than simpler models.[20]

To avoid this kind of overfitting, we ensure that firm-years for a single AAER are kept in the same fold. In fact, we might want to keep *firms* in the same cross-validation fold as an extra level of protection against overfitting.[21]

Another issue related to timing that we might consider in the cross-validation sample is whether it is appropriate to evaluate a model trained on data that includes AAERs at Enron in 1998–2000 in predicting fraud at another firm in, say 1993. We argue the concern about having a feasible forecasting approach that requires us to only forecast future fraud does not preclude training a model using seemingly impossible "prediction" exercises. It is important to recognize that the goals of validating and testing models are very different. In validating the model, we are trying to maximize *expected* out-of-sample performance (e.g., what it will be when used with a test set), while in testing the model, we are trying to provide a realistic measure of the performance of the model on truly novel data.

Using cross-validation rather than the single fiscal-year validation sample approach used by Bao et al. (2020) offers some benefits, including effectively using the full model-building data set for validation and also validating using data from multiple fiscal years, which reduces the potential for models to be excessively calibrated for models that work well only in fiscal 2001.

To make it easier to implement this for the approaches we discuss below, we first organize our data into folds that can be used below. We explain the role of `pick(everything())` in Chapter 13.

```
n_folds <- 5
folds <- 1:n_folds

set.seed(2024)

sample_splits <-
  data_train |>
  select(gvkey) |>
  distinct() |>
  mutate(fold = sample(folds, nrow(pick(everything())), replace = TRUE))
```

The code above implements the folds for 5-fold cross-validation by assigning each GVKEY to one of 5 folds that will use below. In essence, for fold i, we will estimate the model using data in the other folds $(j \neq i)$ and then evaluate model performance by comparing the predicted classification for each observation in fold i with its actual classification. We then

---

[20]This idea is also related to p-hacking as discussed in Chapter 19, whereby a hypothesis that appears to apply in an estimation sample does not have predictive power in a new sample.

[21]Note that we are relying on the selection of the meta-parameter to provide some degree of protection against this kind of overfitting, but it may be that the models will gravitate to the features that allow them to overfit "serial frauds" no matter the meta-parameter values. In such a case, we could still rely on cross-validation performance as a reliable indicator of the resultant weakness of the models we train.

repeat that for each fold i. Later in this chapter, we will also use cross-validation to choose values for meta-parameters.

26.3 Model foundations

There are broadly two basic building blocks with which many statistical learning models are constructed. The first kind of building block fits models using linear combinations of features. For example, in regression-based models, we might use ordinary least squares as the basic unit. With a binary classification problem such as the one we have here, a standard building block is the logistic regression model.

The second kind of building block is the classification tree. We consider each of these building blocks in turn, including their weaknesses and how these weaknesses can be addressed using approaches that build on them.

26.3.1 Logistic regression

We begin with logistic regression, also known as **logit**. Logistic regression models the log-odds of an event using a function linear in its parameters:

$$\log\left(\frac{p}{1-p}\right) = \beta_0 + \beta_1 x_1 + \cdots + \beta_P x_P.$$

Here p is the modelled probability of the event's occurrence and P denotes the number of predictors. Kuhn and Johnson (2013, p. 280) suggest that "the logistic regression model is very popular due to its simplicity and ability to make inferential statements about model terms." While logit can be shown to be the maximum likelihood estimator for certain structural models, it is not always optimized for prediction problems (see Kuhn and Johnson, 2013, p. 281). While we will not use logit to "make inferential statements" in this chapter, we explore it given its popularity and simplicity.

Having identified a model-building data set, we first fit a logit model and store the result in `fm1`. As discussed in Chapter 23, logit falls in the class of models known as **generalized linear models** (GLMs), and thus can be estimated using `glm()` provided with base R. If we specify `family = binomial`, `glm()` will use logit as the **link function**, as `link = "logit"` is the default for the `binomial` family.

```
formula <- str_c(y_var, " ~ ", str_c(X_vars, collapse = " + "))
fm1 <- glm(formula, data = data_train, family = binomial)
```

We can assess within-sample performance of the logit model by comparing the predicted outcomes with the actual outcomes for `data_train`. Here we use the fitted probability (`prob`) to assign the predicted classification as `TRUE` when `prob > 0.5` and `FALSE` otherwise. (As discussed below, there may be reasons to choose a different cut-off from `0.5`.)

```
within_sample1 <-
  data_train |>
  mutate(score = predict(fm1,
                         newdata = pick(everything()), type = "response"),
         predicted = as.integer(score > 0.5))
```

We can produce a **confusion matrix**, which is a tabulation of number of observations in the sample by predicted and true outcome values.

```
table(predicted = within_sample1$predicted,
      response = within_sample1$misstate)
```

```
          response
predicted FALSE   TRUE
        0 66965    208
        1     0      1
```

Here we see 1 true positive case, 66,965 true negative cases, 208 false negative cases, and 0 false positive cases.

While there are no meta-parameters to select with the basic logistic regression model, we can use cross-validation as an approach for estimating out-of-sample prediction performance.

The following function takes the argument `fold`, fits a logit model for observations not in that fold, then classifies observations that *are* in that fold, and returns the values. As we did in Chapter 14, we use `!!` to distinguish the `fold` supplied to the function from `fold` found in `sample_splits`.[22]

```
logit_predict <- function(fold) {
  dft <-
    data_train |>
    inner_join(sample_splits, by = "gvkey")

  fm1 <-
    dft |>
    filter(fold != !!fold) |>
    glm(formula, data = _, family = binomial)

  dft |>
    filter(fold == !!fold) |>
    mutate(score = predict(fm1, pick(everything()), type = "response"),
           predicted = as.integer(score > 0.5)) |>
    select(gvkey, fyear, score, predicted)
}
```

We can call `logit_predict()` for each fold, combine the results into a single data frame, and then join with data on `misstate` from `data_train`.

```
logit_fit <-
  folds |>
  map(logit_predict) |>
  list_rbind() |>
  inner_join(data_train, by = c("gvkey", "fyear")) |>
  select(gvkey, fyear, predicted, score, misstate)
```

The predicted value for each observation is based on a model that was trained using data excluding the firm for which we are predicting.

---

[22]Chapter 19 of Wickham (2019) has more details on this "unquote" operator `!!`.

```
head(logit_fit)
```

```
# A tibble: 6 x 5
  gvkey  fyear predicted    score misstate
  <chr>  <dbl>     <int>    <dbl> <lgl>
1 001253  1994         0 0.000270 TRUE
2 001253  1995         0 0.000194 TRUE
3 008062  1995         0 0.00331  TRUE
4 025475  1993         0 0.00313  TRUE
5 009538  1995         0 0.00824  TRUE
6 009538  1996         0 0.00877  TRUE
```

Again we can produce the confusion matrix using the predictions from cross-validation. Now we see 0 true positive cases, 66,956 true negative cases, 209 false negative cases, and 9 false positive cases.

```
table(predicted = logit_fit$predicted, response = logit_fit$misstate)
```

```
         response
predicted FALSE  TRUE
        0 66956   209
        1     9     0
```

As poor as the within-sample classification performance was, it seems it was actually optimistic relative to performance assessed using cross-validation.

26.3.2 Classification trees

The second fundamental building block that we consider here is the classification tree. According to Kuhn and Johnson (2013, p. 173), "tree-based models consist of one or more nested `if-then` statements for the predictors that partition the data. Within these partitions, a model is used to predict the outcome." The models are often called simply **trees** for reasons that are clear from their typical graphical representation (see Figure 26.2). Other terminology follows from the tree metaphor: the splits give rise to branches and a terminal node of a tree is called a **leaf**.[23]

Trees can be categorized as either regression trees or as classification trees. A typical output for a regression tree is a predicted numerical value of an outcome given the covariates. In contrast, a typical output for a classification tree is a predicted class value (e.g., "hot dog" or "not hot dog"). Given our focus on binary classification problems, we focus on classification trees.

There are many approaches for building trees. One of the oldest and most popular techniques is the classification and regression tree (CART) of Breiman et al. (1984). The CART algorithm is a form of **recursive partitioning**, where the partitioning refers to the division of the sample into disjoint sets and recursive refers to the fact that the partitioning can be performed recursively (i.e., each set formed by partition can be partitioned in turn).

Algorithms for fitting CART models are provided by the `rpart()` function offered by the `rpart` package. We now fit a model (`fm2`) using recursive partitioning. We set two control parameters: `cp` and `minbucket`.[24]

---

[23]The metaphor is not perfect in that leaves are generally not found at the bottom of real trees.

[24]The meaning of these parameters is deferred to the discussion questions below.

```
fm2 <- rpart(formula, data = data_train,
             control = rpart.control(cp = 0.001, minbucket = 5),
             method = "class")
```

While the logistic regression model stored in **fm1** is described by coefficients on each of the 28 features, the tree stored in **fm2** can be represented by the graph depicted in Figure 26.2, which we create using the **rpart.plot()** function from the **rpart.plot** package.[25]

```
rpart.plot(fm2, extra = 1)
```

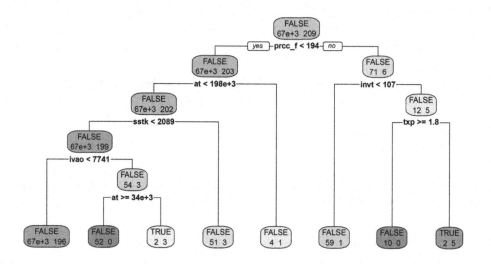

FIGURE 26.2
Fitted tree produced by recursive partitioning.

Again we can assess within-sample performance of **fm2** as we did with **fm1**, albeit with slightly different commands to extract predicted probabilities and classes.

```
within_sample2 <-
  data_train |>
  mutate(prob = predict(fm2, newdata = pick(everything()))[, 2],
         predicted = prob > 0.5)

table(predicted = within_sample2$predicted,
      response = within_sample2$misstate)
```

```
          response
predicted FALSE   TRUE
    FALSE 66961    201
    TRUE      4      8
```

Here we see 8 true positive cases, 66,961 true negative cases, 201 false negative cases, and 4 false positive cases.

---

[25]Each node contains the predicted class for observations falling in that node, as well as the number of observations of each class.

26.4 Performance metrics

While it is intuitively clear that the performance of our models is rather unimpressive, we need ways to measure model performance more formally and precisely. A number of evaluation metrics are available to measure the performance of prediction models. In this section, we discuss some of these metrics with a focus on two-class classification problems.[26]

26.4.1 Accuracy

One standard classification performance metric is **accuracy**, defined as $\frac{TP+TN}{TP+FN+FP+TN}$, where TP (true positive) is (in our current context) the number of fraudulent firm-years that are correctly classified as fraud; FN (false negative) is the number of fraudulent firm-years that are misclassified as non-fraud; TN (true negative) is the number of non-fraudulent firm-years that are correctly classified as non-fraud; and FP (false positive) is the number of non-fraudulent firm-years that are misclassified as fraud.

Unfortunately, accuracy is a not a very useful metric in this setting because of the severe class imbalance. With fewer than 1% of firms in a typical year being affected by accounting fraud, simply classifying all firm-years as non-fraud would yield accuracy of more than 99%.

Two other statistics that are often considered are **sensitivity** (also known as the true positive rate, or TPR) and **specificity** (also known as the **true negative rate** or TNR)

$$Sensitivity = TPR = \frac{TP}{TP+FN}$$

$$Specificity = TNR = \frac{TN}{TN+FP}$$

Often we focus on the complement to specificity, which is know as the **false positive rate** (FPR) and calculated as follows:

$$1 - Specificity = FPR = 1 - \frac{TN}{TN+FP} = \frac{FP}{TN+FP}$$

26.4.2 AUC

There is an inherent trade-off between sensitivity and specificity. A classification model will typically return something like an estimated probability that an observation is a positive case. We can choose a cut-off probability and classify all observations with an estimated probability at or above the cut-off as a positive case, and classify all observations with estimated probabilities below the cut-off as negative cases. Starting at a cut-off of zero, every observation would be classified as a positive and the TPR would be 1, but the FPR would also be 1. As the cut-off increases, both the TPR and the FPR will decrease.

If we collect the pairs of values of TPR and FPR as we increase the cut-off, we can plot the TPR against the FPR to form what is called the **response operator curve** (or **ROC**).

In terms of the ROC, the worst-case model just randomly assigns probabilities to each observation. So with a cut-off of 0.25, we'd classify 75% of observations as positive and 25%

---

[26]See Chapters 5 and 11 of Kuhn and Johnson (2013) for more details on performance metrics for regression problems and classification problems, respectively.

as negative. In expectation, we'd have 75% of the positive cases classified as positive, and 75% of the negative cases classified as positive, so the *TPR* and *FPR* values would be equal in expectation. This would yield an ROC close to the 45-degree line.

The best-case model would be a perfect classifier that yielded zero *FNR* and 100% *TPR*. This would yield an ROC that rose vertically along the y-axis and horizontally at $y = 100\%$.

This suggests the area under the ROC (or **AUC**) as a measure of the quality of a model, with values of 0.5 for the random-guess model and 1 for the perfect classifier. Below, we follow Bao et al. (2020) (and many others) in using AUC as the performance measure we use to evaluate models.

26.4.3 NDCG@k

Bao et al. (2020) also consider an alternative performance evaluation metric called the normalized discounted cumulative gain at position k (NDCG@k).

The NDCG@k metric can be motivated using the idea of budget constraints. Suppose that a regulator (or other party interested in unearthing frauds) has a limited budget and can only investigate k frauds in a given year. Denoting the number of actual frauds in a year as j, if $j \geq k$, then the best that the regulator can do is to investigate k cases, all of which are frauds. If $j < k$, then the regulator can do no better than investigating k cases with all j frauds being among those investigated.

The assumption is that the regulator investigates the k cases with the highest estimated probability of being frauds (scores). This suggests a performance measure that scores a model higher the greater the number of true frauds ranked in the top k scores.

NDCG@k is such a measure. NDCG@k also awards more "points" the higher in the list (based on scores) each true fraud is placed. This is implemented by weighting each correct guess by $1/\log_2(i+1)$. Formally, the **discounted cumulative gain** at position k is defined as follows:

$$\text{DCG@k} = \sum_{i=1}^{k} rel_i / \log_2(i+1)$$

The measure NDCG@k is DCG@k normalized by the DCG@k value obtained when all the true instances of fraud are ranked at the top of the ranking list. Hence, the values of NDCG@k are bounded between 0 and 1, and a higher value represents better performance.

Bao et al. (2020) select k so that the number of firm-years in the ranking list represents 1% of all the firm-years in a test year. Bao et al. (2020) "select a cut-off of 1% because the average frequency of accounting fraud punished by the SEC's AAERs is typically less than 1% of all firms in a year."

Bao et al. (2020) report modified versions of sensitivity and specificity: Sensitivity@k = $\frac{TP@k}{TP@k+FN@k}$ and Precision@k = $\frac{TP@k}{TP@k+FP@k}$ where TP@k, FN@k and FP@k are analogous to TP, FN and FP, but based on classifying as frauds only the $k\%$ of observations with the highest scores.

26.4.4 Exercises

1. One claimed benefit of classification trees is their ease of interpretation. Can you provide an intuitive explanation for the fitted tree in Figure 26.2? If so, outline your explanation. If not, what challenges prevent you from doing so?

2. Use 5-fold cross-validation to assess the performance of the classification tree approach using the parameters above.

3. What do the parameters `cp` and `minbucket` represent?

4. How does the classification tree change if you increase the parameter `cp`?

5. Use 5-fold cross-validation and AUC to assess the performance of the classification tree approach with three different parameters for `minbucket` (the one used above and two others).

6. Calculate AUC for the data stored in `within_sample2` above (this is for the tree stored in `fm2` and shown in Figure 26.2).

26.5   Penalized models

Having a number of ways to measure model performance alone does not address the basic problem of poor model performance using the two fundamental approaches above. In this section and the next, we describe two strategies that can be used to improve model performance.

The first approach uses **penalized models**. Perhaps the easiest way to understand penalized models is to recognize that many common modelling approaches seek to minimize the sum of squared errors

$$\text{SSE} = \sum_{i=1}^{n} (y_i - \hat{y}_i)^2.$$

With penalized models, the objective is modified by adding a penalty term. For example, in **ridge regression** the following objective function is minimized.[27]

$$\text{SSE}_{L_2} = \sum_{i=1}^{n} (y_i - \hat{y}_i)^2 + \lambda \sum_{j=1}^{P} \beta_j^2.$$

An alternative approach is **least absolute shrinkage and selection operator** (or **lasso**) model, which uses the following objective function:

$$\text{SSE}_{L_1} = \sum_{i=1}^{n} (y_i - \hat{y}_i)^2 + \lambda \sum_{j=1}^{P} |\beta_j|.$$

While both ridge regression and lasso will **shrink** coefficients towards zero (hence the alternative term **shrinkage methods**), some coefficients will be set to zero by lasso for some values of λ. In effect, lasso performs both shrinkage and feature selection.

A third shrinkage method worth mentioning is **elastic net** (Kuhn and Johnson, 2013, p. 126), which includes the penalties from both ridge regression and lasso

$$\text{SSE}_{\text{ENet}} = \sum_{i=1}^{n} (y_i - \hat{y}_i)^2 + \lambda_1 \sum_{j=1}^{P} \beta_j^2 + \lambda_2 \sum_{j=1}^{P} |\beta_j|.$$

---

[27]See Kuhn and Johnson (2013, p. 123) for more discussion of penalized models.

One aspect of these penalized objective functions that we have not discussed is λ. Implicitly, $\lambda > 0$ (if $\lambda = 0$, then the penalty term is zero), but we have said nothing about how this parameter is selected.[28] In fact, λ is a **meta-parameter** of the kind alluded to above, as it is not estimated directly as part of the estimation process.

A common approach to setting meta-parameters like λ uses cross-validation. With cross-validation, a number of candidate values of λ are considered. For each candidate value and each cross-validation fold, the model is estimated omitting the data in the fold. The performance of the model is evaluated using predicted values for the cross-validation folds (e.g., the average mean-squared error, or MSE, across all folds). Finally, the value of λ that maximizes performance (e.g., minimizes MSE) is chosen as the value to be used when applying the model to new data (e.g., test data when applicable).

To illustrate, we use the lasso analogue to the logit regression we estimated above.[29] For this purpose, we use the `glmnet` library and specify `alpha = 1` to focus on the lasso model.

The `glmnet` library contains built-in functionality for cross-validation. We simply need to specify the performance metric to use in evaluating models (here we set `type.measure = "auc"` to use AUC) and—because we constructed our folds to put all observations on a given firm in the same fold—the `foldid` for each observation

```
dft <-
  data_train |>
  inner_join(sample_splits, by = "gvkey")

fm_lasso_cv <-
  cv.glmnet(x = as.matrix(dft[X_vars]),
            y = dft[[y_var]],
            family = "binomial",
            alpha = 1,
            type.measure = "auc",
            foldid = dft[["fold"]],
            keep = TRUE)
```

The `glmnet()` function will automatically consider a range of λ values. Note that the scale of the features matters with penalized regression, as the penalties increase with the magnitude of the coefficients, which in turn depends on the scale of the variables. For example, a logit coefficient on `at` expressed in millions of dollars of 0.123 is equivalent to a coefficient of 123 on `at` expressed in thousands of dollars. Thus `glmnet()` will by default scale the features (see `? glmnet` for additional details).

```
tibble(lambda = fm_lasso_cv$lambda, auc = fm_lasso_cv$cvm) |>
  ggplot(aes(x = lambda, y = auc)) +
  geom_line()
```

In Figure 26.3, we see that AUC is maximized with $\lambda \approx 0.00032$.

Predicted values for each value of λ considered are stored in `fit.preval` of `fm_lasso_cv` as log-odds ratios. We can convert these to probabilities using the function $f(x) =$

---

[28] If $\lambda < 0$, the model would not be penalizing models, but "rewarding" them.

[29] The knowledgeable reader might note that logit is estimated by maximum likelihood estimation, but the basic idea of penalization is easily extended to such approaches, as discussed by Kuhn and Johnson (2013, p. 303).

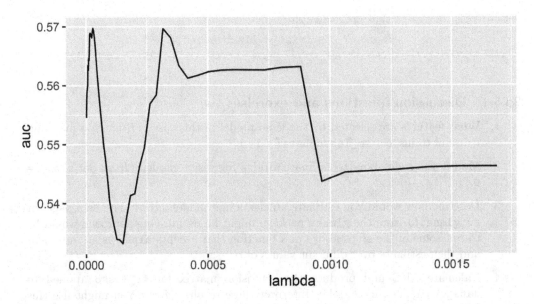

FIGURE 26.3
AUC for lasso for different λ values

$\frac{\exp(x)}{1+\exp(x)}$. Here we select the value of λ that maximizes AUC (this is stored in `fm_lasso_cv$lambda.min`).[30]

```
idmin <- match(fm_lasso_cv$lambda.min, fm_lasso_cv$lambda)

fit_lasso_cv <-
  dft |>
  select(misstate) |>
  mutate(logodds = fm_lasso_cv$fit.preval[, idmin],
         prob = exp(logodds) / (1 + exp(logodds)),
         predicted = logodds > 0)
```

From this we can calculate the confusion matrix based on the models fitted using cross-validation.

```
table(predicted = fit_lasso_cv$predicted,
      response = fit_lasso_cv$misstate)
```

```
         response
predicted FALSE  TRUE
    FALSE 66960   209
    TRUE      5     0
```

We can produce the fitted model based on our chosen λ by estimating a single model for the entire training data set. We store this model in `fm_lasso`.

```
fm_lasso <- glmnet(x = as.matrix(data_train[X_vars]),
                   y = data_train[[y_var]],
```

---

[30]A popular alternative is to use the largest value of λ that yields AUC within one standard error of the maximum AUC. This value is stored in `fm_lasso_cv$lambda.1se`.

```
              family = "binomial",
              lambda = fm_lasso_cv$lambda.min,
              alpha = 1)
```

26.5.1 Discussion questions and exercises

1. What features are selected by the lasso model in this case? (*Hint*: You may find it useful to inspect `fm_lasso$beta[, 1]`.)

2. Modify the code above to estimate a ridge regression model. (*Hint*: Set `alpha = 0`.)

3. Describe how would you estimate an elastic net model. (*Hint*: The help page for `cv.glmnet()` from the `glmnet` package might be useful. One approach would be to put some of the steps above in a function that accepts `alpha` as an argument and then estimate over different values of α.)

4. Calculate AUC and produce the confusion matrix for `fm_lasso` applied to `data_train` (i.e., in sample). Interpret these results. (*Hint*: You might use this code: `predict(fm_lasso, newx = as.matrix(data_train[X_vars]))` and the `auc` function supplied with the `farr` package.)

26.6 Ensemble methods

A second approach to addressing the weaknesses of building models directly from the building blocks uses **ensemble methods**, which are methods that combine the predictions from many models to generate a single prediction for each observation. The ensemble methods we cover here are all based on trees.

One of the earliest ensemble methods is **bagging**, which is short for *bootstrap aggregation*. With bagging a value M is selected and, for each $i \in (1, \dots, M)$, a bootstrap sample of the original data is taken and a model is fit. The prediction for any observation is an average of the predictions of the M models.[31]

One benefit of bagging is that for any given observation, several of the M models will not have used that observation in the fitting process, so the predictions from these models are, in a sense, out of sample for that observation. Such observations are termed **out-of-bag** and provide a natural measure of the predictive performance much like cross-validation measures.

One additional aspect of bagging is that there are no dependencies between the fitted models. So one could in principle divide the estimation of the M models between M computers (or cores) and (assuming sufficient computing resources) generate a prediction almost as quickly as one could from a single model.

One refinement of bagging is the **random forest**, which not only uses bootstrap samples of the observations, but also randomly draws a subset of $k < P$ features in fitting each model.

---

[31]What this averaging means is straightforward when the models are regression models. With classification models, a common approach is to take the classification that has the most "votes"; if more models predict "hot dog" than predict "not hot dog" for an observation, then "hot dog" would be the prediction for the ensemble.

By randomly selecting features in addition to randomly selecting observations, random forests generate trees with lower correlation than bagging does. This can reduce the risk that the model overfits the sample using certain features and also produces more robust models when features are highly correlated.

The final set of ensemble methods we discuss are known as **boosting** methods. The basic idea of boosting is that each observation has a weight that varies from one iteration to the next: observations that are incorrectly classified in iteration k receive more weight in iteration $k + 1$. In this way, the observations that are difficult to classify can receive increasingly larger weight until the algorithm is able to identify a model that classifies them correctly. Below we describe two boosting algorithms in more detail: AdaBoost—which was "the first practical implementation of boosting theory" (Kuhn and Johnson, 2013, p. 389)—and RUSBoost—which was introduced to accounting research by Bao et al. (2020).[32] Because the outcome of iteration k affects the model fit in iteration $k + 1$, boosting is not as easily implemented in parallel as bagging or random forests.

26.6.1   AdaBoost

Boosting is perhaps best understood in the context of an actual algorithm implementing it. Hastie et al. (2013, p. 339) outline the AdaBoost algorithm as follows:

1. Initialize the observation weights $w_i = \frac{1}{N}$, $i = 1, 2, ..., N$.
2. For $m = 1 \to M$:
    (a) Fit a classifier $G_m(x)$ to the training data using weights w_i.
    (b) Compute

$$\text{err}_m = \frac{\sum_{i=1}^{N} w_i I(y_i \neq G_m(x_i))}{\sum_{i=1}^{N} w_i}$$

    (c) Compute $\alpha_m = \log((1 - \text{err}_m)/\text{err}_m)$.
    (d) Set $w_i \leftarrow w_i \exp\left[\alpha_m I(y_i \neq G_m(x_i))\right]$, $i = 1, 2, ..., N$.
3. Output $G(x) = \text{sign}\left[\sum_{m=1}^{M} \alpha_m G_m(x)\right]$

A couple of observations on this algorithm are in order. First, step 3 implicitly assumes that the classes are coded as either $+1$ or -1 so that the sign indicates the class. Second, we have not described what classifier is represented by $G_m(x)$.

We provide an implementation of AdaBoost in the `farr` package. Because the RUSBoost algorithm we discuss below is a variation on AdaBoost, we use one function (`rusboost()`) to cover both AdaBoost (`rus = FALSE`) and RUSBoost (`rus = TRUE`). We use `rpart()` discussed above as the basic building block for `rusboost()` (i.e., $G_m(x)$). This means any arguments that can be passed to `rpart()` via the `control` argument can be used with `rusboost()`.

In addition, a user of `rusboost()` can specify the number of iterations (M) using the `size` argument and a learning rate ($L \in (0, 1]$) using the `learn_rate` argument. The learning rate modifies step 2(c) of the AdaBoost algorithm above to:

    (c) Compute $\alpha_m = L \log((1 - \text{err}_m)/\text{err}_m)$.

(Hence, the default `learn_rate = 1` implements the algorithm above.)

---

[32] For more on boosting, see pp. 389–392 of Kuhn and Johnson (2013).

The following convenience function `fit_rus_model()` allows us to set `maxdepth` and `minbucket` values passed to `rpart()` directly.

```
fit_rus_model <- function(df, size = 30, rus = TRUE, learn_rate = 1,
                          maxdepth = NULL, minbucket = NULL,
                          ir = 1) {
  if (!is.null(maxdepth)) control <- rpart.control(maxdepth = maxdepth)
  if (!is.null(minbucket)) control <- rpart.control(minbucket = minbucket)

  fm <- rusboost(formula, df, size = size, ir = ir, learn_rate = learn_rate,
             rus = rus, control = control)
  return(fm)
}
```

We next create `rus_predict()`, a convenience function for calculating fitted probabilities and predicted classes for data in a fold.

```
rus_predict <- function(fold, ...) {

  dft <-
    data_train |>
    inner_join(sample_splits, by = "gvkey") |>
    mutate(misstate = as.integer(misstate))

  fm <-
    dft |>
    filter(fold != !!fold) |>
    fit_rus_model(...)

  res <-
    dft |>
    filter(fold == !!fold) |>
    mutate(prob = predict(fm, pick(everything()), type = "prob"),
           predicted = predict(fm, pick(everything()), type = "class")) |>
    select(gvkey, fyear, prob, predicted)

  res
}
```

We next create `get_auc()`, which fits models by fold and returns the resulting cross-validated AUC. Note that `get_auc()` uses `future_map()` from the `furrr` library.

```
get_auc <- function(...) {
  set.seed(2021)
  rus_fit <-
    folds |>
    future_map(rus_predict,
               .options = furrr_options(seed = 2021),
               ...) |>
    list_rbind() |>
    inner_join(data_train, by = c("gvkey", "fyear")) |>
    select(gvkey, fyear, predicted, prob, misstate)
```

TABLE 26.1

AdaBoost cross-validation results for different meta-parameters

size	learn_rate	maxdepth	auc
30	0.100	5	0.646
30	0.100	7	0.644
50	0.100	7	0.642
50	0.100	5	0.624
50	0.100	3	0.624
50	1.000	5	0.611

```
  auc(score = rus_fit$prob, response = as.integer(rus_fit$misstate))
}
```

The meta-parameters that we evaluate are `size` (M, the number of iterations), `learn_rate` (L, the learning rate), and `maxdepth` (the maximum depth of trees).

```
params <- expand_grid(rus = FALSE,
                      size = c(30, 50),
                      learn_rate = c(0.1, 0.5, 1),
                      maxdepth = c(3, 5, 7))
```

We can now use our `get_auc()` function to evaluate the model performance for each of the possible parameter values in `params`.

```
plan(multisession)

adaboost_results <-
  params |>
  mutate(auc = pmap_dbl(params, get_auc)) |>
  select(-rus)
```

Table 26.1 provides the first few rows of the results.

```
adaboost_results |>
  arrange(desc(auc)) |>
  head()
```

We store the row (set of parameter values) that maximizes AUC in `optim_ada`.

```
optim_ada <- adaboost_results |> filter(auc == max(auc))
```

26.6.2 RUSBoost

RUSBoost is a variant of AdaBoost that makes use of random undersampling (RUS) to address the problem of class imbalance learning (Seiffert et al., 2008). When there are very few positive cases relative to negative cases, RUSBoost allows a model to train using samples with substantially more positive cases than found in the broader population. Given the relatively small number of fraud firm-years, the setting of Bao et al. (2020) seems to be a good one for RUSBoost. Seiffert et al. (2008) lay out the RUSBoost algorithm as follows:[33]

---

[33]We modify the algorithm slightly to include a learning-rate parameter, L in step 2(d).

1. Initialize the observation weights $w_i = \frac{1}{N}$, $i = 1, 2, ..., N$.
2. For $m = 1 \rightarrow M$:
    (a) Create temporary training data set S'_t with distribution D'_t using random under-sampling.
    (b) Fit a classifier $G_m(x)$ to the training data using weights w_i.
    (c) Compute

    $$\text{err}_m = \frac{\sum_{i=1}^{N} w_i I(y_i \neq G_m(x_i))}{\sum_{i=1}^{N} w_i}$$

    (d) Compute $\alpha_m = L \log((1 - \text{err}_m)/\text{err}_m)$.
    (e) Set $w_i \leftarrow w_i \exp\left[\alpha_m I(y_i \neq G_m(x_i))\right]$, $i = 1, 2, ..., N$.
3. Output $G(x) = \text{sign}\left[\sum_{m=1}^{M} \alpha_m G_m(x)\right]$.

In each iteration, the RUS algorithm samples from the training data differently for fraudulent and non-fraudulent firm-years. RUSBoost requires setting the ratio between the number of undersampled majority class (i.e., non-fraud) observations and the number of minority class (i.e., fraud) observations. The `rusboost()` function allows the user to specify this ratio using the `ir` argument. Bao et al. (2020) estimate RUSBoost models using a 1:1 ratio (i.e., `ir = 1`), which means each sample they used in estimation contained the same number of fraud and non-fraudulent observations.

Fraudulent firm-years—which are much rarer—are randomly sampled with replacement up to their number in the training sample (e.g., if there are 25 fraudulent observations in the training data, 25 such observations will be in the training sample used). Non-fraudulent firm-years—which are more numerous—are randomly sampled without replacement up to a number that is typically expressed as a multiple of the number of fraudulent observations in the training sample (e.g., if there are 25 fraudulent observations in the training data and `ir = 1`, 25 such observations will be in the training sample used).

In addition to `ir = 1`, we consider `ir = 2` (the latter selecting two non-fraud observations for each fraud observation).

We use a different set of meta-parameter values for `size`, `learn_rate`, and `maxdepth` for RUSBoost than we did for AdaBoost, as the relation between meta-parameters and performance differs.[34]

```
params_rus <- expand_grid(rus = TRUE,
                          size = c(30, 100, 200, 300),
                          learn_rate = c(0.01, 0.1, 0.5, 1),
                          maxdepth = c(2, 3, 5, 10, 15),
                          ir = c(1, 2))
```

Again, we can now use the `get_auc()` function to evaluate the model performance for each of the possible parameter values (now in `params_rus`).

```
plan(multisession)

rusboost_results <-
  params_rus |>
  mutate(auc = pmap_dbl(params_rus, get_auc))
```

---

[34]In writing this chapter, we actually considered a much wider range of meta-parameter values, but we limit the grid of values here because of computational cost.

TABLE 26.2
RUSBoost cross-validation results for different
meta-parameters

rus	size	learn_rate	maxdepth	ir	auc
True	300	0.010	15	2	0.694
True	200	0.010	15	2	0.693
True	200	0.010	5	2	0.691
True	200	0.010	10	2	0.691
True	300	0.010	10	2	0.690
True	100	0.010	15	2	0.690

Table 26.2 provides the first few rows of the results.

```
rusboost_results |>
  arrange(desc(auc)) |>
  head()
```

We store the row that maximizes AUC in `optim_rus`.

```
optim_rus <- rusboost_results |> filter(auc == max(auc))
```

Compared with AdaBoost, we see that RUSBoost performance is maximized with larger
models (`size` is 300 rather than 30) and deeper trees (`maxdepth` is 15 rather than 5). We
see that both models use a `learn_rate` less than 1.

Before testing our models, it is noteworthy that the AUCs for both AdaBoost and RUSBoost
are appreciably higher than those for the logit model, the single tree, and the penalized logit
models from above.

26.7   Testing models

We now consider the out-of-sample performance of our models. Given the apparent su-
periority of the boosted ensemble approaches (AdaBoost and RUSBoost) over the other
approaches, we focus on those two approaches here.

We follow Bao et al. (2020) in a number of respects. First, we use fiscal years 2003 through
2008 as our test years. Second, we fit a separate model for each test year using data on
AAERs that had been released on the posited test date for each test year.

Third, we exclude the year prior to the test year from the model-fitting data set (we do
this by using `filter(fyear <= test_year - gap)` where `gap = 2`). But note that the
rationale provided by Bao et al. (2020) for omitting this year differs from ours: Bao et
al. (2020) wanted some assurance that the AAERs were observable when building the test
model; in contrast, we simply want to reduce measurement error in the `misstate` variable
when using too-recent a sample.[35]

Fourth, rather than using `gap` to get assurance that the AAERs were observable when
building the test model, we use data on AAER dates to ensure that the AAERs we use

---

[35]As discussed above many of the apparent non-fraud observations will later be revealed to be frauds.

had been revealed by our test dates. For each `test_year`, we assume that we are fitting a
model on 30 September in the year after `test_year`, so as to allow at least four months
for financial statement data to be disclosed (fiscal years for, say, 2001 can end as late as 31
May 2002, so 30 September gives four months for these).

```
test_years <- 2003:2008

test_dates <-
  tibble(test_year = test_years) |>
  mutate(test_date = as.Date(paste0(test_year + 1, "-09-30")))
```

We now create `train_model()`, which allows us to pass a set of features, specify the
`test_year` and the parameters to be used in fitting the model (e.g., `learn_rate` and `ir`)
and which returns a fitted model.

```
train_model <- function(features, test_year, gap = 2,
                        size, rus = TRUE, learn_rate, ir, control) {

  test_date <-
    test_dates |>
    filter(test_year == !!test_year) |>
    select(test_date) |>
    pull()

  aaers <- get_aaers(test_date = test_date)

  data_train <-
    aaers |>
    right_join(features, by = c("gvkey", "fyear")) |>
    filter(fyear <= test_year - gap) |>
    mutate(misstate = as.integer(coalesce(misstate, FALSE)))

  fm <- rusboost(formula, data_train, size = size, learn_rate = learn_rate,
                rus = rus, ir = ir, control = control)
  return(fm)
}
```

Note that in training our final model, we can use all available data in the training data
set. For a test year of 2003, our training data set is the same as that used for selecting
the meta-parameters above, but there we generally only used a subset of the data in fitting
models (i.e., in cross-validation).

We next create `test_model()`, which allows us to pass a set of features (`features`), a fitted
model (`fm`) and to specify the test year (`test_year`) and which returns the predicted values
and estimated probabilities for the test observations.

```
test_model <- function(features, fm, test_year) {

  target <- as.character(stats::as.formula(fm[["formula"]])[[2]])

  df <-
    features |>
    left_join(aaer_long, by = c("gvkey", "fyear")) |>
```

```
    select(-max_year, -min_year) |>
    mutate(misstate = as.integer(!is.na(p_aaer)))

  data_test <-
    df |>
    filter(fyear == test_year)

  return(list(score = predict(fm, data_test, type = "prob"),
              predicted = as.integer(predict(fm, data_test,
                                          type = "class") == 1),
              response = as.integer(data_test[[target]] == 1)))
}
```

We next make `fit_test()`, a convenience function that calls `train_model()` then `test_model()` in turn.[36]

```
fit_test <- function(x, features, gap = 2,
                     size, rus = TRUE, learn_rate, ir = 1, control) {
  set.seed(2021)
  fm <- train_model(features = features, test_year = x,
                    size = size, rus = rus, ir = ir,
                    learn_rate = learn_rate, control = control, gap = gap)
  test_model(features = features, fm = fm, test_year = x)
}
```

We first test AdaBoost using the meta-parameter values stored in `optim_ada` and store the results in `results_ada`. We have included the values generated from code in Section 26.6 here so that you do not need to run the code from that section.

```
optim_ada <- tibble(size = 30, learn_rate = 0.1, maxdepth = 5)
```

Again, we use `future_map()` from the `furrr` library here.

```
plan(multisession)

results_ada <- future_map(test_years, fit_test,
                          features = features,
                          size = optim_ada$size,
                          rus = FALSE,
                          learn_rate = optim_ada$learn_rate,
                          control =
                            rpart.control(maxdepth = optim_ada$maxdepth),
                          .options = furrr_options(seed = 2021))
```

We then test RUSBoost using the meta-parameter values stored in `optim_rus` and store the results in `results_rus`. We have included the values generated from code in Section 26.6 here so that you do not need to run the code from that section.

```
optim_rus <- tibble(size = 300, learn_rate = 0.01, maxdepth = 15, ir = 2)
```

---

[36]It also sets the seed of the random-number generator, but this likely has little impact on the results apart from replicability.

```
plan(multisession)

results_rus <- future_map(test_years, fit_test,
                          features = features,
                          size = optim_rus$size,
                          rus = TRUE,
                          ir = optim_rus$ir,
                          learn_rate = optim_rus$learn_rate,
                          control = rpart.control(
                            maxdepth = optim_rus$maxdepth),
                          .options = furrr_options(seed = 2021))
```

To make it easier to tabulate results, we create **get_stats()**, which collects performance statistics for a set of results that can be used to tabulate the results.

```
get_stats <- function(results) {
  response <- results$response
  score <- results$score

  ndcg_k <- function(k) {
    c(ndcg(score, response, k = k))
  }

  res <- list(auc(score, response),
              unlist(map(seq(0.01, 0.03, 0.01), ndcg_k)))
  unlist(res)
}
```

Bao et al. (2020) appear to evaluate test performance using the means across test years, so we make **add_means()** to add a row of means to the bottom of a data frame.

```
add_means <- function(df) {
  bind_rows(df, summarise(df, across(everything(), mean)))
}
```

We use **add_means()** function in **table_results()**, which extracts performance statistics from a set of results, adds means, then formats the resulting data frame for display.

```
table_results <- function(results, ...) {
  results |>
    map(get_stats) |>
    bind_rows() |>
    add_means() |>
    mutate(test_year = c(test_years, "Mean")) |>
    select(test_year, everything())
}
```

We can now simply call **table_results()** for **results_ada** and **results_rus** in turn, with results being shown in Tables 26.3 and 26.4.

```
table_results(results_ada)
```

```
table_results(results_rus)
```

TABLE 26.3
Results: AdaBoost

test_year	auc	ndcg_0.01	ndcg_0.02	ndcg_0.03
2003	0.666	0.027	0.063	0.063
2004	0.598	0.021	0.021	0.031
2005	0.678	0.000	0.000	0.023
2006	0.637	0.038	0.038	0.038
2007	0.570	0.000	0.000	0.000
2008	0.592	0.030	0.030	0.030
Mean	0.623	0.019	0.025	0.031

TABLE 26.4
Results: RUSBoost

test_year	auc	ndcg_0.01	ndcg_0.02	ndcg_0.03
2003	0.729	0.054	0.060	0.085
2004	0.728	0.029	0.039	0.059
2005	0.716	0.020	0.032	0.032
2006	0.728	0.000	0.015	0.015
2007	0.647	0.000	0.000	0.000
2008	0.639	0.023	0.023	0.023
Mean	0.698	0.021	0.028	0.036

Comparing Table 26.3 with Table 26.4, it does seem that RUSBoost is superior to AdaBoost in the current setting using the features drawn from Bao et al. (2020). Note that Bao et al. (2020) do not separately consider AdaBoost as an alternative to RUSBoost. Bao et al. (2020) argue that RUSBoost is superior to the all-features logit model, which seems consistent with our evidence above, though we did not consider the logit model beyond the cross-validation analysis given its poor performance (but see exercises below).

26.8 Further reading

There are many excellent books for learning more about statistical learning and prediction. Kuhn and Johnson (2013) provide excellent coverage and intuitive explanations. The text by Hastie et al. (2013) is widely regarded as the standard text. James et al. (2022) provide a more accessible introduction.

26.9 Discussion questions and exercises

1. In asserting that RUSBoost is superior to AdaBoost, we did not evaluate the statistical significance of the difference between the AUCs for the two approaches. Describe how you might evaluate the statistical significance of this difference.

2. Compare the performance statistics from cross-validation and out-of-sample testing. What do these numbers tell you? Do these provide evidence on whether using (say) 2001 data to train a model that is evaluated on (say) 1993 frauds is problematic?

3. Bao et al. (2022) report an AUC in the test sample for an all-features logit model of 0.7228 (see Panel B of Table 1). What differences are there between our approach here and those in Bao et al. (2020) and Bao et al. (2022) that might account for the differences in AUC?

4. Bao et al. (2022) report an AUC in the test sample for an all-features logit model of 0.6842 (see Panel B of Table 1). Calculate an equivalent value for AUC from applying logistic regression to data from above. You may find it helpful to adapt the code above for RUSBoost in the following steps:

   a. Create a function `train_logit()` analogous to `train_model()` above, but using `glm()` in place of `rusboost()`.
   b. Create a function `test_logit()` analogous to `test_model()` above, but using calls to `predict()` like those used above for logistic regression.
   c. Create a function `fit_test_logit()` analogous to `fit_test()` above, but calling the functions you created above.
   d. Use `map()` or `future_map()` with `test_years` and `fit_test_logit()` and store the results in `results_logit`.
   e. Tabulate results using `table_results(results_logit)`. (Table 26.5 contains our results from following these steps.)

TABLE 26.5
Performance statistics for logit models

test_year	auc	ndcg_0.01	ndcg_0.02	ndcg_0.03
2003	0.668	0.000	0.029	0.046
2004	0.673	0.000	0.000	0.020
2005	0.654	0.000	0.013	0.024
2006	0.646	0.000	0.016	0.016
2007	0.598	0.000	0.000	0.000
2008	0.635	0.060	0.060	0.077
Mean	0.646	0.010	0.020	0.031

5. Do the AUC results obtained from logit (either from your answer to the previous question or Table 26.5) surprise you given the cross-validation and in-sample AUCs calculated (in exercises) above? How does the average AUC compare with the 0.6842 of Bao et al. (2022)? What might account for the difference?

6. Provide an intuitive explanation of the out-of-sample (test) NDCG@k results for the RUSBoost model. (Imagine you are trying to explain these to a user of prediction models like those discussed earlier in this chapter.) Do you believe that these results are strong, weak, or somewhere in between?

7. We followed Bao et al. (2020) in using `gap = 2` (i.e., training data for fiscal years up to two years before the test year), but discussed the trade-offs in using a shorter gap period (more data, but more measurement error). How would you evaluate alternative gap periods?

8. Which results are emphasized in Bao et al. (2020)? What are the equivalent values in Bao et al. (2022)? Which results are emphasized in Bao et al. (2022)?

9. Which audience (e.g., practitioners, researchers) would be most interested in the results of Bao et al. (2020) and Bao et al. (2022)? What limitations of Bao et al. (2020) and Bao et al. (2022) might affect the interest of this audience?

10. Provide reasons why the tree-based models outperform the logit-based models in this setting? Do you think this might be related to the set of features considered? How might you alter the features if you wanted to improve the performance of the logit-based models?

Part V

Appendices

A

Linear algebra

Matrices come to us from a branch of mathematics known as linear algebra. A full course in linear algebra would provide a lot more detail than we can cover here. Our goal is merely to provide enough background to make sense of the material in the text that uses matrices. We focus on some basic details such as what matrices are, some of the operations we can perform with them, and some useful results.

In the canonical setting in econometrics, we have n observations on k variables. Each observation might be a person or a firm, or even a firm at a particular point in time. In earlier chapters of the book, we considered the possibility that the variables for observation i are related in the following way:

$$y_i = \beta_0 + x_{1i}\beta_1 + x_{2i}\beta_2 + \cdots + x_{ki}\beta_k + \epsilon_i$$

For example, y_i might represent the profitability of a firm in a given year and the various x variables are factors assumed to affect that profitability, such as capital stock and market concentration. The error term (ϵ_i) allows the equation to hold when the variables x_{1i} through x_{ki} do not determine the exact value of y_i. Given we have n observations, we actually have n equations.

$$y_1 = x_{11}\beta_1 + x_{21}\beta_2 + \cdots + x_{k1}\beta_k + \epsilon_1$$
$$y_2 = x_{12}\beta_1 + x_{22}\beta_2 + \cdots + x_{k2}\beta_k + \epsilon_2$$
$$\vdots = \quad \vdots \quad\quad \vdots \quad\quad \vdots \quad\quad \vdots \quad\quad \vdots$$
$$y_n = x_{ni}\beta_1 + x_{ni}\beta_2 + \cdots + x_{ni}\beta_k + \epsilon_n$$

As we shall see, matrices allow us to write this system of equations in a succinct fashion that allows manipulations to be represented concisely.

A.1   Vectors

For an observation, we might have data on sales, profit, R&D spending, and fixed assets. We can arrange these data as a **vector**: $y = $ (sales, profit, R&D, fixed assets). This y_i is an n-tuple (here $n = 4$), which is a finite ordered list of elements. A more generic representation of a y would be $y = (y_1, y_2, \dots, y_n)$.

A.1.1   Operations on vectors

Suppose we have two vectors $x = (x_1, x_2, \dots, x_n)$ and $y = (y_1, y_2, \dots, y_n)$.

Vectors of equal length can be added:

$$x + y = (x_1 + y_1, x_2 + y_2, \dots, x_n + y_n)$$

and subtracted:
$$x - y = (x_1 - y_1, x_2 - y_2, \dots, x_n - y_n)$$
Vectors can also be multiplied by real number:

$$\lambda y = (\lambda y_1, \lambda y_2, \dots, \lambda y_n)$$

Definition A.1 (Dot product). The *dot product* (or *scalar product*) of two n-vectors x and y is denoted as $x \cdot y$ and is defined as:

$$x \cdot y = x_1 y_1 + x_2 y_2 + \dots + x_n y_n = \sum_{i=1}^{n} x_i y_i.$$

A.2   Matrices

A **matrix** is a rectangular array of real numbers.

$$A = \begin{bmatrix} a_{11} & a_{12} & \cdots & a_{1k} \\ a_{21} & a_{22} & \cdots & a_{2k} \\ \cdot & \cdot & \cdots & \cdot \\ a_{m1} & a_{m2} & \cdots & a_{mk} \end{bmatrix}$$

Matrices are typically denoted with capital letters (e.g., A) and the generic element of a matrix is denoted as a_{ij}. We can also express a matrix in terms of its generic element and its dimensions as $\left[a_{ij} \right]_{m \times k}$.

Two important matrices are the **null matrix**, **0**, which contains only zeros, and the **identity matrix** of size n, I_n or simply I, which has diagonal elements equal to one ($i_{kk} = 1, \forall k$) and all other elements equal to zero ($i_{jk} = 0$ for $\forall j \neq k$). Provided that I and A are conformable for multiplication (e.g., they are both $n \times n$ square matrices), then $AI = A$ and $IA = A$, hence the term *identity* matrix (in some ways, I is the matrix equivalent of the number 1).

Each row or column of a matrix can be considered a vector, so that an $m \times k$ matrix can be viewed as m k-vectors (the rows) or k m-vectors (the columns).

A.2.1   Operations on matrices

Suppose we have two matrices $A = \left[a_{ij} \right]_{m \times k}$ and $B = \left[b_{ij} \right]_{m \times k}$, then we can add these matrices

$$A + B = \left[a_{ij} + b_{ij} \right]_{m \times k}$$

We can multiply a matrix by a real number λ

$$\lambda A = \left[\lambda a_{ij} \right]_{m \times k}$$

Matrix multiplication is defined for two matrices if the number of columns for the first is equal to the number of rows of the second.

Given matrices $A = \left[a_{ij}\right]_{m \times l}$ and $B = \left[b_{jk}\right]_{l \times n}$, the $m \times n$ matrix AB with typical element c_{ik} is defined as

$$AB = \left[c_{ik} := \sum_{j=1}^{l} a_{ij}b_{jk}\right]_{m \times n}$$

Alternatively $c_{ik} = a_i \cdot b_k$, where a_i is the i-th row of A and b_k is the k-th column of B. Not that multiplication of A and B requires that they be **conformable* for multiplication. In particular, the number of columns of A must equal the number of rows of B for AB to exist. If the number of rows of A does not equal the number of columns of B, then BA will not exist (let alone equal AB).

Definition A.2 (Transpose). The matrix $B = \left[b_{ij}\right]_{n \times m}$ is called the *transpose* of a matrix $A = \left[a_{ij}\right]_{m \times n}$ (and denoted A^{T}) if $b_{ij} = a_{ji}$ for all $i \in \{1, 2, ..., m\}$ and all $j \in \{1, 2, ..., n\}$.

Definition A.3 (Square matrix). A matrix that has the same number of rows and columns is called a *square matrix*.

Definition A.4 (Symmetric). A square matrix is *symmetric* if $a_{ij} = a_{ji}, \forall i, j$. Clearly if A is a symmetric matrix, then $A = A^{\mathsf{T}}$.

Definition A.5 (Matrix inverse). An $m \times m$ square matrix A has an **inverse**, if there exists a matrix denoted as A^{-1} such that $A^{-1}A = I_m$ and $AA^{-1} = I_m$, where I_m denotes the $m \times m$ identity matrix. A matrix that has an inverse is said to be *invertible* or *non-singular*.

Properties of inverses.

1. If an inverse of A exists, it is unique.
2. If $\alpha \neq 0$ and A is invertible, then $(\alpha A)^{-1} = 1/\alpha A^{-1}$.
3. If A and B are both invertible $m \times m$ matrices, then $(AB)^{-1} = B^{-1}A^{-1}$.

Here we show a couple of useful results about transposes. First, for two square, invertible matrices A and B, we have $(AB)^{\mathsf{T}} = B^{\mathsf{T}}A^{\mathsf{T}}$.

$$(AB)^{\mathsf{T}} = [ab_{ij}]^{\mathsf{T}}$$
$$= [ab_{ji}]$$
$$= \left[\sum_{k=1}^{n} a_{jk}b_{ki}\right]$$
$$= \left[\sum_{k=1}^{n} (B^{\mathsf{T}})_{ik}(A^{\mathsf{T}})_{kj}\right]$$
$$= B^{\mathsf{T}}A^{\mathsf{T}}$$

Second, that for a square, invertible matrix A, we have $(A^{\mathsf{T}})^{-1} = (A^{-1})^{\mathsf{T}}$:

$$AA^{-1} = I$$
$$(A^{-1})^{\mathsf{T}} A^{\mathsf{T}} = I$$
$$(A^{-1})^{\mathsf{T}} A^{\mathsf{T}} (A^{\mathsf{T}})^{-1} = (A^{\mathsf{T}})^{-1}$$
$$(A^{-1})^{\mathsf{T}} = (A^{\mathsf{T}})^{-1}$$

Definition A.6 (Diagonal matrix). A square matrix A is a *diagonal matrix* if $a_{ij} = 0, \forall i \neq j$. In words, all off-diagonal elements of a diagonal matrix are zero.

Definition A.7 (Linear independence). Let $\{x_1, x_2, \ldots, x_r\}$ be a set of $n \times 1$ vectors. We say that these vectors are *linearly independent* if and only if

$$\alpha_1 x_1 + \alpha_2 x_2 + \cdots + \alpha_r x_r = 0 \tag{A.1}$$

implies that $\alpha_1 = \alpha_2 = \cdots = \alpha_r = 0$. If Equation A.1 holds for a set of scalars that are not all zero, then $\{x_1, x_2, \ldots, x_r\}$ is *linearly dependent*.

Definition A.8 (Matrix rank). Let A be an $m \times k$ matrix. The *rank* of a matrix A is the maximum number of linearly independent columns of A. If A is $m \times k$ and the rank of A is k, then A has *full column rank*. If A is $m \times k$ and $m \geq k$, then its rank can be at most k.

Some properties of rank include:

1. The rank of A equals the rank of A^T.
2. If A is a $k \times k$ square matrix with rank k, then it is **non-singular**.

Definition A.9 (Idempotent). A matrix A is *idempotent* if it has the property that $AA = A$.

Definition A.10 (Projection matrix). Given a matrix X, the **projection matrix** for X is denoted as P_X and is defined as

$$P_X = X(X^\mathsf{T}X)^{-1}X^\mathsf{T}$$

The following shows that P_X is idempotent:

$$P_X P_X = X(X^\mathsf{T}X)^{-1}X^\mathsf{T}X(X^\mathsf{T}X)^{-1}X^\mathsf{T} = X(X^\mathsf{T}X)^{-1}X^\mathsf{T} = P_X$$

Note also that P_X is **symmetric**, which means that it and its transpose are equal, as the following demonstrates:

$$\begin{aligned}
P_X^\mathsf{T} &= \left(X(X^\mathsf{T}X)^{-1}X^\mathsf{T}\right)^\mathsf{T} \\
&= \left(X(X^\mathsf{T}X)^{-1}X^\mathsf{T}\right)^\mathsf{T} \\
&= X\left((X^\mathsf{T}X)^{-1}\right)^\mathsf{T} X^\mathsf{T} \\
&= X\left((X^\mathsf{T}X)^\mathsf{T}\right)^{-1} X^\mathsf{T} \\
&= X\left(X^\mathsf{T}X\right)^{-1} X^\mathsf{T} \\
&= P_X
\end{aligned}$$

In this analysis, we used two results about transposes discussed above.

A.3 The OLS estimator

The **classical linear regression model** assumes that the data-generating process has $y = X\beta + \epsilon$ where $\epsilon \sim IID(0, \sigma^2 I)$, where y and ϵ are n-vectors, X is an $n \times k$ matrix (including the constant term), β is a k-vector, and I is the $n \times n$ identity matrix.[1]

[1] Here $IID(0, \sigma^2 I)$ means independent and identically distributed with mean 0 and variance $\sigma^2 I$.

As discussed in Chapter 3, the ordinary least-squares (OLS) estimator is given by:

$$\hat{\beta} = \left(X^{\mathsf{T}}X\right)^{-1} X^{\mathsf{T}}y$$

Here we can see that we can only calculate $\hat{\beta}$ if $X^{\mathsf{T}}X$ is invertible, which requires that it be of rank k. This requires that no one column of X is a linear combination of the other columns of X.

Assuming $\mathbb{E}[\epsilon|X] = 0$, we can derive the following result:

$$\begin{aligned}
\mathbb{E}\left[\hat{\beta}\right] &= \mathbb{E}\left[\mathbb{E}\left[\hat{\beta}|X\right]\right] \\
&= \mathbb{E}\left[\left(X^{\mathsf{T}}X\right)^{-1} X^{\mathsf{T}}(X\beta + \epsilon)|X\right] \\
&= \mathbb{E}\left[\left(X^{\mathsf{T}}X\right)^{-1} X^{\mathsf{T}}X\beta|X\right] + \mathbb{E}\left[\left(X^{\mathsf{T}}X\right)^{-1} X^{\mathsf{T}}\epsilon|X\right] \\
&= \beta + \left(X^{\mathsf{T}}X\right)^{-1} X^{\mathsf{T}}\mathbb{E}\left[\epsilon|X\right] \\
&= \beta
\end{aligned}$$

This demonstrates that $\hat{\beta}$ is unbiased given these assumptions. But note that the assumption that $\mathbb{E}[\epsilon|X] = 0$ can be a strong one in some situations. For example, Davidson and MacKinnon point out that "in the context of time-series data, [this] assumption is a very strong one that we may often not feel comfortable making." As such, many textbook treatments replace $\mathbb{E}[\epsilon|X] = 0$ with weaker assumptions and focus on the asymptotic property of **consistency** instead of unbiasedness.

A.4 Further reading

This appendix barely scratches the surface of matrices and linear algebra. Many econometrics textbooks have introductory sketches of linear algebra that go beyond what we have provided here. Chapter 1 of Davidson and MacKinnon (2004) and Appendix D of Wooldridge (2000) cover the results provided here and more. Standard introductory texts for mathematical economics, such as Chiang (1984) and Simon and Blume (1994), provide introductions to linear algebra.

B

SQL primer

This brief appendix aims to serve two groups of users. The first group comprises those who have followed the material in the main part of the book and would like a quick introduction to SQL. The second group comprises those who know SQL (say, SAS's `PROC SQL`) would like a quick introduction to the `dplyr`-based approach to R that we use in this book.

B.1 What are SQL and `dplyr`?

SQL is a specialized language for manipulating and retrieving tabular data used by almost all modern database systems.

The R package `dplyr` is a core part of the Tidyverse and perhaps the package we use the most in this book. From the Tidyverse website:[1]

> `dplyr` is a grammar of data manipulation, providing a consistent set of **verbs** that help you solve the most common data manipulation challenges:
>
> - `mutate()` adds new variables that are functions of existing variables
> - `select()` picks variables based on their names
> - `filter()` picks cases based on their values
> - `summarize()` reduces multiple values down to a single summary
> - `arrange()` changes the ordering of the rows

Prior to the advent of the `dplyr` in 2014,[2] most users of R would have used base R[3] functions and operators, such as `subset()`, `$`, and `[`. However, `dplyr` provides a much more consistent framework for manipulating data that is easier to learn, especially for users familiar with SQL. This is unsurprising given that SQL provides something like a "grammar of data manipulation" of its own. In fact, each `dplyr` verb has an SQL equivalent keyword or concept, as seen in Table B.1, which provides a translation table of sorts between `dplyr` and SQL.

In this appendix, we use the following packages:

```
library(DBI)
library(dplyr)
```

As in earlier chapters, we set up a database connection that we can use within R. We also create remote data frames for each of the two tables from WRDS that we use in this appendix: `crsp.dsf` and `crsp.dsi`.

---

[1] https://dplyr.tidyverse.org
[2] https://blog.rstudio.com/2014/01/17/introducing-dplyr/
[3] https://dplyr.tidyverse.org/articles/base.html

DOI: 10.1201/9781003456230-B

TABLE B.1
SQL translations of key `dplyr` verbs

dplyr verb (R)	SQL equivalent
|>	FROM
select()	SELECT
filter()	WHERE
group_by()	GROUP BY
arrange()	ORDER BY
mutate()	used-defined columns
summarize()	used-defined aggregate columns

```
db <- dbConnect(RPostgres::Postgres())

crsp.dsf <- tbl(db, Id(table = "dsf", schema = "crsp"))
crsp.dsi <- tbl(db, Id(table = "dsi", schema = "crsp"))
```

B.2 SQL terms `SELECT` and `FROM`

Let's begin with a basic SQL query. (Note that, throughout this appendix, we only display the first five records when there are more than five records returned by a query. Note that a **record** is SQL terminology for what we might call a row in the context of an R data frame.)

```
SELECT date, vwretd, ewretd
FROM crsp.dsi
```

date	vwretd	ewretd
1925-12-31		
1926-01-02	0.005689	0.009516
1926-01-04	0.000706	0.005780
1926-01-05	−0.004821	−0.001927
1926-01-06	−0.000423	0.001182

The query above

- extracts the data in three columns (`date`, `vwretd`, `ewretd`) (the first line)
- from the table named `crsp.dsi` (the second line)

While the syntax differs, SQL's `SELECT` operates very much like `select()` from `dplyr`.

Translating this into `dplyr` code using the pipe operator (`|>`), it's easy to see that the order of presentation is one of the big differences between SQL and `dplyr`.

To use `dplyr` we first need to set up the table on which to operate; once we've done so we can see that SQL's `FROM` is implicit in the `|>` operator.

```
crsp.dsi |>
  select(date, vwretd, ewretd)
```

date	vwretd	ewretd
1925-12-31		
1926-01-02	0.005689	0.009516
1926-01-04	0.000706	0.005780
1926-01-05	−0.004821	−0.001927
1926-01-06	−0.000423	0.001182

B.3 SQL WHERE

The filter() verb from dplyr corresponds to WHERE in SQL. Note that WHERE goes *after* the FROM clause in SQL, though in practice the query optimizer[4] will execute the filter implied by the WHERE clause before executing other elements of a query (it would be wasteful to perform calculations on data that are going to be filtered out later on).

```
SELECT date, vwretd, ewretd
FROM crsp.dsi
WHERE date = '2015-01-02'
```

date	vwretd	ewretd
2015-01-02	−0.000242	0.001470

In dplyr, this query could be written as follows:

```
crsp.dsi |>
  select(date, vwretd, ewretd) |>
  filter(date == '2015-01-02')
```

date	vwretd	ewretd
2015-01-02	−0.000242	0.001470

But it could also be written with the filter coming first.

```
crsp.dsi |>
  filter(date == '2015-01-02') |>
  select(date, vwretd, ewretd)
```

date	vwretd	ewretd
2015-01-02	−0.000242	0.001470

---

[4]https://www.postgresql.org/docs/current/planner-optimizer.html

Note that each of these `dplyr` queries is implemented in the same way by the PostgreSQL query optimizer, as can be seen by examining the output from `explain()`.

```
crsp.dsi |>
  select(date, vwretd, ewretd) |>
  filter(date == '2015-01-02') |>
  explain()
```

```
<SQL>
SELECT "date", "vwretd", "ewretd"
FROM "crsp"."dsi"
WHERE ("date" = '2015-01-02')


<PLAN>
                                                          QUERY PLAN
1 Index Scan using dsi_date_idx on dsi  (cost=0.29..8.30 rows=1 width=20)
2                                    Index Cond: (date = '2015-01-02'::date)
```

```
crsp.dsi |>
  filter(date == '2015-01-02') |>
  select(date, vwretd, ewretd) |>
  explain()
```

```
<SQL>
SELECT "date", "vwretd", "ewretd"
FROM "crsp"."dsi"
WHERE ("date" = '2015-01-02')


<PLAN>
                                                          QUERY PLAN
1 Index Scan using dsi_date_idx on dsi  (cost=0.29..8.30 rows=1 width=20)
2                                    Index Cond: (date = '2015-01-02'::date)
```

By adding EXPLAIN to the SQL above, we can see also that the `dplyr` queries are viewed by the query optimizer as equivalent to that SQL.

```
EXPLAIN
SELECT date, vwretd, ewretd
FROM crsp.dsi
WHERE date = '2015-01-02'
```

```
                                                          QUERY PLAN
1 Index Scan using dsi_date_idx on dsi  (cost=0.29..8.30 rows=1 width=20)
2                                    Index Cond: (date = '2015-01-02'::date)
```

This illustrates a nice feature of SQL engines, namely that you can leave some of the details of the query to the software and focus on the higher-level requirements of your query (i.e., which observations to keep, etc.).

B.4 SQL ORDER BY

An R data frame can be considered a list of vectors, where each vector has a well-defined order. In contrast, SQL tables should be not considered to have any particular order when there is no ORDER BY clause.[5]

An ORDER BY clause is placed at the end of an SQL query (reflecting in this case the order of operations) followed by the columns that the query is to be sorted on.

```
SELECT date, vwretd, ewretd
FROM crsp.dsi
ORDER BY vwretd
```

date	vwretd	ewretd
1987-10-19	−0.171346	−0.103897
1929-10-29	−0.119539	−0.134856
2020-03-16	−0.118168	−0.107631
1929-10-28	−0.113022	−0.095202
1929-11-06	−0.097020	−0.083819

Thus, ORDER BY works very much like `arrange()`.

```
crsp.dsi |>
  select(date, vwretd, ewretd) |>
  arrange(vwretd)
```

date	vwretd	ewretd
1987-10-19	−0.171346	−0.103897
1929-10-29	−0.119539	−0.134856
2020-03-16	−0.118168	−0.107631
1929-10-28	−0.113022	−0.095202
1929-11-06	−0.097020	−0.083819

To reverse the order, use the DESC keyword after the relevant variable.

```
SELECT date, vwretd, ewretd
FROM crsp.dsi
ORDER BY vwretd DESC, date
```

date	vwretd	ewretd
1925-12-31		
1933-03-15	0.156838	0.203020
1929-10-30	0.122155	0.125714
2008-10-13	0.114918	0.107422
1931-10-06	0.111278	0.088742

---

[5]See https://stackoverflow.com/questions/20050341 on this point.

Thus, DESC is very much equivalent to desc() in dplyr apart from the minor syntactical difference that desc() is written as a function of the relevant variable.

```
crsp.dsi |>
  select(date, vwretd, ewretd) |>
  arrange(desc(vwretd), date)
```

date	vwretd	ewretd
1925-12-31		
1933-03-15	0.156838	0.203020
1929-10-30	0.122155	0.125714
2008-10-13	0.114918	0.107422
1931-10-06	0.111278	0.088742

It is important to note that some differences in the implementation details between data frames in R and tables in SQL are seen with arrange(). While a data frame in R is fundamentally a list of *ordered* vectors, SQL tables are best thought of as sets of rows without a natural order. As such, one needs to pay attention to when arrange() is implemented in a query that depends on it.

To illustrate this issue, suppose we get stock returns for Apple and order them from largest to smallest returns.[6]

```
apple <-
  crsp.dsf |>
  filter(permno == 14593L) |>
  select(permno, date, ret)

apple |>
  arrange(desc(ret))
```

permno	date	ret
14593	1980-12-12	
14593	1997-08-06	0.332278
14593	1998-01-02	0.238095
14593	1996-07-18	0.237037
14593	1998-01-06	0.192913

The following code illustrates that the order created above is lost when we merge with crsp.dsi. If you are using window queries with SQL data sources, use window_order() to order data within windows.

```
apple |>
  inner_join(crsp.dsi, by = "date") |>
  select(permno, date, ret, vwretd)
```

---

[6]Note that the NA value from Apple's first day of trading is viewed as being the largest value due to the way that SQL collates NULL values.

permno	date	ret	vwretd
14593	1980-12-12		0.014859
14593	1980-12-15	−0.052061	0.001605
14593	1980-12-16	−0.073227	0.007485
14593	1980-12-17	0.024691	0.016168
14593	1980-12-18	0.028916	0.004134

B.5 SQL approach to `mutate()`

Creating new variables in SQL is quite straightforward. In addition to basic mathematical operators, such as + and *, PostgreSQL and DuckDB have an extensive array of functions available for use in queries.[7]

The names of calculated variables can be specified using AS, as in the following query:

```
SELECT date,
  vwretd - ewretd AS ret_diff,
  ln(1 + vwretd) AS log_vwretd,
  date_part('year', date) AS year
FROM crsp.dsi
```

date	ret_diff	log_vwretd	year
1925-12-31			1925
1926-01-02	−0.003827	0.005673	1926
1926-01-04	−0.005074	0.000706	1926
1926-01-05	−0.002894	−0.004833	1926
1926-01-06	−0.001605	−0.000423	1926

The translation of this query into `dplyr` requires the use of the `mutate()` verb, but this translation is straightforward, as can be seen below:

```
crsp.dsi |>
  mutate(ret_diff = vwretd - ewretd,
         log_vwretd = log(1 + vwretd),
         year = year(date)) |>
  select(date, ret_diff, log_vwretd, year)
```

date	ret_diff	log_vwretd	year
1925-12-31			1925
1926-01-02	−0.003827	0.005673	1926
1926-01-04	−0.005074	0.000706	1926
1926-01-05	−0.002894	−0.004833	1926
1926-01-06	−0.001605	−0.000423	1926

---

[7]See https://www.postgresql.org/docs/current/functions.html.

We see that the `dplyr` code is *slightly* more verbose than the SQL because the SQL equivalent of `mutate()` is an implicit part of the `SELECT` portion of the statement.

Note that we use the `ln()` in SQL, but use `log()` in the `dplyr` code. In SQL, `log()` returns the base-10 logarithm, but `dplyr` will translate common functions such as `log()` to their SQL equivalents.[8] An alternative approach would be to use `ln()` with `dplyr`. Because `ln()` is not a recognized R function, `ln()` is passed to SQL unmodified. Similarly, `year()` is translated by `dbplyr` to its nearest equivalent in SQL (for `year()` something using SQL's `EXTRACT` that is equivalent to `date_part()`). We can use `show_query()` to see the generated SQL.

```
crsp.dsi |>
  mutate(ret_diff = vwretd - ewretd,
         log_vwretd = log(1 + vwretd),
         year = year(date)) |>
  select(date, ret_diff, log_vwretd, year) |>
  show_query()
```

```
<SQL>
SELECT
  "date",
  "vwretd" - "ewretd" AS "ret_diff",
  LN(1.0 + "vwretd") AS "log_vwretd",
  EXTRACT(year FROM "date") AS "year"
FROM "crsp"."dsi"
```

B.6 SQL `GROUP BY` and aggregates

One use for `group_by()` in R is to create summary statistics and the like for each of the groups implied by the `group_by()` variables. In SQL, `GROUP BY` is used with **aggregate** functions, which are like the functions used with `summarize()`.

Suppose we want to get the maximum and minimum returns, as well as a measure of the volatility of the returns, for each year. In SQL, we could do the following:

```
SELECT date_part('year', date) AS year,
  max(vwretd) AS max_ret,
  min(vwretd) AS min_ret,
  stddev(ln(1 + vwretd)) AS sd_ret
FROM crsp.dsi
GROUP BY year
ORDER BY sd_ret DESC
```

---

[8]See https://www.postgresql.org/docs/current/functions-math.html.

year	max_ret	min_ret	sd_ret
1925			
1932	0.109846	−0.073118	0.027422
1933	0.156838	−0.093272	0.025666
2008	0.114918	−0.089818	0.025397
2020	0.091556	−0.118168	0.021380

And the `dplyr` equivalent would look like this:

```
crsp.dsi |>
  mutate(year = year(date)) |>
  group_by(year) |>
  summarize(max_ret = max(vwretd, na.rm = TRUE),
            min_ret = min(vwretd, na.rm = TRUE),
            sd_ret = sd(log(1 + vwretd), na.rm = TRUE)) |>
  arrange(desc(sd_ret))
```

year	max_ret	min_ret	sd_ret
1925			
1932	0.109846	−0.073118	0.027422
1933	0.156838	−0.093272	0.025666
2008	0.114918	−0.089818	0.025397
2020	0.091556	−0.118168	0.021380

Note that `na.rm = TRUE` could be omitted, as SQL always omits these values. The inclusion of this argument serves only as a reminder to the user of the code that `NA` values are omitted.[9]

---

[9]Setting `na.rm = TRUE` also suppresses warnings from `dplyr` regarding the handling of missing values in SQL.

C

Research computing overview

In this appendix, we provide a brief overview of the research computing landscape with a focus on the statistical programming languages most commonly used in empirical accounting research. We also discuss some fundamental ideas associated with data management that help to explain the approach we emphasize in this book.

C.1 Languages

A downside of developing a course that involves hands-on data analysis is that material in at least one statistical programming language is needed. In this course, we focus on R and, somewhat indirectly, on PostgreSQL as tools. We believe that this choice is much less restrictive than it may seem. However, before explaining this choice, it makes sense to discuss the primary alternatives.

C.1.1 SAS

According to Wikipedia[1], "SAS is a statistical software suite developed by SAS Institute for data management, advanced analytics, multivariate analysis, business intelligence, criminal investigation, and predictive analytics." SAS has long been a staple of research in accounting and finance, in part because it is used by Wharton Research Data Services[2] (WRDS, pronounced "words"). WRDS provides its data sets in SAS data format and historically WRDS offered SAS as the primary programming language on its research computing servers.

SAS offers a number of advantages to researchers.

- Rich set of statistical tools
- Ability to use SQL to prepare data sets (via `PROC SQL`)
- Ability to process large data sets due to an on-disk orientation to data processing
- Availability to most researchers via WRDS servers

However, SAS has some disadvantages.

- Specialized programming language is relatively foreign to users of other languages
- Proprietary software limited to Windows and Linux (i.e., not MacOS)
- Limited set of data types
- Limitations in handling textual data
- Some important functionality (e.g., graphs) seems fairly inaccessible to most users

---

[1] https://en.wikipedia.org/wiki/SAS_(software)
[2] https://wrds-www.wharton.upenn.edu

C.1.2 Stata

According to Wikipedia[3], "Stata is a general-purpose statistical software package created in 1985 by StataCorp. Most of its users work in research, especially in the fields of economics, sociology, political science, biomedicine, and epidemiology."

Stata offers a number of advantages to researchers.

- Rich set of statistical tools, especially for economists and those in downstream disciplines, such as finance and accounting
- Availability on all three end-user operating systems (Windows, MacOS, and Linux)
- Friendly interface coupled with command-based orientation suits many workflows

However, Stata has some disadvantages.

- Proprietary system (i.e., it costs money)
- In-memory orientation makes it unsuitable for large data sets
- Historical limitation to one in-memory data set makes data preparation challenging
- Tendency for Stata users to produce code that is opaque

C.1.3 Python

According to Wikipedia[4], "Python is an interpreted, high-level and general-purpose programming language." While not designed specifically for statistical or econometric analyses, Python has become a mainstay of the data science community, and there exist many packages that make it possible to perform research computing tasks for accounting research in Python.

Python offers a number of advantages to researchers.

- As a full-featured computer programming language, pretty much *anything* can be done using Python
- Rich set of statistical tools, especially in areas such as statistical learning
- Availability on all three end-user operating systems (Windows, MacOS, and Linux)

However, Python has some disadvantages.

- Due to its origins as a general-purpose computing language, some design choices are not optimized for data analysis
- A little more difficult to set up for a relatively non-technical user

C.1.4 PostgreSQL

PostgreSQL is in some ways out of place in this list, as it is not a programming language or statistical software package. But given that much of this course uses PostgreSQL, we discuss it briefly here. According to its website[5], "PostgreSQL is a powerful, open source object-relational database system with over 35 years of active development that has earned it a strong reputation for reliability, feature robustness, and performance." For our purposes, the relevant fact is that PostgreSQL is a powerful system for storing data in a way that it can be processed, combined, and used in analysis.

---

[3]https://en.wikipedia.org/wiki/Stata
[4]https://en.wikipedia.org/wiki/Python_(programming_language)
[5]https://www.postgresql.org

There are a number of similar database systems, including SQLite, MySQL, and Oracle. We use PostgreSQL in part because we make extensive use of WRDS data, which is made available in a PostgreSQL database, and also because PostgreSQL is a very robust database system offering perhaps the strongest feature set among the main open-source alternatives.

Starting from Chapter 6, we make extensive use of data sources from the PostgreSQL[6] supplied by WRDS. WRDS essentially provides data in two forms: as SAS data files (accessible via SFTP[7]) and as tables in a PostgreSQL database.

Chapter 2 provides an extended tutorial on R with an emphasis on the so-called Tidyverse[8]. Using `dbplyr`, tables in a (possibly remote) database can be represented as a kind of data frame that allows us "to use remote database tables as if they are in-memory data frames by automatically converting `dplyr` code into SQL" (we explain what a data frame is in Chapter 2). In this way, the `dbplyr` package allows us to deploy the skills we learn in Chapter 2 to the manipulation and analysis of data stored in relational databases almost seamlessly. The best way to understand these ideas is through hands-on use, for which we provide guidance in the chapters beginning with Chapter 6.

C.1.5 R

According to Wikipedia[9], "R is a programming language and free software environment for statistical computing and graphics".

Compared to SAS and Stata, R has the benefit of being open-source and free. Like SAS, there is a seemingly endless list of packages to support almost any statistical procedure or data management task that one can think of. A disadvantage of R is that, like Stata, its default approach uses in-memory data analysis. However, as we will see, this disadvantage can be overcome with packages that facilitate on-disk data manipulation and analysis.

C.1.6 Why R?

This book contains a lot of R code and a reader might ask why we chose R for this book and we provide an answer in this section.

A more basic question might be: Why choose any software package? We believe that there are significant pedagogical benefits to covering not only research papers and their methods, but in studying in detail the computing steps required to apply those methods to produce the results in those papers. For example, having actual code that a reader can tweak makes it easy for the reader to appreciate how the results might have been different if different approaches had been used. Producing the numerous replications found in this book required a significant investment of time that a reader is unlikely to be able to make for everything covered in this book, especially for a reader who is new to research computing.

Having decided to include code in the book, why did we choose R in particular? An alternative approach would have been to include code in multiple languages. For example, "The Effect: An Introduction to Research Design and Causality"[10] (Huntington-Klein, 2021) provides code in Python, R, and Stata. And "Causal Inference: The Mixtape"[11] (Cunningham, 2021) provides code in R and Stata.

---

[6]https://www.postgresql.org

[7]https://en.wikipedia.org/wiki/SSH_File_Transfer_Protocol

[8]https://www.tidyverse.org

[9]https://en.wikipedia.org/wiki/R_(programming_language)

[10]https://theeffectbook.net

[11]https://mixtape.scunning.com

There are two critical differences between these two books and ours. First, we felt it was important to include code to reproduce *every* analysis in the book. In contrast, "The Effect" and "The Mixtape" only include code to produce some of the analyses they contain. Including code for every analysis in those books would likely have resulted in significant greater length and complexity, and doing so in multiple languages even more so.

Second, "The Effect" and "The Mixtape" largely skip complicated data steps and use small data sets. In contrast, we generally go through all the data steps required to produce our analyses.[12] And also we use (relatively!) large data sets, such as CRSP and Compustat that require paying attention to the "logistics" of getting the data into whatever statistical analysis package is being used. In this regard, the approach we use in this book provides a particularly low-cost way to access these data sets, as we discuss in Section 6.1. Achieving similar results with (say) Stata would require in most cases a significant "data step" (to use SAS terminology) using SQL followed by data analysis of the resulting data frames using native tools. We believe that the `dplyr`-based approach used here results in code that is easier to understand.[13]

Finally, this book has been written with Quarto, a software package that works most seamlessly with R code. Producing this book using Stata or SAS would have been much more challenging.[14]

Even though this book uses R, we believe it should be of interest to dedicated users of SAS, Stata, and Python. First, one can probably get a lot of out this book with a minimal investment in learning R. We have endeavoured to make the code understandable even if viewed as a kind of pseudo-code (in this regard, we think R—especially with `dplyr`—is superior to, say, SAS or Stata). While one could read the book without running code, we strongly encourage you to run code in some places, even if just to be able to do some of the exercises. After some initial set-up (see Section 1.2), *every* analysis in the book can be produced by copying and pasting the code into R and the book has been written so that the code in any chapter can be run independently of that in other chapters (though code later in a chapter may depend on that earlier in the chapter).

Second, we believe that some investment in R is worthwhile even if you will remain a dedicated user of, say, Stata. Even a dedicated Stata user is likely to benefit from the ability to read SAS, as co-authors may use it, and authors of papers may supply SAS code. As R's popularity increases, the ability to read R is likely to become more important. Already R is a kind of *lingua franca* in much of statistics, including (along with Python) machine learning.

Finally, the researcher who is a "dedicated user of X" is probably an endangered species. Most emerging researchers will likely need to use multiple statistical languages over their careers and R is a good candidate to be one of those.[15]

---

[12]The few exceptions are those where we use data from `farr`, the companion package for this book, and in general the data steps are made available in the source code for that package: https://github.com/iango w/farr.

[13]While there will often be an implicit SQL-based step, `dplyr` handles most of the SQL-related details. SAS is an alternative with excellent native handling of WRDS data, but our understanding is that nowadays few accounting researchers exclusively use SAS for their research, suggesting it is not sufficient.

[14]Recent developments have made it easier to work with remote data sources using Python and it is possible that a future (alternative) edition of this book will use Python.

[15]For example, the Accounting Coding Camp provided by three accounting professors covers SAS, Stata, and Python: https://accountingcodingcamp.com.

C.2   Data management

Researchers use data stored in a variety of forms. In this section, we outline some desired attributes for a data storage approach. We then describe some common approaches to data storage, before evaluating them against the criteria outlined below.

C.2.1   Desiderata for data storage

In this section, we discuss features that are desired for a data storage approach. Not all features will be relevant to any given use case. Similarly, some features will be more important than others in a given situation. Nonetheless, these are important attributes across a wide portfolio of possible research uses.

1. **Fast random access to on-disk data.** Occasionally you will see benchmarks that compare alternative approaches to processing in-memory data. However, the reality is that data mostly "live" on disk and we want the ability to access these data quickly. In many cases, we will want only a small sample of the observations in a data set. For example, we may want stock-return data for Microsoft for a few dates.[16] These data are found on `crsp.dsf` in the WRDS PostgreSQL database and can be retrieved from that system very quickly. In contrast, some data storage systems would require us to load (or at least scan) the entirety of `crsp.dsf` to get these rows.

2. **Data accessible by any software package.** Ideally, our data would be accessible from R, Stata, Python, Perl, ... pretty much any software. Being able to do this has subtle but important advantages for data collaboration and multilingual programming. For example, you may be a Stata devotee who needs help preparing a data set from Web data. Finding people with the skills to do this is going to be a lot easier if you don't need them to have Stata skills so you can work with a Python guru. So long as the data can be accessed through Stata, a Stata user is unlikely to care much if the data were created using Python or Stata. Alternatively, you may have written some complicated Perl code to generate some data sets, and occasionally need to tweak the code to accommodate changes in the source data. So long as the data sets are stored in a form that can be accessed by any system, you will have no need to translate the entire code into something you still know how to code in.

3. **Data accessible from anywhere.** Ideally, data should be able to be accessed from anywhere with an internet connection (in the 2020s, this essentially means anywhere). For example, it should be easy for a researcher in Boston to share data with co-authors in California, Australia, or Singapore.

4. **Possibility of centralized processing.** While less important than it once was, the ability to move data processing and analysis to a computer more powerful than one's laptop can still be useful today. In the text, there are a number of cases where the heavy part of data processing is accomplished on the WRDS PostgreSQL server even when executing R code locally. SAS offers similar functionality with `RSUBMIT`.

---

[16]Here "random" means that we may not know which observations we want ahead of time.

5. **Support for a wide range of data types.** Ideally, available data types will include strings, floating-point, integers, large integers, dates, timestamps with time zones, JSON, XML, and other types. At one end, text files are simply text, with data types either being inferred when reading the data or specified in a separate file. At the other end are systems such as PostgreSQL, which offer all the data types listed above and more.

6. **Ability to handle textual data with ease.** Textual data have emerged as an important data source in research in recent years. In practice, good text-handling capability often means supporting Unicode and encoding as UTF-8. According to Stata[17], "Unicode is the modern way that computers encode characters such as the letters in the words you are now reading." Additionally, text fields can be "wide" and the data storage approach should not impose restrictions on these that limit its usefulness.

C.2.2   Data storage approaches

We can think of four basic approaches to data storage:

1. Text files
2. Package-specific binary data formats
3. General-purpose binary data formats
4. Relational databases

One way to store data is in text files, such as comma-separated values (CSV) files. These text files, which may be compressed to reduce storage space requirements and increase read speeds, can then be read into pretty much any statistical programming package. So long as one addresses issues such as encoding (e.g., UTF-8 versus Windows-1252) and quote characters (e.g., " or ') embedded in values, text files are a robust approach to data storage and arguably the best way to share certain kinds of data with random strangers (e.g., posting `.csv` files on a webpage).

However, text files are not ideal for fast random access to on-disk data. This disadvantage matters when one is interested in using only small slivers of data at a time, which is very common in accounting research. For example, suppose we are interested in getting the stock returns for Microsoft Corporation from 3 January 2018 to 6 January 2018. If the data source is a text file version of CRSP's daily stock file (`crsp.dsf`), then we would typically need to read in the entire text file, which would take several seconds, even with a very fast storage medium (such as solid-state drives), and require several gigabytes of memory (`crsp.dsf` is about 16 GB in SAS format).

Additionally, text files do not easily incorporate information about data types. This means it is generally necessary to provide additional information about a data set for statistical packages read them correctly. For example, should a field with values like 991201, 120305, 231130 be read as text, as floating-point numerical values, as integers, or even as dates? Or if a field contains date-time values, what is the format? And what is the time zone?

Package-specific storage means storing data as, say, SAS or Stata data files. This approach can address some weaknesses of text files.

First, most statistical software packages have a richer set of data types and these are generally embedded into the native data storage formats for each package. Nonetheless, some

---

[17]https://www.stata.com/stata14/unicode/

packages have limitations. For example, everything in SAS is fundamentally of either a floating-point numerical type or a fixed-width character type. Things such as dates and times are essentially formatting overlays over these types, and such formatting is easy to break, sometimes causing serious issues in practice. Stata is fairly similar. For example, dates and times in Stata are stored as integers and special formatting is applied to these to display them so that they are recognized as dates or times by a human.

Second, some statistical software packages have the ability to facilitate random on-disk access by indexing data files. For example, SAS allows the user to index data so that small portions of the data can be loaded quickly without having to process the entire data file (e.g., the SAS data file for `crsp.dsf` on WRDS is indexed by `permno`, `permco`, and `date`; so accessing the stock-return data for Microsoft for those few dates is quite fast).

While package-specific data formats do address some of the deficiencies of text files, they do so at the cost of locking users into the applicable package. While `.dta` (Stata) data files can be read into R, they are unlikely to be the best approach to storing data unless all analyses are run using Stata (in which case, limitations of Stata are likely to arise). Either of the following approaches is likely to be superior to such an approach.

A third approach uses a binary format that is accessible to multiple software packages or systems. Perhaps the leading example of this for on-disk storage is the parquet format. The parquet format is provided by Apache Arrow, "a software development platform for building high performance applications that process and transport large data sets."[18] The parquet format offers much of the versatility of CSV files, while also allowing for a rich system of data types and random access to data. As seen in Chapter 23, parquet files can be combined with DuckDB to provide a lightweight, high-performance alternative to PostgreSQL. While this approach does not facilitate the centralization of data processing as easily as running a database server does, the performance of approaches based on DuckDB and parquet files on even limited hardware is often more than adequate. Appendix E provides guidance on setting up your own parquet-based data repository for research.

A relational database provides a more rigid, structured way of storing data. Almost all database packages use Structured Query Language (SQL) as the primary language for the user to interact with the data and a brief primer on SQL is provided in Appendix B. This course focuses on data stored in the PostgreSQL database offered by WRDS.

The WRDS PostgreSQL server offers a number of advantages. First, much of the data processing can be handled by the WRDS server. Second, the WRDS database offers fast random access to on-disk data.

PostgreSQL has excellent support for data types without requiring detailed understanding from users. While PostgreSQL likely uses a similar approach to storing dates as is used by SAS or Stata, a user is never forced to be aware of these details. Most of the commonly used types in PostgreSQL (e.g., timestamps with time zones) are recognized fairly seamlessly and converted to appropriate types when data are brought into R or Python. PostgreSQL also handles text data well.

PostgreSQL is easily accessed from other languages, including Python, R, and Stata. This benefit is perhaps a significant driver of WRDS's decision to add a PostgreSQL database to its offerings. With more users wanting to use Python or R for analysis, sticking to a SAS-format-only approach was likely not viable. Finally, if you have a WRDS account, the WRDS PostgreSQL database is accessible from anywhere with an internet connection.

---

[18]See the Apache Arrow website at https://arrow.apache.org/overview/.

While the WRDS PostgreSQL database provides many of these benefits, it does not provide any mechanism to manage non-WRDS data and most significant research projects will involve such data. Using a relational database as the central data repository for your research offers a number of benefits, and Appendix D describes how to set up your own PostgreSQL server.

D

Running PostgreSQL

Throughout the book, we have used the WRDS PostgreSQL server as the primary source of data for our analysis. While WRDS PostgreSQL server offers access to a large menagerie of data with very little set-up, there are limitations to using the WRDS server alone for research. For example, if you are using `crsp.dsf` intensively on a near-daily basis, then you may be downloading a lot of data over time. As such, it might be more efficient to have a local copy of `crsp.dsf`.

The purpose of this appendix is to provide guidance on one possible solution, namely running your own PostgreSQL server. An alternative approach—creating a repository of parquet files—is discussed in Appendix E.

D.1 Setting up a personal server

Setting up your own PostgreSQL server overcomes some of the disadvantages of using the WRDS server.

One disadvantage is that, because WRDS only allows read-only connections, one cannot store computed results on the WRDS PostgreSQL server. There are two types of computed tables that one can create. Temporary tables, which are created using the `compute()` function from `dplyr`, can dramatically simplify queries and increase performance, but disappear once you disconnect from the database. Permanent tables, which can be created using the `compute()` function with `temporary = FALSE` argument, can be useful to break the research process into several discrete steps.

Another disadvantage of the WRDS server is the limits it imposes on using our own data. While we can use the `copy_inline()` function from the `dbplyr` package to create something similar to a temporary table on the server using local data, this will not be the best approach if our local data sets are large.[1] And, while we could download `crsp.dsf` as an R data file and do the analysis on our own computer, this is likely to be painful even if we have enough RAM to load `crsp.dsf`.

An alternative approach that builds on the framework used in this book is to create your own PostgreSQL database and use that. While there are some set-up and maintenance costs associated with this, it's actually fairly manageable using the steps we describe below.

---

[1] In fact, `copy_inline()` was added to `dbplyr` after a request related to this book: https://github.com/tidyverse/dbplyr/issues/628.

DOI: 10.1201/9781003456230-D

D.1.1   Installation steps

1. Install Python. You may already have Python installed. If not, Miniconda[2] is one option.
2. Install the **wrds2pg** package.

```
pip install wrds2pg --upgrade
```

3. Install PostgreSQL and initialize the server

Installation of PostgreSQL varies by platform (and on some platforms there are multiple approaches).[3] A good place to start is the PostgreSQL webpage[4].

4. Create a database.

You may have already done this in the previous step. If not, and you have the server running, you could do this using SQL:

```
CREATE DATABASE wrds;
```

5. Set up environment variables for **wrds2pg**.

See here[5] for details.

```
export PGHOST="localhost"
export PGDATABASE="wrds"
export WRDS_ID="iangow"
export PGUSER="igow"
export PGPORT=5432
```

You might also use the environment variable PGPASSWORD:

```
export PGPASSWORD="password"
```

But it is better to use a password file[6].

6. Run Python.

7. Within Python, use the **wrds2pg** module to get data.

```
from wrds2pg import wrds_update
wrds_update("dsi", "crsp", dbname="wrds", host="localhost")
```

D.1.2   Getting the tables used in this book

Following the steps above, the script below can be used to get the tables used in this book. Note that a few tables require special handling.

---

[2]https://docs.conda.io/en/latest/miniconda.html

[3]On Ubuntu, we install using say **sudo apt install postgresql-16**; on MacOS, Postgres.app provides a very easy-to-manage PostgreSQL implementation: https://postgresapp.com.

[4]https://www.postgresql.org/download/

[5]https://github.com/iangow/wrds2pg/

[6]https://www.postgresql.org/docs/current/libpq-pgpass.html

Some SAS data files on CRSP and Compustat have special missing values that SAS's PROC EXPORT function turns into simple character values, which PostgreSQL cannot accept in fields that are not text types. The fix_missing = True argument to wrds_update converts such missing values into regular missing values.

The script also fixes variable types that are not well-formatted in the original SAS files (e.g., permno should be integer).

Finally, the script creates **indexes**, which dramatically increase performance of table joins or filter() operations.

Note that the script below likely takes a few hours to run, primarily because of crsp.dsf, which is about 20 GB of data. However, subsequent runs of the script will only download data if the SAS data file on WRDS has been updated and so will usually run much faster.

```python
#!/usr/bin/env python3
from wrds2pg import wrds_update, make_engine, process_sql

engine = make_engine()

# CRSP
wrds_update('ccmxpf_lnkhist', 'crsp', fix_missing=True,
            col_types= {'lpermno': 'integer', 'lpermco': 'integer'})

dsf = wrds_update('dsf', 'crsp', fix_missing=True,
                  col_types={'permno': 'integer',
                             'permco': 'integer'})
if dsf:
    process_sql('CREATE INDEX ON crsp.dsf (permno, date)', engine)
    process_sql('CREATE INDEX ON crsp.dsf (permco)', engine)

erdport1 = wrds_update('erdport1', 'crsp',
                       keep='permno date decret')
if erdport1:
    process_sql('CREATE INDEX ON crsp.erdport1 (permno, date)', engine)

dsi = wrds_update('dsi', 'crsp')
if dsi:
    process_sql('CREATE INDEX ON crsp.dsi (date)', engine)

wrds_update('comphist', 'crsp', fix_missing=True)

dsedelist = wrds_update('dsedelist', 'crsp', fix_missing=True,
                        col_types={'permno': 'integer',
                                   'permco': 'integer'})
if dsedelist:
    process_sql('CREATE INDEX ON crsp.dsedelist (permno)', engine)

dseexchdates = wrds_update('dseexchdates', 'crsp',
                           col_types= {'permno': 'integer',
                                       'permco': 'integer'})
if dseexchdates:
    process_sql('CREATE INDEX ON crsp.dseexchdates (permno)', engine)
```

```python
msf = wrds_update('msf', 'crsp', fix_missing=True,
                  col_types= {'permno': 'integer',
                              'permco': 'integer'})
if msf:
    process_sql('CREATE INDEX ON crsp.msf (permno, date)', engine)

msi = wrds_update('msi', 'crsp')
if msi:
    process_sql('CREATE INDEX ON crsp.msi (date)', engine)

wrds_update('mse', 'crsp', fix_missing=True,
            col_types= {'permno': 'integer',
                        'permco': 'integer'})

wrds_update('stocknames', 'crsp',
            col_types= {'permno': 'integer',
                        'permco': 'integer'})

dsedist = wrds_update('dsedist', 'crsp', fix_missing=True,
                      col_types={'permno': 'integer',
                                 'permco': 'integer'})
if dsedist:
    process_sql('CREATE INDEX ON crsp.dsedist (permno)', engine)

# Fama-French data
updated = wrds_update('factors_daily', 'ff')

# Compustat
company = wrds_update('company', 'comp')
if company:
    process_sql('CREATE INDEX ON comp.company (gvkey)', engine)

funda = wrds_update('funda', 'comp', fix_missing=True)
if funda:
    process_sql('CREATE INDEX ON comp.funda (gvkey)', engine)

wrds_update('funda_fncd', 'comp')

fundq = wrds_update('fundq', 'comp', fix_missing=True)
if fundq:
    process_sql('CREATE INDEX ON comp.fundq (gvkey, datadate)', engine)

updated = wrds_update('r_auditors', 'comp')

idx_daily = wrds_update('idx_daily', 'comp')
if idx_daily:
    process_sql('CREATE INDEX ON comp.idx_daily (gvkeyx)', engine)

aco_pnfnda = wrds_update('aco_pnfnda', 'comp')
if aco_pnfnda:
```

```
    process_sql('CREATE INDEX ON comp.aco_pnfnda (gvkey, datadate)', engine)

wrds_update('seg_customer', 'compseg')
wrds_update('names_seg', 'compseg')
```

D.2 Setting up a shared server

While setting up a personal PostgreSQL server gives you more flexibility, greater potential can be realized if you can share your server with collaborators. For example, you might write Python code to perform textual analysis on SEC filings, then have a co-author pull the data from the database directly into Stata for regression analysis. Having data in a single database is in many ways superior to solutions such as Dropbox that are not really set up for sharing multi-gigabyte data files. This short appendix describes some details of sharing a database.

D.2.1 Opening up the database

By default, a PostgreSQL cluster is not open to connections from other computers and we need to edit two configuration files to change this.

1. Issue two SQL commands to locate the two configuration files (pg_hba.conf and postgresql.conf) that we need to edit. Here we use the command-line program for accessing PostgreSQL (psql), but anything that allows you to execute SQL against the server should work.

```
(base) brontegow@brontegow-ubuntu-mate:~$ psql
psql (13.2)
SSL connection (protocol: TLSv1.3)
Type "help" for help.

crsp=# SHOW hba_file ;
              hba_file
-----------------------------------
 /etc/postgresql/13/main/pg_hba.conf
(1 row)

crsp=# SHOW config_file ;
               config_file
-----------------------------------------
 /etc/postgresql/13/main/postgresql.conf
(1 row)
```

2. Add the following line to the end of pg_hba.conf

```
host all all 0.0.0.0/0 md5
```

You may need to edit the file as a super user (e.g., sudo vim /etc/ postgresql/13/main/pg_hba.conf).

3. Change listen_addresses in postgresql.conf

There should already be a line related to `listen_addresses` in the file, but it is likely commented out (i.e., starts with #). Uncomment the line and edit it so that PostgreSQL listens from all addresses. After editing, the line should look like this:

```
listen_addresses = '*'
```

Again, you may need to edit the file as a super user (e.g., `sudo vim /etc/postgresql/13/main/postgresql.conf`).

After making these two edits, it will be necessary to reboot the server. The details of how to reboot the server will depend on how you installed it, so consult instructions specific to your installation (in the worst case, restarting your computer should do the trick).

D.2.2   Creating user logins

Suppose that Bronte has a co-author named Gemma to whom she wants grant access to the database. Note that `wrds_update` function automatically creates two roles for each schema. For example, in creating a local copy of `crsp.dsf`, which is stored in schema `crsp`, it creates the role `crsp`, which owns the data (and therefore can delete or modify it), and the role `crsp_access`, which only has read access to the data. Similar roles are created for data in the `comp`, `compseg`, and `ff` schemas.

In this case, we want to create the role `gemma` and give her read-only access to the data in the four schemas: `crsp`, `comp`, `compseg`, and `ff`. The following commands achieve this:

```
(base) brontegow@brontegow-ubuntu-mate:~$ psql
psql (13.2 (Ubuntu 13.2-1))
SSL connection (protocol: TLSv1.3)
Type "help" for help.

crsp=# CREATE USER gemma ENCRYPTED PASSWORD 'bronte2006';
CREATE ROLE
crsp=# GRANT crsp_access TO gemma;
GRANT ROLE
crsp=# GRANT comp_access TO gemma;
GRANT ROLE
crsp=# GRANT compseg_access TO gemma;
GRANT ROLE
crsp=# GRANT ff_access TO gemma;
GRANT ROLE
```

D.2.3   Testing access

Now Gemma should have access to the database. We need to give Gemma the IP address for the server and with that information, she should be able to access the data. In the following example, the IP address for the server is assumed to be `108.26.187.8`.

```
psql -h 108.26.187.8 -d wrds -U gemma
Password for user gemma:
psql (13.2 (Ubuntu 13.2-1.pgdg20.04+1))
SSL connection (protocol: TLSv1.3)
Type "help" for help.

crsp=> SELECT date, vwretd FROM crsp.dsi LIMIT 10;
```

```
     date    |  vwretd
-------------+-----------
 1925-12-31 |
 1926-01-02 |  0.005689
 1926-01-04 |  0.000706
 1926-01-05 | -0.004821
 1926-01-06 | -0.000423
 1926-01-07 |  0.004988
 1926-01-08 | -0.003238
 1926-01-09 |  0.002209
 1926-01-11 |  -0.00854
 1926-01-12 | -0.000929
(10 rows)
```

Note that the above assumed that the machine on which the database is hosted can be accessed via an IP address (or a URL). If you are using a home computer, there is a good chance that your IP address is not fixed. So, the following steps may be necessary to give access to others outside your home network:

1. Fix the local IP address of your server. For example, I have one computer set to `192.168.1.3`.
2. Direct all traffic on the port for your database to this computer. This is accomplished via the port-forwarding settings of your internet router. For example, the default port for PostgreSQL is **5432** and we might forward all traffic on this port to `192.168.1.3`.
3. Get a domain name. For example, a service such as GoDaddy[7] can tell us what domains are available to acquire. At the time of writing `brontesdb.com` is available and would cost \$12 for the first year.
4. Use a dynamic DNS service (such as `dyn.com`[8]) to make sure that your chosen domain name (`brontesdb.com` in our example) always points to your IP address.

After taking the steps above, Gemma could now connect to our database like this:

```
psql -h brontesdb.com -d wrds -U gemma
```

Note that Gemma could change her password to something more to her liking using SQL such as the following:

```
ALTER ROLE gemma PASSWORD 'gemma2007';
```

Because the password is encrypted on the database server, Bronte cannot see the password that Gemma has chosen.

---

[7]http://godaddy.com
[8]http://dyn.com

E

Making a parquet repository

Throughout the book, we have relied on the WRDS PostgreSQL server as our primary source of larger data sets, primarily from CRSP and Compustat. However, there are downsides to relying on the WRDS PostgreSQL server to this degree. One (minor) downside is that one needs to be online to get data. Another downside is that we quickly run into limits when we have large amounts of non-WRDS data that need to be combined with WRDS data for analysis. Finally, many researchers would prefer to have greater clarity on the underlying data sets than is provided by an approach that relies on the ever-changing data sets on WRDS.

In Appendix D, we considered an approach that uses a user-created PostgreSQL server. While this has the merit of using the same underlying approach used by WRDS and the rock-solid backbone provided by PostgreSQL, it does require the user to invest in understanding a technology at a deeper level than many researchers would be comfortable with.

In Chapter 23, we introduced an approach that used parquet data files as a foundation for analysis. This approach is likely to be more transparent to many researchers accustomed to storing data in individual data files. This may lead many researchers to ask if such an approach could be extended to handle data from CRSP and Compustat more generally.

The answer is "yes" and this appendix outlines some options for researchers interested in pursuing such an approach.

The code in this appendix uses the following packages. For instructions on how to set up your computer to use the code found in this book, see Section 1.2.

```
library(DBI)
library(dplyr)
library(farr)
library(ggplot2)
library(stringr)
library(dbplyr)      # window_order()
```

E.1   Data management approaches

Once upon a time, research computing was largely done on shared computers managed by others. Data of the kind provided by CRSP and Compustat would have resided on these shared computers and often researchers would have had limited allocations for data storage. For example, many academics would have relied on the servers provided by WRDS to conduct research. If WRDS updated Compustat or CRSP, then the results from running a researcher's code might change.

DOI: 10.1201/9781003456230-E   548

When computing power and storage became greater and cheaper, many researchers came to conduct research on their personal computers. Today, it is quite feasible to maintain local copies of major CRSP and Compustat data tables on the hard drive of one's laptop. In fact, one could keep multiple copies, say one per project.

E.2 Organizing data

For brevity, we assume you have elected to maintain a core repository of data files shared across projects (e.g., Compustat, CRSP) as well as project-specific data files. The shared data might also include data from non-WRDS sources, such as conference call files from StreetEvents or FactSet or data obtained from SEC EDGAR. The project-specific data files might be hand-collected or scraped data relevant to a specific project, as well as the results of processing other data for your project.

This approach to data organization is easily implemented using **schemas** in a relational database, such as PostgreSQL or MariaDB. An obvious choice for schema names for WRDS-derived data is the name of the corresponding schema in the WRDS PostgreSQL database (e.g., `crsp` for CRSP data and `comp` for Compustat data).

But if we want to maintain a file system–based approach rather than setting up a relational database, we might choose to organize our data files into directories. With such an approach, we could mimic the schemas discussed above, we could put tables from CRSP in `crsp`, tables from Compustat in `comp`, and so on. If we wanted a project-specific version of `crsp.dsf` for a project stored in `mega_paper`, we could just put that in (say) a `crsp` directory in the `mega_paper` directory.

Having elected to choose a data-file approach to storing data, a natural question arises about the format to use. Many Stata users would use Stata data files (`.dta`). R users might use RDS files (`.rds`). SAS users typically keep SAS data files (`.sas7bdat`). However, `.dta` and `.rds` files are a bad choice for large files.[1] Loading these files into Stata or R means reading the whole file off disk into RAM. This takes time and uses up memory. Storing `crsp.dsf` as `dsf.dta` or `dsf.rds` is likely to result in a file that is several gigabytes on disk and even more in memory.

The approach in this chapter uses parquet files, which we first saw in Chapter 23. Parquet files have a rich type system and are created in a way that makes it easier to get just the data we need. Additionally, parquet files are **column-oriented** which typically leads to superior performance for data analysis than formats that are **row-oriented**, such as CSV files or database tables. Finally, parquet files are typically much smaller than the equivalent CSV files.

Relative to text files or most package-specific storage formats, parquet files offer most of the benefits of relational databases:

1. Fast random access to on-disk data.
2. Data accessible by any software package.[2]
3. Data can be properly typed.
4. Textual data are handled with ease.

---

[1]SAS works in a way that makes SAS data files a reasonable choice.
[2]This is almost true for parquet files, but not quite. Stata cannot really work with on-disk data and has no native support for parquet data.

A couple of potential advantages of a PostgreSQL server are lost with this approach. First, it's slightly more difficult to centralize computing in the way that can often be achieved with a database server. Second, data are not "accessible from anywhere" in quite the same way. However, parquet files offer benefits without requiring a database server to be set up, a complex step for many users.

E.3 Canonical WRDS data

There are two underlying sources of WRDS data:

1. The SAS files on the WRDS cloud server
2. The WRDS PostgreSQL database

For many years (or even decades), the canonical WRDS data were those in SAS files on the WRDS servers. These SAS files formed the basis of the web queries familiar to many researchers.[3]

In recent years, WRDS has offered a PostgreSQL server that provides a more platform-neutral offering for researchers using more modern languages, such as Python or R. While the PostgreSQL server was initially constructed from the SAS data files, it seems possible that some data are ingested by the PostgreSQL server directly without an intermediate step using SAS data formats.

WRDS highlights support for four languages: R, Python, Stata, and SAS. For R, Python, and Stata, it is assumed that users will access WRDS data via the PostgreSQL server. Only for SAS is it assumed that the SAS data files will be used directly.

Initially, the SAS data files appeared to be the canonical data sources. Some data sets were not covered by the PostgreSQL database (e.g., TAQ data). Other data sets were in the PostgreSQL server, but with some data loss. For example, non-ASCII text in SAS data files was lost in early versions of the PostgreSQL database; this appears no longer to be the case.

At this stage, it seems that both the WRDS SAS data files and the WRDS PostgreSQL database can be viewed as canonical sources for WRDS data. Nonetheless there are minor differences between these data sources. First, the underlying schema names can differ in some cases.

Second, there may be differences in the underlying data types. As discussed in Appendix C, SAS has only two underlying data types and is therefore much less strict about what data goes into each column. The more specific data types in SAS are actually implemented as presentations of the underlying data. Nonetheless some data may be presented as one type in SAS and another type in PostgreSQL. For example, `permno` is presented as a whole number in SAS, but as a floating-point value in PostgreSQL. Practically, this makes little difference.[4]

One data source that is perhaps incorrectly viewed as a canonical data source is the web query interface offered by WRDS. The web query interface leaves much to be desired from a

---

[3]It seems plausible that these web queries still draw on SAS data files to this day.

[4]SAS stores the data as floating-point values in any case.

reproducibility perspective as manual steps are needed to extract data and often it is necessary to eliminate variables or periods to manage the size of the resulting files. Additionally, the relation between the data returned by some queries and the underlying tables in SAS or PostgreSQL is sometimes unclear. Some web queries appear to join data from multiple underlying tables in ways that are not always transparent.

E.4 Converting WRDS data to parquet

In this section, we discuss three approaches that readers could use to create their own parquet-based repositories of WRDS data. To keep the repository we create organized, we put it under a single directory, which we flag using the `DATA_DIR` environment variable. As we need to interact with WRDS, we need to communicate our WRDS ID to the code, which we can also do using an environment variable.

```
Sys.setenv(DATA_DIR = "~/Dropbox/pq_data/",
           WRDS_ID = "iangow")
```

E.4.1 Approach 1: Get WRDS PostgreSQL data using DuckDB

One approach to getting WRDS data is to connect to the PostgreSQL server much as we have done throughout the book. Having made this connection, we can easily copy the data to a local DuckDB database, and from there, it is easy to write the data to parquet files on our local storage.

For the reasons discussed in Section 6.1, we use environment variables to give R the information it needs to connect to the WRDS PostgreSQL server.

```
Sys.setenv(PGHOST = "wrds-pgdata.wharton.upenn.edu",
           PGDATABASE = "wrds",
           PGUSER = "iangow",
           PGPORT = 9737)
```

We can now connect to the WRDS PostgreSQL server (`pg` here) and to a DuckDB instance (`db`). If we want to copy the `crsp.msi` table from WRDS to our local machine, we simply create `msi`, a remote data table, then copy this to `db` using the `copy_to()` function provided by `dplyr`. We name the table (`name = "msi"`) so that we can refer to it in subsequent queries. Then, we save the data to a file named `msi.parquet` (DuckDB will infer that we want a parquet file from the `.parquet` file extension). Finally, we disconnect from both databases (`pg` and `db`).

```
pg <- dbConnect(RPostgres::Postgres())
db <- dbConnect(duckdb::duckdb())
msi <- tbl(pg, Id(schema = "crsp", table = "msi"))
copy_to(db, df = msi, name = "msi", overwrite = TRUE)

if (!dir.exists(file.path(Sys.getenv("DATA_DIR"), "crsp"))) {
  dir.create(file.path(Sys.getenv("DATA_DIR"), "crsp"))
}
file_name <- file.path(Sys.getenv("DATA_DIR"), "crsp/msi.parquet")
dbExecute(db, str_c("COPY msi TO '", file_name, "'"))
```

```
[1] 1177
```

```
dbDisconnect(pg)
dbDisconnect(db)
```

This basic approach is taken by the `pg_to_parquet()` function provided as part of the `farr` package. As we have the PostgreSQL connection details set up for WRDS and also have the `DATA_DIR` environment variable set up, we can issue a single line to get a local copy of `crsp.msi`.

```
pg_to_parquet("msi", "crsp")
```

The main downside of this approach is that it is very memory-intensive with large tables.

E.4.2 Approach 2: Get WRDS PostgreSQL data using Python

In Appendix D, we saw how we can use the `wrds2pg` Python library to import WRDS SAS data into a local PostgreSQL server with the `wrds_update()` function. The `db2pq` Python library offers `wrds_update_pq()`, a function modelled on `wrds_update()` that creates parquet files using data stored in the WRDS PostgreSQL database. To install the required version of `db2pq`, use the following command from the command line:

```
pip install db2pq --upgrade
```

Assuming we have the necessary environment variables set up (e.g., `WRDS_ID` and `DATA_DIR`), creating a parquet file is straightforward. (Note that the following lines are Python code.)

```
from db2pq import wrds_update_pq
wrds_update_pq("msi", "crsp")
```

One benefit of this approach is that we can specify variable types. Additionally, it places much less of a burden than the approach using `pg_to_parquet()` above. The main downside is the need to use Python. A script to obtain all tables needed to run the code in this book using this approach is provided in Section E.6.1.

E.4.3 Approach 3: Get WRDS SAS data using Python

The `wrds2pg` Python library also offers `wrds_update_pq()`, a function modelled after `wrds_update()` that uses SAS code to generate data that are saved as a parquet file. To install the required version of `wrds2pg`, use the following command from the command line.[5]

```
pip install wrds2pg --upgrade
```

Assuming we have the necessary environment variables set up (e.g., `WRDS_ID` and `DATA_DIR`), creating a parquet file from a WRDS SAS file is straightforward. (Note that the following lines are Python code.)

```
from wrds2pg import wrds_update_pq
wrds_update_pq("msi", "crsp")
```

If you don't have the `DATA_DIR` environment variable set, you can supply it to the `wrds_update_pq()` function directly.

---

[5]If you installed `wrds2pg` before December 2023, then you will need to update to the current version.

```
wrds_update_pq("msi", "crsp", data_dir = "~/Dropbox/pq_data/")
```

One benefit of this approach is that we can leverage the features of `wrds2pg`, such as the ability to specify variable types, rename and drop variables, and so on. Additionally, importing from WRDS SAS data is fast and often results in better typed data than that provided by the WRDS PostgreSQL database. Finally, while `wrds_update_pq()` does require a significant amount of RAM, it places less of a burden than the R approach using `pg_to_parquet()` above. The main downside is the need to use Python. A script to obtain all tables needed to run the code in this book using this approach is provided in Section E.6.2.

E.5 Working with parquet files

To illustrate how we can work with these parquet files, we retrace some of the steps we took in Chapter 21. We will assume that you have parquet versions of the three data tables in your repository, perhaps as a result of running one of the three approaches to getting the data listed here.

> **❗ Important**
>
> Downloading `comp.funda` will take a few minutes using any of the three approaches below.

E.5.1 Getting the data

E.5.1.1 Approach 1: Get WRDS PostgreSQL data using DuckDB

While this approach is simpler than the other two, it will take longer and it requires more RAM. It is not recommended for large files.

```
pg_to_parquet("seg_customer", "compseg")
pg_to_parquet("names_seg", "compseg")
pg_to_parquet("funda", "comp")
```

E.5.1.2 Approach 2: Get WRDS PostgreSQL data using Python

This is the most robust and fastest approach. Also, it uses less RAM than either of the other two approaches.

```
from db2pq import wrds_update_pq

wrds_update_pq("seg_customer", "compseg")
wrds_update_pq("names_seg", "compseg")
wrds_update_pq("funda", "comp")
```

E.5.1.3 Approach 3: Get WRDS SAS data using Python

As the `comp.funda` SAS data file includes special missing value codes that are exported by SAS as text values, we need to use `fix_missing = True` to convert these to missing values.

```
from wrds2pg import wrds_update_pq

wrds_update_pq("seg_customer", "compseg")
wrds_update_pq("names_seg", "compseg")
wrds_update_pq("funda", "comp", fix_missing = True)
```

E.5.2 Replicating analysis using parquet files

Assuming you now have the three data tables, we can proceed with some analysis from
Chapter 21. We start by creating an in-memory DuckDB database.

```
db <- dbConnect(duckdb::duckdb())
```

We next use `load_parquet()` from the `farr` package to load the tables we created above into
our DuckDB database, `db`. By default, `load_parquet()` will use the environment variable
`DATA_DIR` to locate the repository of parquet files. If `DATA_DIR` has *not* been set, we set it
now.

```
Sys.setenv(DATA_DIR = "~/Dropbox/pq_data/")
```

```
seg_customer <- load_parquet(db, "seg_customer", "compseg")
names_seg <- load_parquet(db, "names_seg", "compseg")
funda <- load_parquet(db, "funda", "comp")
```

As an alternative to setting the `DATA_DIR` environment variable, you can supply the location
of the repository of parquet files to the `load_parquet()` function directly as the `data_dir`
argument.

```
load_parquet(db, "funda", "comp", data_dir = "~/Dropbox/pq_data/")
```

The next two tables come with the `farr` package, but need to be copied to `db`.

```
undisclosed_names <- copy_to(db, undisclosed_names)
llz_2018 <- copy_to(db, llz_2018)
```

The next block of code is essentially unchanged from the code in Chapter 21. The two
changes are:

- We do *not* use `collect()`, as we want the data to remain in our database as long as possible. In the original code, we needed to `collect()` data to merge with `undisclosed_names`
  and `llz_2018`. These are now all in a single database.
- We add `na.rm = TRUE` in a couple of places to suppress warnings. SQL always uses `na.rm`
  `= TRUE`.

```
customers <-
  seg_customer |>
  filter(ctype == "COMPANY")

disclosure_raw <-
  customers |>
  filter(between(datadate, "1994-01-01", "2010-12-31")) |>
  left_join(undisclosed_names, by = "cnms") |>
  mutate(disclosed = coalesce(disclosed, TRUE)) |>
  select(gvkey, datadate, cnms, salecs, disclosed)
```

```
sales <-
  funda |>
  filter(indfmt == "INDL", datafmt == "STD",
         consol == "C", popsrc == "D") |>
  select(gvkey, datadate, sale)

prin_cust_df <-
  disclosure_raw |>
  inner_join(sales, by = c("gvkey", "datadate")) |>
  group_by(gvkey, datadate) |>
  filter(!is.na(salecs), sale > 0) |>
  summarize(prin_cust = max(salecs / sale, na.rm = TRUE),
            .groups = 'drop') |>
  mutate(has_prin_cust = prin_cust >= 0.1)

disclosure <-
  disclosure_raw |>
  inner_join(sales, by = c("gvkey", "datadate")) |>
  semi_join(llz_2018, by = "gvkey") |>
  group_by(gvkey, datadate) |>
  summarize(ratio = mean(as.double(!disclosed), na.rm = TRUE),
            ratio_sale = sum(as.double(!disclosed) * salecs, na.rm = TRUE) /
              sum(salecs, na.rm = TRUE),
            .groups = "drop") |>
  mutate(year = year(datadate))
```

```
disclosure |>
  summarize(across(c(ratio, ratio_sale),
                   \(x) mean(x, na.rm = TRUE))) |>
  collect()
```

```
# A tibble: 1 x 2
  ratio ratio_sale
  <dbl>      <dbl>
1 0.459      0.464
```

```
dbDisconnect(db)
```

E.5.3 Working with larger parquet files

The power of parquet files becomes even more apparent when working with larger data sets, such as `crsp.dsf`. Here we assume that you have a copy of `crsp.dsf` in your repository, perhaps as a result of running code like the following Python code. The resulting data file will be over 3 GB (this is much compressed relative to the original SAS or PostgreSQL data).

> **❗ Important**
>
> Downloading `crsp.dsf` can take 30 minutes or more depending on the speed of your connection to WRDS and the approach used.

```
from db2pq import wrds_update_pq
wrds_update_pq("dsf", "crsp")
```

The following code does a number of analytical tasks. First, we count the number of rows in the table. Second, we count the number of rows by year. Third, we extract the data for a particular date. Next, we identify the PERMNO for Apple, then use that to plot cumulative returns for Apple over time.

Completing all these tasks takes about 3 seconds!

```
now <- Sys.time()
db <- dbConnect(duckdb::duckdb())

dsf <- load_parquet(db, "dsf", "crsp")
stocknames <- load_parquet(db, "stocknames", "crsp")

dsf |>
  count() |>
  collect()
```

```
# A tibble: 1 x 1
          n
      <dbl>
1 105258380
```

```
dsf |>
  mutate(year = year(date)) |>
  count(year) |>
  arrange(desc(year)) |>
  collect()
```

```
# A tibble: 99 x 2
    year       n
   <dbl>   <dbl>
 1  2023 2353668
 2  2022 2389874
 3  2021 2187044
 4  2020 1948489
 5  2019 1911581
 6  2018 1869102
 7  2017 1827464
 8  2016 1828303
 9  2015 1820106
10  2014 1770410
# i 89 more rows
```

```
dsf_subset <-
  dsf |>
  filter(date == "1986-01-07") |>
  collect()

apple_permno <-
  stocknames |>
```

```
  filter(str_detect(comnam, "^APPLE COM")) |>
  pull(permno)

dsf |>
  filter(permno == apple_permno) |>
  group_by(permno) |>
  window_order(date) |>
  mutate(cumret = exp(cumsum(log(1 + coalesce(ret, 0))))) |>
  ggplot(aes(x = date, y = cumret)) +
  geom_line()
```

```
Sys.time() - now
```

```
Time difference of 2.14948 secs
```

```
dbDisconnect(db)
```

E.6   Creating a parquet library

In this section, we describe two approaches to creating a local repository of the data used in this book in parquet format.[6] Either approach requires installation of Python, but the first (db2pq) uses the WRDS PostgreSQL database server as the data source while the second (wrds2pg) gets the data from the SAS files provided by WRDS.

Note that the db2pq approach does inspect the WRDS SAS data files to determine whether the current WRDS data is more recent than that in the local repository. So a benefit of the wrds2pg approach is that it relies only on one data source (SAS data files). However, the db2pq approach will generally be faster and also avoids the significant demands on RAM imposed by wrds2pg in converting data to parquet format and therefore we recommend the db2pq approach.[7]

E.6.1   Using db2pq

Successfully executing the following steps will result in a local repository of parquet files comprising all WRDS data used in this book.

1. Install Python. You may already have Python installed. If not, Miniconda[8] is one option.

2. Install the db2pq package.

```
pip install db2pq --upgrade
```

3. Set up environment variables for db2pq. See here[9] for details.

---

[6]Above we discussed three approaches, but "Approach 1" will be problematic with larger data files. The two approaches here are "Approach 2" and "Approach 3".

[7]This concern with RAM does not apply to the wrds_update() function used in Appendix D.

[8]https://docs.conda.io/en/latest/miniconda.html

[9]https://github.com/iangow/wrds2pg/

```
export WRDS_ID="iangow"
export DATA_DIR="~/Dropbox/pq_data"
```

4. Run Python.

5. Within Python, use the **db2pq** module to get data. Running the following script creates a repository comprising all the data sets used in this book. (If you have already downloaded some tables, you might want to comment those out so that you don't download them again.)

```python
#!/usr/bin/env python3
from db2pq import wrds_update_pq

# CRSP
wrds_update_pq('ccmxpf_lnkhist', 'crsp',
               col_types={'lpermno': 'int32',
                          'lpermco': 'int32'})
wrds_update_pq('dsf', 'crsp',
               col_types={'permno': 'int32',
                          'permco': 'int32'})
wrds_update_pq('dsi', 'crsp')
wrds_update_pq('erdport1', 'crsp')
wrds_update_pq('comphist', 'crsp')
wrds_update_pq('dsedelist', 'crsp',
               col_types={'permno': 'int32',
                          'permco': 'int32'})
wrds_update_pq('dseexchdates', 'crsp', col_types={'permno': 'int32',
                                                  'permco': 'int32'})
wrds_update_pq('msf', 'crsp', col_types={'permno': 'int32',
                                         'permco': 'int32'})
wrds_update_pq('msi', 'crsp')
wrds_update_pq('mse', 'crsp',
               col_types={'permno': 'int32',
                          'permco': 'int32'})
wrds_update_pq('stocknames', 'crsp',
               col_types={'permno': 'int32',
                          'permco': 'int32'})
wrds_update_pq('dsedist', 'crsp',
               col_types={'permno': 'int32',
                          'permco': 'int32'})

# Fama-French library
wrds_update_pq('factors_daily', 'ff')

# Compustat
wrds_update_pq('company', 'comp')
wrds_update_pq('funda', 'comp')
wrds_update_pq('funda_fncd', 'comp')
wrds_update_pq('fundq', 'comp')
```

```
wrds_update_pq('r_auditors', 'comp')
wrds_update_pq('idx_daily', 'comp')
wrds_update_pq('aco_pnfnda', 'comp')

# The segment data is in comp_segments_hist_daily in PostgreSQL,
# but in compsegd in SAS, so we need to use sas_schema to find the SAS data.
wrds_update_pq('seg_customer', 'compseg')
wrds_update_pq('names_seg', 'compseg')
```

E.6.2 Using `wrds2pg`

Successfully executing the following steps will result in a local repository of parquet files comprising all WRDS data used in this book.

1. Install Python. You may already have Python installed. If not, Miniconda[10] is one option.

2. Install the `wrds2pg` package.

```
pip install wrds2pg --upgrade
```

3. Set up environment variables for `wrds2pg`. See here[11] for details.

```
export WRDS_ID="iangow"
export DATA_DIR="~/Dropbox/pq_data"
```

4. Run Python.

5. Within Python, use the `wrds2pg` module to get data. Running the following script creates a repository comprising all the data sets used in this book. (If you have already downloaded some tables, you might want to comment those out so that you don't download them again.)

```
#!/usr/bin/env python3
from wrds2pg import wrds_update_pq

# CRSP
wrds_update_pq('ccmxpf_lnkhist', 'crsp', fix_missing=True,
               col_types={'lpermno': 'integer',
                          'lpermco': 'integer'})
wrds_update_pq('dsf', 'crsp', fix_missing=True,
               col_types={'permno': 'integer',
                          'permco': 'integer'})
wrds_update_pq('dsi', 'crsp')
wrds_update_pq('erdport1', 'crsp', fix_missing=True)
wrds_update_pq('comphist', 'crsp', fix_missing=True)
wrds_update_pq('dsedelist', 'crsp', fix_missing=True,
               col_types={'permno': 'integer',
```

[10]https://docs.conda.io/en/latest/miniconda.html
[11]https://github.com/iangow/wrds2pg/

```
                                        'permco': 'integer'})
wrds_update_pq('dseexchdates', 'crsp',
               col_types={'permno': 'integer',
                          'permco': 'integer'})
wrds_update_pq('msf', 'crsp', fix_missing=True,
               col_types={'permno': 'integer',
                          'permco': 'integer'})
wrds_update_pq('msi', 'crsp')
wrds_update_pq('mse', 'crsp', fix_missing=True,
               col_types={'permno': 'integer',
                          'permco': 'integer'})
wrds_update_pq('stocknames', 'crsp',
               col_types={'permno': 'integer',
                          'permco': 'integer'})
wrds_update_pq('dsedist', 'crsp', fix_missing=True,
               col_types={'permno': 'integer',
                          'permco': 'integer'})

# Fama-French library
wrds_update_pq('factors_daily', 'ff')

# Compustat
wrds_update_pq('company', 'comp',)
wrds_update_pq('funda', 'comp', fix_missing=True)
wrds_update_pq('funda_fncd', 'comp')
wrds_update_pq('fundq', 'comp', fix_missing=True)
wrds_update_pq('r_auditors', 'comp')
wrds_update_pq('idx_daily', 'comp')
wrds_update_pq('aco_pnfnda', 'comp')

# compseg
wrds_update_pq('seg_customer', 'compseg')
wrds_update_pq('names_seg', 'compseg')
```

References

Abel, A.B., Mishkin, F.S., 1983. On the econometric testing of rationality-market efficiency. The Review of Economics and Statistics 65, 318–323. https://doi.org/10.2307/1924498

Aboody, D., Barth, M.E., Kasznik, R., 2004. Firms' voluntary recognition of stock-based compensation expense. Journal of Accounting Research 42, 123–150. https://doi.org/10.1111/j.1475-679X.2004.00132.x

Adams, C.P., 2020. Learning microeconometrics with R, Chapman & Hall/CRC The R Series. CRC Press. https://doi.org/10.1201/9780429288333

Adler, J., 2012. R in a nutshell, In a nutshell. O'Reilly Media, Incorporated.

Ahern, K.R., Dittmar, A.K., 2012. The changing of the boards: The impact on firm valuation of mandated female board representation. The Quarterly Journal of Economics 127, 137–197. https://doi.org/10.1093/qje/qjr049

Akobeng, A.K., 2005. Understanding randomised controlled trials. Archives of Disease in Childhood 90, 840–844. https://doi.org/10.1136/adc.2004.058222

Alexander, G.J., Peterson, M.A., 2008. The effect of price tests on trader behavior and market quality: An analysis of Reg SHO. Journal of Financial Markets 11, 84–111. https://doi.org/10.1016/j.finmar.2007.06.002

Amihud, Y., 2002. Illiquidity and stock returns: Cross-section and time-series effects. Journal of Financial Markets 5, 31–56. https://doi.org/10.1016/s1386-4181(01)00024-6

Angrist, J.D., 1990. Lifetime earnings and the Vietnam era draft lottery: Evidence from Social Security administrative records. American Economic Review 80, 313–336.

Angrist, J.D., Pischke, J.-S., 2014. Mastering 'metrics: The path from cause to effect. Princeton University Press.

Angrist, J.D., Pischke, J.-S., 2010. The credibility revolution in empirical economics: How better research design is taking the con out of econometrics. Journal of Economic Perspectives 24, 3–30. https://doi.org/10.1257/jep.24.2.3

Angrist, J.D., Pischke, J.-S., 2008. Mostly harmless econometrics: An empiricist's companion. Princeton University Press.

Appel, I.R., Gormley, T.A., Keim, D.B., 2024. Identification using Russell 1000/2000 index assignments: A discussion of methodologies. Critical Finance Review 13, 151–224. https://doi.org/10.1561/104.00000139

Armstrong, C., Kepler, J.D., Samuels, D., Taylor, D., 2022. Causality redux: The evolution of empirical methods in accounting research and the growth of quasi-experiments. Journal of Accounting and Economics 74, 101521. https://doi.org/10.1016/j.jacceco.2022.101521

Armstrong, C.S., Core, J.E., Guay, W.R., 2014. Do independent directors cause improvements in firm transparency? Journal of Financial Economics 113, 383–403. https://doi.org/10.1016/j.jfineco.2014.05.009

Armstrong, C.S., Gow, I.D., Larcker, D.F., 2013. The efficacy of shareholder voting: Evidence from equity compensation plans. Journal of Accounting Research 51, 909–950. https://doi.org/10.1111/1475-679X.12023

Aronow, P.M., Miller, B.T., 2019. Foundations of agnostic statistics. Cambridge University Press. https://doi.org/10.1017/9781316831762

Ball, R., Brown, P., 2019. Ball and Brown (1968) after fifty years. Pacific-Basin Finance Journal 53, 410–431. https://doi.org/10.1016/j.pacfin.2018.12.008

Ball, R., Brown, P., 1968. An empirical evaluation of accounting income numbers. Journal of Accounting Research 6, 159–178. https://doi.org/10.2307/2490232

Ball, R., Kothari, S.P., Nikolaev, V.V., 2013. Econometrics of the Basu asymmetric timeliness coefficient and accounting conservatism. Journal of Accounting Research 51, 1071–1097. https://doi.org/10.1111/1475-679x.12026

Ball, R., Kothari, S.P., Robin, A., 2000. The effect of international institutional factors on properties of accounting earnings. Journal of Accounting and Economics 29, 1–51. https://doi.org/10.1016/s0165-4101(00)00012-4

Ball, R., Shivakumar, L., 2008. How much new information is there in earnings? Journal of Accounting Research 46, 975–1016. https://doi.org/10.1111/j.1475-679X.2008.00299.x

Bamber, L.S., Christensen, T.E., Gaver, K.M., 2000. Do we really "know" what we think we know? A case study of seminal research and its subsequent overgeneralization. Accounting, Organizations and Society 25, 103–129. https://doi.org/10.1016/S0361-3682(99)00027-6

Bao, Y., Ke, B., Li, B., Yu, Y.J., Zhang, J., 2022. Erratum. Journal of Accounting Research 60, 1635–1646. https://doi.org/10.1111/1475-679X.12454

Bao, Y., Ke, B., Li, B., Yu, Y.J., Zhang, J., 2020. Detecting accounting fraud in publicly traded US firms using a machine learning approach. Journal of Accounting Research 58, 199–235. https://doi.org/10.1111/1475-679X.12292

Basu, S., 1997. The conservatism principle and the asymmetric timeliness of earnings. Journal of Accounting and Economics 24, 3–37. https://doi.org/10.1016/s0165-4101(97)00014-1

Beaver, W.H., 1998. Financial reporting: An accounting revolution, 3rd ed. Prentice-Hall, Inc.

Beaver, W.H., 1968. The information content of annual earnings announcements. Journal of Accounting Research 6, 67–92. https://doi.org/10.2307/2490070

Bebchuk, L.A., Cohen, A., Hirst, S., 2017. The agency problems of institutional investors. Journal of Economic Perspectives 31, 89–102. https://doi.org/10.1257/jep.31.3.89

Bennedsen, M., Nielsen, K.M., Perez-Gonzalez, F., Wolfenzon, D., 2007. Inside the family firm: The role of families in succession decisions and performance. The Quarterly Journal of Economics 122, 647–691. https://doi.org/10.1162/qjec.122.2.647

Bernard, V.L., Schipper, K., 1994. Recognition and disclosure in financial reporting.

Bernard, V.L., Thomas, J.K., 1989. Post-earnings-announcement drift: Delayed price response or risk premium? Journal of Accounting Research 27, 1–36. https://doi.org/10.2307/2491062

Bertrand, M., Schoar, A., 2003. Managing with style: The effect of managers on firm policies. The Quarterly Journal of Economics 118, 1169–1208. https://doi.org/10.1162/003355303322552775

Beyer, A., Cohen, D.A., Lys, T.Z., Walther, B.R., 2010. The financial reporting environment: Review of the recent literature. Journal of Accounting and Economics 50, 296–343. https://doi.org/10.1016/j.jacceco.2010.10.003

Bird, A., Karolyi, S.A., 2017. Governance and taxes: Evidence from regression discontinuity. The Accounting Review 92, 29–50. https://doi.org/10.2308/accr-51520

Black, B.S., Desai, H., Litvak, K., Yoo, W., Yu, J.J., 2024. The SEC's short-sale experiment: Evidence on causal channels and reassessment of indirect effects. Management Science 70, 5131–5156. https://doi.org/10.1287/mnsc.2023.4918

Black, B.S., Desai, H., Litvak, K., Yoo, W., Yu, J.J., 2019. Pre-analysis plan for the Reg SHO reanalysis project. SSRN Electronic Journal. https://doi.org/10.2139/ssrn.3415529

Black, F., 1976. The dividend puzzle. The Journal of Portfolio Management 2, 5–8. https://doi.org/10.3905/jpm.1976.408558

Black, F., Scholes, M., 1973. The pricing of options and corporate liabilities. Journal of Political Economy 81, 637–654. https://doi.org/10.1086/260062

Bloomfield, M.J., 2021. The asymmetric effect of reporting flexibility on priced risk. Journal of Accounting Research 59, 867–910. https://doi.org/10.1111/1475-679X.12346

Bloomfield, R., Rennekamp, K., Steenhoven, B., 2018. No system is perfect: Understanding how registration-based editorial processes affect reproducibility and investment in research quality. Journal of Accounting Research 56, 313–362. https://doi.org/10.1111/1475-679X.12208

Boone, A.L., White, J.T., 2015. The effect of institutional ownership on firm transparency and information production. Journal of Financial Economics 117, 508–533. https://doi.org/10.1016/j.jfineco.2015.05.008

Breiman, L., Friedman, J., Stone, C.J., Olshen, R.A., 1984. Classification and regression trees. Taylor & Francis.

Brown, L.D., Lee, Y.-J., 2011. Changes in option-based compensation around the issuance of SFAS 123R. Journal of Business Finance & Accounting 38, 1053–1095. https://doi.org/10.1111/j.1468-5957.2011.02247.x

Brown, N.C., Stice, H., White, R.M., 2015. Mobile communication and local information flow: Evidence from distracted driving laws. Journal of Accounting Research 53, 275–329. https://doi.org/10.1111/1475-679X.12077

Brown, P., 1989. Ball and Brown [1968]. Journal of Accounting Research 27, 202–216. https://doi.org/10.2307/2491072

Burks, J.J., Cuny, C., Gerakos, J., Granja, J., 2018. Competition and voluntary disclosure: Evidence from deregulation in the banking industry. Review of Accounting Studies 23, 1471–1511. https://doi.org/10.1007/s11142-018-9463-1

Burt, R.A.P., 2000. The randomization failed—discard the study. International Journal of Pharmaceutical Medicine 14, 319–321. https://doi.org/10.2165/00124363-200012000-00009

Call, A.C., Martin, G.S., Sharp, N.Y., Wilde, J.H., 2018. Whistleblowers and outcomes of financial misrepresentation enforcement actions. Journal of Accounting Research 56, 123–171. https://doi.org/10.1111/1475-679X.12177

Call, A.C., Sharp, N.Y., Wilde, J.H., 2019. A response to "Are a few huge outcomes distorting financial misconduct research?" Econ Journal Watch 16, 35–37.

Cameron, A.C., Gelbach, J.B., Miller, D.L., 2011. Robust inference with multiway clustering. Journal of Business & Economic Statistics 29, 238–249. https://doi.org/10.1198/jbes.2010.07136

Cameron, A.C., Gelbach, J.B., Miller, D.L., 2008. Bootstrap-based improvements for inference with clustered errors. The Review of Economics and Statistics 90, 414–427. https://doi.org/10.1162/rest.90.3.414

Cameron, A.C., Trivedi, P.K., 2009. Microeconometrics using Stata. Stata Press.

Cameron, A.C., Trivedi, P.K., 2005. Microeconometrics: Methods and applications. Cambridge University Press.

Card, D., Krueger, A.B., 1994. Minimum wages and employment: A case study of the fast-food industry in New Jersey and Pennsylvania. The American Economic Review 84, 772–793. https://doi.org/10.2307/2118030

Chava, S., Roberts, M.R., 2008. How does financing impact investment? The role of debt covenants. The Journal of Finance 63, 2085–2121. https://doi.org/10.1111/j.1540-6261.2008.01391.x

Chen, S., Huang, Y., Li, N., Shevlin, T., 2019. How does quasi-indexer ownership affect corporate tax planning? Journal of Accounting and Economics 67, 278–296. https://doi.org/10.1016/j.jacceco.2018.01.001

Chen, W., Hribar, P., Melessa, S., 2018. Incorrect inferences when using residuals as dependent variables. Journal of Accounting Research 56, 751–796. https://doi.org/10.1111/j.1540-6261.2008.01391.x

Chiang, A.C., 1984. Fundamental methods of mathematical economics. McGraw-Hill.

Choudhary, P., Rajgopal, S., Venkatachalam, M., 2009. Accelerated vesting of employee stock options in anticipation of FAS 123-R. Journal of Accounting Research 47, 105–146. https://doi.org/10.1111/j.1475-679X.2008.00316.x

Christensen, H.B., Floyd, E., Liu, L.Y., Maffett, M., 2017. The real effects of mandated information on social responsibility in financial reports: Evidence from mine-safety records. Journal of Accounting and Economics 64, 284–304. https://doi.org/10.1016/j.jacceco.2017.08.001

Cochrane, J.H., 2009. Asset pricing: Revised edition. Princeton University Press.

Cohen, D.A., Dey, A., Lys, T.Z., 2013. Corporate governance reform and executive incentives: Implications for investments and risk taking. Contemporary Accounting Research 30, 1296–1332. https://doi.org/10.1111/j.1911-3846.2012.01189.x

Cunningham, S., 2021. Causal inference: The mixtape. Yale University Press.

Davidson, R., MacKinnon, J.G., 2004. Econometric theory and methods. Oxford University Press.

DeAngelo, L.E., 1986. Accounting numbers as market valuation substitutes: A study of management buyouts of public stockholders. The Accounting Review 61, 400–420.

Dechow, P., Khimich, N., Sloan, R., 2011. The accrual anomaly, in: The Handbook of Equity Market Anomalies: Translating Market Inefficiencies into Effective Investment Strategies. Wiley Finance, pp. 23–49.

Dechow, P.M., Sloan, R.G., Sweeney, A.P., 1995. Detecting earnings management. The Accounting Review 70, 193–225.

Dechow, P.M., Sloan, R.G., Zha, J., 2014. Stock prices and earnings: A history of research. Annual Review of Financial Economics 6, 343–363. https://doi.org/10.1146/annurev-financial-110613-034522

DeFond, M., Erkens, D.H., Zhang, J., 2017. Do client characteristics really drive the Big N audit quality effect? New evidence from propensity score matching. Management Science 63, 3628–3649. https://doi.org/10.1287/mnsc.2016.2528

Diether, K.B., Lee, K.-H., Werner, I.M., 2009. It's SHO time! Short-sale price tests and market quality. The Journal of Finance 64, 37–73. https://doi.org/10.1111/j.1540-6261.2008.01428.x

Dietrich, J.R., Muller, K.A., Riedl, E.J., 2022. On the validity of asymmetric timeliness measures of accounting conservatism. Review of Accounting Studies 28, 2150–2195. https://doi.org/10.1007/s11142-022-09684-2

Dietrich, J.R., Muller, K.A., Riedl, E.J., 2007. Asymmetric timeliness tests of accounting conservatism. Review of Accounting Studies 12, 95–124. https://doi.org/10.1007/s11142-006-9023-y

Doll, R., Peto, R., 1976. Mortality in relation to smoking: 20 years' observations on male British doctors. British Medical Journal 2, 1525–1536. https://doi.org/10.1136/bmj.2.6051.1525

Dunn, P.K., Smyth, G.K., 2018. Generalized linear models with examples in R, Springer Texts in Statistics. Springer New York. https://doi.org/10.1007/978-1-4419-0118-7

Dunning, T., 2012. Natural experiments in the social sciences: A design-based approach. Cambridge University Press.

Dyck, A., Morse, A., Zingales, L., 2010. Who blows the whistle on corporate fraud? The Journal of Finance 65, 2213–2253. https://doi.org/10.1111/j.1540-6261.2010.01614.x

Erkens, D.H., Subramanyam, K.R., Zhang, J., 2014. Affiliated banker on board and conservative accounting. The Accounting Review 89, 1703–1728. https://doi.org/10.2308/accr-50798

Ertimur, Y., Ferri, F., Oesch, D., 2015. Does the director election system matter? Evidence from majority voting. Review of Accounting Studies 20, 1–41. https://doi.org/10.1007/s11142-014-9284-9

Fama, E.F., 1991. Efficient capital markets: II. The Journal of Finance 46, 1575–1617. https://doi.org/10.1111/j.1540-6261.1991.tb04636.x

Fama, E.F., 1970. Efficient capital markets: A review of theory and empirical work. The Journal of Finance 25, 383–417. https://doi.org/10.2307/2325486

Fama, E.F., Fisher, L., Jensen, M.C., Roll, R., 1969. The adjustment of stock prices to new information. International Economic Review 10, 1–21. https://doi.org/10.2307/2525569

Fama, E.F., French, K.R., 2008. Dissecting anomalies. The Journal of Finance 63, 1653–1678. https://doi.org/10.1111/j.1540-6261.2008.01371.x

Fama, E.F., MacBeth, J.D., 1973. Risk, return, and equilibrium: Empirical tests. Journal of Political Economy 81, 607–636. https://doi.org/10.1086/260061

Fang, V.W., Huang, A.H., Karpoff, J.M., 2019. Reply to "the Reg SHO reanalysis project: Reconsidering Fang, Huang and Karpoff (2016) on Reg SHO and earnings management" by Black et al. (2019). SSRN Electronic Journal. https://doi.org/10.2139/ssrn.3507033

Fang, V.W., Huang, A.H., Karpoff, J.M., 2016. Short selling and earnings management: A controlled experiment. The Journal of Finance 71, 1251–1294. https://doi.org/10.1111/jofi.12369

Fisher, L., 1966. Some new stock-market indexes. The Journal of Business 39, 191–225. https://doi.org/10.1086/294848

Fisher, R.A., 1935. The design of experiments. Oliver and Boyd.

Foster, G., 1977. Quarterly accounting data: Time-series properties and predictive-ability results. The Accounting Review 52, 1–21.

Foster, G., Olsen, C., Shevlin, T., 1984. Earnings releases, anomalies, and the behavior of security returns. The Accounting Review 59, 574–603.

Frank, K.A., 2000. Impact of a confounding variable on a regression coefficient. Sociological Methods & Research 29, 147–194. https://doi.org/10.1177/0049124100029002001

Freedman, D.A., 2009. Statistical models and causal inference: A dialogue with the social sciences. Cambridge University Press.

Friedl, J.E.F., 2006. Mastering regular expressions: Understand your data and be more productive. O'Reilly Media.

Frison, L., Pocock, S.J., 1992. Repeated measures in clinical trials: Analysis using mean summary statistics and its implications for design. Statistics in Medicine 11, 1685–1704. https://doi.org/10.1002/sim.4780111304

Gelman, A., Imbens, G., 2019. Why high-order polynomials should not be used in regression discontinuity designs. Journal of Business & Economic Statistics 37, 447–456. https://doi.org/10.1080/07350015.2017.1366909

Gillies, D., 2011. The Russo-Williamson thesis and the question of whether smoking causes heart disease, in: Illari, P.M., Russo, F., Williamson, J. (Eds.), Causality in the Sciences. Oxford University Press, pp. 110–125.

Glaeser, S., Guay, W.R., 2017. Identification and generalizability in accounting research: A discussion of Christensen, Floyd, Liu, and Maffett (2017). Journal of Accounting and Economics 64, 305–312. https://doi.org/10.1016/j.jacceco.2017.08.003

Glymour, M.M., Greenland, S., 2008. Causal diagrams, in: Rothman, K.J., Greenland, S., Lash, T.L. (Eds.), Modern Epidemiology. Lippincott, Williams; Wilkins.

Goldberger, A.S., 1972. Structural equation methods in the social sciences. Econometrica 40, 979–1001. https://doi.org/10.2307/1913851

Gow, I.D., 2022. farr: Data and code for financial accounting research. CRAN: Contributed Packages. https://doi.org/10.32614/cran.package.farr

Gow, I.D., Kells, S.S., 2018. The Big Four: The curious past and perilous future of the global accounting monopoly. Berrett-Koehler Publishers.

Gow, I.D., Larcker, D.F., McCall, A., Tayan, B., 2013. Sneak preview: How ISS dictates equity plan design. Stanford Closer Look Series.

Gow, I.D., Larcker, D.F., Reiss, P.C., 2016. Causal inference in accounting research. Journal of Accounting Research 54, 477–523. https://doi.org/10.1111/1475-679X.12116

Gow, I.D., Ormazabal, G., Taylor, D.J., 2010. Correcting for cross-sectional and time-series dependence in accounting research. The Accounting Review 85, 483–512. https://doi.org/10.2308/accr.2010.85.2.483

Goyvaerts, J., Levithan, S., 2009. Regular expressions cookbook. O'Reilly Media.

Green, J., Hand, J.R.M., Soliman, M.T., 2011. Going, going, gone? The apparent demise of the accruals anomaly. Management Science 57, 797–816. https://doi.org/10.1287/mnsc.1110.1320

Grolemund, G., 2014. Hands-on programming with R, Safari books online. O'Reilly Media, Incorporated.

Guay, W.R., Samuels, D., Taylor, D.J., 2016. Guiding through the fog: Financial statement complexity and voluntary disclosure. Journal of Accounting and Economics 62, 234–269. https://doi.org/10.1016/j.jacceco.2016.09.001

Guo, S., Atkinson, P., Delamont, S., Cernat, A., Sakshaug, J.W., Williams, R.A., 2020. Propensity score analysis. SAGE Publications Limited.

Hagopian, K., 2006. Point of view: Expensing employee stock options is improper accounting. California Management Review 48, 136–156. https://doi.org/10.2307/41166364

Hail, L., Tahoun, A., Wang, C., 2018. Corporate scandals and regulation. Journal of Accounting Research 56, 617–671. https://doi.org/10.1111/1475-679X.12201

Harvey, C.R., 2017. Presidential address: The scientific outlook in financial economics. The Journal of Finance 72, 1399–1440. https://doi.org/10.1111/jofi.12530

Hastie, T., Tibshirani, R., Friedman, J., 2013. The elements of statistical learning: Data mining, inference, and prediction, Springer series in statistics. Springer New York. https://doi.org/10.1007/978-0-387-84858-7

Healy, K., 2018. Data visualization: A practical introduction. Princeton University Press.

Healy, P.M., 1985. The effect of bonus schemes on accounting decisions. Journal of Accounting and Economics 7, 85–107. https://doi.org/10.1016/0165-4101(85)90029-1

Heckman, J.J., Singer, B., 2017. Abducting economics. American Economic Review 107, 298–302. https://doi.org/10.1257/aer.p20171118

Hoekstra, M., 2009. The effect of attending the flagship state university on earnings: A discontinuity-based approach. The Review of Economics and Statistics 91, 717–724. https://doi.org/10.1162/rest.91.4.717

Hopkins, J.J., Lang, M.H., Zhao, J.D., 2022. The rise of VIEs in China: Balancing state control and access to foreign capital. Journal of Financial Reporting 7, 105–130. https://doi.org/10.2308/jfr-2021-017

Hribar, P., Collins, D.W., 2002. Errors in estimating accruals: Implications for empirical research. Journal of Accounting Research 40, 105–134. https://doi.org/10.1111/1475-679X.00041

Huntington-Klein, N., 2021. The effect: An introduction to research design and causality. Chapman; Hall/CRC Press.

Iliev, P., 2010. The effect of SOX Section 404: Costs, earnings quality, and stock prices. The Journal of Finance 65, 1163–1196. https://doi.org/10.1111/j.1540-6261.2010.01564.x

Imbens, G.W., Kalyanaraman, K., 2012. Optimal bandwidth choice for the regression discontinuity estimator. The Review of Economic Studies 79, 933–959. https://doi.org/10.1093/restud/rdr043

Imbens, G.W., Rubin, D.B., 2015. Causal inference for statistics, social and biomedical sciences: An introduction. Cambridge University Press. https://doi.org/10.1017/CBO9781139025751

Irarrazabal, A., Moxnes, A., Opromolla, L.D., 2013. The margins of multinational production and the role of intrafirm trade. Journal of Political Economy 121, 74–126. https://doi.org/10.1086/669877

Jackson, S.B., Liu, X.K., Cecchini, M., 2009. Economic consequences of firms' depreciation method choice: Evidence from capital investments. Journal of Accounting and Economics 48, 54–68. https://doi.org/10.1016/j.jacceco.2009.06.001

Jame, R., Johnston, R., Markov, S., Wolfe, M.C., 2016. The value of crowdsourced earnings forecasts. Journal of Accounting Research 54, 1077–1110. https://doi.org/10.1111/1475-679x.12121

James, G., Witten, D., Hastie, T., Tibshirani, R., 2022. An introduction to statistical learning: With applications in R, Springer texts in statistics. Springer US. https://doi.org/10.1007/978-1-0716-1418-1

Jayaraman, S., Shivakumar, L., 2012. Agency-based demand for conservatism: Evidence from state adoption of antitakeover laws. Review of Accounting Studies 18, 95–134. https://doi.org/10.1007/s11142-012-9205-8

Jones, J.J., 1991. Earnings management during import relief investigations. Journal of Accounting Research 29, 193–228. https://doi.org/10.2307/2491047

Kajüter, P., Klassmann, F., Nienhaus, M., 2018. The effect of mandatory quarterly reporting on firm value. The Accounting Review 94, 251–277. https://doi.org/10.2308/accr-52212

Karolyi, G.A., Taboada, A.G., 2015. Regulatory arbitrage and cross-border bank acquisitions. The Journal of Finance 70, 2395–2450. https://doi.org/10.1111/jofi.12262

Kelly, B., Ljungqvist, A., 2012. Testing asymmetric-information asset pricing models. Review of Financial Studies 25, 1366–1413. https://doi.org/10.1093/rfs/hhr134

Kennedy, P., 2008. A guide to econometrics. Wiley.

Khan, M., Srinivasan, S., Tan, L., 2016. Institutional ownership and corporate tax avoidance: New evidence. The Accounting Review 92, 101–122. https://doi.org/10.2308/accr-51529

Khan, U., Li, B., Rajgopal, S., Venkatachalam, M., 2017. Do the FASB's standards add shareholder value? The Accounting Review 93. https://doi.org/10.2308/accr-51840

Klasa, S.J., Ortiz-Molina, H., Serfling, M.A., Srinivasan, S., 2018. Protection of trade secrets and capital structure decisions. Journal of Financial Economics 128, 266–286. https://doi.org/10.1016/j.jfineco.2018.02.008

Koh, P.-S., Reeb, D.M., 2015. Missing R&D. Journal of Accounting and Economics 60, 73–94. https://doi.org/10.1016/j.jacceco.2015.03.004

Koller, M., Stahel, W.A., 2017. Nonsingular subsampling for regression S estimators with categorical predictors. Computational Statistics 32, 631–646. https://doi.org/10.1007/s00180-016-0679-x

Kothari, S.P., 2001. Capital markets research in accounting. Journal of Accounting and Economics 31, 105–231. https://doi.org/10.1016/S0165-4101(01)00030-1

Kothari, S.P., Leone, A.J., Wasley, C.E., 2005. Performance matched discretionary accrual measures. Journal of Accounting and Economics 39, 163–197. https://doi.org/10.1016/j.jacceco.2004.11.002

Kraft, A., Leone, A.J., Wasley, C.E., 2007. Regression-based tests of the market pricing of accounting numbers: The Mishkin test and ordinary least squares. Journal of Accounting Research 45, 1081–1114. https://doi.org/10.1111/j.1475-679X.2007.00261.x

Kreps, D.M., 1990. A course in microeconomic theory. Princeton University Press.

Kuhn, M., Johnson, K., 2013. Applied predictive modeling, SpringerLink: Bücher. Springer New York. https://doi.org/10.1007/978-1-4614-6849-3

Kuhn, M., Silge, J., 2022. Tidy modeling with R. O'Reilly Media.

Kuvvet, E., 2019. Are a few huge outcomes distorting financial misconduct research? Econ Journal Watch 16, 1–34.

Larcker, D.F., Ormazabal, G., Taylor, D.J., 2011. The market reaction to corporate governance regulation. Journal of Financial Economics 101, 431–448. https://doi.org/10.1016/j.jfineco.2011.03.002

Larcker, D.F., Richardson, S.A., Tuna, İ., 2007. Corporate governance, accounting outcomes, and organizational performance. The Accounting Review 82, 963–1008. https://doi.org/10.2308/accr.2007.82.4.963

Larcker, D.F., Rusticus, T.O., 2010. On the use of instrumental variables in accounting research. Journal of Accounting and Economics 49, 186–205. https://doi.org/10.1016/j.jacceco.2009.11.004

Lawrence, A., Minutti-Meza, M., Zhang, P., 2011. Can Big 4 versus non-Big 4 differences in audit-quality proxies be attributed to client characteristics? The Accounting Review 86, 259–286. https://doi.org/10.2308/accr.00000009

Lee, D.S., Lemieux, T., 2010. Regression discontinuity designs in economics. Journal of Economic Literature 48, 281–355. https://doi.org/10.1257/jel.48.2.281

Leftwich, R.W., Zmijewski, M.E., 1994. Contemporaneous announcements of dividends and earnings. Journal of Accounting, Auditing and Finance 725–762. https://doi.org/10.1177/0148558X9400900406

Lennox, C.S., Francis, J.R., Wang, Z., 2012. Selection models in accounting research. The Accounting Review 87, 589–616. https://doi.org/10.2308/accr-10195

Leone, A.J., Minutti-Meza, M., Wasley, C.E., 2019. Influential observations and inference in accounting research. The Accounting Review 94, 337–364. https://doi.org/10.2308/accr-52396

Leuz, C., 2007. Was the Sarbanes–Oxley Act of 2002 really this costly? A discussion of evidence from event returns and going-private decisions. Journal of Accounting and Economics 44, 146–165. https://doi.org/10.1016/j.jacceco.2007.06.001

Li, S.X., Sandino, T., 2018. Effects of an information sharing system on employee creativity, engagement, and performance. Journal of Accounting Research 56, 713–747. https://doi.org/10.1111/1475-679X.12202

Li, Y., Lin, Y., Zhang, L., 2018. Trade secrets law and corporate disclosure: Causal evidence on the proprietary cost hypothesis. Journal of Accounting Research 56, 265–308. https://doi.org/10.1111/1475-679X.12187

Li, Y., Zhang, L., 2015. Short selling pressure, stock price behavior, and management forecast precision: Evidence from a natural experiment. Journal of Accounting Research 53, 79–117. https://doi.org/10.1111/1475-679X.12068

Lin, Y., Mao, Y., Wang, Z., 2017. Institutional ownership, peer pressure, and voluntary disclosures. The Accounting Review 93, 283–308. https://doi.org/10.2308/accr-51945

Listokin, Y., 2008. Management always wins the close ones. American Law and Economics Review 10, 159–184. https://doi.org/10.1093/aler/ahn010

Loughran, T., McDonald, B., 2014. Measuring readability in financial disclosures. The Journal of Finance 69, 1643–1671. https://doi.org/10.1111/jofi.12162

MacKinlay, A.C., 1997. Event studies in economics and finance. Journal of Economic Literature 35, 13–39.

Mahase, E., 2021. Covid-19: Pfizer's paxlovid is 89% effective in patients at risk of serious illness, company reports. BMJ 375. https://doi.org/10.1136/bmj.n2713

Manchiraju, H., Rajgopal, S., 2017. Does corporate social responsibility (CSR) create shareholder value? Evidence from the Indian Companies Act 2013. Journal of Accounting Research 55, 1257 1300. https://doi.org/10.1111/1475-679X.12174

Maronna, R.A., Martin, R.D., Yohai, V.J., Salibián-Barrera, M., 2019. Robust statistics: Theory and methods (with R). John Wiley & Sons.

Matloff, N., 2011. The art of R programming: A tour of statistical software design. No Starch Press.

McCrary, J., 2008. Manipulation of the running variable in the regression discontinuity design: A density test. Journal of Econometrics 142, 698–714. https://doi.org/10.1016/j.jeconom.2007.05.005

McCrum, D., 2022. Money men: The inspiration for Netflix's Skandal! A hot startup, a billion dollar fraud, a fight for the truth. Transworld.

McNichols, M., Wilson, G.P., 1988. Evidence of earnings management from the provision for bad debts. Journal of Accounting Research 26, 1–31. https://doi.org/10.2307/2491176

Michels, J., 2017. Disclosure versus recognition: Inferences from subsequent events. Journal of Accounting Research 55, 3–34. https://doi.org/10.1111/1475-679X.12128

Miller, G.S., Skinner, D.J., 2015. The evolving disclosure landscape: How changes in technology, the media, and capital markets are affecting disclosure. Journal of Accounting Research 53, 221–239. https://doi.org/10.1111/1475-679X.12075

Mishkin, F.S., 1983. A rational expectations approach to macroeconomics: Testing policy ineffectiveness and efficient-markets models. University of Chicago Press.

Morgan, S.L., Winship, C., 2014. Counterfactuals and causal inference. Cambridge University Press. https://doi.org/10.1017/cbo9781107587991

Morrow, J.D., Frei, B., Longmire, A.W., Gaziano, J.M., Lynch, S.M., Shyr, Y., Strauss, W.E., Oates, J.A., Roberts, L.J., 1995. Increase in circulating products of lipid peroxidation F_2-isoprostanes in smokers: Smoking as a cause of oxidative damage. New England Journal of Medicine 332, 1198–1203. https://doi.org/10.1056/NEJM199505043321804

Newey, W.K., West, K.D., 1987. A simple, positive semi-definite, heteroskedasticity and autocorrelation consistent covariance matrix. Econometrica 55, 703–708. https://doi.org/10.2307/1913610

Neyman, J., 1923. On the application of probability theory to agricultural experiments. Essay on principles. Section 9. Statistical Science 5, 465–472.

Nichols, D.C., Wahlen, J.M., 2004. How do earnings numbers relate to stock returns? A review of classic accounting research with updated evidence. Accounting Horizons 18, 263–286. https://doi.org/10.2308/acch.2004.18.4.263

Pearl, J., 2009b. Causal inference in statistics: An overview. Statistics Surveys 3, 96–146. https://doi.org/10.1214/09-SS057

Pearl, J., 2009a. Causality: Models, reasoning and inference. Cambridge University Press.

Pearl, J., Mackenzie, D., 2019. The book of why: The new science of cause and effect, An allen lane book. Penguin Books.

R Core Team, 2021. R: A language and environment for statistical computing. R Foundation for Statistical Computing, Vienna, Austria.

Reiss, P.C., Wolak, F.A., 2007. Structural econometric modeling: Rationales and examples from industrial organization, in: Heckman, J.J., Leamer, E.E. (Eds.), Handbook of Econometrics. Elsevier, pp. 4277–4415. https://doi.org/10.1016/S1573-4412(07)06064-3

Roberts, M.R., Whited, T.M., 2013. Endogeneity in empirical corporate finance, in: Handbook of the Economics of Finance. Elsevier, pp. 493–572. https://doi.org/10.1016/B978-0-44-453594-8.00007-0

Russo, F., Williamson, J., 2007. Interpreting causality in the health sciences. International Studies in the Philosophy of Science 21, 157–170. https://doi.org/10.1080/02698590701498084

Salkever, D.S., 1976. The use of dummy variables to compute predictions, prediction errors, and confidence intervals. Journal of Econometrics 4, 393–397. https://doi.org/10.1016/0304-4076(76)90027-0

Santos Silva, J.M.C., Tenreyro, S., 2011. Further simulation evidence on the performance of the poisson pseudo-maximum likelihood estimator. Economics Letters 112, 220–222. https://doi.org/10.1016/j.econlet.2011.05.008

Santos Silva, J.M.C., Tenreyro, S., 2006. The log of gravity. The Review of Economics and Statistics 88, 641–658. https://doi.org/10.1162/rest.88.4.641

Schipper, K., Thompson, R., 1983. The impact of merger-related regulations on the shareholders of acquiring firms. Journal of Accounting Research 21, 184–221. https://doi.org/10.2307/2490943

Seiffert, C., Khoshgoftaar, T.M., Van Hulse, J., Napolitano, A., 2008. RUSBoost: Improving classification performance when training data is skewed, in: 2008 19th International Conference on Pattern Recognition. pp. 1–4.

Shadish, W.R., Cook, T.D., Campbell, D.T., 2002. Experimental and quasi-experimental designs for generalized causal inference. Houghton Mifflin.

Shiller, R., 1984. Stock prices and social dynamics (Cowles Foundation Discussion Papers No. 719R). Cowles Foundation for Research in Economics, Yale University.

Shipman, J.E., Swanquist, Q.T., Whited, R.L., 2016. Propensity score matching in accounting research. The Accounting Review 92, 213–244. https://doi.org/10.2308/accr-51449

Simmons, J.P., Nelson, L.D., Simonsohn, U., 2011. False-positive psychology: Undisclosed flexibility in data collection and analysis allows presenting anything as significant. Psychological Science 22, 1359–1366. https://doi.org/10.1177/0956797611417632

Simon, C.P., Blume, L., 1994. Mathematics for economists. W. W. Norton & Company.

Sloan, R.G., 1996. Do stock prices fully reflect information in accruals and cash flows about future earnings? The Accounting Review 71, 289–315.

Stock, J.H., Wright, J.H., Yogo, M., 2002. A survey of weak instruments and weak identification in generalized method of moments. Journal of Business & Economic Statistics 20, 518–529. https://doi.org/10.1198/073500102288618658

Tan, L., 2013. Creditor control rights, state of nature verification, and financial reporting conservatism. Journal of Accounting and Economics 55, 1–22. https://doi.org/10.1016/j.jacceco.2012.08.001

Tenreyro, S., 2007. On the trade impact of nominal exchange rate volatility. Journal of Development Economics 82, 485–508. https://doi.org/10.1016/j.jdeveco.2006.03.007

Thaler, R.H., 2015. Misbehaving: The making of behavioral economics. W. W. Norton.

Thistlethwaite, D.L., Campbell, D.T., 1960. Regression-discontinuity analysis: An alternative to the ex post facto experiment. Journal of Educational Psychology 51, 309–317. https://doi.org/10.1037/h0044319

Thompson, S.B., 2011. Simple formulas for standard errors that cluster by both firm and time. Journal of Financial Economics 99, 1–10. https://doi.org/10.1016/j.jfineco.2010.08.016

Tinbergen, J., 1962. The world economy: Suggestions for an international economic policy. Twentieth Century Fund, New York, NY.

Vashishtha, R., 2014. The role of bank monitoring in borrowers' discretionary disclosure: Evidence from covenant violations. Journal of Accounting and Economics 57, 176–195. https://doi.org/10.1016/j.jacceco.2014.04.002

Verrecchia, R.E., 1983. Discretionary disclosure. Journal of Accounting and Economics 5, 179–194. https://doi.org/10.1016/0165-4101(83)90011-3

Walker, S., 2021. Post-split underreaction: The importance of prior split history. International Review of Financial Analysis 78, 101945. https://doi.org/10.1016/j.irfa.2021.101945

Watts, R.L., 1992. Accounting choice theory and market-based research in accounting. The British Accounting Review 24, 235–267. https://doi.org/10.1016/S0890-8389(05)80023-X

White, H., 1984. Asymptotic theory for econometricians, Economic theory, econometrics, and mathematical economics. Academic Press, Harcourt Brace Jovanovich, Orlando, FL.

White, H., 1980. A heteroskedasticity-consistent covariance matrix estimator and a direct test for heteroskedasticity. Econometrica 48, 817–838. https://doi.org/10.2307/1912934

Wickham, H., 2019. Advanced R, second edition, Chapman & Hall/CRC The R Series. CRC Press. https://doi.org/10.1201/9781351201315

Wickham, H., Çetinkaya-Rundel, M., Grolemund, G., 2023. R for data science. O'Reilly Media, Sebastopol, CA.

Wilkinson, L., 2005. The grammar of graphics, Statistics and Computing. Springer-Verlag. https://doi.org/10.1007/0-387-28695-0

Wooldridge, J.M., 2010. Econometric analysis of cross section and panel data, 2nd ed. MIT Press.

Wooldridge, J.M., 2000. Introductory econometrics. South-Western Thomson Learning.

Xie, H., 2001. The mispricing of abnormal accruals. The Accounting Review 76, 357–373. https://doi.org/10.2308/accr.2001.76.3.357

Young, A., 2018. Will the real specification please stand up? A comment on Andrew Bird and Stephen Karolyi. Econ Journal Watch 15, 35–48.

Zhang, I.X., 2007. Economic consequences of the Sarbanes-Oxley Act of 2002. Journal of Accounting and Economics 44, 74–115. https://doi.org/10.1016/j.jacceco.2007.02.002

Index

R functions that appear very frequently, such as `mutate()` and `select()`, are omitted from this index.

Printed in the United States
by Baker & Taylor Publisher Services

Printed in the United States
by Baker & Taylor Publisher Services